STUDENT'S SOLUTIONS MANUAL

Calculus

Part I *Graphical, Numerical, Algebraic*

FINNEY THOMAS DEMANA WAITS

STUDENT'S SOLUTIONS MANUAL

Calculus

Part I *Graphical, Numerical, Algebraic*

DAVID L. WINTER
MARY JEAN WINTER

MICHIGAN STATE UNIVERSITY

Addison-Wesley Publishing Company
Reading, Massachusetts • Menlo Park, California • New York
Don Mills, Ontario • Wokingham, England • Amsterdam • Bonn
Sydney • Singapore • Tokyo • Madrid • San Juan • Milan • Paris

Reprinted with corrections, March 1995.
ISBN 0-201-56907-8

3 4 5 6 7 8 9 10-BAH-979695

We would like to acknowledge the important help provided by our solution checkers, Alyson E. Elliott and Harriet E. Winter. We would like to thank Tammy A. Hatfield and Catherine E. Friess for their advice and great technical skill in preparing the manuscript.

TABLE OF CONTENTS

CHAPTER 1

PREREQUISITES FOR CALCULUS

1.1 COORDINATES AND GRAPHS IN THE PLANE

1. In the given viewing rectangle $-5 \leqq x \leqq 5$ and $-10 \leqq y \leqq 10$. The only points satisfying these conditions are $(0, 6)$ and $(5, -5)$.

3. One choice: $[-17, 21]$ by $[-12, 76]$

5. For graph reading in the entire viewing rectangle we want (sometimes but not always) to have as many scale marks as possible but still have them far enough apart to be easily distinguishable. For this purpose we suggest 25 or fewer scale marks in each direction. At the same time the distance between them should be a convenient number for measurement, for example, $m10^n$ where m is a positive integer and n is an integer, positive, negative or 0. Here we have $\frac{x\,\text{Max}-x\,\text{Min}}{25} = \frac{50-(-10)}{25} = 2.4$. $x\,\text{Scl} = 5$ would be convenient. $\frac{y\,\text{Max}-y\,\text{Min}}{25} = \frac{50-(-50)}{25} = 4$. Again $y\,\text{Scl} = 5$ would be convenient.

7. $\frac{x\,\text{Max}-x\,\text{Min}}{25} = 0.04$ so we could take $x\,\text{Scl} = 0.05$, $\frac{y\,\text{Max}-y\,\text{Min}}{25} = \frac{200}{25} = 8$ so $y\,\text{Scl} = 10$ would be convenient.

9. $y = x + 1$.
x-intercept: $0 = x + 1$, $x = -1$.
y-intercept: $y = 0 + 1 = 1$.

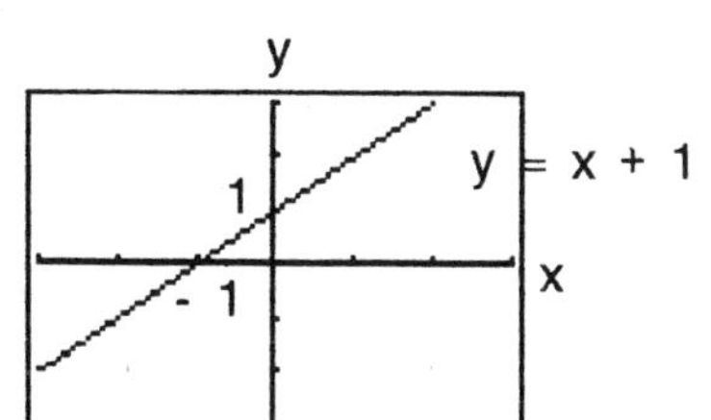

11. $y = -x^2$.
$(0, 0)$ gives the only intercepts.
y-intercept: $y = 4$.

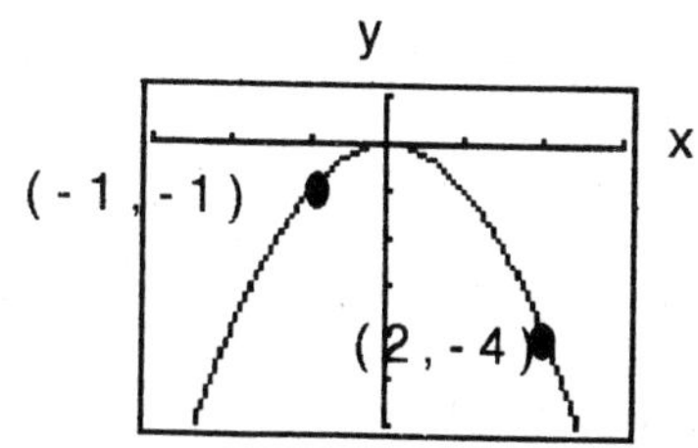

13. $x = -y^2$.
(0, 0) gives the only intercepts.

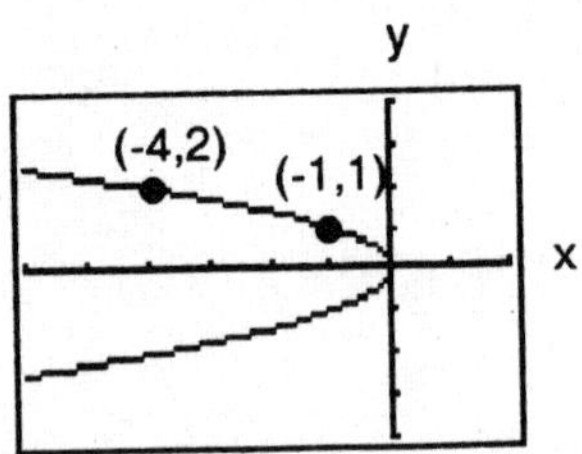

15. (e) **17.** (e) **19.** (e)

21. A complete graph of $y = 3x - 5$ may be viewed in the rectangle $[-2, 4]$ by $[-10, 5]$. x-intercept: $0 = 3x - 5$, $x = 5/3$. y-intercept: $y = 0 - 5 = -5$.

23. A graph of $y = 10 + x - 2x^2$ may be viewed in the rectangle $[-6, 6]$ by $[-30, 15]$. x-intercepts: $0 = 10 + x - 2x^2$ leads to $2x^2 - x - 10 = 0$, $(2x - 5)(x + 2) = 0$, $x = -2,\ 5/2$. y-intercept: $y = 10$.

25. A graph of $y = 2x^2 - 8x + 3$ may be viewed in the rectangle $[-7, 10]$ by $[-20, 70]$. x-intercepts: $2x^2 - 8x + 3 = 0$ leads to $x = (4 \pm \sqrt{10})/2$ or $x \approx 0.42, 3.58$. y-intercept: $y = 3$.

27. Using TRACE on the graph of $y = x^2 + 4x + 5 = (x + 2)^2 + 1$ in the rectangle $[-8, 4]$ by $[-2, 20]$, we see there are no x-intercepts, $y = 5$ is the y-intercept and $(-2, 1)$ is the low point.

29. We graph $y = 12x - 3x^3$ in the rectangle $[-4, 4]$ by $[-15, 15]$. $x = \pm 2, 0$ are the x-intercepts, $y = 0$ is the y-intercept. Using TRACE, $(-1.14, -9.24)$ and $(1.14, 9.24)$ are the approximate local low and high points, respectively.

31. A complete graph of $y = -x^3 + 9x - 1$ can be obtained in the viewing rectangle $[-5, 5]$ by $[-35, 35]$. The y-intercept is -1. Use of TRACE yields the following approximations. x-intercepts: $-3.02, 0.08, 2.93$; low point: $(-1.75, -11.39)$; high point: $(1.75, 9.39)$.

33. An idea of a complete graph of $y = x^3 + 2x^2 + x + 5$ can be obtained by using the viewing rectangles $[-3, 2]$ by $[-2, 10]$ and $[-2, 1]$ by $[4, 6]$. The y-intercept is 5 and using TRACE in the first rectangle, we obtain -2.44 as the approximate x-intercept. In the second rectangle, we obtain $(-1, 5)$ and $(-0.33, 4.85)$ as the approximate local high and low points.

35. Every pixel in the first column of pixels has x-coordinate $x\,\mathrm{Min}$. Every pixel in the second column has x-coordinate $x\,\mathrm{Min} + \delta x$. Every pixel in the third column has screen x-coordinate $x\,\mathrm{Min} + 2\delta x$. Continuing inductively, we see that every pixel in the ith column has screen x-coordinate $x\,\mathrm{Min} + (i-1)\delta x$. Similarly every pixel in the jth row, starting from the bottom, has screen y-coordinate $y\,\mathrm{Max} - (j - i)\delta y$.

37. We assume $N = 127$ and $M = 63$.

$\delta x = \frac{x\,\mathrm{Max} - x\,\mathrm{Min}}{N-1} = \frac{a-(-10)}{126} = 1, \quad a = 116.$

$\delta y = \frac{y\,\mathrm{Max} - y\,\mathrm{Min}}{M-1} = \frac{b-(-10)}{62} = 1, \quad b = 52.$

39. We take $N = 127$ and $M = 63$. $\delta x = \frac{a-(-10)}{126} = 0.5$, giving $a = 53$. $\delta y = \frac{b-(-10)}{62} = 2$, giving $b = 114$.

41. Yes, the point $(100, 60)$ is above the curve at $x = 100$.

1.2 SLOPE, AND EQUATIONS FOR LINES

1. $\Delta x = -1 - 1 = -2, \ \Delta y = -1 - 2 = -3$

3. $\Delta x = -8 - (-3) = -5, \ \Delta y = 1 - 1 = 0$

5. $m = \frac{\Delta y}{\Delta x} = \frac{1-(-2)}{2-1} = 3$. The slope of lines perpendicular to AB is $-1/3$.

7. $m = \frac{3-3}{-1-2} = 0$. The perpendicular lines are vertical and have no slope.

9. $\sqrt{(x_2 - x_1)^2 + (y_2 - y_1)^2} = \sqrt{(0-1)^2 + (1-0)^2} = \sqrt{2}$

11. $\sqrt{(-\sqrt{3} - 2\sqrt{3})^2 + (1-4)^2} = \sqrt{(-3\sqrt{3})^2 + (-3)^2} = \sqrt{9 \cdot 3 + 9} = 6$

13. $\sqrt{(0-a)^2 + (0-b)^2} = \sqrt{a^2 + b^2}$

15. $|-3| = 3$

17. $|-2 + 7| = |5| = 5$

19. $|(-2)3| = |-2||3| = 2 \cdot 3 = 6$

21. a) $x = 2$ b) $y = 3$

23. a) $x = 0$ b) $y = -\sqrt{2}$

25. $y - 1 = (1)(x - 1)$ or $y = x$

27. $y - 1 = (1)[x - (-1)] = x + 1$ or $y = x + 2$

29. $y = 2x + b$

31. $m = \frac{3-0}{2-0} = \frac{3}{2}$, $y - 0 = \frac{3}{2}(x - 0)$ or $y = \frac{3}{2}x$

33. $x = 1$

35. $m = \frac{-2-1}{2-(-2)} = \frac{-3}{4}$. $y - 1 = -\frac{3}{4}[x - (-2)]$ leads to $3x + 4y + 2 = 0$

37. $y = mx + b = 3x - 2$. Answer: $y = 3x - 2$

39. $y = x + \sqrt{2}$

41. $y = -5x + 2.5$

43. $3x + 0 = 12$, $x = 4$ is x-intercept. $0 + 4y = 12$, y-intercept is 3.

45. $4x - 3y = 12$. x-intercept: $4x - 0 = 12$, $x = 3$.
y-intercept: $0 - 3y = 12$, $y = -4$.

47. $y = 2x + 4$, x-intercept: $0 = 2x + 4$, $x = -2$. y-intercept: $y = 4$.

49. For the x-intercept we set $y = 0$: $\frac{x}{3} + 0 = 1$, $x = 3$. For the y-intercept we set $x = 0$: $0 + \frac{y}{4} = 1$, $y = 4$.

51. $\frac{x}{a} + \frac{y}{b} = 1$. For the x-intercept set $y = 0$: $\frac{x}{a} + 0 = 1$, $x = a$. For the y-intercept set $x = 0$: $0 + \frac{y}{b} = 1$, $y = b$. The point is that if we write a linear equation in the form $\frac{x}{a} + \frac{y}{b} = 1$, we can immediately read off the intercepts, or, if we know the intercepts, we can immediately write down an equation of the line.

53. $P(0,0)$, $L : y = -x + 2$. For L, $m = -1$. Hence for a perpendicular line the slope is $-(1/-1) = 1$. The line through $(0,0)$ perpendicular to L is $y - 0 = (1)(x - 0)$ or $y = x$. A complete graph of $y = -x + 2$ and $y = x$ may be seen in the viewing rectangle $[-6,6]$ by $[-4,4]$. For the point of intersection of the two lines $y = -x + 2 = x$, $2x = 2$, $x = 1$, $y = x = 1$ and $(1,1)$ is the point of intersection. The distance from P to L is the distance from P to $(1,1)$: $\sqrt{(1-0)^2 + (1-0)^2} = \sqrt{2}$.

55. $P(1,2)$, $L : x+2y = 3$. $y = -\frac{1}{2}x + \frac{3}{2}$, L has slope $-\frac{1}{2}$. The line through $(1,2)$ perpendicular to L has equation $y - 2 = 2(x - 1)$ or $y = 2x$. We may graph $y = -.5x + 1.5$ and $y = 2x$ in the viewing rectangle $[-6,6]$ by $[-4,4]$. For the point of intersection $y = 2x = -\frac{1}{2}x + \frac{3}{2}$, $4x = -x + 3$, $x = \frac{3}{5}$ and $(\frac{3}{5}, \frac{6}{5})$ is the point of intersection. The distance is $\sqrt{(1 - \frac{3}{5})^2 + (2 - \frac{6}{5})^2} = \frac{2\sqrt{5}}{5}$.

57. $P(3,6)$, $L : x + y = 3$. $y = -x + 3$ so L has slope -1. The required perpendicular has equation $y - 6 = (1)(x-3)$ or $y = x+3$. Graph $y = -x+3$ and $y = x + 3$ in the viewing rectangle $[-7,8]$ by $[-3,7]$. The point of intersection is the y-intercept $(0,3)$. The distance is $\sqrt{(0-3)^2 + (3-6)^2} = 3\sqrt{2}$

59. $P(2,1)$, $L : y = x + 2$. L has slope 1 so any line parallel to L has slope 1. The line requested has equation $y - 1 = (1)(x-2)$ or $y = x - 1$. The graphs of $y = x + 2$ and $y = x - 1$ may viewed in the standard viewing rectangle $[-10,10]$ by $[-10,10]$.

61. $P(1,0)$, $L : 2x + y = -2$ or $y = -2x - 2$. Parallel line through $P : y - 0 = -2(x-1)$ or $y = -2x + 2$.

63. a) $|x-3|$ b) $|x-(-2)| = |x+2|$

65. The distance between 5 and $x = 1$.

67. a) $a = -2$ b) $a \geqq 0$

69. a) From $(0, 69°)$ to $(0.4, 68°)$, $\frac{68°-69°}{0.4-0} = -2.5°/\text{in.}$

b) From $(0.4, 68°)$ to $(4, 10°)$, $\frac{10°-68°}{4-0.4} \approx -16.1°/\text{in.}$

c) From $(4, 10°)$ to $(4.6, 5°)$, $\frac{5°-10°}{4.6-4} = -8\frac{1}{3}°/\text{in.}$

71. $p = kd + 1$ and $10.94 = k100 + 1$. Hence $k = 0.0994$ and $p = 0.0994d + 1$. Letting $d = 50$ meters, we find $p = 5.97$ atmospheres.

73. $F = \frac{9}{5}C + 32$. Setting $F = C$, we have $C = \frac{9}{5}C + 32$, $-\frac{4}{5}C = 32$, $C = -40°$. $-40°$ Celcius is equivalent to $-40°$ Fahrenheit.

75. a) $d(t) = 45t + d_0$ where d_0 is the distance of the car from the point P at time $t = 0$.

c) $m = 45$

d) If $t = 0$ corresonds to a specific time, say 1:00 p.m., then negative values of t would correspond to times before 1:00 p.m.

e) The initial distance (from point P) $d_0 = 30$ miles.

77. Let A, B, C be, respectively, $(-1,1),(2,0),(2,3)$. For one side of the parallelogram we may take the vertical line segment BC of length 3. Thus for the fourth vertex D we take either $(-1,4)$ or $(-1,-2)$. The third possibility for D, $(5,2)$, is found as the intersection of line $BD(y=\frac{2}{3}(x-2))$ and line $CD(y=3-\frac{1}{3}(x-2))$.

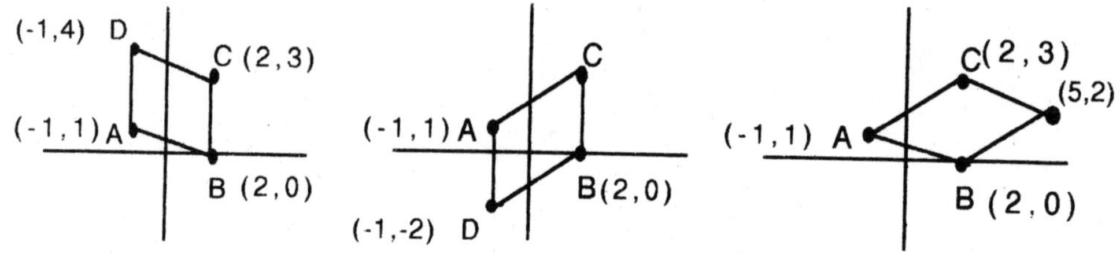

79. The point of intersection is $(1,1)$. An equation of the line through $(1,1)$ and $(1,2)$ is $x=1$.

1.3 RELATIONS, FUNCTIONS, AND THEIR GRAPHS

1. This relation is not a function. The graph does not pass the vertical line test. There are x-values that correspond to two y-values.

3. This is a function. No value of x corresponds to two y-values.

5. a) $(3,-1)$ b) $(-3,1)$ c) $(-3,-1)$

7. a) $(-2,-1)$ b) $(2,1)$ c) $(2,-1)$

9. a) $(1,\sqrt{2})$ b) $(-1,-\sqrt{2})$ c) $(-1,\sqrt{2})$

11. a) $(0,-\pi)$ b) $(0,\pi)$ c) $(0,-\pi)$

13. Domain: $[1,\infty)$, range: $[2,\infty)$

15. Domain: $(-\infty,0]$, range: $(-\infty,0]$

17. Domain: $(-\infty,3]$, range: $[0,\infty)$

19. Domain: $(-\infty,2)\cup(2,\infty)$, range: $(-\infty,0)\cup(0,\infty)$; any non-zero number is a reciprocal of another number.

21. Domain: $(-\infty,\infty)$, range: $[-9,\infty)$. Symmetric about the y-axis.

23. Domain = range = $(-\infty,\infty)$. No symmetry.

25. Domain = range = $(-\infty, \infty)$. No symmetry.

27. Domain = range = $(-\infty, 0) \cup (0, \infty)$. Symmetric about the origin.

29. Domain: $(-\infty, 0) \cup (0, \infty)$. Since $\frac{1}{x}$ can have any value except 0, $1 + \frac{1}{x}$ can have any value except 1. Range: $(-\infty, 1) \cup (1, \infty)$. No symmetry.

31. a) No b) No c) $(0, \infty)$

33. Odd **35.** Neither **37.** Even **39.** Even **41.** Odd

43. Symmetric about the y-axis. Graph $y = -x^2$ in the viewing rectangle $[-10, 10]$ by $[-10, 10]$.

45. Symmetric about the y-axis. Graph $y = 1/x^2$ in the viewing rectangle $[-5, 5]$ by $[0, 3]$.

47. Since $(-x)(-y) = xy$, the equation is unchanged if x is replaced by $-x$ and y by $-y$. Thus the graph is symmetric about the origin. A complete graph of $y = 1/x$ can be obtained in $[-4, 4]$ by $[-4, 4]$.

49. $x^2y^2 = 1$ has graph symmetric about both axes and the origin. Graph $y = 1/|x|$ and $y = -1/|x|$ in the viewing rectangle $[-4, 4]$ by $[-4, 4]$.

51. The graph of $y = |x + 3|$ can be obtained by translating the graph of $y = |x|$ three units to the left. Graph $y = \text{abs}(x + 3)$ in the viewing rectangle $[-7, 1]$ by $[0, 4]$.

53. Graph $y = \frac{|x|}{x}$ in the viewing rectangle $[-2, 2]$ by $[-2, 2]$. There is not point on the graph when $x = 0$.

55. $y = \frac{x-|x|}{2}$. $y = x$ when $x \leqq 0$ and $y = 0$ when $x \geqq 0$.

57. a) Graph $y = 3 - x + 0\sqrt{1 - x}$ and $y = 2x + 0\sqrt{x - 1}$ in the viewing rectangle $[-10, 10]$ by $[0, 20]$. b) $f(0) = 3$, $f(1) = 2$, $f(2.5) = 5$.

59. a) Graph $y = 1 + 0\sqrt{5 - x}$ in the viewing rectangle $[-3, 10]$ by $[0, 2]$. It is understood that the x-axis for $x \geqq 5$ is part of the graph and the point $(5, 1)$ is not. b) $f(0) = 1$, $f(5) = 0$, $f(6) = 0$.

61. a) Graph $y = 4 - x^2 + 0\sqrt{1 - x}$, $y = \frac{3}{2}x + \frac{3}{2} + 0\sqrt{x - 1} + 0\sqrt{3 - x}$ and $y = x + 3 + 0\sqrt{x - 3}$ in the viewing rectangle $[-3, 7]$ by $[-5, 10]$. b) $f(0.5) = 3.75$, $f(1) = 3$, $f(3) = 6$, $f(4) = 7$.

63. a) $1 - |x-1|,\ 0 \leqq x \leqq 2$ b) $f(x) = \begin{cases} 2, & 0 \leqq x < 1 \\ 0, & 1 \leqq x < 2 \\ 2, & 2 \leqq x < 3 \\ 0, & 3 \leqq x \leqq 4 \end{cases}$

65. a) $0 \leq x < 1$ b) $-1 < x \leqq 0$

67. a) Graph of $y = x - [x],\ -3 \leqq x \leqq 3$

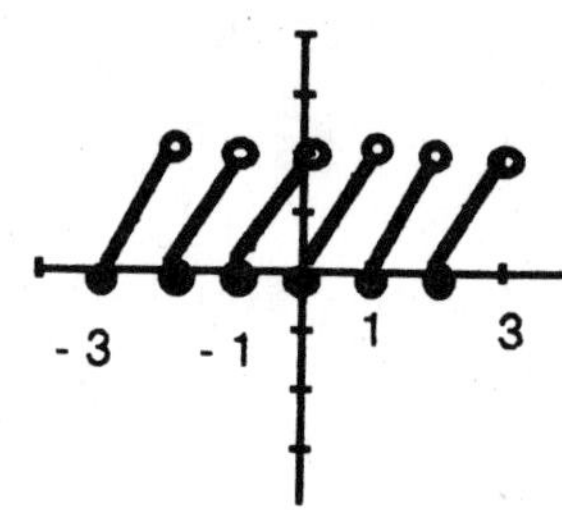

b) Graph of $y = [x] - \lceil x \rceil,\ -3 \leqq x \leqq 3$

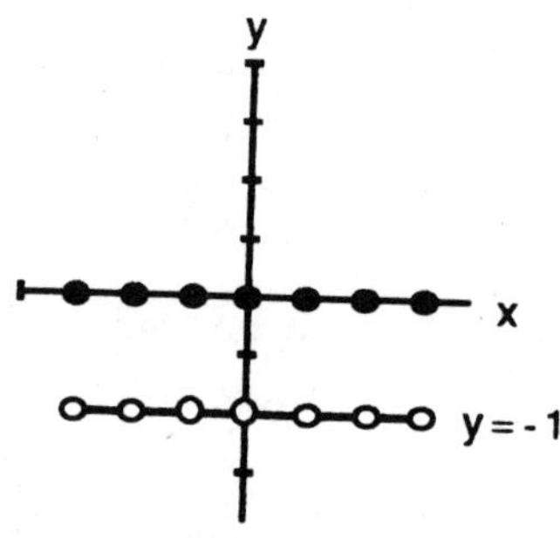

69. Graph $y = abs(x+1) + 2abs(x-3)$ in the viewing rectangle $[-2,5]$ by $[0,12]$. For $x \leqq -1$, both $x+1$ and $x-3$ are non-positive and $f(x) = |x+1| + 2|x-3| = -(x+1) - 2(x-3) = -3x+5$. For $-1 < x \leq 3$, $x+1$ is positive and $x-3$ is nonpositive and $f(x) = x+1-2(x-3) = -x+7$. For $x > 3$ both $x+1$ and $x-3$ are positive and $f(x) = x+1+2(x-3) = 3x-5$. Thus

$$f(x) = \begin{cases} -3x+5, & x \leqq -1 \\ -x+7, & -1 < x \leqq 3 \\ 3x-5, & x > 3 \end{cases}$$

71. Graph $y = |x| + |x-1| + |x-3|$ in the viewing rectangle $[-1,4]$ by $[0,7]$. $y = -3x+4,\ x \leqq 0,\ y = -x+4,\ 0 < x \leqq 1,\ y = x+2,\ 1 < x \leqq 3,\ y = 3x-4,\ x > 3$.

73. $f(x) = x$, domain: $(-\infty, \infty)$. $g(x) = \sqrt{x-1}$, domain: $[1, \infty)$. $f(x)+g(x) = x + \sqrt{x-1}$, domain $(f+g)$: $[1,\infty)$, complete graph in $[0,5]$ by $[0,10]$. $f(x) - g(x) = x - \sqrt{x-1}$, domain $(f-g)$: $[1,\infty)$, complete graph in $[0,5]$

by $[0,4]$. $f \circ g(x) = f(g(x)) = f(\sqrt{x-1}) = \sqrt{x-1}$, domain $f \circ g : [1,\infty)$, complete graph in $[0,5)$ by $[0,2]$. $f(x)/g(x) = x/\sqrt{x-1}$, domain: $(1,\infty)$, complete graph in $[0,10]$ by $[0,5]$. $g(x)/f(x) = \sqrt{x-1}/x$, domain: $[1,\infty)$, complete graph in $[0,10]$ by $[0,0.5]$, graph starts at $(1,0)$.

75. a) $f(g(0)) = f(-3) = 2$ b) $g(f(0)) = g(5) = 22$ c) $f(g(x)) = f(x^2-3) = (x^2-3)+5 = x^2+2$ d) $g(f(x)) = g(x+5) = (x+5)^2-3 = x^2+10x+22$ e) $f(f(-5)) = f(0) = 5$ f) $g(g(2)) = g(1) = -2$ g) $f(f(x)) = f(x+5) = (x+5)+5 = x+10$ h) $g(g(x)) = g(x^2-3) = (x^2-3)^2-3 = x^4-6x^2+6$

77. a) $f \circ g(x) = f(g(x)) = f(x-7) = \sqrt{x-7}$
b) $f(g(x)) = f(x+2) = 3(x+2)$
c) $f \circ g(x) = f(g(x)) = \sqrt{g(x)-5} = \sqrt{x^2-5}$ so $g(x) = x^2$
d) $f(g(x)) = f(\frac{x}{x-1}) = \frac{(\frac{x}{x-1})}{(\frac{x}{x-1})-1} = \frac{x}{x-(x-1)} = x$
e) $f(g(x)) = 1 + \frac{1}{g(x)} = x$, $\frac{1}{g(x)} = x-1$, $g(x) = \frac{1}{x-1}$
f) $f(g(x)) = f(\frac{1}{x}) = x$ so $f(x) = \frac{1}{x}$.

79. a) $C(10) = 72$ b) $C(30) - C(20)$ is the increase of cost if the production level is raised from 20 to 30 items daily.

81. The two functions are identical.

83. $g(f(x)) = g(x^2+2x+1) = g((x+1)^2) = |x+1|$. $g(x) = \sqrt{x}$ is one possibility.

85. Graph $y = abs(x+3) + abs(x-2) + abs(x-4)$ in the viewing rectangle $[-4,5]$ by $[0,15]$. We see that $d(x)$ is minimized when $x = 2$ so you would put the table next to Machine 2.

87. Graph $y = 2|x+3| + |x+1| + 3|x-2| + |x-6|$ in the viewing rectangle $[-4,7]$ by $[15,40]$. $d(x)$ now has minimum value 17 when $x = 2$. The table should be placed next to Machine 3.

1.4 GEOMETRIC TRANSFORMATIONS: SHIFTS, REFLECTIONS, STRETCHES, AND SHRINKS

1. a) $y = (x+4)^2$ b) $y = (x-7)^2$

3. a) Position 4 b) Position 1 c) Position 2 d) Position 3

5. Shift the graph of $|x|$ to the left 4 units. Then shift the resulting graph down 3 units.

7. Reflect the graph of $y = \sqrt{x}$ over the y-axis and then stretch the resulting graph vertically by a factor of 3.

9. Stretch the graph of $y = \frac{1}{x}$ vertically by a factor of 2 and shift the resulting graph down 3 units.

11. Shift the graph of $y = x^3$ right 3 units. Shrink the resulting graph by a factor of 0.5. Reflect the last graph over the x-axis and shift the last graph up 1 unit.

13. Shift the graph of $y = \frac{1}{x}$ to the right 2 units (obtaining the graph of $\frac{1}{x-2}$). Shift the resulting graph up 3 units.

15. Start with the graph of $y = \sqrt[3]{x}$. Stretch vertically by a factor of 4 and then reflect over the y-axis. We now have the graph of $y = 4\sqrt[3]{-x}$. Now shift to right 2 units obtaining the graph of $y = 4\sqrt[3]{-(x-2)} = 4\sqrt[3]{2-x}$. Finally shift the last graph 5 units down. Viewing rectangle: $[-8, 12]$ by $[-15, 5]$. Domain = range = $(-\infty, \infty)$.

17. $y = 5\sqrt[3]{-x} - 1$. Start with the graph of $y = \sqrt[3]{x}$, reflect through the y-axis, stretch vertically by a factor of 5, shift vertically down one unit. Check your result by graphing y in $[-5, 5]$ by $[-10, 10]$. Since $\sqrt[3]{x}$ has domain = range = $(-\infty, \infty)$, the same is true of the present function.

19. Start with the graph of $y = \frac{1}{x^2}$. Shift to left 3 units, reflect over the x-axis and shift up 2 units. Viewing rectangle: $[-10, 5]$ by $[-3, 3]$. Domain = $(-\infty, -3) \cup (-3, \infty)$, range = $(-\infty, 2)$.

21. Check your result by graphing $y = -2((x-1)^{(1/3)})^2 + 1$ in the viewing rectangle $[-1, 3]$ by $[-2, 1]$. Domain = $(-\infty, \infty)$, range = $(-\infty, 1]$.

23. Check your result by graphing $y = 2[1-x] = 2\ \mathrm{Int}(1-x)$ in the viewing window $[-3, 4]$ by $[-6, 8]$. (Graph this in Dot Mode if possible.) Domain = $(-\infty, \infty)$, range = $\{2n : n = 0, \pm 1, \pm 2, \ldots\}$.

25. $y = x^2 \to y = 3x^2 \to y = 3x^2 + 4$

27. $y = \frac{1}{x} \to y = \frac{1}{x} - 2 \to y = 0.2(\frac{1}{x} - 2)$

29. $y = |x| \to y = |x+2| \to y = 3|x+2| \to y = 3|x+2| + 5$

31. $y = x^3 \to y = -x^3 \to y = -0.8x^3 \to y = -0.8(x-1)^3 \to y = -0.8(x-1)^3 - 2$

33. $y = \sqrt{x} \to y = \sqrt{-x} \to y = 5\sqrt{-x} \to y = 5\sqrt{-(x+y)} \to y = 5\sqrt{-(x+6)} + 5$

35. $y = \sqrt{3x} \to y = \sqrt{3(\frac{1}{2}x)} \to y = \sqrt{\frac{3}{2}x} + 1$

37. Vertical stretch by 4 : $y = x^2 \to y = 4x^2$. Horizontal shrink by 0.5 : $y = x^2 \to y = (2x)^2 = 4x^2$. The resulting curve is the same in both cases. Let $c > 1$ be given. A vertical stretch by c applied to $y = x^n : y = x^n \to y = cx^n$ and a horizontal shrink by $\frac{1}{\sqrt[n]{c}} : y = x^n \to y = (\sqrt[n]{c}x)^n = cx^n$ have the same end result.

39. In #25 and #26 we obtain, respectively, $y = 3x^2 + 4$ and $y = 3(x^2 + 4) = 3x^2 + 12$. We obtain different geometric results by reversing the order of the transformations. The second graph is 8 units above the first graph.

41. a) Reflect across the y-axis. b) Reflect across the x-axis. c) They are the same: $y = \sqrt[3]{-x} = -\sqrt[3]{x}$.

43. $y = mx + b$

45. $(-1, 1) \to (2, 1) \to (2, 3)$. $(0, 0) \to (3, 0) \to (3, 2)$. $(1, 1) \to (4, 1) \to (4, 3)$. The points are $(2, 3), (3, 2), (4, 3)$.

47. $(-1, 1) \to (-1, 2) \to (-1, -2)$. $(0, 0) \to (0, 0) \to (0, 0)$. $(1, 1) \to (1, 2) \to (1, -2)$. The points are $(-1, -2), (0, 0), (1, -2)$.

49.

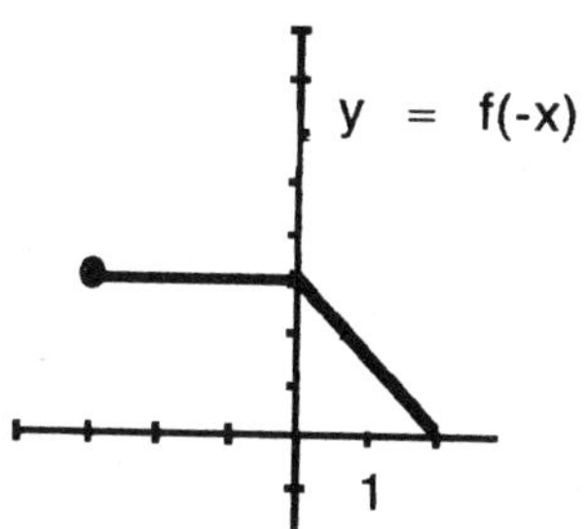

51.

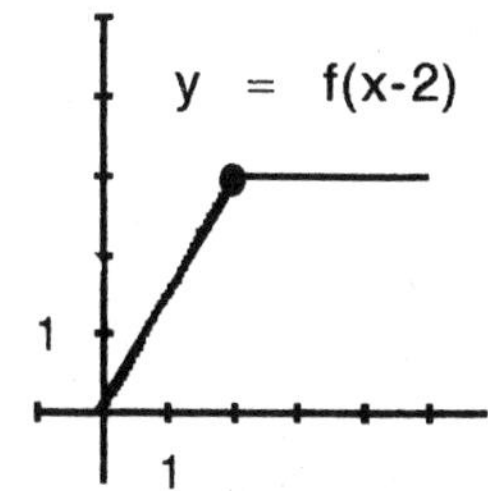

53.

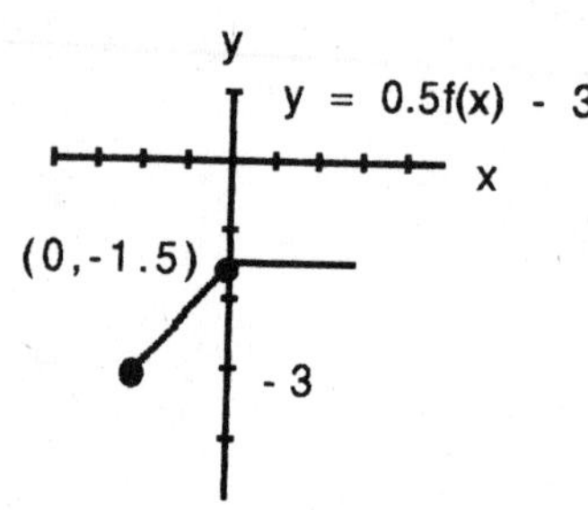

55.

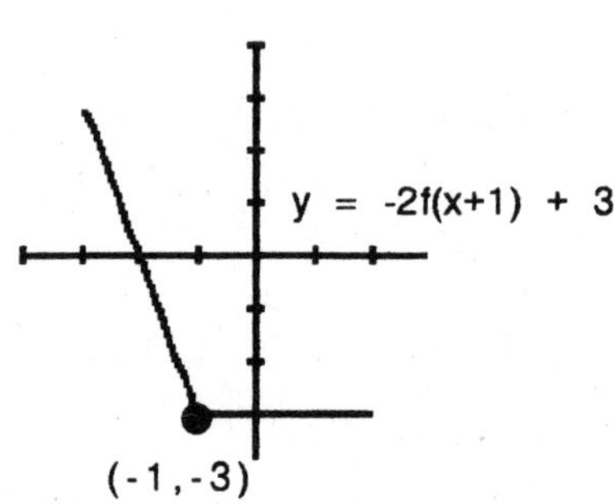

57. Using $y = 2 + \frac{5}{x-2}$, we start with the graph of $y = \frac{1}{x}$. Stretch vertically by a factor of 5, shift right 2 units, shift up 2.

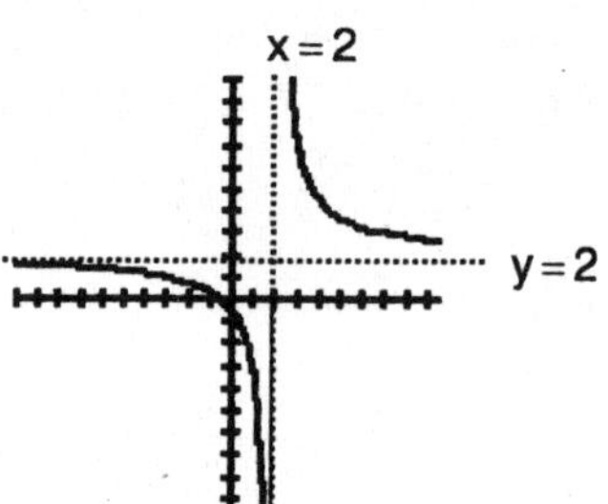

59. $y = -2x^2 + 12x - 11 = -2(x^2 - 6x) - 11 = -2(x^2 - 6x + 9) + 2(9) - 11 = -2(x-3)^2 + 7$. Vertex $= (3, 7)$, axis of symmetry: $x = 3$. Check your result by graphing $y = -2x^2 + 12x - 11$ in the viewing rectangle $[-1, 7]$ by $[-16, 7]$.

61. $y = 4x^2 + 20x + 19 = 4[x^2 + 5x + (\frac{5}{2})^2] - 4(\frac{5}{2})^2 + 19 = 4(x + \frac{5}{2})^2 - 6$. Vertex: $(-\frac{5}{2}, -6)$, axis of symmetry: $x = -\frac{5}{2}$. Check your result by graphing the function in $[-5, -0]$ by $[-6, 10]$.

63. $x = y^2 - 6y + 11 = y^2 - 6y + 9 + 2 = (y-3)^2 + 2$. Start with the graph of $x = y^2$. Shift up 3, shift right 2. $(y-3)^2 = x - 2$, $y = 3 \pm \sqrt{x-2}$. Check your sketch by graphing $y = 3 + \sqrt{x-2}$ and $y = 3 - \sqrt{x-2}$ in the viewing rectangle $[2, 20]$ by $[-5, 11]$. In parametric mode use $x = t^2 - 6t + 11$, $y = t$, $-5 \leqq t \leqq 11$.

65. $x = 2y^2+4y+1 = 2(y^2+2y)+1 = 2(y^2+2y+1)-2\cdot1+1 = 2(y+1)^2-1$. Shift down 1, stretch horizontally by 2, shift left 1. $\frac{x+1}{2} = (y+1)^2$, $y = -1\pm\sqrt{\frac{x+1}{2}}$. Check your sketch by graphing $y = -1+\sqrt{\frac{x+1}{2}}$ and $y = -1-\sqrt{\frac{x+1}{2}}$ in $[-1, 10]$ by $[-4, 2]$. In parametric mode $x = 2t^2 + 4t + 1$, $y = t$, $-4 \leqq t \leqq 2$.

67. $x = -2y^2 + 12y - 13 = -2(y^2 - 6y + 9) + 18 - 13 = -2(y - 3)^2 + 5$. Shift up 3, horizontal stretch by 2, reflection across y-axis, shift right 5. $(y-3)^2 = \frac{x-5}{-2}$, $y = 3\pm\sqrt{\frac{5-x}{2}}$. Check your sketch by graphing $y = 3+\sqrt{\frac{5-x}{2}}$ and $y = 3-\sqrt{\frac{5-x}{2}}$ in $[-9, 5]$ by $[0, 7]$. In parametric mode $x = -2t^2 + 12t - 13$, $y = t$, $0 \leqq t \leqq 7$.

69. Two. Only the graphs of $g(x) = (x + 2)^2 + 3$ and $g(x) = (2 - x)^2 + 3$ are possible.

1.5 SOLVING EQUATIONS AND INEQUALITIES GRAPHICALLY

1. $6x^2+5x-6 = 0$. By the quadratic formula or by factoring $((2x+3)(3x-2) = 0)$, we see that $\{-3/2, 2/3\}$ is the solution set. The graph of $y = 6x^2+5x-6$ is a parabola opening upwards with $-3/2$, $2/3$ as x-intercepts.

3. $4x^3 - 16x^2 + 15x + 2 = 0$. By inspection 2 is a root and so $x - 2$ must be a factor: $(x - 2)(4x^2 - 8x - 1) = 0$ by long division. The solution set is $\{1 - \sqrt{5}/2, 2, 1 + \sqrt{5}/2\}$.

5. $2x^2 + 7x - 4 = (2x - 1)(x + 4) = 0$. The solution set is $\{-4, \frac{1}{2}\}$.

7. $18x^3 - 3x^2 - 14x - 4 = 0$. We determine that $-\frac{1}{2}$ is a root and factor: $2(x+\frac{1}{2})(9x^2-6x-4) = 0$. Using the preceding exercise, we find the solution set to be $\{-0.5, -0.41, 1.08\}$ or $\{-\frac{1}{2}, \frac{1\pm\sqrt{5}}{3}\}$

9. We give a sequence for the smallest solution only. $[-2, -1]$ by $[-1, 1]$, $x\,Scl = 0.1$; $[-2, -1.9]$ by $[-0.1, 0.1]$, $x\,Scl = 0.01$; $[-1.94, -1.93]$ by $[-0.01, 0.01]$, $x\,Scl = 0.001$; $[-1.931, -1.930]$ by $[-0.001, 0.001]$, $x\,Scl = 0.0001$.

11. $9x^2 - 6x + 5 = 0$ has no real solution since the discriminant $b^2 - 4ac = (-6)^2 - 4(9)(5) < 0$.

13. $2x^3-8x^2+3x+9 = 0$. Graphing the cubic in $[-10, 10]$ by $[-10, 10]$, $x\,Scl = 1$ suggests $x = 3$ is a solution. This is verified and leads to factoring $(x - 3)(2x^2 - 2x - 3) = 0$. The solution set is $\{3, \frac{1\pm\sqrt{7}}{2}\}$

15. Obtain a complete graph of $y = x^3 - 21x^2 + 111x - 71$ in $[0, 15]$ by $[-70, 120]$. To solve the problem we zoom-in to the 3 x-intercepts and obtain $\{0.74, 7.56, 12.70\}$.

17. We assume that the number x satisfies $2 < x < 6$. The statement a) $0 < x < 4$ about x does not contain precise enough information about x to be able to determine whether the statement is true or false. The same goes for the statement g) which is equivalent to $-2 < x < 6$. Use of rules of inequalities shows that the remaining statements are all equivalent to $2 < x < 6$ whence they are true.

19. $\{-2, 2\}$

21. $|2x + 5| = 4$ leads to $2x + 5 = -4$ and $2x + 5 = 4$. These have solutions $x = -9/2$ and $x = -1/2$. The solution set is $\{-9/2, -1/2\}$.

23. $|8 - 3x| = 9$. $8 - 3x = -9$ and $8 - 3x = 9$. $x = 17/3$ and $x = -1/3$.

25. The distance between y and 1 is at most 2: $-1 \leqq y \leqq 3$.

27. $|3y - 7| < 2$ is equivalent to $-2 < 3y - 7 < 2$, $5 < 3y < 9$, $5/3 < y < 3$.

29. The distance between 1 and y is less than $1/10$. $0.9 < y < 1.1$

31. The midpoint of the interval is $(3 + 9)/2 = 6$ and it is 3 units from the midpoint to an endpoint. Thus $|x - 6| < 3$.

33. The midpoint is $(-5 + 3)/2 = -1$ and from there to an endpoint is 4 units. $|x + 1| < 4$.

35. $|x - 5| < 2$ is equivalent to $-2 < x - 5 < 2$, $3 < x < 7$. Graph $y = abs(x - 5)$ and $y = 2$ and verify that the first curve is below the second in the interval.

37. $|\frac{4}{x-1}| \leqq 2$ is equivalent to $\frac{|x-1|}{4} \geqq \frac{1}{2}$, $x \neq 1$. $|x - 1| \geq 2$, distance between x and 1 at least 2, leads to $x \leqq -1$ or $x \geqq 3$ but $x \neq 1$. Thus $(-\infty, -1] \cup [3, \infty)$ is the solution set.

39. We observe that the graph of $y = |x + 3|$ is below the graph of $y = 5$ in the interval $[-8, 2]$. $|x + 3| = |x - (-3)| \leqq 5$, the distance between x and -3 is at most 5. Hence $-8 \leqq x \leqq 2$.

41. $|\frac{2}{x+3}| < 1$. $\frac{|x+3|}{2} > 1$, $x \neq -3$, $|x + 3| > 2$. We obtain $x + 3 < -2$ or $x + 3 > 2$, i.e., $x < -5$ or $x > -1$. $(-\infty, -5) \cup (-1, \infty)$ is the solution set.

43. $|2-3x|<4$. $-4<2-3x<4$, $-6<-3x<2$, $2>x>-2/3$. Answer: $(-2/3,2)$

45. $x^2+3x-10 \leqq 0$, $(x+5)(x-2) \leqq 0$. The graph of the left-hand side is a parabola that opens upward. Thus the solution set will be the interval between the two zeros including the endpoints. Answer: $[-5,2]$

47. Let $y=x^3-6x^2+5x+6$. The graph suggests, and we verify, that 2 is a root. $y=(x-2)(x^2-4x-3)$. The other roots are $2\pm\sqrt{7}$. Answer: $(-\infty,2-\sqrt{7}]\cup[2,2+\sqrt{7}]$

49. We graph $y=x^3-4x^2+3.99x$ in $[-2,5]$ by $[-2,2]$. We know this is a complete graph because we a dealing with a cubic function. We determine by Zoom-In those intervals where the graph is above the x-axis. Answer: $(0,1.9)\cup(2.1,\infty)$.

51. $y=30x-x^2=x(30-x)$. The vertex will occur for $x=15$ midway between the two zeros. Thus $(15,225)$ is the vertex.

53. b) If the y-range is too large, the graphing utility cannot distinguish between very close values of y.

55. We may multiply both sides of the equation by the least common multiple of the denominators of the coefficients. This produces an equivalent equation with integer coefficients. For the given example the l.c.m. is 6 and we obtain the equation $12x^3+3x^2-4x+6=0$. The possible rational roots of this equation are $\pm\frac{p}{q}$ where $p=1,2,3$ or 6 and $q=1,2,3,4,6$ or 12.

57. a) Let y be the length of the adjacent side. $2x+2y=100$, $x+y=50$, $y=50-x$, $A=xy$, $A=x(50-x)$. b) Check you sketch by graphing $y=x(50-x)$ in $[-10,60]$ by $[-100,700]$. c) Domain $=(-\infty,\infty)$, range $=(-\infty,625]$ d) Both x and y must be positive so $0<x<50$. e) Using TRACE for the graph in b) we get the approximate dimensions 13.8ft by 36.2ft. Algebraically we have to solve the system $x+y=50$, $x(50-x)=500$. This gives the dimensions $25-5\sqrt{5}$ft by $25+5\sqrt{5}$ft. f) In the problem situation the graph is below $y=500$ for $0<x<25-5\sqrt{5}$ and for $25+5\sqrt{5}<x<50$.

59. a) $A(x)=(8.5-2x)(11-2x)$ b) Graph $y=A(x)$ in $[0,10]$ by $[-2,50]$ c) Domain is $(-\infty,\infty)$. The vertex occurs midway between the x-intercepts at $((8.5/2)+(11/2))/2=4.875$. $A(4.875)=-1.5625$. The range is $[-1.5625,\infty)$. d) We must have $8.5-2x>0$ or $0<x<8.5/2$, $0<x<4.25$. The graph in this interval is the graph of the problem situation. e) $y = (8.5 - 2x)(11 - 2x)$, $y = 60$ has one solution for $0 < x < 4.25$. Zooming in we find its x-coordinate to be $x = 0.95$in.

61. a) $V(x) = 0.1x + 0.25(50 - x)$ b) Graph $y = V(x)$ in $[-20, 100]$ by $[-2, 20]$ c) Domain = range = $(-\infty, \infty)$ d) $0 \leqq x \leqq 50$, x an integer e) $0.1x + 0.25(50 - x) = 9.2$ leads to $10x + 25(50 - x) = 920$, $2x + 5(50 - x) = 184$, $-3x = -66$, $x = 22$. There are 22 dimes and 28 quarters. f) $0.1x + 0.25(50 - x) = 6.25$ does not have an integral solution.

63. a) $V = x(20 - 2x)(25 - 2x)$ b) Graph $y = V$ in $[-2, 17]$ by $[-100, 900]$. c) Domain = range = $(-\infty, \infty)$ d) We must have $x > 0$ and $20 - 2x > 0$ so $0 < x < 10$. Graphing $y = V$ in $[0, 10]$ by $[0, 900]$ gives a graph of the problem situation. e) We zoom in to find the high point to be $(3.68, 820.53)$. For the maximum volume of 820.53in^3 we take $x = 3.68$in.

65. a) Let L be the length (horizontal as pictured) of the lid. $2x + 2L = 11$, $2L = 11 - 2x$, $L = 5.5 - x$. b) $V(x) = x(5.5 - x)(8.5 - 2x)$ c) Graph V in $[-1, 7]$ by $[-10, 50]$. The domain and range are both the set of all real numbers. d) The width must be positive: $8.5 - 2x > 0$ which leads to $0 < x < 4.25$. Graph V in $[0, 4.25]$ by $[-10, 50]$. e) In the same viewing window we also graph $y = 25$ and find the x-coordinates of the points of intersection: 0.753in and 2.592in f) $x = 1.585$in. $V_{\text{max}} = V(1.585) = 33.074\text{in}^3$.

1.6 RELATIONS, FUNCTIONS, AND THEIR INVERSES

1. Not one-to-one. Does not pass the Horizontal Line Test.

3. One-to-one

5. The given line is determined by $(-3, 0)$ and $(0, 2)$. The inverse relation has graph a line through $(0, -3)$ and $(2, 0)$.

7.

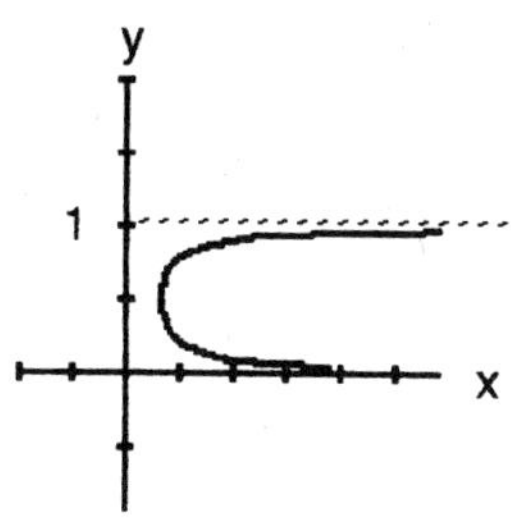

9. $x^2 + (y-2)^2 = 4$. Graph $y = 2 + \sqrt{4-x^2}$ and $y = 2 - \sqrt{4-x^2}$ in $[-3.4, 3.4]$ by $[0, 4]$.

11. $(x-3)^2 + (y+4)^2 = 25$. Graph $y = -4 + \sqrt{25-(x-3)^2}$ and $y = -4 - \sqrt{25-(x-3)^2}$ in $[-5.5, 11.5]$ by $[-9, 1]$.

13. $x^2 + y^2 = 4$

15. $(x-3)^2 + (y-3)^2 = 9$

17. $x^2 + y^2 - 6x + 8y = -16$. $(x^2 - 6x) + (y^2 + 8y) = -16$, $(x^2 - 6x + 9) + (y^2 + 8y + 16) = -16 + 9 + 16$, $(x-3)^2 + (y+4)^2 = 9$. Center: $(3, -4)$. Radius: 3. Graph $x = 3 + 3\cos t$, $y = -4 + 3\sin t$, $0 \leqq t \leqq 2\pi$, in $[-2.1, 8.1]$ by $[-7, -1]$. Domain $= [0, 6]$, range $= [-7, -1]$.

19. $x^2+y^2+4x+6y+8 = 0$, $x^2+4x+4+y^2+6y+9 = -8+4+9$, $(x+2)^2+(y+3)^2 = 5$. $C = (-2, -3)$, $r = \sqrt{5}$. Domain $= [-2-\sqrt{5}, -2+\sqrt{5}]$, range $= [-3-\sqrt{5}, -3+\sqrt{5}]$. Graph $x = -2+\sqrt{5}\cos t$, $y = -3+\sqrt{5}\sin t$, $0 \leqq t \leqq 2\pi$, in $[-8.15, 4.15]$ by $[-5.7, 1.5]$.

21. $16x^2 - 9y^2 = 144$. $9y^2 = 16x^2 - 144 = 16(x^2 - 9)$, $y^2 = \frac{16}{9}(x^2 - 9)$. We graph $y = \frac{4}{3}\sqrt{x^2-9}$ and $y = -\frac{4}{3}\sqrt{x^2-9}$ in $[-10, 10]$ by $[-10, 10]$. The calculator screen shows incorrect gaps in the curve when the slopes of tangent lines become numerically very large.

23. a) All points outside the boundary of the unit circle. b) All points in the interior of the circular disk with center $(0, 0)$ and radius 2. c) The ring between the two circles.

25. $(x+2)^2 + (y+1)^2 < 6$

27. The inverse relation is given by $x = \frac{3}{y-2} - 1$. Its graph is obtained by graphing $x_1 = 3/(t-2) - 1$, $y_1 = t$, $-10 \leqq t \leqq 10$, tstep $= 0.1$, in $[-10, 10]$ by $[-10, 10]$. We see that the inverse is a function since it passes the vertical line test.

29. Graph $x = t^3 - 4t + 6$, $y = t$, $-10 \leqq t \leqq 10$, tstep $= 0.1$, in $[-10, 20]$ by $[-15, 15]$. This is not a function.

31. The inverse is given by $x = \ell n\, y^2$ or $e^x = y^2$ which we can graph by graphing $y = e^{x/2}$ and $y = -e^{x/2}$ in $[-2, 5]$ by $[-10, 10]$. This is not a function.

33. $y = f(x) = 2x + 3$. To find f^{-1} we interchange x and $y : x = 2y + 3$, $y = (x-3)/2 = f^{-1}(x)$. $f \circ f^{-1}(x) = f(f^{-1}(x)) = f[(x-3)/2] = 2[(x-3)/2] + 3 = (x-3) + 3 = x$. $f^{-1} \circ f(x) = f^{-1}(2x+3) = [(2x+3) - 3]/2 = x$. Graph $y = 2x + 3$ and $y = (x-3)/2$ in $[-10, 10]$ by $[-10, 10]$.

35. $y = f(x) = x^3 - 1$. $x = y^3 - 1$, $y^3 = x + 1$, $y = \sqrt[3]{x+1} = f^{-1}(x)$.

37. $y = f(x) = x^2 + 1$, $x \geqq 0$. $x = y^2 + 1$, $y = \pm\sqrt{x-1}$. The domain of f, $x \geqq 0$, must be the range of f^{-1} so we take $f^{-1}(x) = \sqrt{x-1}$.

39. $y = f(x) = -(x-2)^2$, $x \leqq 2$. $x = -(y-2)^2$, $y - 2 = \pm\sqrt{-x}$, $y = f^{-1}(x) = 2 - \sqrt{-x}$, $x \leqq 0$. We make sure that the domain $x \leqq 2$ and the range $(-\infty, 0]$ of f become the range $(\infty, 2]$ and the domain $(-\infty, 0]$ of f^{-1}.

41. $y = f(x) = \frac{1}{x^2}$, $x \geqq 0$. $x = \frac{1}{y^2}$, $y = \pm\frac{1}{\sqrt{x}}$. We have $f^{-1}(x) = \frac{1}{\sqrt{x}}$ because the domain $x \geqq 0$ of f must be the range of f^{-1}.

43. $y = f(x) = \frac{2x+1}{x+3}$, $x \neq -3$. $x = \frac{2y+1}{y+3}$, $(y+3)x = 2y + 1$, $xy + 3x = 2y + 1$, $xy - 2y = 1 - 3x$, $y(x-2) = 1 - 3x$, $y = f^{-1}(x) = \frac{1-3x}{x-2}$, $x \neq 2$.

45. Check your sketch by graphing $y = 2[\ell n(x-4)/\ell n(3)] - 1$ in $[3, 14]$ by $[-10, 4]$ noting that $x = 4$ is a vertical asymptote.

47. Check your sketch by graphing $y = -3[\ell n(x+2)/\ell n(0.5)] + 2$ in $[-2, 4]$ by $[-18, 10]$.

49. Graph $y = 5(e^{3x}) + 2$ in $[-2, 1]$ by $[0, 20]$.

51. Graph $y = -2(3^x) + 1$ in $[-4, 2]$ by $[-10, 2]$.

53. The graph of $\log x = \log_{10} x$ has the same shape as that in Figure 1.91. We start with the graph of $y = \log x$, reflect it across the x-axis and stretch vertically by a factor of 3. (We now have the graph of $y = -3\log x$.) We then shift left 2 and then up 1. The domain comes from $x + 2 > 0$ so it is $(-2, \infty)$ and the range is $(-\infty, \infty)$.

55. Start with the graph of $y = 3^x$, shift left 1 ($y = 3^{x+1}$), reflect across the y-axis ($y = 3^{-x+1}$), stretch vertically by a factor of 2 and shift up 1.5. Domain $= (-\infty, \infty)$, range $= (1.5, \infty)$.

57. Shift the graph of $x^2 + y^2 = 9$ three units left and 5 units up. Endpoints of domain are $x = -3 \pm 3$ and the endpoints of the range are $y = 5 \pm 3$. Thus domain $= [-6, 0]$, range $= [2, 8]$.

59. $f(x) = 2^x$. $f^{-1}(x) = \log_2 x$. Graph $y_1 = 2^x$, $y_2 = \frac{\ell n\, x}{\ell n\, 2}$, $y_3 = x$ in $[-7.5, 14.6]$ by $[-5, 8]$.

61. Graph $y_1 = \log_3 x = \frac{\ell n\, x}{\ell n\, 3}$, $y_2 = 3^x$, $y_3 = x$ in $[-7.5, 14.6]$ by $[-5, 8]$.

63. $e^x + e^{-x} = 3$. Let $u = e^x$. Then $e^{-x} = \frac{1}{e^x} = \frac{1}{u}$. $u + \frac{1}{u} = 3$. $u^2 + 1 = 3u$, $u^2 - 3u + 1 = 0$, $u = \frac{3 \pm \sqrt{9-4}}{2}$, $u = \frac{3 \pm \sqrt{5}}{2} = e^x$. Hence $x = \ell n(\frac{3-\sqrt{5}}{2})$ and $\ell n(\frac{3+\sqrt{5}}{2})$.

65. $\log_2 x + \log_2(4 - x) = 0$. $\log_2[x(4 - x)] = 0$, $x(4 - x) = 2^0 = 1$, $4x - x^2 = 1$, $x^2 - 4x + 1 = 0$, $x = \frac{4 \pm \sqrt{16-4}}{2} = \frac{4 \pm 2\sqrt{3}}{2} = 2 \pm \sqrt{3}$. In the original equation we must have $x > 0$ and $4 - x > 0$, i.e., $0 < x < 4$. Both these solutions qualify. $\{2 - \sqrt{3}, 2 + \sqrt{3}\}$

67. Since $R^{-1} = \{(b, a) : (a, b) \text{ is in } R\}$, the hint is the entire proof.

69. $\mathrm{Log}_b a = \frac{\log_a a}{\log_a b} = \frac{1}{\log_a b}$

1.7 A REVIEW OF TRIGONOMETRIC FUNCTIONS

1. $510 \cdot \frac{\pi}{180} = \frac{17\pi}{6} = 8.901$ **3.** $-42\frac{\pi}{180} = -0.73$ **5.** $6.2(\frac{180}{\pi}) = 355.234°$

7. $-2(\frac{180}{\pi}) = -114.592°$ **9.** $s = r\theta = (2)\frac{5\pi}{8} = \frac{5\pi}{4}$

11. $r = \frac{s}{\theta} = \frac{3\pi}{(\pi/6)} = 18$ **13.** $\theta = \frac{s}{r} = \frac{7}{14} = \frac{1}{2}$

15. a) $\sin\frac{\pi}{3}=\frac{\sqrt{3}}{2}$, $\cos\frac{\pi}{3}=\frac{1}{2}$, $\tan\frac{\pi}{3}=\sqrt{3}$, $\cot\frac{\pi}{3}=\frac{\sqrt{3}}{3}$, $\sec\frac{\pi}{3}=2$, $\csc\frac{\pi}{3}=\frac{2\sqrt{3}}{3}$

b) $\sin(-\frac{\pi}{3})=-\frac{\sqrt{3}}{2}$, $\cos(-\frac{\pi}{3})=\frac{1}{2}$, $\tan(-\frac{\pi}{3})=-\sqrt{3}$, $\cot(-\frac{\pi}{3})=-\frac{\sqrt{3}}{3}$, $\sec(-\frac{\pi}{3})=2$, $\csc(-\frac{\pi}{3})=-\frac{2\sqrt{3}}{3}$

17. a) $\sin(6.5)=0.2151$, $\cos(6.5)=0.9766$, $\tan(6.5)=0.2203$, $\cot(6.5)=4.5397$, $\sec(6.5)=1.0240$, $\csc(6.5)=4.6486$

b) $\sin(-6.5)=-0.2151$, $\cos(-6.5)=0.9766$, $\tan(-6.5)=-0.2203$, $\cot(-6.5)=-4.5397$, $\sec(-6.5)=1.0240$, $\csc(-6.5)=-4.6486$

19. a) $\sin\frac{\pi}{2}=1$, $\cos\frac{\pi}{2}=0$, $\tan\frac{\pi}{2}$ is undefined, $\cot(\frac{\pi}{2})=0$, $\sec\frac{\pi}{2}$ is undefined, $\csc\frac{\pi}{2}=1$

b) $\sin\frac{3\pi}{2}=-1$, $\cos\frac{3\pi}{2}=0$, $\tan\frac{3\pi}{2}$ is undefined, $\cot\frac{3\pi}{2}=0$, $\sec\frac{3\pi}{2}$ is undefined, $\csc\frac{3\pi}{2}=-1$

21. $\sin^{-1}(0.5)=\frac{\pi}{6}$, $30°$

23. $\tan^{-1}(-5)=-1.3734$, $-78.6901°$

25. In parametric mode graph $x_1(t)=\cos t$, $y_1(t)=\sin t$, t Min $=0$, t Max $=2\pi$, t Step $=0.1$. Then use TRACE. Then for a given t value, the displayed $x=\cos t$ and $y=\sin t$.

27. $[-\pi,2\pi]$ by $[-1,1]$, $[-\pi,2\pi]$ by $[-1,1]$, $[-1.5\pi,1.5\pi]$ by $[-2,2]$, respectively.

29. $[-270°,450°]$ by $[-3,3]$, $[-360°,360°]$ by $[-3,3]$, $[-180°,180°]$ by $[-3,3]$, respectively.

31. Check by graphing the functions in $[-\pi,\pi]$ by $[-3,3]$.

33. Graph the functions in $[0,2\pi]$ by $[-1,1]$.

35. $y=2\cos\frac{x}{3}$. Amplitude $=2$. Period $=(2\pi)/(1/3)=6\pi$. To see one period of the function graph it in $[0,6\pi]$ by $[-2,2]$. There is a horizontal stretch by a factor of 3.

37. $y=\cot(2x+\frac{\pi}{2})=\cot[2(x+\frac{\pi}{4})]$. A horizontal shrinking by a factor of $\frac{1}{2}$ is applied to the graph of $y=\cot x$ followed by a horizontal shift left $\frac{\pi}{4}$ units. The period is $\frac{\pi}{2}$. Graphing the function in $[-\frac{\pi}{4},\frac{3\pi}{4}]$ by $[-2,2]$ shows two periods of the function.

39. Start with the graph of $y=\csc x$, shrink horizontally by a factor of $1/3$, shift horizontally left $\pi/3$ units ($y=\csc[3(x+\frac{\pi}{3})]$), stretch vertically by a factor of 3, shift vertically downward 2 units. The period is $2\pi/3$. The domain is the set of all real numbers except the solutions of $\sin(3x+\pi)=0$, that is except

$3x+\pi = n\pi$, $3x = (n-1)\pi$, $x = m\frac{\pi}{3}$, m an integer. The range of $\csc(3x+\pi)$ is $(-\infty,-1) \cup (1,\infty)$. The range of $3\csc(3x+\pi)$ is $(-\infty,-3) \cup (3,\infty)$ and that of $3\csc(3x+\pi) - 2$ is $(-\infty,-5) \cup (1,\infty)$. One period of the graph may be viewed in $[-\frac{\pi}{3}, \frac{\pi}{3}]$ by $[-11, 7]$.

41. Start with the graph of $y = \tan x$, shrink horizontally by a factor of $1/3$, shift horizontally left $\pi/3$ units, stretch vertically by a factor of 3, reflect across the x-axis, shift vertically upward 2 units. The period is $\pi/3$. The domain is all real numbers except the solutions of $\cos(3x+\pi) = 0$, $3x+\pi = (2m+1)\pi/2$, $3x = (2m-1)\pi/2$, $x = (2m-1)\pi/6$, the odd multiples of $\pi/6$. The range is $(-\infty,\infty)$. One period of the graph may be viewed in $[-\frac{\pi}{6}, \frac{\pi}{6}]$ by $[-11, 15]$.

43. $\cos x = -0.7$. Two solutions are $\pm\cos^{-1}(-0.7)$, the only solutions in the interval $[-\pi, \pi]$. The solution set is $\{\pm\cos^{-1}(-0.7) + 2n\pi\}$.

45. $\tan x = 4$ has solution set $\{(\tan^{-1} 4) + n\pi\}$

47. To solve $\sin x = 0.2x$, we may first graph $y = 0.2x - \sin x$ in $[-10, 10]$ by $[-2, 2]$. We see that there are 3 solutions (x-intercepts) and we Zoom-In to find them: $\{-2.596, 0, 2.596\}$.

49. We graph $y = \ell n\ x - \sin x$ in $[0, 4]$ by $[-2, 2]$ and Zoom-In to the x-intercept: $x = 2.219$.

51. We may use the work in Exercise 47 and solve the equivalent inequality $0.2x - \sin x < 0$. The set of x for which the graph is below the x-axis is $(-\infty, -2.596) \cup (0, 2.596)$.

53. Graph $x_1 = \sqrt{5}\cos t$, $y_1 = \sqrt{5}\sin t$, $0 \leqq t \leqq 2\pi$ in $[-3.8, 3.8]$ by $[-\sqrt{5}, \sqrt{5}]$. We use the "squaring" device on the calculator to produce a viewing window in which the curve appears to be a circle.

55. Graph $x_1 = 2+3\cos t$, $y_1 = -3+3\sin t$, $0 \leqq t \leqq 2\pi$, in $[-3.1, 7.1]$ by $[-6, 0]$.

57. $y = 2\sin x + 3\cos x$. $a = 2, b = 3$. $A = \sqrt{a^2+b^2} = \sqrt{13}$. $\cos\alpha = a/A = 2/\sqrt{13}$, $\sin\alpha = 3/\sqrt{13}$. Thus α is in the first quadrant and we may take $\alpha = \sin^{-1}(3/\sqrt{13})$. $y = A\sin(x+\alpha) = \sqrt{13}\sin(x+\alpha)$, $\alpha = \sin^{-1}(3/\sqrt{13}) = 0.9828$.

59. $y = \sin 2x + \cos 2x = \sqrt{2}[\sin 2x \cos(\pi/4) + \cos 2x \sin(\pi/4)] = \sqrt{2}\sin(2x + \frac{\pi}{4})$.

61. b) and d); c) and e)

63. $1 + \tan^2\theta = 1 + (\frac{y}{x})^2 = \frac{x^2+y^2}{x^2} = \frac{1}{x^2} = \sec^2\theta$.

65. The equation $y = \cos(-x)$ is the same as $y = \cos x$ so the graph of the cosine is symmetric with respect to the y-axis. $-y = \sin(-x) = -\sin x$ is the same as $y = \sin x$ so the graph of the sine is symmetric with respect to the origin. Similarly, the graph of the tangent is symmetric with respect to the origin.

67. a) yes, b) $-1 \leqq \cos 2x \leqq 1$, c) $0 \leqq 1 + \cos 2x \leqq 1 + 1$, $0 \leqq \frac{1+\cos 2x}{2} \leqq 1$, d) $0 \leqq \sqrt{\frac{1+\cos 2x}{2}} \leqq \sqrt{1} = 1$. The domain is the set of all reals; the range is the interval $[0, 1]$. Alternative approach: $y = \sqrt{(1+\cos 2x)/2} = \sqrt{\cos^2 x} = |\cos x|$.

69. a) 37 b) period $= 2\pi/(2\pi/365) = 365$ c) 101 units to the right d) 25 units upward

71. We obtain $\cos(A - A) = \cos A \cos A + \sin A \sin A$ or $1 = \cos^2 A + \sin^2 A$ for the first equation. For the second equation we obtain $\sin(A - A) = \sin A \cos A - \cos A \sin A$ or $0 = 0$.

73. $\cos(A + \frac{\pi}{2}) = \cos A \cos \frac{\pi}{2} - \sin A \sin \frac{\pi}{2} = -\sin A$. If we start the cosine curve, reflect it across the x-axis and shift horizontally $\frac{\pi}{2}$ units to the left, we obtain the sine curve. $\sin(A + \frac{\pi}{2}) = \sin A \cos \frac{\pi}{2} + \cos A \sin \frac{\pi}{2} = \cos A$.

75. $\cos 15° = \cos(45° - 30°) = \cos 45° \cos 30° + \sin 45° \sin 30° = \frac{\sqrt{2}}{2}\frac{\sqrt{3}}{2} + \frac{\sqrt{2}}{2}\frac{1}{2} = \frac{\sqrt{2}}{4}(\sqrt{3}+1)$ or $\frac{\sqrt{6}+\sqrt{2}}{4}$

77. $\sin \frac{7\pi}{12} = \sin(\frac{\pi}{4} + \frac{\pi}{3}) = \sin \frac{\pi}{4} \cos \frac{\pi}{3} + \cos \frac{\pi}{4} \sin \frac{\pi}{3} = \frac{\sqrt{2}}{2}\frac{1}{2} + \frac{\sqrt{2}}{2}\frac{\sqrt{3}}{2} = \frac{\sqrt{2}+\sqrt{6}}{4}$

79. $\cos^2 \frac{\pi}{8} = \frac{1+\cos\frac{\pi}{4}}{2} = \frac{1+\frac{\sqrt{2}}{2}}{2} = \frac{2+\sqrt{2}}{4}$ **81.** $\sin^2 \frac{\pi}{12} = \frac{1-\cos\frac{\pi}{6}}{2} = \frac{1-\frac{\sqrt{3}}{2}}{2} = \frac{2-\sqrt{3}}{4}$

83. We discuss only $\cos 2\theta = \cos^2\theta - \sin^2\theta$. We may graph $y_1 = \cos 2x$ and $y_2 = \cos^2 x - \sin^2 x$ in $[0, 2\pi]$ by $[-2, 2]$ and see that we get just one curve. Alternatively we can shift one curve up slightly and see that we get two parallel curves: In the same viewing rectangle graph $y_1 = \cos 2x$ and $y_2 = \cos^2 x - \sin^2 x + 0.5$.

85. $\tan(A + B) = \frac{\sin(A+B)}{\cos(A+B)} = \frac{\sin A \cos B + \cos A \sin B}{\cos A \cos B - \sin A \sin B} = \frac{\tan A + \tan B}{1 - \tan A \tan B}$. The last step was obtained by dividing all terms by $\cos A \cos B$.

87. a) Let $f(x) = \cot x$. $f(-x) = \frac{\cos(-x)}{\sin(-x)} = \frac{\cos x}{-\sin x} = -f(x)$, proving $f(x)$ is odd.

b) Let $g(x) = \frac{h(x)}{k(x)}$ where $h(x)$ is even and $k(x)$ is odd. $g(-x) = \frac{h(-x)}{k(-x)} = \frac{h(x)}{-k(x)} = -g(x)$ proving $g(x)$ is odd where defined.

c) The graph of $y = \cot(-x) = -\cot x$ can be obtained by reflecting the graph of $y = \cot x$ across the x-axis.

89. Use the method indicated in the solution of Exercise 87.

91. $f(x) = \sin(60x)$. $60x$ will range between 0 and 2π if x ranges between 0 and $\frac{2\pi}{60}$. Thus $f(x)$ has fundamental period $\frac{\pi}{30}$. Graph $y = \sin(60x)$ in the window $[0, \frac{\pi}{30}]$ by $[-1, 1]$.

93. Since parallel lines have the same slope m and the same angle of inclination α, we may assume that the line passes through the origin. Let (a, b) be a point on the line, $a \neq 0$, $b \geqq 0$. Then $m = \frac{b}{a} = \tan\alpha$.

95. $m = \frac{5.5-2}{3+1} = 0.875$. In degrees, $\alpha = \tan^{-1} 0.875 = 41.186°$.

97. $2x - 6y = 7$, $6y = 2x - 7$, $y = \frac{1}{3}x - \frac{7}{6}$, $m = \frac{1}{3}$. In degrees, $\alpha = \tan^{-1}\frac{1}{3} = 18.435°$.

PRACTICE EXERCISES, CHAPTER 1

1. a) $(1, -4)$ b) $(-1, 4)$ c) $(-1, -4)$

3. a) $(-4, -2)$ b) $(4, 2)$ c) $(4, -2)$

5. a) origin b) y-axis

7. a) both axes and the origin b) none of the mentioned symmetries

9. $x = 1$, $y = 3$ **11.** $x = 0$, $y = -3$

13. Using $y - y_0 = m(x - x_0)$, we get $y - 3 = 2(x - 2)$ or $y = 2x - 1$. y-intercept is $y = -1$. x-intercept: $0 = 2x - 1$, $x = \frac{1}{2}$. We may graph the line by drawing the line through $(\frac{1}{2}, 0)$ and $(0, -1)$.

15. $y - 0 = -(x - 1)$. Intercepts: $x = 1$, $y = 1$

17. $y + 6 = 3(x - 1)$ or $y = 3x - 9$. Intercepts: $x = 3$, $y = -9$.

19. $y - 2 = -\frac{1}{2}(x + 1)$ or $y = -\frac{1}{2}x + \frac{3}{2}$. Intercepts: $x = 3$, $y = \frac{3}{2}$.

21. $m = \frac{y_2 - y_1}{x_2 - x_1} = \frac{3+2}{1+2} = \frac{5}{3}$. $y + 2 = \frac{5}{3}(x + 2)$, $3y + 6 = 5(x + 2) = 5x + 10$, $3y = 5x + 4$.

23. $m = \frac{4+1}{4-2} = \frac{5}{2}$. $y + 1 = \frac{5}{2}(x - 2) = \frac{5}{2}x - 5$, $y = \frac{5}{2}x - 6$

25. $y = \frac{1}{2}x + 2$ **27.** $y = -2x - 1$

29. $2x - y = -2$ or $y = 2x + 2$ so $m = 2$. a) $y = 2(x - 6)$ or $2x - y = 12$ b) $y = -\frac{1}{2}(x - 6)$ or $2y = -x + 6$ or $x + 2y = 6$. c) For the point of intersection of the two lines: $y = 2x + 2 = -\frac{1}{2}x + 3$, $\frac{5}{2}x = 1$, $x = \frac{2}{5}$, $y = \frac{14}{5}$. Distance $= \sqrt{(6 - \frac{2}{5})^2 + (0 - \frac{14}{5})^2} = \frac{14}{5}\sqrt{5}$

31. $L : y = -\frac{4}{3}x + 12$, $m = -\frac{4}{3}$. a) $y + 12 = -\frac{4}{3}(x - 4)$ or $4x + 3y = -20$ b) $y + 12 = \frac{3}{4}(x - 4)$. c) For the point of intersection of L and the perpendicular we obtain $(\frac{228}{25}, -\frac{204}{25})$. The distance is $\frac{32}{5}$.

33. Domain = range = $(-\infty, \infty)$

35. The range of $y = 2|x - 1|$ is $[0, \infty)$ so the range of $y = 2|x - 1| - 1$ is $[-1, \infty)$. The domain is $(-\infty, \infty)$. Check your sketch by graphing $y = 2\,\text{abs}(x - 1) - 1$ in $[-2, 4]$ by $[-1, 5]$.

37. Domain = $(-\infty, \infty)$, range = $[-1, 1]$.

39. Domain = $(-\infty, 0]$, range = $(-\infty, \infty)$. Check your sketch by graphing $y = -\sqrt{-x}$ and $y = \sqrt{-x}$ in $[-9, 0]$ by $[-3, 3]$.

41. Domain = range = $(-\infty, \infty)$. Graph $f(x)$ in $[-9, 4]$ by $[-100, 100]$.

43. Domain = $(1, \infty)$, range = $(-\infty, \infty)$. Graph $y = \ell n(x - 1)/\ell n(7) + 1$ in $[1, 3]$ by $[-3, 3]$, recalling that $x = 1$ is a vertical asymptote.

45. $$f(x) = \begin{cases} -2x - 1, & x < -3 \\ 5, & -3 \leqq x < 2 \\ 2x + 1, & x \geqq 2 \end{cases}.$$

Domain = $(-\infty, \infty)$, range = $[5, \infty)$. Graph $y = \text{abs}(x - 2) + \text{abs}(x + 3)$ in $[-5, 4]$ by $[4, 9]$.

47. Stretch vertically by a factor of 2, reflect across the x-axis, shift horizontally right one unit, shift vertically 5 units upward.

49. $f(x) = 3\sin(3x + \pi) = 3\sin[3(x + \frac{\pi}{3})]$. Stretch vertically by a factor of 3, shrink horizontally by a factor of 1/3, shift horizontally $\pi/3$ units left.

51. $y = x^2$, $y = 2x^2$, $y = -2x^2$, $y = -2(x - 2)^2$, $y = -2(x - 2)^2 + 3$.

53. $y = \frac{1}{x}$, $y = \frac{3}{x}$, $y = \frac{3}{x+2}$, $y = \frac{3}{x+2} + 5$

55.

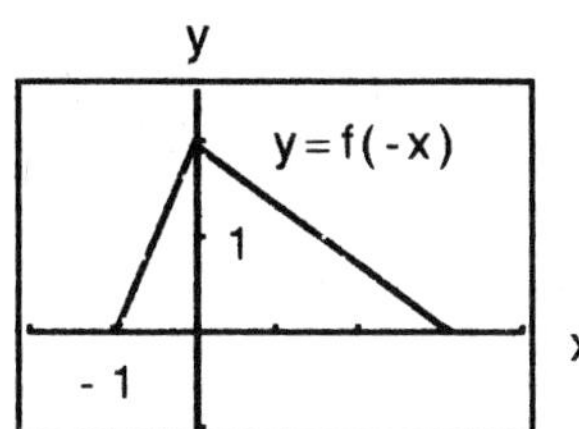

57.

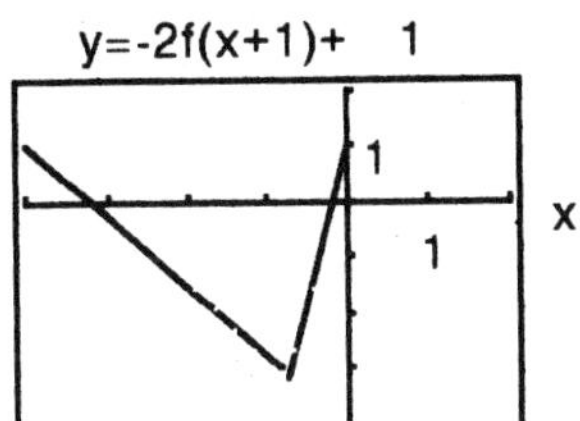

59. $y = -x^2 + 4x - 1 = -(x^2 - 4x) - 1 = -(x^2 - 4x + 4) + 4 - 1 = 3 - (x - 2)^2$. $(2, 3)$ is the vertex, $x = 2$ is the axis of symmetry. Check your sketch by graphing y in $[-2, 5]$ by $[-6, 3]$.

61. a) $\cos(-x) = \cos x$, even b) even c) even

63. a) even b) odd c) odd

65. a) even b) odd c) odd

67. Graph y in $[-2, 2]$ by $[0, 2]$. The function is periodic of period 1.

69. Graph $y_1 = \sqrt{-x}$ and $y_2 = \sqrt{x}$ at the same time in $[-4, 4]$ by $[0, 2]$.

71. The graph consists of one period of the sine function on $[0, 2\pi]$ together with all points on the x-axis larger than 2π.

73. $y = \begin{cases} 1 - x, & 0 \leqq x < 1 \\ 2 - x, & 1 \leqq x \leqq 2 \end{cases}$

75. For $f(x)$, domain = range = all real numbers except 0. For the remaining functions, domain = range = all positive real numbers.

77. $(x - 1)^2 + (y - 1)^2 = 1$

79. $(x - 2)^2 + (y + 3)^2 = \frac{1}{4}$

81. $(3, -5)$, 4

83. $(-1, 7)$, 11

85. a) $x^2 + y^2 < 1$ b) $x^2 + y^2 \leqq 1$

87. $|x - 1| = \frac{1}{2}$ leads to $x - 1 = -\frac{1}{2}$ and $x - 1 = \frac{1}{2}$. The solution set is $\{\frac{1}{2}, \frac{3}{2}\}$.

89. $\frac{2x}{5} + 1 = -7$ and $\frac{2x}{5} + 1 = 7$. $\{-20, 15\}$

91. $-\frac{1}{2} \leqq x + 2 \leqq \frac{1}{2}$, $-\frac{5}{2} \leqq x \leqq -\frac{3}{2}$

93. $-\frac{3}{5} < y - \frac{2}{5} < \frac{3}{5}$, $-\frac{1}{5} < y < 1$

95. $\{0.19, 2.47, 4.34\}$

97. $2 + \log_3(x - 2) + \log_3(3 - x) = 2 + \log_3[(x - 2)(3 - x)] = 0$, $\log_3(-x^2 + 5x - 6) = -2$, $-x^2 + 5x - 6 = 3^{-2} = \frac{1}{9}$, $-9x^2 + 45x - 54 = 1$, $9x^2 - 45x + 55 = 0$. From the quadratic formula we get the solution set $\{\frac{15 - \sqrt{5}}{6}, \frac{15 + \sqrt{5}}{6}\}$. Solving with zoom-in, $\{2.127, 2.873\}$.

99. $|1-2x|<3$, $-3<1-2x<3$, $-4<-2x<2$, $2>x>-1$ or $-1<x<2$.

101. $|\frac{3}{x-2}|<1$ is equivalent to $\frac{|x-2|}{3}>1$, $|x-2|>3$ which has solution set $(-\infty,-1)\cup(5,\infty)$.

103. The inequality is equivalent to $x^3-7x^2+12x-2<0$. We seek the set of all x for which the graph of the left-hand side lies below the x-axis. Using the graph and Exercise 95 we get $(-\infty,0.19)\cup(2.47,4.34)$.

105. a) $30(\frac{\pi}{180})=\frac{\pi}{6}$ b) $22(\frac{\pi}{180})=0.122\pi$,
c) $-130(\frac{\pi}{180})=-0.722\pi$ d) $-150(\frac{\pi}{180})=-\frac{5\pi}{6}$

107. a) 0.891, 0.454, 1.965, 0.509, 2.205, 1.122

b) −0.891, 0.454, −1.965, −0.509, 2.205, −1.122

c) $\sqrt{3}/2$, $-1/2$, $-\sqrt{3}$, $-\sqrt{3}/3$, -2, $2\sqrt{3}/3$

d) $-\sqrt{3}/2$, $-1/2$, $\sqrt{3}$, $\sqrt{3}/3$, -2, $-2\sqrt{3}/3$

109. Graph the functions in $[0,2\pi]$ by $[-1,2]$.

111. a) $(\cos\frac{\pi}{6})^2=(\frac{\sqrt{3}}{2})^2=\frac{3}{4}$ b) $\cos^2\frac{\pi}{6}=\frac{1+\cos\frac{\pi}{3}}{2}=\frac{1+\frac{1}{2}}{2}=\frac{3}{4}$

113. $f(x)=2-3x$. $x=2-3y$, $3y=2-x$, $f^{-1}(x)=\frac{2-x}{3}$. $f\circ f^{-1}(x)=f(\frac{2-x}{3})=2-3(\frac{2-x}{3})=2-2+x=x$. $f^{-1}\circ f(x)=f^{-1}(2-3x)=\frac{2-(2-3x)}{3}=x$. Graph $y=2-3x$ and $y=\frac{2-x}{3}$ in $[-10,10]$ by $[-5,5]$.

115. $y=x^3-x$. The inverse relation is determined by $x=y^3-y$. This may be graphed in parametric mode using $x_1=t^3-t$, $y_1=t$, $-2\leqq t\leqq 2$ in $[-6,6]$ by $[-2,2]$. The inverse relation is not a function.

117. $\sin^{-1}(0.7)=0.775$, $44.427°$

119. Graph $y=|\cos x|$ in $[-\frac{\pi}{2},\frac{\pi}{2}]$ by $[0,1]$. The graph is complete because the function has period π.

121. For x in the interval $[-\frac{\pi}{2},\frac{3\pi}{2}]$, $y=\begin{cases}0, & -\frac{\pi}{2}\leqq x<\frac{\pi}{2}\\ -\cos x, & \frac{\pi}{2}\leqq x\leqq\frac{3\pi}{2}\end{cases}$. Graphing this part gives a complete graph because the function has period 2π.

123. a) $A(x)=(\frac{x}{4})^2+(\frac{100-x}{4})^2$ b) Graph this function in $[-50,150]$ by $[300,1000]$ c) The domain is the set of all real numbers. Since $A(0)=A(100)$, $A(50)=2(\frac{50}{4})^2=312.5$ is the minimum of A and the range is $[312.5,\infty)$. d) $0<x<100$ e) We must solve $(\frac{x}{4})^2+(\frac{100-x}{4})^2=400$, $x^2+10000-200x+x^2=$

6400, $2x^2 - 200x + 3600 = 0$, $x^2 - 100x + 1800 = 0$, $(x-50)^2 = 700$, $x = 50 \pm 10\sqrt{7}$. The two lengths are $50 - 10\sqrt{7}$in and $50 + 10\sqrt{7}$in f) We can only approach the maximum obtained by not cutting and getting $(\frac{100}{4})^2 = 625\text{in}^2$. As noted in c) if both pieces are 50in, we get the minimum area of 312.5in^2.

125. In the first quadrant $|x| + |y| = 1$ becomes $x + y = 1$ and the graph is the line segment connecting $(0,1)$ and $(1,0)$. Since the graph is symmetric with respect to both axes, its graph is the square with vertices $(0,1), (1,0), (0,-1)$, $(-1,0)$.

CHAPTER 2

LIMITS AND CONTINUITY

2.1 LIMITS

1. $\lim_{x\to 2} 2x = 2\cdot 2 = 4$ **3.** $\lim_{x\to 1}(3x-1) = (3\cdot 1 - 1) = 2$

5. $\lim_{x\to -1} 3x(2x-1) = 3(-1)[2(-1)-1] = -3[-3] = 9$

7. $\lim_{x\to -2}(x+3)^{171} = (-2+3)^{171} = 1^{171} = 1$

9. $\lim_{x\to 1}(x^3+3x^2-2x-17)$
$= 1^3 + 3(1)^2 - 2(1) - 17$
$= 1+3-2-17$
$= 4-19$
$= -15$

11. $\lim_{x\to -1} \frac{x+3}{x^2+3x-1} = \frac{-1+3}{(-1)^2+3(-1)+1} = \frac{2}{1-3+1} = \frac{2}{-1} = -2$

13. $\lim_{y\to -3} \frac{y^2+4y+3}{y^2-3} = \frac{(-3)^2+4(-3)+3}{(-3)^2-3}$
$= \frac{9-12+3}{9-3}$
$= \frac{0}{6}$
$= 0$

15. $\lim_{x\to -2}\sqrt{x-2}$. $\sqrt{x-2}$ is defined only for $x \geqq 2$ so it is not defined for values of x near -2. Thus the limit does not exist.

17. $\lim_{x\to 0}\frac{|x|}{x}$. Since $\frac{|x|}{x}$ is not defined when $x = 0$, we cannot use substitution. $\lim_{x\to 0^-}\frac{|x|}{x} = \lim_{x\to 0^-}\frac{-x}{x} = \lim_{x\to 0^-}(-1) = -1$. But $\lim_{x\to 0^+}\frac{|x|}{x} = \lim_{x\to 0^+}\frac{x}{x} = 1$. Since the two one-sided limits are not equal, the limit does not exist.

19. $\lim_{x\to 1}\frac{x-1}{x^2-1} = \lim_{x\to 1}\frac{x-1}{(x-1)(x+1)} = \lim_{x\to 1}\frac{1}{x+1} = \frac{1}{1+1} = \frac{1}{2}$

21. $\lim_{t\to 1}\frac{t^2-3t+2}{t^2-1} = \lim_{t\to 1}\frac{(t-1)(t-2)}{(t-1)(t+1)} = \lim_{t\to 1}\frac{t-2}{t+1} = \frac{1-2}{1+1} = -\frac{1}{2}$

23. $\lim_{x\to 2}\frac{2x-4}{x^3-2x^2} = \lim_{x\to 2}\frac{2(x-2)}{x^2(x-2)} = \lim_{x\to 2}\frac{2}{x^2} = \frac{2}{2^2} = \frac{1}{2}$

25. $\lim_{x\to 0} \frac{\frac{1}{2+x}-\frac{1}{2}}{x} = \lim_{x\to 0} \frac{\frac{2-(2+x)}{2(2+x)}}{x}$
$= \lim_{x\to 0} \frac{\frac{-x}{2(2+x)}}{x}$
$= \lim_{x\to 0} \frac{-1}{2(2+x)} = -\frac{1}{4}$

27. $\lim_{x\to 0} x \sin \frac{1}{x} = 0$

29. The limit does not exist.

31. $\lim_{x\to 0} \frac{2^x-1}{x} = 0.693\ldots$ We will later be able to show that the limit is $\ell n\, 2$.

33. We graph $y = 100(1 + 0.06/x)^x$ in $[0, 1000]$ by $[106, 106.5]$. As x increases, using TRACE, the limiting value of y, rounded to two decimal places, is \$106.18. After compounding 8 times, the value of the investment has already reached \$106.15. So frequent compounding within one year is not much of an advantage to the investor of \$100.

35. There appear to be no points of the graph of the function very near to $(0, 4)$. The actual graph is a straight line with one point, $(0, 4)$, missing. $y = (x^2 + 4x + 4 - 4)/x = x + 4$ when $x \neq 0$.

37. $f(x) = \frac{x^3-1}{x-2}$. The graphs indicate $\lim_{x\to 2^+} f(x) = \infty$ and $\lim_{x\to 2^-} f(x) = -\infty$. When $x > 2$, the numerator $\to 7$ while the denominator $\to 0$ but is always positive. When $x < 2$, the denominator is negative and $f(x) \to -\infty$ as $x \to 2^-$.

39. Only the following relevant one-sided limits exist: $\lim_{x\to -1^+} f(x) = 1$, $\lim_{x\to 0^-} f(x) = 0$, $\lim_{x\to 0^+} f(x) = 0$, $\lim_{x\to 1^-} f(x) = 1$, $\lim_{x\to 1^+} f(x) = 0$, $\lim_{x\to 2^-} f(x) = 0$. We do therefore have $\lim_{x\to 0} f(x) = 0$. The true statements are a), b), d), e), f).

41. a)

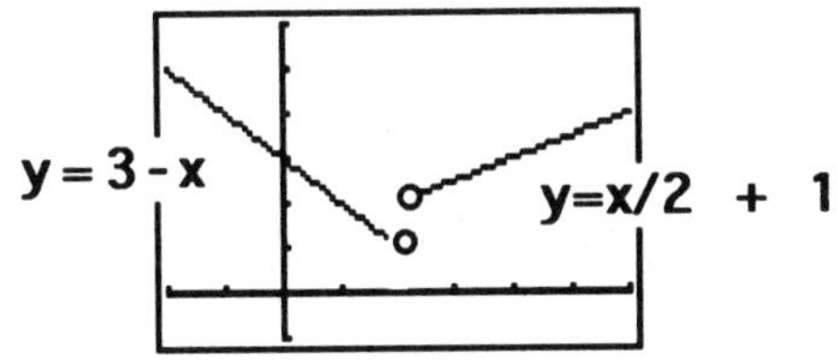

b) $\lim_{x\to 2^+} f(x) = 2$, $\lim_{x\to 2^-} f(x) = 1$

c) Does not exist because right-hand and left-hand limits are not equal.

43. a) A complete graph of $f(x)$ can be obtained on a graphing calculator by graphing both $y = (x-1)^{-1} + 0\sqrt{(1-x)}$ and $y = x^3 - 2x + 5 + 0\sqrt{(x-1)}$ in the viewing rectangle $[-3, 5]$ by $[-25, 25]$.

b) $\lim_{x\to 1^+} f(x) = 4$ and $\lim_{x\to 1^-} f(x)$ does not exist.

c) No. For this limit to exist, the two limits in b) must be equal to the same finite number.

45. $\text{Lim}_{x\to 2^+} f(x) = \lim_{x\to 2^+}(x^2 + 5x - 3) = 11.$

$\text{Lim}_{x\to 2^-} f(x) = \lim_{x\to 2^-}(a - x^2) = a - 4.$

$\text{Lim}_{x\to 2} f(x)$ will exist if and only if $a - 4 = 11$ or $a = 15$.

47. a)

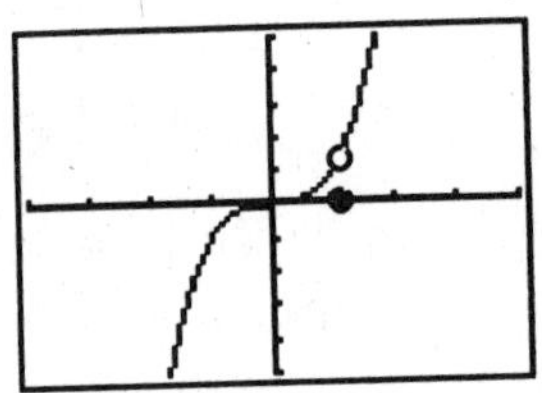

$[-4, 4]$ by $[-6, 6]$

b) $\text{Lim}_{x\to 1^-} f(x) = 1 = \lim_{x\to 1^+} f(x)$

c) $\text{Lim}_{x\to 1} f(x) = 1.$

49.

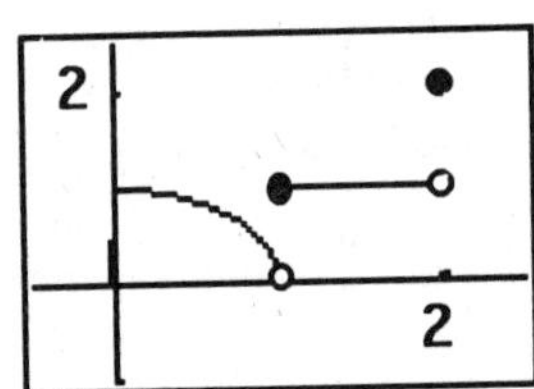

a) All points c except $c = 0, 1, 2$. b) $x = 2$ c) $x = 0$

51. $\lim_{x\to 0+}[x] = \lim_{x\to 0+} 0 = 0$

53. $\lim_{x\to 0.5}[x] = \lim_{x\to 0.5} 0 = 0$ because when x is sufficiently close to 0.5, $\lfloor x \rfloor = 0$.

55. $\lim_{x\to 0+} \frac{x}{|x|} = \lim_{x\to 0+} \frac{x}{x} = \lim_{x\to 0+} 1 = 1$ because $|x| = x$ when $x > 0$.

57. When $x > a$, $x - a > 0$ and $|x - a| = x - a$. Hence $\lim_{x\to a+} \frac{|x-a|}{x-a} = \lim_{x\to a+} \frac{x-a}{x-a} = \lim_{x\to a+} 1 = 1$.

59. $\lim_{x\to c} f(x) = 5$ and $\lim_{x\to c} g(x) = 2$

a) $\lim_{x\to c} f(x)g(x) = \lim_{x\to c} f(x) \cdot \lim_{x\to c} g(x) = 5 \cdot (2) = 10$

b) $\lim_{x\to c} 2f(x)g(x) = 2\lim_{x\to c} f(x)g(x) = 2(10) = 20$

61. a) $\lim_{x\to b}(f(x) + g(x)) = \lim_{x\to b} f(x) + \lim_{x\to b} g(x) = 7 + (-3) = 4$

b) $\lim_{x\to b} f(x) \cdot g(x) = \lim_{x\to b} f(x) \cdot \lim_{x\to b} g(x) = 7 \cdot (-3) = -21$

c) $\lim_{x\to b} 4g(x) = 4\lim_{x\to b} g(x) = 4(-3) = -12$

d) $\lim_{x\to b} f(x)/g(x) = \lim_{x\to b} f(x)/\lim_{x\to b} g(x) = 7/-3 = -7/3$

63. $\lim_{x\to 0} x \sin x = \lim_{x\to 0} x \lim_{x\to 0} \sin x = 0 \cdot 0 = 0$. The graph of $y = x \sin x$ in the viewing rectangle $[-10, 10]$ by $[-10, 10]$ suggests this limit.

65. A view of the graph of $f(x) = x^2 \sin x$ in the square $[-1, 1]$ by $[-1, 1]$ supports $\lim_{x\to 0} f(x) = 0$. $\lim_{x\to 0} x^2 \sin x = \lim_{x\to 0} x^2 \lim_{x\to 0} \sin x = 0 \cdot 0 = 0$.

67. The magnified graph of the function in the viewing rectangle $[-0.03, 0.03]$ by $[-0.001, 0.001]$ strongly suggests the limit is 0 as $x \to 0$.

69. $\lim_{x\to 0}(1 + x)^{4/x} = 54.598$ with error less than 0.01.

71. Graph $y = \frac{2^x-1}{x}$ in $[-9, 7]$ by $[-0.3, 3.5]$. Then zoom in to the y-intercept. We see that $\lim_{x\to 0} \frac{2^x-1}{x} = 0.6931$ with error less than 0.0001.

75. Graph the function in $[1.49\pi, 1.51\pi]$ by $[-0.00001, 0.00001]$ and use TRACE. This suggests the limit is about 9.536×10^{-7}. This is close to the actual limit which can be shown to be $\frac{1}{2^{20}}$. Let $u = x - \frac{3\pi}{2}$, $x = u + \frac{3\pi}{2}$. Then $u \to 0$ as $x \to \frac{3\pi}{2}$ and $\sin x = \sin(u + \frac{3\pi}{2}) = -\cos u$. $\frac{(1+\sin x)^{20}}{(x-\frac{3\pi}{2})^{40}} = \frac{(1-\cos u)^{20}}{u^{40}} \frac{(1+\cos u)^{20}}{(1+\cos u)^{20}} = \frac{(1-\cos^2 u)^{20}}{u^{40}} \frac{1}{(1+\cos u)^{20}} = (\frac{\sin u}{u})^{40} \frac{1}{(1+\cos u)^{20}} \to \frac{1}{2^{20}}$

77. The midpoint of OP is $(\frac{a}{2}, \frac{a^2}{2})$ and OP has slope $\frac{a^2}{a} = a$. Thus the perpendicular bisector has equation $y - \frac{a^2}{2} = -\frac{1}{a}(x - \frac{a}{2}) = -\frac{1}{a}x + \frac{1}{2}$ or $y = -\frac{1}{a}x + (\frac{1}{2} + \frac{a^2}{2})$. Hence $b = \frac{1}{2} + \frac{a^2}{2} \to \frac{1}{2}$ as $a \to 0$ or as $P \to 0$. In the viewing window $[-1, 1]$ by $[-.5, 1]$ graph $y_1 = -A^{-1}x + .5 + \frac{A^2}{2}$, $y_2 = x^2$ and $y_3 = Ax$, successively storing to A the values $A = 0.4, 0.1, 0.01$ and using TRACE to find the y-intercept of y_1. We find the vlaues to be $0.58, 0.505, 0.50005$.

2.2 CONTINUOUS FUNCTIONS

1. a) Yes, $f(-1) = 0$ b) Yes, $\lim_{x \to -1^+} f(x) = 0$ c) Yes d) Yes

3. Since f is not defined at $x = 2$, it can't be continuous at $x = 2$.

5. a) $\lim_{x \to 2} f(x) = 0$ b) Define $g(x) = f(x)$, $x \neq 2$, $g(2) = 0$. Then g is an extension of f which is continuous at $x = 2$.

7. $f(x)$ is continous at all points of $[-1, 2]$ except $x = 0$ and $x = 1$. $\lim_{x \to 0} f(x)$ exists but it is not equal to $f(0)$. $\lim_{x \to 1} f(x)$ does not exist.

9. $f(x)$ is continuous at all points except $x = 2$. $\lim_{x \to 2} f(x)$ does not exist.

11. $f(x)$ is continuous at all points except $x = 1$. $\lim_{x \to 1^-} f(x)$ does not exist.

13. $f(x)$ is continuous at all points except $x = 1$. $\lim_{x \to 1} f(x) = 1 \neq f(1) = 0$.

15. a)

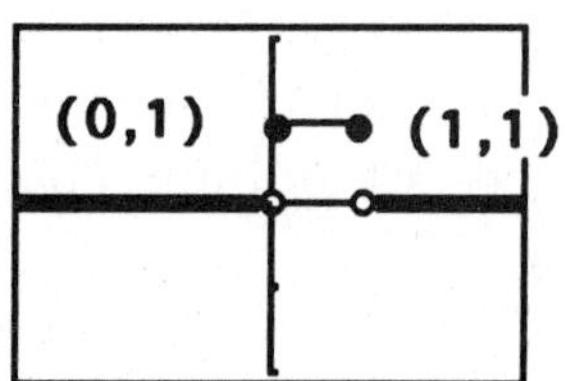

b) f is continuous at all points except $x = 0$ and $x = 1$ where there are jump discontinuities.

17. $x = 2$

19. $y = \frac{x+1}{x^2 - 4x + 3} = \frac{x+1}{(x-1)(x-3)}$ is undefined at $x = 1$ and $x = 3$.

21. $y = \frac{x^3-1}{x^2-1}$ is undefined when $x = \pm 1$. The factor $x - 1$ can be cancelled only when $x \neq 1$; $x = 1$ is a removable discontinuity. y is not continuous only at $x = \pm 1$.

23. $|x - 1|$ is the composite of two continuous functions $|x|$ (Example 6) and $x - 1$. It is therefore continuous by Theorem 6.

25. The quotient of two continuous functions is discontinuous only at points where the denominator is 0. Thus $x = 0$ is the only point of discontinuity.

27. $\sqrt{2x+3}$ will be discontinuous at x if and only if $2x+3 < 0$, that is, $x < -3/2$.

29. The cube root function $x^{1/3}$ is continuous for all x and the same is true of $2x - 1$. Thus $y = \sqrt[3]{2x-1}$ is the composite of two continuous functions. By Theorem 6 y is continuous at all points. Answer: No discontinuities.

31. $\text{Lim}_{x\to 1} \frac{x^2-1}{x-1} = \lim_{x\to 1}(x+1) = 2$. Hence $f(1) = \lim_{x\to 1} f(x)$ and f is continuous at $x = 1$.

33. $\text{Lim}_{x\to 2} h(x) = \lim_{x\to 2} \frac{(x-2)(x+5)}{x-2} = \lim_{x\to 2}(x+5) = 7$. Thus if $h(2) = 7$, $h(x)$ will be continuous at $x = 2$.

35. $\text{Lim}_{x\to 4} g(x) = \lim_{x\to 4} \frac{(x-4)(x+4)}{(x-4)(x+1)} = \lim_{x\to 4} \frac{x+4}{x+1} = \frac{8}{5}$. If we assign $g(4) = \frac{8}{5}$, the extended function will be continuous at $x = 4$.

37. $\text{Lim}_{x\to 3^-} f(x) = \lim_{x\to 3^-}(x^2 - 1) = 8$ and $\lim_{x\to 3^+} f(x) = \lim_{x\to 3^+} 2ax = 2a(3) = 6a$. We must have $6a = 8$ or $a = 4/3$. Then we have

$$f(x) = \begin{cases} x^2 - 1, & x < 3 \\ \frac{8}{3}x, & x \geqq 3. \end{cases}$$

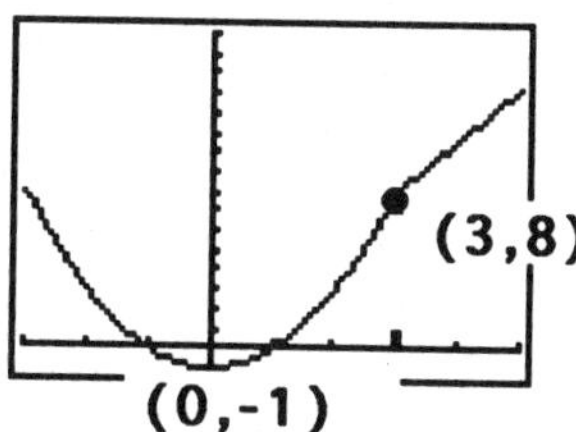

39. $\lim_{x\to 0} \sec x = \lim_{x\to 0} \frac{1}{\cos x} = \frac{\lim_{x\to 0} 1}{\lim_{x\to 0} \cos x} = \frac{1}{\cos 0} = \frac{1}{1} = 1$

41. $\text{Lim}_{x\to 0}\frac{1+\cos x}{2} = \frac{1+\cos 0}{2} = \frac{1+1}{2} = 1$. We are able to replace x by 0 because $\cos x$ is a continuous function.

43. $\lim_{x\to 0}\cos(1-\frac{\sin x}{x}) = \cos(1-1) = \cos 0 = 1$

45. $f(x) = x^3+4 = 2,\ x^3 = -2,\ x = -\sqrt[3]{2} \approx -1.2599$

47. Zooming in to the x-intercept as much as possible we obtain $x = 1.324717957$ with error at most 10^{-9}. In part (b) we obtain the same result.

49. The maximum value 2 of f is taken on at $x = 2$ and $x = 3$. $\text{Lim}_{x\to 1^-} f(x) = 0$ but the value 0 is not taken on by f for any x; the minimum 0 can be approached arbitrarily closely but cannot be attained. Since f has discontinuities in $[0,4]$, this does not contradict Theorem 7.

51. The maximum value 1 is not attained but is only approached as x approaches ± 1. The minimum value 0 is attained at $x = 0$. Theorem 7 is not contradicted because the interval $(-1,1)$ is not closed.

53. We are given $f(0) < 0 < f(1)$. By Theorem 8 there exists some c in $[0,1]$ such that $f(c) = 0$. A possible graph is

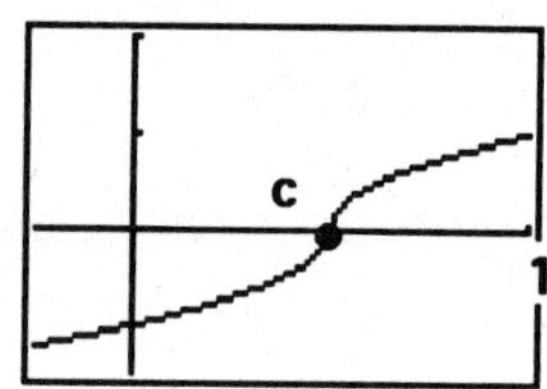

55. Let $f(x) = e^{-x} - x$. Then $f(-1) = e+1 > 0$ and $f(1) = \frac{1}{e} - 1 < 0$. By the Intermediate Value Theorem there is a number c, $-1 < c < 1$, such that $f(c) = 0$, i.e., $e^{-c} = c$. Thus c is a solution of the original equation.

2.3 THE SANDWICH THEOREM AND $(\sin\theta)/\theta$

1. $\lim_{x\to 0}\frac{1}{\cos x} = \frac{1}{\cos 0} = \frac{1}{1} = 1$

3. $\lim_{x\to 0}\frac{1+\sin x}{1+\cos x} = \frac{1+\sin 0}{1+\cos 0} = \frac{1+0}{1+1} = \frac{1}{2}$

5. $\lim_{x\to 0}\frac{x}{\sin x} = \lim_{x\to 0}\frac{1}{\frac{\sin x}{x}} = \frac{1}{1} = 1$

7. $\lim_{x\to 0} \frac{\sin 2x}{x} = \lim_{x\to 0} 2\frac{\sin 2x}{2x} = 2\lim_{x\to 0} \frac{\sin 2x}{2x} = 2\lim_{\theta\to 0} \frac{\sin\theta}{\theta} = 2\cdot 1 = 2$. Here $\theta = 2x$.

9. $\lim_{x\to 0} = \frac{\tan 2x}{2x} = \lim_{x\to 0} \frac{1}{\cos 2x}\frac{\sin 2x}{2x} = 1\cdot 1 = 1$

11. $\lim_{x\to 0} \frac{\sin x}{2x^2-x} = \lim_{x\to 0} \frac{\sin x}{x}\frac{1}{2x-1} = 1\cdot(-1) = -1$

13. $\lim_{x\to 0} \frac{\sin^2 x}{x} = \lim_{x\to 0} \sin x \frac{\sin x}{x} = 0\cdot 1 = 0$

15. $\lim_{x\to 0} \frac{3\sin 4x}{\sin 3x} = 3\lim_{x\to 0} \frac{\sin 4x}{\sin 3x} = 3\lim_{x\to 0} \frac{(4x)\frac{\sin 4x}{4x}}{(3x)\frac{\sin 3x}{3x}} = 3\lim_{x\to 0} \frac{4}{3}\frac{\frac{\sin 4x}{4x}}{\frac{\sin 3x}{3x}} = 3\cdot\frac{4}{3}\cdot\frac{1}{1} = 4$.

17. a) Approximately 0.6 b) Very close

c) $\lim_{x\to 0} \frac{\tan 3x}{\sin 5x} = \lim_{x\to 0} \frac{\frac{\sin 3x}{\cos 3x}}{\sin 5x} = \lim_{x\to 0} \frac{\frac{1}{\cos 3x}\frac{\sin 3x}{3x}3x}{\frac{\sin 5x}{5x}5x} = 1\cdot 1\cdot\frac{3}{5} = \frac{3}{5}$.

19. Since $\lim_{x\to 0} 1 - \frac{x^2}{6} = 1$ and $\lim_{x\to 0} 1 = 1$, it follows from the inequality and the Sandwich Theorem that $\lim_{x\to 0} \frac{\sin x}{x} = 1$.

21. a) We graph $y_1 = \frac{1}{2} - \frac{x^2}{24}$, $y_2 = \frac{1-\cos x}{x^2}$ and $y_3 = \frac{1}{2}$ (y_1 and y_2 are extremely close near $x = 0$) in the window $[-6, 6]$ by $[-0.1, 0.6]$ supporting the inequality.

b) $\frac{1}{2}$ c) Since both y_1 and $y_3 \to \frac{1}{2}$ as $x \to 0$, $y_2 = f(x) \to \frac{1}{2}$ by the Sandwich Theorem.

23. The numerator $\cos x$ approaches 1 as $x \to 0$ while the denominator approaches 0. Thus the fraction can be made arbitrarily large in absolute value if x is sufficiently close to 0. Therefore the fraction cannot approach any finite number and so the limit does not exist.

25. Let $f(\theta) = \frac{\sin\theta}{\theta}$ where the sine function is in degree mode. Then $f(\theta) = \frac{\sin(\frac{\pi}{180}\theta)}{\theta}$ if sine is in radian mode. $\frac{\sin(\frac{\pi}{180}\theta)}{\theta} = \frac{(\frac{\pi}{180})\sin(\frac{\pi}{180}\theta)}{(\frac{\pi}{180}\theta)} \to (\frac{\pi}{180})(1) = \frac{\pi}{180}$ as $\theta \to 0$.

27. $f(x) = \frac{\tan 3x}{x}$.

x	± 0.1	± 0.01	± 0.001
$f(x)$	3.0934	3.0009	3.0000

Conjecture: $\lim_{x\to 0} f(x) = 3$.

29. $f(x) = \frac{x-\sin x}{x^2}$.

x	-0.01	0.01	-0.001	0.001
$f(x)$	-0.00167	0.00167	-0.00017	0.00017

Conjecture: $\lim_{x\to 0} f(x) = 0$.

2.4 LIMITS INVOLVING INFINITY

1. a) $\lim_{x\to\infty} \frac{2x+3}{5x+7} = \lim_{x\to\infty} \frac{2+\frac{3}{x}}{5+\frac{7}{x}} = \frac{2+0}{5+0} = \frac{2}{5}$

b) $\lim_{x\to-\infty} \frac{2x+3}{5x+3} = \frac{2}{5}$ by the same reasoning.

3. $\lim_{x\to\infty} \frac{x+1}{x^2+3} = \lim_{x\to\infty} \frac{\frac{1}{x}+\frac{1}{x^2}}{1+\frac{3}{x^2}} = \frac{0+0}{1+0} = \frac{0}{1} = 0$

b) $\lim_{x\to-\infty} \frac{x+1}{x^2+3} = 0$ by the same reasoning.

5. a) $\lim_{x\to\infty} \frac{3x^2-6x}{4x-8} = \lim_{x\to\infty} \frac{3x-6}{4-\frac{8}{x}} = \infty$

b) $\lim_{x\to-\infty} \frac{3x^2-6x}{4x-8} = -\infty$ by the same method.

7. a) $\lim_{x\to\infty} \frac{1}{x^3-4x+1} = \lim_{x\to\infty} \frac{\frac{1}{x^3}}{1-\frac{4}{x^2}+\frac{1}{x^3}} = \frac{0}{1-0+0} = 0$

b) $\lim_{x\to-\infty} \frac{1}{x^3-4x+1} = 0$ by the same method.

9. a) $\lim_{x\to\infty} \frac{-2x^3-2x+3}{3x^3+3x^2-5x} = \lim_{x\to\infty} \frac{-2-\frac{2}{x^2}+\frac{3}{x^3}}{3+\frac{3}{x}-\frac{5}{x^2}} = \frac{-2-0+0}{3+0-0} = -\frac{2}{3}$

b) $\lim_{x\to-\infty} \frac{-2x^3-2x+3}{3x^3+3x^2-5x} = -\frac{2}{3}$ by the same method.

11. a) $\lim_{x\to\infty} \left(\frac{-x}{x+1}\right)\left(\frac{x^2}{5+x^2}\right) = \lim_{x\to\infty} \left(\frac{-1}{1+\frac{1}{x}}\right)\left(\frac{1}{\frac{5}{x^2}+1}\right) = (-1)(1) = -1$

b) $\lim_{x\to-\infty} \left(\frac{-x}{x+1}\right)\left(\frac{x^2}{5+x^2}\right) = -1$ by the same proof.

13. $\lim_{x\to2^+} \frac{1}{x-2} = \infty$ because $x-2$ is always positive in this process.

15. $\lim_{x\to2^+} \frac{x}{x-2} = \infty$ because the numerator approaches 2 and the denominator approaches 0 but is always positive in the process.

17. $\lim_{x\to-3^+} \frac{1}{x+3} = \infty$ (the denominator is always positive)

19. $\lim_{x\to-3^+} \frac{x}{x+3} = \lim_{x\to-3^+} x\frac{1}{x+3} = -\infty$ by Theorem 11 or it can be argued that the numerator is negative and approaching -3 while the denominator is positive and approaching 0.

21. $\lim_{x\to\pm\infty} f(x) = \lim_{x\to\pm\infty} \frac{x-2}{2x^2+3x-5} = \lim_{x\to\pm\infty} \frac{\frac{1}{x}-\frac{2}{x^2}}{2+\frac{3}{x}-\frac{5}{x^2}} = \frac{0-0}{2+0-0} = 0$. Therefore $y = 0$ is the end behavior asymptote. $2x^2+3x-5 = (2x+5)(x-1)$ so there are vertical asymptotes at $x = -\frac{5}{2}$ and $x = 1$.

23. $\lim_{x\to\pm\infty} g(x) = \lim_{x\to\pm\infty} \frac{3x^2-x+5}{x^2-4} = \lim_{x\to\pm\infty} \frac{3-\frac{1}{x}+\frac{5}{x^2}}{1-\frac{4}{x^2}} = 3$ so $y = 3$ is the end behavior asymptote. VA: $x = \pm 2$.

25. $f(x) = \frac{x^2-2x+3}{x+2} = (x-4) + \frac{11}{x+2}$ so $x-4$ is the end behavior asymptote. VA: $x = -2$.

27. $g(x) = x^2+2x+2+\frac{5}{x-2}$ so x^2+2x+2 is the end behavior asymptote. VA: $x = 2$.

29. $\lim_{x\to\pm\infty}\left(\frac{x}{x^2+3} - \frac{1+\frac{2}{x^2}}{\frac{1}{x^2}+\frac{1}{x}-1}\right) = 0-(-1) = 1$ so $y = 1$ is the end behavior asymptote. The quadratic formula yields the roots of $1+x-x^2$ and so the vertical asymptotes: $x = \frac{1\pm\sqrt{5}}{2}$.

31. To study $y = \frac{1}{x^2-4}$, a convenient viewing rectangle for its graph is $[-3,3]$ by $[-5,5]$. From this graph we see a) $\lim_{x\to 2^+} \frac{1}{x^2-4} = \infty$, b) $\lim_{x\to 2^-} \frac{1}{x^2-4} = -\infty$, c) $\lim_{x\to -2^+} \frac{1}{x^2-4} = -\infty$, and d) $\lim_{x\to -2^-} \frac{1}{x^2-4} = \infty$

33. The graph of $f(x) = \frac{x^2-1}{2x+4}$ in the viewing window $[-10,10]$ by $[-10,10]$ indicates that a) $\lim_{x\to -2^+} f(x) = \infty$, and b) $\lim_{x\to -2^-} f(x) = -\infty$.

35. $\lim_{x\to 0^+} \frac{[x]}{x} = \lim_{x\to 0^+} \frac{0}{x} = \lim_{x\to 0^+} 0 = 0$. Here we use the fact that $[x] = 0$ if $0 < x < 1$.

37. $\lim_{x\to\infty} \frac{|x|}{|x|+1} = \lim_{x\to\infty} \frac{x}{x+1} = \lim_{x\to\infty} \frac{1}{1+\frac{1}{x}} = \frac{1}{1+0} = 1$. Here we use the fact that $|x| = x$ for $x \geqq 0$.

39. $\lim_{x\to 0^+} \frac{1}{\sin x} = \infty$ because $\sin x$ is a positive number if x is a small positive number $0 < x < \pi$.

41. $\lim_{x\to\frac{\pi}{2}^+} \frac{1}{\cos x} = -\infty$ because $\cos x < 0$ for $\frac{\pi}{2} < x < \frac{3\pi}{2}$.

43. $\lim_{x\to -\infty} f(x) = \lim_{x\to -\infty} \frac{1}{x} = 0$. $\lim_{x\to 0^-} f(x) = \lim_{x\to 0^-} \frac{1}{x} = -\infty$. $\lim_{x\to 0^+} f(x) = \lim_{x\to 0^+}(-1) = -1$. $\lim_{x\to\infty} f(x) = \lim_{x\to\infty}(-1) = -1$.

45. We first prove $\lim_{x\to\infty} \frac{\sin x}{x} = 0$. We know $-1 \leqq \sin x \leqq 1$ for all x. Hence for $x > 0$, $-\frac{1}{x} \leqq \frac{\sin x}{x} \leqq \frac{1}{x}$. But $\lim_{x\to\infty}(-\frac{1}{x}) = 0 = \lim_{x\to\infty} \frac{1}{x}$. By the Sandwich Theorem $\lim_{x\to\infty} \frac{\sin x}{x} = 0$. Therefore $\lim_{x\to\infty}(2+\frac{\sin x}{x}) = 2+0 = 2$.

47. $\lim_{x\to\infty}(1+\cos\frac{1}{x}) = \lim_{\theta\to 0^+}(1+\cos\theta) = 1+\cos 0 = 2$.

49. $-1 \leqq \sin 2x \leqq 1$ for all x. Hence for $x > 0$, $-\frac{1}{x} \leqq \frac{\sin 2x}{x} \leqq \frac{1}{x}$. But $\lim_{x\to\infty}(-\frac{1}{x}) = 0 = \lim_{x\to\infty} \frac{1}{x}$. By the Sandwich Theorem, $\lim_{x\to\infty} \frac{\sin 2x}{x} = 0$.

51. $\lim_{x\to\pm\infty} \frac{2x^2}{x^2+1} = \lim_{x\to\pm\infty} \frac{2}{1+\frac{1}{x^2}} = \frac{2}{1+0} = 2$ and $\lim_{x\to\pm\infty} \frac{2x^2+5}{x^2} = \lim_{x\to\pm\infty} (2+\frac{5}{x^2}) = 2+0 = 2$. By the Sandwich Theorem, $\lim_{x\to\infty} f(x) = 2$ and $\lim_{x\to-\infty} f(x) = 2$.

53. Each graph satisfies $y\to\infty$ as $x\to\infty$ and $y\to-\infty$ as $x\to-\infty$. As the power of x increases, the vertical steepness of the graph increases for $|x|>1$.

55. $\lim_{x\to\pm\infty} \frac{f(x)}{-\frac{1}{7}} = \lim_{x\to\pm\infty} -7f(x) = -7\lim_{x\to\pm\infty} f(x) = -7(-\frac{1}{7}) = 1$ using the result of Example 8 (the same method can be used for $x\to-\infty$).

57. One such function is $f(x) = \begin{cases} x+1, & x \leqq 2 \\ \frac{1}{5-x}, & 2<x<5 \\ -1, & x \geqq 5 \end{cases}$. Graph $y_1 = x+1+0\sqrt{2-x}$, $y_2 = \frac{1}{5-x} + 0\sqrt{x-2} + 0\sqrt{5-x}$ and $y_3 = -1 + 0\sqrt{x-5}$ in $[-5,10]$ by $[-10,10]$.

59. $\lim_{x\to 0} f(x) = \lim_{x\to 0} \frac{1}{x}$ does not exist. $\lim_{x\to 0} g(x) = \lim_{x\to 0} x = 0$. $\lim_{x\to 0} f(x)g(x) = \lim_{x\to 0} \frac{x}{x} = 1$.

61. $\lim_{x\to 2} \frac{3}{x-2}$ does not exist, $(x-2)^3 \to 0$ as $x\to 2$, $f(x)g(x) = \frac{3}{x-2}(x-2)^3 = 3(x-2)^2 \to 0$ as $x\to 2$.

63. Let $f(x) = x^3$, $g(x) = \frac{1}{x^2}$. Then $\lim_{x\to 0}(fg) = \lim_{x\to 0} x = 0$.
Let $f(x) = 5x^2$, $g(x) = \frac{1}{x^2}$. Then $\lim_{x\to 0}(fg) = \lim_{x\to 0} 5 = 5$.
Let $f(x) = x^2$, $g(x) = \frac{1}{x^4}$. Then $\lim_{x\to 0}(fg) = \lim_{x\to 0} \frac{1}{x^2} = \infty$.
In each case $\lim_{x\to 0} f(x) = 0$ and $\lim_{x\to 0} g(x) = \infty$.

65. $\lim_{x\to\pm\infty} \frac{f(x)}{a_n x^n} = \lim_{x\to\pm\infty} \frac{a_n x^n + a_{n-1}x^{n-1} + a_{n-2}x^{n-2} + \cdots + a_1 x + a_0}{a_n x^n} = \lim_{x\to\pm\infty}(1 + \frac{a_{n-1}}{a_n}\frac{1}{x} + \frac{a_{n-2}}{a_n}\frac{1}{x^2} + \cdots + \frac{a_1}{a_n}\frac{1}{x^{n-1}} + \frac{a_0}{a_n}\frac{1}{x^n}) = 1+0+0+\cdots+0+0 = 1$

67. Using graphs, support $(1+\frac{1}{x})^x \to e$ as $x\to\pm\infty$.

69. Using graphs, we can support $(1+\frac{0.07}{x})^x \to e^{0.07}$ as $x\to\pm\infty$.

71. $xe^{-x} = \frac{x}{e^x} \to 0$ as $x\to\infty$. $xe^{-x} \to -\infty$ as $x\to-\infty$. This is supported by graphing $y = xe^{-x}$ in the window $[-3,5]$ by $[-5,1]$.

73. $y = xe^x \to \infty$ as $x\to\infty$. $y\to 0$ as $x\to-\infty$. This may be supported by graphing y in $[-5,3]$ by $[-1,5]$.

75. As θ increases $\sin\theta$ steadily oscillates between the values -1 and $+1$ passing through all intermediate values. No matter how large θ gets, $\sin\theta$ thereafter does not stay arbitrarily close to any fixed value.

2.5 CONTROLLING FUNCTION OUTPUTS

1. a) $0 < x < 6$. Not equivalent. b) $1 < x - 1 < 7$. Adding 1, we get $2 < x < 8$. Equivalent. c) $1 < \frac{x}{2} < 4$. Multiplying by 2, we get $2 < x < 8$. Equivalent. d) $\frac{1}{8} < \frac{1}{x} < \frac{1}{2}$. Taking reciprocals, we get $8 > x > 2$. Equivalent. e) $x > 8$. Not equivalent. f) $|x - 5| < 3$. $-3 < x - 5 < 3$. Adding 5, we get $2 < x < 8$. Equivalent. g) $4 < x < 10$. Not equivalent. h) $-8 < -x < -2$. Multiplying by -1, we get $8 > x > 2$. Equivalent.

3. Change $|x + 3| < 1$ to $-1 < x + 3 < 1$ to $-4 < x < -2$. Answer: g). Equivalently, we can read $|x + 3| < 1$ as $|x - (-3)| < 1$, that is, the distance between x and -3 is less than 1 and thus $-4 < x < -1$.

5. Change $|\frac{x}{2}| < 1$ to $-1 < \frac{x}{2} < 1$ to $-2 < x < 2$. Answer: e)

7. Change $|2x - 5| \leqq 1$ to $-1 \leqq 2x - 5 \leqq 1$ to $4 \leqq 2x \leqq 6$ to $2 \leqq x \leqq 3$. Answer: h)

9. $|\frac{x-1}{5}| \leqq 1$ leads to $|x - 1| \leqq 5$, $-5 \leqq x - 1 \leqq 5$, $-4 \leqq x \leqq 6$. Answer: i)

11. $|y - 2| \leqq 5$. $-5 \leqq y - 2 \leqq 5$, $-3 \leqq y \leqq 7$.

13. $|2y - 5| < 1$. $-1 < 2y - 5 < 1$, $4 < 2y < 6$, $2 < y < 3$.

15. Change $|\frac{y}{2} - 1| \leqq 1$ to $-1 \leqq \frac{y}{2} - 1 \leqq 1$ to $0 \leqq \frac{y}{2} \leqq 2$ to $0 \leqq y \leqq 4$. Answer: $0 \leqq y \leqq 4$

17. $|2 - y| < \frac{1}{5}$. $-\frac{1}{5} < 2 - y < \frac{1}{5}$, $-\frac{11}{5} < -y < -\frac{9}{5}$, $\frac{9}{5} < y < \frac{11}{5}$.

19. The midpoint of the interval is $\frac{1+8}{2} = \frac{9}{2}$ and it has radius $\frac{9}{2} - 1 = \frac{7}{2}$. Answer: $|x - \frac{9}{2}| < \frac{7}{2}$

21. The midpoint of the interval is $\frac{-4+1}{2} = -\frac{3}{2}$ and it has radius $1 - (-\frac{3}{2}) = \frac{5}{2}$. Answer: $|x + \frac{3}{2}| < \frac{5}{2}$

23. $0.5 < x^2 < 1.5$ yields $\sqrt{0.5} < |x| < \sqrt{1.5}$ or after appropriate rounding $0.71 < |x| < 1.22$. Thus we obtain $-1.22 < x < -0.71$

25. For x in the interval $[0, \frac{\pi}{2}]$, $\cos x$ is decreasing and so $0.2 < \cos x < 0.6$ yields $\cos^{-1} 0.6 < x < \cos^{-1} 0.2$, or rounding appropriately, $0.93 < x < 1.36$.

27. The graph of $y = \cos x$ is symmetric with respect to the y-axis. We need only reflect the corresponding interval in $[0, \frac{\pi}{2}]$ found in Exercise 25 over the y-axis. We obtain $-1.36 < x < -0.93$.

29. $99.9 < x^2 < 100.1$ yields $\sqrt{99.9} < x < \sqrt{100.1}$, or rounding to thousandths appropriately, $9.995 < x < 10.004$.

31. Change $3.9 < \sqrt{x-7} < 4.1$ to $3.9^2 < x - 7 < 4.1^2$ to $22.21 < x < 23.81$.

33. Change $4 < \frac{120}{x} < 6$ to $\frac{1}{4} > \frac{x}{120} > \frac{1}{6}$ to $30 > x > 20$ or $20 < x < 30$.

35. The graph of $y = \frac{3-2x}{x-1}$ is steadily falling as it passes through the horizontal channel between $y = -3.1$ and $y = -2.9$. Next we solve for x in terms of $y: (x-1)y = 3-2x$, $xy - y = 3 - 2x$, $xy + 2x = y + 3$, $x(y+2) = y+3$, $x = (y+3)/(y+2)$. Thus when $y = -2.9$, $x = (-2.9+3)/(-2.9+2) = -\frac{0.1}{0.9} = -\frac{1}{9}$ and, similarly, when $y = -3.1$, $x = \frac{1}{11}$. Answer: $-\frac{1}{9} < x < \frac{1}{11}$

37. $10.5 < x^2 - 5 < 11.5$ leads to $15.5 < x^2 < 16.5$, $\sqrt{15.5} < x < \sqrt{16.5}$ since $x_0 > 0$. Thus $3.94 < x < 4.06$ rounding to hundredths appropriately.

39. We graph $y = 4.8$, $y = 5.2$ and $y = x^3 - 9x$ in the same viewing rectangle, $[-4, 5]$ by $[-10, 10]$, for example. We then zoom in on the two points of intersection near $x = -3$. Rounding to hundredths appropriately, we obtain $-2.68 < x < -2.66$.

41. $0.4 < e^x < 0.6$ leads to $\ell n\, 0.4 < \ell n\, e^x < \ell n\, 0.6$ since $\ell n\, x$ is an increasing function. Because $\ell n\, e^x = x$, we obtain after rounding appropriately to hundredths $-0.91 < x < -0.52$.

43. $|f(x) - y_0| < E = 0.5$ is equivalent to $|x + 1 - 4| < 0.5$ or $|x - 3| < 0.5$ which is the desired inequality.

45. In the same viewing retangle we graph $y = 2.8$, $y = 3.2$ and $y = 2x^2 + 1$. The latter curve is rising near $x = 1$ and we zoom in on its points of intersection with the horizontal lines near $x = 1$. The curve is in the channel for $0.95 < x < 1.04$ after appropriate rounding. Thus we may take $|x - 1| < 0.04$.

47. $\lim_{x\to 1} x^2 = 1$, $\lim_{x\to 1} x^2 = 1$, $\lim_{x\to \pi/6} \sin x = 0.5$, $\lim_{x\to 3} \frac{x+1}{x-2} = 4$, respectively.

49. $\lim_{x\to 10} x^2 = 100$, $\lim_{x\to -10} x^2 = 100$, $\lim_{x\to 23} \sqrt{x-7} = 4$, $\lim_{x\to 10} \sqrt{19-x} = 3$, $\lim_{x\to 24}(\frac{120}{x}) = 5$, $\lim_{x\to 1/4}(\frac{1}{4x}) = 1$, $\lim_{x\to 0} \frac{3-2x}{x-1} = -3$, $\lim_{x\to -3} \frac{3x+8}{x+2} = 1$, respectively.

51. $8.99 < \pi(x/2)^2 < 9.01$ leads to $\frac{8.99}{\pi} < (\frac{x}{2})^2 < \frac{9.01}{\pi}$, $\frac{4(8.99)}{\pi} < x^2 < \frac{4(9.01)}{\pi}$ and to $\sqrt{\frac{4(8.99)}{\pi}} < x < \sqrt{\frac{4(9.01)}{\pi}}$. Rounding appropriately to thousandths, we obtain $3.384 < x < 3.387$ or, in symmetric form, $|x - x_0| < 0.001$.

53. $f(x) = \frac{3x+1}{x-2} = 3 + \frac{7}{x-2} \to 3$ as $x \to \infty$ (the equality can be obtained by long division). The same equality (or a graph) shows that $f(x) > 3$ for $x > 2$. Thus we want to solve $3 + \frac{7}{x-2} < 3.01$, $x > 2$. This leads to $\frac{7}{x-2} < 0.01$, $x - 2 > \frac{7}{.01}$ (since $x - 2 > 0$), $x > 702$.

55. By long division $f(x) = \frac{2x^2-x+2}{x^2-4} = 2 + \frac{10-x}{x^2-4}$. This shows $f(x) < 2$ for $x > 10$ and $f(x) = 2 + \frac{(10/x^2)-(1/x^2)}{1-(4/x^2)} \to 2 + 0 = 2$ as $x \to \infty$. $f(x) > 1.99$, $x > 10$ leads to $2x^2 - x + 2 > 1.99(x^2 - 4)$, $0.01x^2 - x + 9.96 > 0$, $x^2 - 100x + 996 > 0$. By the quadratic formula, the larger root of the quadratic is $50 + \sqrt{1504}$. Thus we require $x > 50 + \sqrt{1504} \approx 88.781$.

57. In the interval $[0, 2\pi]$, $\sin x = \frac{\sqrt{2}}{2}$ has two solutions $\frac{\pi}{4}$ in the first quadrant and $\pi - \frac{\pi}{4} = \frac{3\pi}{4}$ in the second quadrant, recalling that $\sin(\pi - x) = \sin x$ for all x. We first solve the problem in the first quadrant. We use the result to solve the problem in the second quadrant and then use the periodicity of the sine function to give the complete solution set. Graph $y_1 = \sin x$, $y_2 = \frac{\sqrt{2}}{2} - 0.1$ and $y_3 = \frac{\sqrt{2}}{2} + 0.1$ in $[0.6, 1]$ by $[0.5, 1]$. We then zoom in to the two points of intersection and find that they occur at $x_1 = 0.65241449628$ and $x_2 = 0.93923517764$. Thus in the first quadrant we must have $x_1 < x < x_2$. In the second quadrant we must have $\pi - x_2 < x < \pi - x_1$. The complete solution set is $\{x \mid x_1 + 2n\pi < x < x_2 + 2n\pi \text{ or } (2n+1)\pi - x_2 < x < (2n+1)\pi - x_1\}$.

59. By long division $f(x) = \frac{x-3}{2x+1} = \frac{1}{2} - \frac{7}{2(2x+1)} > \frac{1}{2}$ for $x < -\frac{1}{2}$, and $f(x) \to \frac{1}{2}$ as $x \to -\infty$. $f(x) < 0.51$ leads to $\frac{x-3}{2x+1} < 0.51$, $x - 3 > 0.51(2x + 1)$ (since $2x + 1 < 0$), $x - 3 > 1.02x + 0.51$, $-3.51 > 0.02x$, $x < -175.5$.

61. By long division $f(x) = \frac{x-4}{1-3x} = -\frac{1}{3} - \frac{11}{3(1-3x)} < -\frac{1}{3}$. $\frac{x-4}{1-3x} \to -\frac{1}{3}$ as $x \to -\infty$. $\frac{x-4}{1-3x} > -\frac{1}{3} - \frac{1}{100} = -\frac{103}{300}$ leads to $300(x-4) > -103(1-3x)$ (since $1 - 3x > 0$ for $x < \frac{1}{3}$), $9x < -1200 + 103 = -1097$, $x < -\frac{1097}{9}$.

2.6 DEFINING LIMITS FORMALLY WITH EPSILONS AND DELTAS

1. $x_0 - a = 5 - 1 = 4$ and $b - x_0 = 7 - 5 = 2$. Since $2 < 4$, we choose $\delta = 2$. Then $|x - x_0| < 2$ or $|x - 5| < 2$ implies $3 < x < 7$ and so $1 < x < 7$.

3. $x_0 - a = -3 + \frac{7}{2} = \frac{1}{2}$ and $b - x_0 = -\frac{1}{2} + 3 = \frac{5}{2}$. Since $\frac{1}{2} < \frac{5}{2}$, $\delta = \frac{1}{2}$.

5. From the graph we see that if x is between 4.9 and 5.1 on the x-axis, then y on the curve (corresponding to x) is in the range $5.8 < y < 6.2$. Therefore we can take $\delta = 0.1$ because $0 < |x - 5| < \delta = 0.1$ does imply $|f(x) - L| = |y - 6| < 0.2 = \varepsilon$.

7. From the graph we see that $\sqrt{3} < x < \sqrt{5}$ implies that $|f(x) - L| < \varepsilon$. Rounding the first inequality appropriately to hundredths, we get $1.74 < x < 2.23$. Since 2 is closer to 2.23 than to 1.74, we take $\delta = 2.23 - 2 = 0.23$.

9. From the graph we see that $\frac{9}{16} < x < \frac{25}{16}$ implies $|f(x) - L| < \varepsilon = \frac{1}{4}$. $1 - \frac{9}{16} = \frac{7}{16}$ and $\frac{25}{16} - 1 = \frac{9}{16}$. Thus 1 is closer to $\frac{9}{16}$. We take $\delta = \frac{7}{16}$. Note that $|x - x_0| < \delta$ is the same as $|x - 1| < \frac{7}{16}$ which is equivalent to $\frac{9}{16} < x < \frac{23}{16}$. So $\delta = \frac{7}{16}$ works.

11. $\lim_{x\to 1}(2x + 3) = 5$. $|f(x) - L| = |2x + 3 - 5| = |2x - 2| = |2(x - 1)| = |2||x - 1| = 2|x - 1|$. Thus $|f(x) - L| < 0.01$ is $2|x - 1| < 0.01$ which is equivalent to $|x - 1| < \frac{0.01}{2} = 0.005$. Thus $\delta = 0.005$ will do because $0 < |x - 1| < 0.005 \Rightarrow |f(x) - L| < 0.01 = \varepsilon$.

13. $\lim_{x\to 2} \frac{x^2-4}{x-2} = \lim_{x\to 2} \frac{(x-2)(x+2)}{x-2} = \lim_{x\to 2}(x + 2) = 4$. If $x \neq 2$, $\frac{x^2-4}{x-2} = x + 2$ and we may assume this here because x is never equal to 2 as $x \to 2$. Thus $|f(x) - L| = |x + 2 - 4| = |x - 2| < \varepsilon = 0.05$ is equivalent to $0 < |x - 2| < 0.05$ and we may take $\delta = 0.05$.

15. $L = \lim_{x\to 11} \sqrt{x - 7} = \sqrt{11 - 7} = 2$. The inequality $|\sqrt{x - 7} - 2| < 0.01$ leads successively to $-0.01 < \sqrt{x - 7} - 2 < 0.01$, $1.99 < \sqrt{x - 7} < 2.01$, $1.99^2 < x - 7 < 2.01^2$, $7 + 1.99^2 < x < 7 + 2.01^2$, $10.9601 < x < 11.0401$. The distance, 0.0399, between $x_0 = 11$ and the left endpoint is less than the distance between x_0 and the right endpoint. Thus we may take $\delta = 0.0399$.

17. $\lim_{x\to 2} \frac{4}{x} = 2$. $|f(x) - L| < \varepsilon$ is equivalent to each of the following: $|\frac{4}{x} - 2| < 0.4 = \frac{2}{5}$, $2 - \frac{2}{5} < \frac{4}{x} < 2 + \frac{2}{5}$, $\frac{8}{5} < \frac{4}{x} < \frac{12}{5}$, $\frac{5}{12} < \frac{x}{4} < \frac{5}{8}$, $\frac{5}{3} < x < \frac{5}{2}$, $2 - \frac{1}{3} < x < 2 + \frac{1}{2}$. The last inequality is satisfied if $|x - 2| < \frac{1}{3}$. We may therefore take $\delta = \frac{1}{3}$.

19. $|f(x) - 5| = |9 - x - 5| = |4 - x| = |x - 4|$. Thus $|f(x) - 5| < \varepsilon$ is equivalent to $|x - 4| < \varepsilon$. Thus in each case $\delta = \varepsilon$.

21. Let $f(x) = \frac{x+2}{x+1} = \frac{(x+1)+1}{x+1} = 1 + \frac{1}{x+1}$. $\lim_{x\to\infty}(1 + \frac{1}{x+1}) = 1$. Let $\varepsilon > 0$ be given. Then $|f(x) - 1| = \frac{1}{x+1}$ (for $x > -1$) $< \varepsilon$ if and only if $x + 1 > \frac{1}{\varepsilon}$ or $x > \frac{1}{\varepsilon} - 1$. Thus if $N = \frac{1}{\varepsilon} - 1$, $x > N$ implies $|f(x) - 1| < \varepsilon$.

23. Let $f(x) = \frac{x+2}{x+1} = 1 + \frac{1}{x+1}$. Here since $x \to -1^+$, $x > -1$ or $x + 1 > 0$. Let $N > 1$ be given, $f(x) > N$ is equivalent to $1 + \frac{1}{x+1} > N$, $\frac{1}{x+1} > N - 1$, $x + 1 < \frac{1}{N-1}$. Let $\delta = \frac{1}{N-1}$. Then $-1 < x < -1 + \delta$ implies $f(x) > N$. This proves $\lim_{x\to -1^+} f(x) = \infty$.

25. $\lim_{x\to 1} \sin x = 0.84 (= \sin 1)$ rounded to hundredths. In the viewing rectangle $[0.99, 1.01]$ by $[0.83, 0.85]$ we graph $y = 0.84$ and $y = \sin x$ (which appears as a straight line). Using the endpoints of $y = \sin x$ in this rectangle, we calculate the slope $m = \frac{0.8468-0.8360}{1.01-0.99} = 0.54$. As in Example 6, $\delta = \varepsilon|m| = \varepsilon/0.54 = 1.85\varepsilon$ rounding δ down to be safe.

27. $\lim_{x\to 1} \cos x = \cos 1 = 0.54$ after rounding. In the viewing rectangle $[0.99, 1.01]$ by $[0.53, 0.55]$ we graph $y = 0.54$ and $y = \cos x$. Using two points on the graph of $y = \cos x$, we get for the estimate of m, $m = -0.85$. Thus $\delta = \varepsilon/0.85 = 1.17\varepsilon$.

29. $\lim_{x\to 0.5}(x^3 - 4x) = (0.5)^3 - 4(0.5) = -1.88$ after rounding. In the viewing rectangle $[0.49, 0.51]$ by $[-1.89, 1.87]$ we graph $y = -1.88$ and $y = x^3 - 4x$. Using the endpoints to estimate the slope, we get $m = -3.25$. Therefore $\delta = \varepsilon/3.25 = 0.30\varepsilon$ rounding down to be safe.

31. $\lim_{x\to -1} \frac{x}{x^2-4} = \frac{1}{3}$. In the viewing rectangle $[-1.01, -0.99]$ by $[0.32, 0.34]$ we graph $y = 0.33$ and $y = \frac{x}{x^2-4}$. Using the endpoints of the latter graph, we get our estimate of the slope, $m = -0.56$. Thus $\delta = \varepsilon/0.56 = 1.78\varepsilon$ rounding appropriately.

33. $\sqrt{x-5} < \varepsilon$ is equivalent to $x - 5 < \varepsilon^2$ or $x < 5 + \varepsilon^2$ since $x \geqq 5$. Thus $I = (5, 5 + \varepsilon^2)$. Since $5 < x < 5 + \varepsilon^2$ implies $|\sqrt{x-5} - 0| < \varepsilon$, this verifies that $\lim_{x\to 5^+} \sqrt{x-5} = 0$.

35.

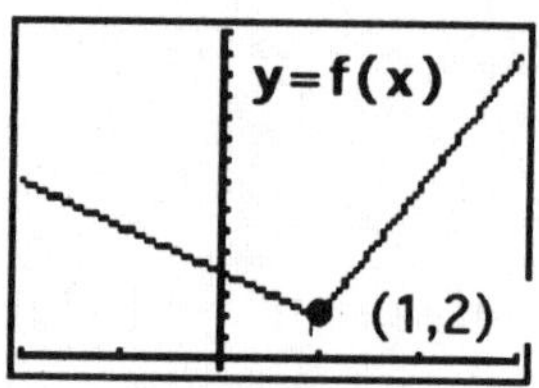

$[-2, 3]$ by $[-1, 16]$

Since $y \geqq 2$, we need only be concerned that $y < 2 + \varepsilon$. For $x < 1$, this is equivalent to $4 - 2x < 2 + \varepsilon$ which leads to $2 < 2x + \varepsilon$ and to $x > 1 - \varepsilon/2$. For $x \geqq 1$, $y < 2+\varepsilon$ is $6x - 4 < 2+\varepsilon$ which leads to $x < 1+\varepsilon/6$. The largest δ can be is the smaller of $\varepsilon/2$, $\varepsilon/6$ which is $\varepsilon/6$ and $I = (1 - \varepsilon/6, 1 + \varepsilon/6)$.

37. $\lim_{x\to 2} f(x) = 5$ means corresponding to any radius $\varepsilon > 0$ about 5, there exists a radius $\delta > 0$ about 2 such that $0 < |x-2| < \delta$ implies $|f(x)-5| < \varepsilon$.

39. Since we need $|x^2 - 4| < \varepsilon < 4$, x cannot be 0 and so $\delta < 2$, $0 < 2 - \delta < x$, i.e., $x > 0$. Now $|x^2 - 4| < \varepsilon$ is equivalent to each of the following: $-\varepsilon < x^2 - 4 < \varepsilon$, $4-\varepsilon < x^2 < 4+\varepsilon$, $\sqrt{4-\varepsilon} < |x| < \sqrt{4+\varepsilon}$, $\sqrt{4-\varepsilon} < x < \sqrt{4+\varepsilon}$ since $x > 0$. The distance from $x = 2$ to the left endpoint is $2 - \sqrt{4-\varepsilon} = (2 - \sqrt{4-\varepsilon})\frac{(2+\sqrt{4-\varepsilon})}{2+\sqrt{4-\varepsilon}} = \frac{\varepsilon}{2+\sqrt{4-\varepsilon}}$, and the distance from $x = 2$ to the right endpoint is $\sqrt{4+\varepsilon} - 2 = (\sqrt{4+\varepsilon} - 2)\frac{(\sqrt{4+\varepsilon}+2)}{\sqrt{4+\varepsilon}+2} = \frac{\varepsilon}{\sqrt{4+\varepsilon}+2}$. Since the second distance has a larger denominator, it is smaller and $\delta = \sqrt{4+\varepsilon} - 2$. This verifies $\lim_{x\to 2} x^2 = 4$ or $\lim_{x\to 2}(x^2 - 4) = 0$. $\delta \to 0$ as $\varepsilon \to 0$. The graph of δ as a function of ε can be viewed by graphing of δ as a function of ε can be viewed by graphing $y = \sqrt{4+x} - 2$ in the rectangle $[0, 4]$ by $[0, 1]$. The endpoints are not included.

41. One need only reverse the order of the steps in Example 3. Each line implies the preceding line in that list of steps.

43. Suppose $\varepsilon < 2$. Then $\delta = \frac{\varepsilon}{2(2+\varepsilon)}$. $|x - 0.5| < \delta$ implies $-\frac{\varepsilon}{2(2+\varepsilon)} < x - \frac{1}{2} < \frac{\varepsilon}{2(2+\varepsilon)} < \frac{\varepsilon}{2(2-\varepsilon)}$. We may now use the steps in Example 4 in reverse order to obtain $|f(x) - 2| < \varepsilon$. Now suppose $\varepsilon \geqq 2$, $\delta = \frac{1}{4}$. $|x - 0.5| < \delta = \frac{1}{4}$ implies $\frac{1}{4} < x < \frac{3}{4}$, $\frac{4}{3} < \frac{1}{x} < 4$, $-\frac{2}{3} < \frac{1}{x} - 2 < 2$ which implies $|\frac{1}{x} - 2| < 2 \leqq \varepsilon$.

45. a)

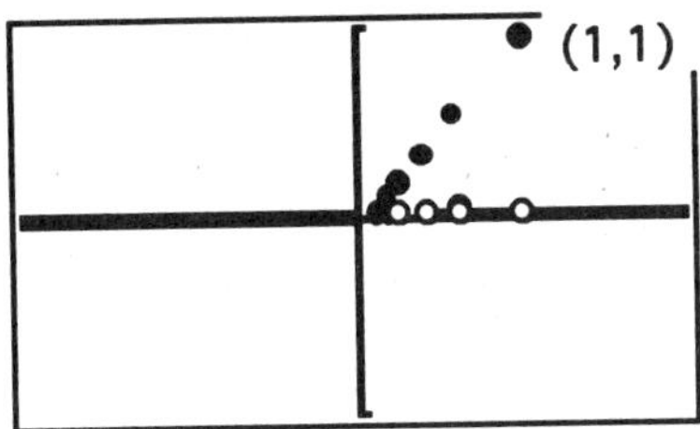

$[-2,2]$ by $[-1,1]$

b) Since $|f(x)-0| = |f(x)| \leqq |x| = |x-0|$, we may take $\delta = \varepsilon$ in the definition of limit to prove $\lim_{x\to 0} f(x) = 0$.

PRACTICE EXERCISES, CHAPTER 2

1. Exists **3.** Exists **5.** Exists **7.** Continuous at $x = a$

9. Not continuous at $x = c$ since $\lim_{x\to c} f(x)$ does not exist.

11. $\text{Lim}_{x\to -2} x^2(x+1) = (-2)^2(-2+1) = 4(-1) = -4$

13. $\text{Lim}_{x\to 3} \frac{x-3}{x^2} = \frac{3-3}{3^2} = \frac{0}{9} = 0$ **15.** $\text{Lim}_{x\to -2} (\frac{x}{x+1})(\frac{3x+5}{x^2+x}) = (\frac{-2}{-1})(\frac{-1}{4-2}) = -1$

17. $\text{Lim}_{x\to 4} \sqrt{1-2x}$ does not exist. **19.** $\lim_{x\to 1} \frac{x^2-1}{x-1} = \lim_{x\to 1} x+1 = 2$

21. $\text{Lim}_{x\to 2} \frac{x-2}{x^2+x-6} = \lim_{x\to 2} \frac{x-2}{(x-2)(x+3)} = \lim_{x\to 2} \frac{1}{x+3} = \frac{1}{5}$

23. $\text{Lim}_{x\to 0} \frac{(1+x)(2+x)-2}{x} = \lim_{x\to 0} \frac{3x+x^2}{x} = \lim_{x\to 0} 3+x = 3$

25. $\text{Lim}_{x\to \infty} \frac{2x+3}{5x+7} = \lim_{x\to \infty} \frac{2+\frac{3}{x}}{5+\frac{7}{x}} = \frac{2+0}{5+0} = \frac{2}{5}$

27. $\text{Lim}_{x\to -\infty} \frac{x^2-4x+8}{3x^3} = \lim_{x\to -\infty} \frac{\frac{1}{x}-\frac{4}{x^2}+\frac{8}{x^3}}{3} = 0$

29. $\text{Lim}_{x\to -\infty} \frac{x^2-7x}{x+1} = \lim_{x\to -\infty} \frac{x(x-7)}{x+1} = \lim_{x\to -\infty} \frac{x(1-\frac{7}{x})}{1+\frac{1}{x}} = -\infty$
because $\lim_{x\to -\infty} \frac{1-\frac{7}{x}}{1+\frac{1}{x}} = 1$.

31. $\text{Lim}_{x\to 3^+}\frac{1}{x-3} = \infty$ **33.** $\text{Lim}_{x\to 0^+}\frac{1}{x^2} = \infty$

35. $\text{Lim}_{x\to 0}\frac{\sin 2x}{4x} = \lim_{x\to 0}\frac{1}{2}\frac{\sin 2x}{2x} = \frac{1}{2}\cdot 1 = \frac{1}{2}$

37. $\text{Lim}_{x\to 0}\frac{\sin^3 2x}{x^3} = \lim_{x\to 0}(\frac{\sin 2x}{x})^3 = \lim_{x\to 0}(2\frac{\sin 2x}{2x})^3 = (2\cdot 1)^3 = 8$

39. a) the y-intercept is approximately 0.78.

b) The values are all close to 0.78

c) f appears to have a minimal value at $x = 0$.

d) $\text{Lim}_{x\to 0}\frac{\sec 2x \csc 9x}{\cot 7x} = \lim_{x\to 0}\frac{1}{\cos 2x}\frac{\sin 7x}{\cos 7x}\frac{1}{\sin 9x}$

$= \lim_{x\to 0}\frac{1}{\cos 2x}\frac{1}{\cos 7x}\frac{\sin 7x}{7x}\frac{1}{\frac{\sin 9x}{9x}}\frac{7x}{9x} = 1\cdot 1\cdot 1\cdot 1\cdot\frac{7}{9} = \frac{7}{9}$

41. a) $\lim_{x\to -2^+}\frac{x+3}{x+2} = \infty$ b) $\lim_{x\to 2^-}\frac{x+3}{x+2} = -\infty$. As $x\to -2$, $x+3$ is positive and $x+2$ is positive in a) but negative in b). The answer may be confirmed by graphing $y = \frac{x+3}{x+2}$.

43. a)

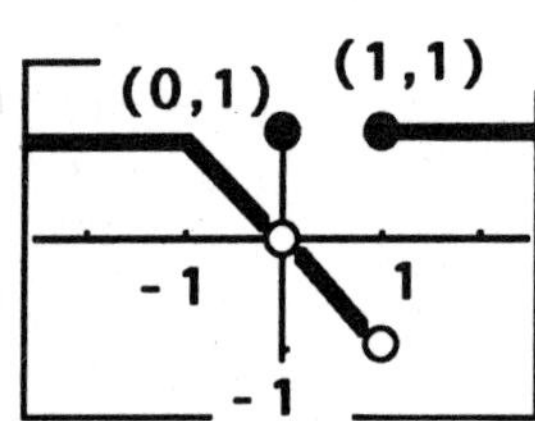

b) $\lim_{x\to -1^+} f(x) = 1$, $\lim_{x\to -1^-} f(x) = 1$, $\lim_{x\to 0^+} f(x) = 0$, $\lim_{x\to 0^-} f(x) = 0$, $\lim_{x\to 1^+} f(x) = 1$, $\lim_{x\to 1^-} f(x) = -1$

c) $\lim_{x\to -1} f(x) = 1$, $\lim_{x\to 0} f(x) = 0$ but $\lim_{x\to 1} f(x)$ does not exist because the right-hand and left-hand limits of f at 1 are not equal.

d) Only at $x = -1$

45. a) A graph of f may be obtained by graphing the functions $y = abs(x^3 - 4x) + 0\sqrt{1-x}$ and $y = x^2 - 2x - 2 + 0\sqrt{x-1}$ in the viewing rectangle $[-5, 7]$ by $[-4, 10]$.

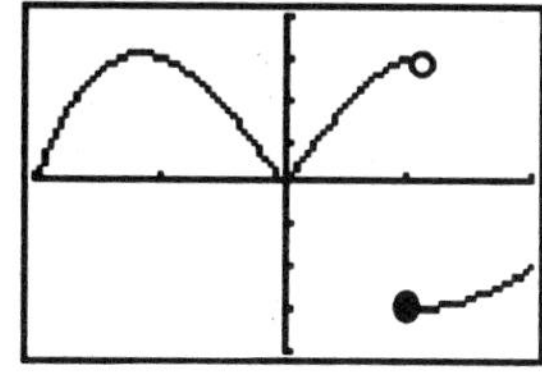

[-2,2] by [-4,4]

b) $\lim_{x\to 1^+} f(x) = \lim_{x\to 1^+}(x^2 - 2x - 2) = -3$.

$\lim_{x\to 1^-} f(x) = \lim_{x\to 1^-} |x^3 - 4x| = 3$

c) f does not have a limit at $x = 1$ because the right-hand and left-hand limits at $x = 1$ are not equal.

d) $x^3 - 4x$ is continuous by 2.2 Example 5 and $|x|$ is continuous by 2.2 Example 8. Thus $|x^3 - 4x|$ is continuous by Theorem 5 and so f is continuous for $x < 1$. For $x > 1$, $f(x) = x^2 - 2x - 2$, a polynomial, is continuous. Thus $f(x)$ is continuous at all points except $x = 1$.

e) f is not continuous at $x = 1$ because the two limits in b) are not equal and so $\lim_{x\to 1} f(x)$ does note exist.

47. a) A graph of f is obtained by graphing $y = -x + 0\sqrt{1-x}$ and $y = x - 1 + 0\sqrt{x-1}$ in the rectangle $[-2, 4]$ by $[-2, 4]$.

b) $\lim_{x\to 1^+} f(x) = \lim_{x\to 1^+}(x-1) = 0$. $\lim_{x\to 1^-} f(x) = \lim_{x\to 1^-} -x = -1$.

c) No value assigned to $f(1)$ makes f continuous at $x = 1$.

49. f is not defined at $x = \pm 2$ so is not continuous at these points. By 2.2 Example 6 f is continuous at all other points.

51. $\lim_{x\to\pm\infty} \frac{2x+1}{x^2-2x+1} = \lim_{x\to\pm\infty} \frac{\frac{2}{x}+\frac{1}{x^2}}{1-\frac{2}{x}+\frac{1}{x^2}} = \frac{0+0}{1-0+0} = 0$.

53. By long division $h(x) = x^2 - x + \frac{3}{x-3}$. Thus $y = x^2 - x$ is the end behavior asymptote of h.

55. a) $\lim_{x\to c} 3f(x) = 3\lim_{x\to c} f(x) = 3(-7) = -21$

b) $\lim_{x\to c}(f(x))^2 = \lim_{x\to c} f(x) \lim_{x\to c} f(x) = (-7)(-7) = 49$

c) $\lim_{x\to c} f(x) \cdot g(x) = \lim_{x\to c} f(x) \lim_{x\to c} g(x) = (-7)(0) = 0$

d) $\lim_{x\to c} \frac{f(x)}{g(x)-7} = \frac{\lim_{x\to c} f(x)}{\lim_{x\to c} g(x)-7} = \frac{-7}{0-7} = 1$

e) $\lim_{x\to c} \cos(g(x)) = \cos[\lim_{x\to c} g(x)] = \cos 0 = 1$.

f) $\lim_{x\to c} |f(x)| = |\lim_{x\to c} f(x)| = |-7| = 7$.

57. Both 0 and $\sqrt{x}$ approach 0 as $x \to 0$. By the Sandwich Theorem $|\sqrt{x}\sin\frac{1}{x}| \to 0$ and hence $\sqrt{x}\sin\frac{1}{x} \to 0$ as $x \to 0$.

59. $\text{Lim}_{x\to\infty} \frac{x+\sin x}{x} = \lim_{x\to\infty}[1 + \frac{\sin x}{x}] = 1 + 0 = 1$

61. $-1 \leqq \sin x \leqq 1$. Hence $-\frac{1}{\sqrt{x}} \leqq \frac{\sin x}{\sqrt{x}} \leqq \frac{1}{\sqrt{x}}$ for $x > 0$. Both $-\frac{1}{\sqrt{x}}$ and $\frac{1}{\sqrt{x}}$ approach 0 as $x \to \infty$. By the Sandwich Theorem $\lim_{x\to\infty} \frac{\sin x}{\sqrt{x}} = 0$.

63. $\text{Lim}_{x\to 3} \frac{x^2+2x-15}{x-3} = \lim_{x\to 3} \frac{(x-3)(x+5)}{x-3} = \lim_{x\to 3}(x+5) = 8$. In order to have $\lim_{x\to 3} f(x) = f(3) = k$, we should set $k = 8$.

65. a) $\lim_{x\to 0^-} f(x) = 0$ b) $\lim_{x\to 0^+} f(x) = \infty$

c) The limit does not exist. To exist both one-sided limits must exist and be equal.

67. This is not a contradiction because $0 < x < 1$ is not a *closed* interval.

69. True because $0 = f(1) < 2.5 < f(2) = 3$ and so by Theorem 7, $2.5 = f(c)$ for some c in $[1, 2]$.

71. Let $f(x) = x + \log x$. $f(\frac{1}{10}) = \frac{1}{10} - 1 < 0$. $f(1) = 1 > 0$. By the Intermediate Value Theorem there is a number c, $\frac{1}{10} < c < 1$ such that $f(c) = 0$. This means that c is a solution of the given equation.

73. $\text{Lim}_{x\to 0} \frac{\sin x}{x} = 1$ means given any radius $\varepsilon > 0$ about 1 there exists a radius $\delta > 0$ about 0 such that, for all x, $0 < |x - 0| < \delta$ implies $|f(x) - 1| < \varepsilon$.

75. This "definition" puts a requirement on $f(x)$ but not necessarily for x near x_0. Thus for any given $\varepsilon > 0$ there is an x such that $|x^2 - 0| < \varepsilon$ but this does not imply $\lim_{x\to x_0} x^2 = 0$ in general.

77. $-\frac{1}{2} < \sqrt{\frac{x+1}{2}} - 1 < \frac{1}{2}$ is equivalent to each of $\frac{1}{2} < \sqrt{\frac{x+1}{2}} < \frac{3}{2}$, $\frac{1}{4} < \frac{x+1}{2} < \frac{9}{4}$, $\frac{1}{2} < x + 1 < \frac{9}{2}$, $-\frac{1}{2} < x < \frac{7}{2}$. The midpoint of the latter interval is $(-\frac{1}{2} + \frac{7}{2})/2 = 3/2$ so that it can be written as $|x - 3/2| < 2$.

79. Let $f(x) = \frac{x-1}{x-3}$. Following the method of 2.5 Example 6, we solve the equations $f(x) = -2.1$ and $f(x) = -1.9$ obtaining 2.35 and 2.31. Thus $f(x)$ is within 0.1 unit of -2 if $2.31 < x < 2.35$ or $|x - 7/3| < 0.02$.

81. Let $f(x) = x^3 - 4x$. We graph $f(x)$, $y = 0.9$ and $y = 1.1$ and use zoom-in to find the x-coordinates of the points of intersection $(-1 < x < 0)$ obtaining -0.28 and -0.228. Thus $f(x)$ is within 0.1 unit of 1 if $-0.280 < x < -0.228$ (rounding appropriately) or if $|x + 0.254| < 0.026$.

83. $|f(x) - 0| + \|x\| = |x| < \varepsilon$ is equivalent to $|x - 0| < \varepsilon$. Thus we must have $0 < \delta \leq \varepsilon$.

85. As $x \to \frac{1}{3}^+$, $1 - 2x \to \frac{1}{3}$ and $3x - 1 \to 0$ through positive values. Hence $\lim_{x\to\frac{1}{3}^+} \frac{1-2x}{3x-1} = \infty$. With $x > \frac{1}{3}$, all of the following are equivalent. $\frac{1-2x}{3x-1} > N$, $1 - 2x > 3xN - N$, $1 + N > x(3N + 2)$, $x < \frac{N+1}{3N+2} = \frac{1}{3} + \frac{1}{3(3N+2)}$, $x - \frac{1}{3} < \frac{1}{3(3N+2)}$. So we choose $\delta = \frac{1}{3(3N+2)}$. Then $0 < x - \frac{1}{3} < \delta$ implies $\frac{1-2x}{3x-1} > N$. Since this is true for any $N > 0$, this confirms the limit.

87. $L = \lim_{x\to 2}(5x - 10) = 0$. $|f(x) - L| = |5x - 10| = 5|x - 2| < \varepsilon$ is equivalent to $|x - 2| < \varepsilon/5$. Thus $\delta = \varepsilon/5 = 0.01$.

89. $L = \lim_{x\to 2}\sqrt{2x - 3} = 1$. $1/2 < \sqrt{2x - 3} < 3/2$ yields $\frac{1}{4} < 2x - 3 < \frac{9}{4}$, $-\frac{3}{4} < 2x - 4 < \frac{5}{4}$, $-\frac{3}{8} < x - 2 < \frac{5}{8}$. Thus $|x - 2| < \frac{3}{8} \Rightarrow |f(x) - L| < \varepsilon$ so $\delta = 3/8$.

91. $L = \lim_{x\to -5} \frac{x-1}{x^2+3x} = \frac{-6}{10} = -0.6$. In the viewing rectangle $[-5.1, -4.9]$ by $[-0.61, -0.59]$ we graph $f(x)$ and using two points on the graph we find $m = -0.32$. Hence we take $\delta = \varepsilon/|m| = 3.12\varepsilon$ rounding down to be safe.

93. $9.9995 < 10 + (t - 70) \times 10^{-4} < 10.0005$ is equivalent to $-0.0005 < (t - 70)10^{-4} < 0.0005$, $-5 < t - 70 < 5$ or $65° < t < 75°$.

CHAPTER 3
DERIVATIVES

3.1 SLOPES, TANGENT LINES, AND DERIVATIVES

1. a) 0 b) -4 because a drawn-in tangent line at $x = 4$ descends 4 y-units as x increases one unit from 4 to 5.

3. a) 1 b) The spacing between horizontal and vertical dots is 0.5 units. The tangent line at $x = 4$ passes through the points $(4, -\frac{3}{4})$ and $(5, -\frac{3}{2})$. Thus $m = \frac{-1.5-(-.75)}{5-4} = -0.75$.

5. a) April 15 (each vertical line of dots corresponds to the first day of a month). Tangent seems to pass through $(120, 30)$ and $(150, 50)$. Hence $m = \frac{50-30}{150-120} = \frac{2}{3}$ degree per day.

b) Yes. Near January 1 and July 1 c) Positive from mid-January until about July 1, negative from about July 2 until mid-December.

7. $f'(x) = \lim_{h\to 0} \frac{f(x+h)-f(x)}{h} = \lim_{h\to 0} \frac{[2(x+h)^2-5]-[2x^2-5]}{h}$
$= \lim_{h\to 0} \frac{2x^2+4xh+2h^2-5-2x^2+5}{h} = \lim_{h\to 0} \frac{4xh+2h^2}{h} = \lim_{h\to 0}(4x+2h) = 4x$

When $x = 3$, $m = f'(3) = 4 \cdot 3 = 12$ and $y = f(3) = 2(3)^2 - 5 = 13$. The tangent at $(3, 13)$ therefore has equation $y - 13 = 12(x - 3)$ or $y = 12x - 23$.

9. $f(x+h) - f(x) = [2(x+h)^2 - 13(x+h) + 5] - [2x^2 - 13x + 5] = h(4x - 13 + 2h)$. $f'(x) = \lim_{h\to 0} \frac{f(x+h)-f(x)}{h} = \lim_{h\to 0}(4x - 13 + 2h) = 4x - 13$. $m = f'(3) = -1$. $f(3) = -16$. Tangent: $y + 16 = (-1)(x - 3)$ or $y = -x - 13$.

11. $f'(x) = \lim_{h\to 0} \frac{\frac{2}{x+h} - \frac{2}{x}}{h} = \lim_{h\to 0} \frac{1}{h} \frac{2x-2x-2h}{(x+h)x} = \lim_{h\to 0} \frac{-2}{(x+h)x} = -\frac{2}{x^2}$.
$f'(3) = -2/3^2 = -2/9$ is the slope and $f(3) = 2/3$. So the tangent at $x = 3$ has equation $y - 2/3 = -(2/9)(x - 3)$ or $2x + 9y = 12$.

13. $\frac{1}{h}[f(x+h)-f(x)] = \frac{1}{h}\left[\frac{x+h}{x+h+1} - \frac{x}{x+1}\right] = \frac{1}{(x+h+1)(x+1)}$.
$f'(x) = \lim_{h\to 0} \frac{1}{(x+h+1)(x+1)} = \frac{1}{(x+1)^2}$. $f'(3) = \frac{1}{16}$. $f(3) = \frac{3}{4}$. Tangent: $y - \frac{3}{4} = \frac{1}{16}(x-3)$ or $16y - x = 9$.

15. $f'(x) = \lim_{h\to 0} \frac{x+h+\frac{9}{x+h}-[x+\frac{9}{x}]}{h} = \lim_{h\to 0}\left\{\frac{h+\frac{9x-9x-9h}{(x+h)x}}{h}\right\} = \lim_{h\to 0}[1+\frac{-9}{(x+h)x}] = 1 - 9/x^2$. $f'(3) = 1 - 9/3^2 = 0$ is the slope and $f(3) = 3 + 9/3 = 6$. An equation of the tangent at $x = 3$ is $y = 6$.

17. $f(x+h) - f(x) = ([1+\sqrt{x+h}] - [1+\sqrt{x}])$
$= \sqrt{x+h} - \sqrt{x} = (\sqrt{x+h} - \sqrt{x})\frac{\sqrt{x+h}+\sqrt{x}}{\sqrt{x+h}+\sqrt{x}}$
$= \frac{x+h-x}{\sqrt{x+h}+\sqrt{x}} = \frac{h}{\sqrt{x+h}+\sqrt{x}}$. $f'(x) = \lim_{h\to 0}\frac{1}{\sqrt{x+h}+\sqrt{x}} = \frac{1}{2\sqrt{x}}$.
$f'(3) = \frac{1}{2\sqrt{3}}$. $f(3) = 1+\sqrt{3}$. Tangent line: $y - (1+\sqrt{3}) = \frac{1}{2\sqrt{3}}(x-3)$.

19. $f'(x) = \lim_{h\to 0}\frac{\sqrt{2(x+h)}-\sqrt{2x}}{h} = \lim_{h\to 0}\frac{\sqrt{2(x+h)}-\sqrt{2x}}{h}\frac{\sqrt{2(x+h)}+\sqrt{2x}}{\sqrt{2(x+h)}+\sqrt{2x}}$
$= \lim_{h\to 0}\frac{2(x+h)-2x}{h(\sqrt{2(x+h)}+\sqrt{2x})} = \lim_{h\to 0}\frac{2}{\sqrt{2(x+h)}+\sqrt{2x}} = \frac{2}{2\sqrt{2x}} = \frac{1}{\sqrt{2x}}$.

$f'(3) = \frac{1}{\sqrt{2\cdot 3}} = \frac{1}{\sqrt{6}}$ is the slope at $x = 3$. $f(3) = \sqrt{6}$ is the slope at $x = 3$. $f(3) = \sqrt{6}$. The tangent at $x = 3$ has equation $y - \sqrt{6} = (1/\sqrt{6})(x-3)$ or $\sqrt{6}y = x + 3$.

21. Using the method of preceding exercises, we find $f'(x) = -2x$. $m = f'(-1) = 2$. Tangent line: $y - 3 = 2(x+1)$ or $y = 2x + 5$. Graph $y_1 = 4 - x^2$ and $y_2 = 2x + 5$ in $[-5, 5]$ by $[-10, 10]$.

23. Proceeding as in Exercise 18 or 19, we find $f'(x) = \frac{1}{2\sqrt{x}}$. $m = f'(1) = \frac{1}{2}$. Tangent line: $y - 1 = \frac{1}{2}(x-1)$ or $y = \frac{1}{2}x + \frac{1}{2}$. Graph $y_1 = \sqrt{x}$ and $y_2 = 0.5x + 0.5$ in $[-3, 5]$ by $[-1, 3]$.

25. $f'(\frac{1}{2}) = \lim_{x\to\frac{1}{2}}\frac{f(x)-f(\frac{1}{2})}{x-\frac{1}{2}} = \lim_{x\to\frac{1}{2}}\frac{x^2-x+1-[(\frac{1}{2})^2-\frac{1}{2}+1]}{x-\frac{1}{2}} = \lim_{x\to\frac{1}{2}}\frac{x^2-(\frac{1}{2})^2-(x-\frac{1}{2})}{x-\frac{1}{2}} = \lim_{x\to\frac{1}{2}}(x+\frac{1}{2}) - 1 = 0$

27. $f(x) - f(-1) = \frac{1}{x+2} - \frac{1}{(-1)+2} = -\frac{(x+1)}{x+2}$. $f'(-1) = \lim_{x\to -1}\frac{f(x)-f(-1)}{x-(-1)} = \lim_{x\to -1}\frac{-1}{x+2} = -1$

29. $f(x) - f(4) = \frac{1}{\sqrt{x}} - \frac{1}{2} = \frac{2-\sqrt{x}}{2\sqrt{x}}\frac{2+\sqrt{x}}{2+\sqrt{x}} = \frac{4-x}{2\sqrt{x}(2+\sqrt{x})}$. $f'(4) = \lim_{x\to 4}\frac{f(x)-f(4)}{x-4} = \lim_{x\to 4}\frac{-1}{2\sqrt{x}(2+\sqrt{x})} = -\frac{1}{16}$

31. The right-hand derivative at $x = 0$ is $\lim_{h\to 0^+} \frac{(0+h)-0}{h} = \lim_{h\to 0^+} 1 = 1$ while the left-hand derivative is $\lim_{h\to 0^-} \frac{(0+h)^2-0^2}{h} = \lim_{h\to 0^-} h = 0$. Since these are unequal, the function is not differentiable at $x = 0$.

33. The left-hand derivative is $\lim_{h\to 0^-} \frac{\sqrt{1+h}-1}{h} = \lim_{h\to 0^-} \frac{\sqrt{1+h}-1}{h} \frac{\sqrt{1+h}+1}{\sqrt{1+h}+1} = \lim_{h\to 0^-} \frac{1+h-1}{h(\sqrt{1+h}+1)} = \lim_{h\to 0^-} \frac{1}{\sqrt{1+h}+1} = \frac{1}{2}$. The right-hand derivative at $x = 1$ is $\lim_{h\to 0^+} \frac{[2(1+h)-1]-1}{h} = \lim_{h\to 0^+} \frac{2h}{h} = 2$. Since the limits are unequal, f is not differentiable at $x = 1$.

35. Let $f(x) = y = -x^2$. a) $f'(x) = -2x$ (see Example 3). b) Graph $y_1 = -x^2$ and $y_2 = -2x$ in $[-3, 3]$ by $[-10, 5]$. c) $y' > 0$ for $x < 0$ and $y' < 0$ for $x > 0$. $f'(0) = 0$. d) y increases on $(-\infty, 0)$ and decreases on $(0, \infty)$. The interval on which $y' > 0 (y' < 0)$ is the interval on which y increases (decreases).

37. a) $f'(x) = x^2$ (See Ex. 5). b) Graph $y_1 = \frac{x^3}{3}$ and $y_2 = x^2$ in $[-3, 3]$ by $[-5, 5]$. c) $f' > 0$ on $(-\infty, 0)$ and $(0, \infty)$. $f'(0) = 0$. d) f is increasing on the same intervals.

3.2 NUMERICAL DERIVATIVES

1. $\text{NDER}(x^2 + 1, 2) = 4$. $f(2) = 5$ and the tangent at $(2, 5)$ has equation $y - 5 = 4(x - 2)$ or $y = 4x - 3$. The graphs of $y = x^2 + 1$ and $y = 4x - 3$ can be viewed in the rectangle $[-10, 10]$ by $[-10, 20]$.

3. $\text{NDER}(\sqrt{4 - x^2}, -1) = 0.58$. $f(-1) = \sqrt{3}$ and the tangent at $(-1, \sqrt{3})$ has equation $y - \sqrt{3} = 0.58(x + 1)$. The graph of $y = \sqrt{4 - x^2}$ and $y = \sqrt{3} + 0.58(x + 1)$ can be viewed in the rectangle $[-6, 6]$ by $[-4, 4]$.

5. $\text{NDER}(\frac{x^2-4}{x^2+1}, 2) = 0.80$. Since $f(2) = 0$, the tangent at $x = 2$ has equation $y = 0.8(x - 2)$. The graph of $f(x)$ and its tangent may be viewed in the rectangle $[-8, 8]$ by $[-8, 8]$.

7. a) only. We can draw the graph without lifting our pencil so the function is continuous. But the function is not differentiable at each of the points which are peaks or low points. At these points the left-hand and right-hand derivatives are not equal (there is not a *unique* tangent line).

9. c). $x = 0$ is not a point of the domain. At every other point there is a unique tangent line so the function is both continuous and differentiable.

11 through 17. In these exercises one may evaluate $D(h)$ and $S(h)$ directly using a calculator, or one may first algebraically simplify $D(h)$ and $S(h)$. If the calculator is used, meaningful results may not be obtained when $h = \pm 10^{-15}$ due to the limits of machine accuracy. This answers part c) of these exercises.

11. b) Conjectures: $f'(2) = 10$, $f'(0) = -2$. $S(h)$ is closer in both cases (in fact exact).

13. b) $f'(2) = -0.25$, $S(h)$ is closer. $f'(0)$ and $D(h)$ for $a = 0$ are not defined but $S(h) = 0$ for all h.

15. a) If $a = 2$, $D(h) = S(h) = 1$ for all the h's considered. If $a = 0$, $D(h) = \begin{cases} -1, & \text{if } h < 0 \\ 1, & \text{if } h > 0 \end{cases}$ and $S(h) = 0$ for all h. b) $f'(2) = 1$; both $S(h)$ and $D(h)$ are exact. $f'(0)$ does not exist.

17. a) If $a = 2$, $D(h) \to 0$ as $h \to 0$ while $S(h) = 0$ for those h's considered. If $a = 0$, $D(h)$ is undefined for $h < 0$ while $D(h) \to \infty$ as $h \to 0^+$. b) $f'(2) = 0$, $S(h)$ is closer. $f'(0)$ does not exist.

19. Even though the derivative may not exist, the values of $S(h)$ may be defined giving meaningless approximations of the derivative.

21. See the comment for Exercise 19.

23. Since $4x - x^2 < 0$ for $x < 0$, $f(x)$ is not defined when $x < 0$ and so the appropriate $S(h)$ is not defined.

25. a) We agree that $f'(a) = \text{NDER}(f(x), a)$ (with $h = 0.01$) rounded to two decimal places and obtain $f'(-1) = 0.14$, $f'(0) = -14.94$, $f'(1.5) = -9.69$ and $f'(3.5) = 0.13$. b) $\text{NDER}(f(x), a)$ may give results even though $f'(a)$ does not exist. This was the case for $f'(0)$. The denominator is zero when $x = 0$ and when $x = x_1$ where $x_1 \approx 1.2$. The numerator is undefined when $\tan x$ is undefined. Hence the domain of f consists of all real numbers except $0, x_1$ and $(2n+1)\pi/2$, n an integer.

27. a) $y_1 = -x^2$, $y_2 = \text{NDER } y_1$. Graph y_1, y_2 in $[-4, 4]$ by $[-10, 10]$. b) y_1' exists for all x. c) y_1' is positive for negative x, negative for positive x and $y_1'(0)$. e) y_1 is increasing over the interval $(-\infty, 0)$ where y_1' is positive. y_1 is decreasing over the interval $(0, \infty)$ where y_1' is negative.

29. a) Graph $y_1 = \sqrt[3]{x-2}$ and $y_2 = \text{NDER } y_1$ in $[0, 3]$ by $[-2.2, 3.8]$. b) y_1' does not exist at $x = 2$ because the tangent line is vertical there. This is suggested by the graph because $y_2 \to \infty$ as $x \to 2$. y_1' is positive, tangent lines have positive slope and y_1 is increasing for all x except $x = 2$.

31. a) Graph $y_1 = \sqrt{1-x}$ and $y_2 = \text{NDER } y_1$ in $[-2, 1]$ by $[-2.2, 2.5]$. b) y_1' does not exist for $x > 1$ because $(1, \infty)$ is not part of the domain of y_1. y_1' does not exist at $x = 1$ because the graph has a vertical tangent there. This is suggested by the graph because $y_1 \to -\infty$ as $x \to 1^-$. y_1' is negative, slopes of tangent lines are negative and y_1 is decreasing on the interval $(-\infty, 1)$.

33. $[0.25895206, 0.25895208]$ by $[0.135, 0.145]$ is one possibility.

35. a) Let $y_1 = -x^2(x < 0) + (4 - x^2)(x \geqq 0)$. Graph NDER$(y_1, x)$ in the given window. b) We see the graph of $y = -2x$, $x \neq 0$. c) With $h = 0.01$, $\text{NDER}(f(x), 0) = \frac{f(0+0.01)-f(0-0.01)}{2(0.01)} = \frac{4-(0.01)^2-[-(0.01)^2]}{0.02} = \frac{4}{0.02} = 200$. $f(x)$ has a jump discontinuity at $x = 0$. So $f'(0)$ cannot exist by Theorem 2.

37. a) Let $y_1 = -\frac{x^2}{2} + 0\sqrt{-x}$, $y_2 = \frac{x^2}{2} + 0\sqrt{x}$, $y_3 = \text{NDER}(y_1, x)$, $y_4 = \text{NDER}(y_2, x)$. Graph y_3, y_4 in the given viewing window. b) This is the graph of $y = \begin{cases} -x, & x < 0 \\ x, & x \geqq 0 \end{cases}$. c) $\text{NDER}(f(x), 0) = \frac{f(0+0.01)-f(0-0.01)}{2(0.01)} = \frac{0.0001}{2(0.01)} = 0.005$. $f'(0)$ does not exist because the left-hand derivative and the right-hand derivative at $x = 0$ are not equal.

39. Graph $y_1 = (x^3 + 6x^2 + 12x)(x < 0) + (-x^2)(x > 0)$ and $y_2 = \text{NDER}(y_1, x)$ in $[-5, 5]$ by $[-10, 10]$. Since the two one-sided derivatives at $x = 0$ are unequal, $f'(0)$ does not exist.

41. Graph $y_1 = -5(abs\, x)^{1/3}$ and $y_2 = \text{NDER}(y_1, x)$ in $[-5, 5]$ by $[-10, 10]$. $f'(0)$ does not exist. For example, $\lim_{h\to0+} \frac{f(0+h)-f(0)}{h} = \lim_{h\to0+} \frac{-5h^{1/3}}{h} = -5 \lim_{h\to0+} \frac{1}{h^{2/3}} = -\infty$.

43. b) $D_x(\sin x) = \cos x$

45. By Example 6 of 3.1, $f'(0)$ does not exist. But $\text{NDER}(|x|, 0) = 0$. $\text{NDER}(f(x), a)$ may exist even if $f'(a)$ does not exist.

47. $0 < \frac{1}{2} < 1$ but $f(x) = \frac{1}{2}$ for no x, $-1 \leqq x \leqq 1$. Therefore f does not have the Intermediate Value Property on the interval. By the theorem alluded to in the text, f cannot be the derivative of any function on the interval.

49. The range of $[x]$ is the set of all integers so it does not have the Intermediate Value Property. Hence it cannot be the derivative of any function on $(-\infty, \infty)$.

3.3 DIFFERENTIATION RULES

1. 1, 0 **3.** $-2x$, -2 **5.** 2, 0 **7.** x^2+x+1, $2x+1$

9. $4x^3 - 21x^2 + 4x$, $12x^2 - 42x + 4$ **11.** $8x - 8$, 8 **13.** $y' = 2x - 1$, $y'' = 2$, $y^{(n)} = 0$ for $n \geqq 3$

15. $y' = 2x^3 - 3x - 1$, $y'' = 6x^2 - 3$, $y''' = 12x$, $y^{(4)} = 12$, $y^{(n)} = 0$ for $n \geqq 5$

17. a) $y' = (x+1)2x + 1 \cdot (x^2+1) = 2x^2 + 2x + x^2 + 1 = 3x^2 + 2x + 1$
b) $y = x^3 + x^2 + x + 1$, $y' = 3x^2 + 2x + 1$

19. a) $y' = (x-1)(2x+1) + 1 \cdot (x^2 + x + 1) = 2x^2 - x - 1 + x^2 + x + 1 = 3x^2$
b) $y = x^3 - 1$, $y' = 3x^2$

21. a) $y' = (3x-1)(2) + 3(2x+5) = 12x + 13$
b) $y = 6x^2 + 13x - 5$, $y' = 12x + 13$

23. a) $y' = x^2(3x^2) + 2x(x^3 - 1) = 5x^4 - 2x$
b) $y = x^5 - x^2$, $y' = 5x^4 - 2x$

25. $\frac{dy}{dx} = \frac{(x+7)\cdot 1-(x-1)\cdot 1}{(x+7)^2} = \frac{x+7-x+1}{(x+7)^2} = \frac{8}{(x+7)^2}$

27. $y = \frac{x^3+7}{x} = x^2 + 7x^{-1}$. $\frac{dy}{dx} = 2x - 7x^{-2} = 2x - \frac{7}{x^2} = \frac{2x^3-7}{x^2}$

29. $y = \frac{(x-1)(x^2+x+1)}{x^3} = \frac{x^3-1}{x^3} = 1 - x^{-3}$. $\frac{dy}{dx} = 3x^{-4} = \frac{3}{x^4}$

31. $y = \frac{1-x}{1+x^2}$. $\frac{dy}{dx} = \frac{(1+x^2)(-1)-(1-x)(2x)}{(1+x^2)^2} = \frac{-1-x^2-2x+2x^2}{(1+x^2)^2} = \frac{x^2-2x-1}{(1+x^2)^2}$

33. $y' = \frac{(1-x^3)2x-x^2(-3x^2)}{(1-x^3)^2} = \frac{2x-2x^4+3x^4}{(1-x^3)^2} = \frac{x^4+2x}{(1-x^3)^2}$

35. $y = (10)\frac{1}{\sqrt{x}-4}$, $y' = (10)\frac{-1}{(\sqrt{x}-4)^2}\left(\frac{1}{2\sqrt{x}}\right) = \frac{-5}{\sqrt{x}(\sqrt{x}-4)^2}$

37. $y' = \frac{(\sqrt{x}+1)(\frac{1}{2\sqrt{x}})-(\sqrt{x}-1)(\frac{1}{2\sqrt{x}})}{(\sqrt{x}+1)^2} = \frac{1}{\sqrt{x}(\sqrt{x}+1)^2}$

39. $y = \frac{1}{(x^2-1)(x^2+x+1)}$. $y' = -\frac{(x^2-1)(2x+1)+2x(x^2+x+1)}{(x^2-1)^2(x^2+x+1)^2} = -\frac{4x^3+3x^2-1}{(x^2-1)^2(x^2+x+1)^2}$

41. $y = 3x^{-2}$. $y' = -6x^{-3} = -\frac{6}{x^3}$. $y' = -6x^{-3}$ so $y'' = 18x^{-4} = \frac{18}{x^4}$.

43. $y = \frac{5}{x^4} = 5x^{-4}$. $y' = -20x^{-5} = -\frac{20}{x^5}$. $y' = -20x^{-5}$ so $y'' = 100x^{-6} = \frac{100}{x^6}$.

45. $y = x + 1 + x^{-1}$ so $y' = 1 - x^{-2}$, $y'' = 2x^{-3}$ or $y' = 1 - \frac{1}{x^2}$, $y'' = \frac{2}{x^3}$.

47. NDER$(x3^{-0.2x}, 1) = 0.63 =$ slope. $f(1) = e^{-0.2}$. The tangent line at $(1, 3^{-0.2})$ is $y - 3^{-0.2} = 0.63(x-1)$. The result is confirmed by viewing $y = x3^{-0.2x}$ and $y = 3^{-0.2} + 0.63(x-1)$ in the rectangle $[-10, 10]$ by $[-10, 10]$.

49. NDER$(\frac{x+3}{x^3-2x+5}, 0) = 0.44$. $f(0) = 3/5$. The tangent line at $(0, 3/5)$ has equation $y - 3/5 = 0.44x$. The result is confirmed by graphing $y = f(x)$ and $y = (3/5) + 0.44x$ in the rectangle $[-1, 1]$ by $[0.3, 0.9]$.

51. The graph of $y =$ NDER$(f(x))$ oscillates, appears to cross the x-axis infinitely often and to be symmetric with respect to the origin. The graph of $y =$ NDER2$(f(x))$ oscillates and appears to cross the x-axis infinitely often and to be symmetric with respect to the y-axis. These graphs can be viewed in $[-10, 10]$ by $[-10, 10]$ and in $[-50, 50]$ by $[-50, 50]$.

53. $y = f'(x)$ or $y =$ NDER$(f(x))$ can be viewed in the rectangle $[-2, 8]$ by $[-4, 4]$. $y = f''(x)$ or $y =$ NDER2$(f(x))$ can be viewed in $[-2, 10]$ by $[-4, 10]$.

55. $y =$ NDER$(f(x))$ can be viewed in the rectangle $[-8, 8]$ by $[-2, 0]$. $y =$ NDER2$(f(x))$ can be viewed in $[-6, 8]$ by $[-1, 1]$.

57. The graph of $y =$ NDER$(f(x))$ oscillates, appears to cross the x-axis infinitely often and to be symmetric with respect to the origin. The three solutions of $f'(x) = 0$ of smallest absolute value are -2.029, 0, 2.029. The graph of $y =$ NDER2$(f(x))$ appears to cross the x-axis infinitely often. The solution set of $f''(x) > 0$ consists of an infinite sequence of intervals. The three closest to the origin are $(-6.578, -3.644)$, $(-1.077, 1.077)$ and $(3.644, 6.578)$ rounding appropriately.

59. We use zoom-in and the graphs of Exercise 53. $f'(x) = 0$ for $x = -0.313$ and $x = 3.198$. $f''(x) > 0$ for x in the set $(-\infty, -1) \cup (1, \infty)$.

61. The graph of $f'(x)$ does not cross the x-axis. Hence there is no solution to $f'(x) = 0$. The solution set of $f''(x) > 0$ (that is, those x for which the graph of $y = f''(x)$ is above the x-axis) is $(-3, -0.333) \cup (5, \infty)$.

63. $f(x) = \frac{2x-5}{3x^2+4}$. $f'(x) = \frac{(3x^2+4)2-(2x-5)6x}{(3x^2+4)^2} = \frac{-6x^2+30x+8}{(3x^2+4)^2} = \frac{-2(3x^2-15x-4)}{(3x^2+4)^2}$. $f''(x) = (-2)\frac{[(3x^2+4)^2(6x-15)-(3x^2-15x-4)2(3x^2+4)6x]}{(3x^2+4)^4} = (-2)\frac{[(3x^2+4)(6x-15)-12x(3x^2-15x-4)]}{(3x^2+4)^3} = \frac{6[6x^3-45x^2-24x+20]}{(3x^2+4)^3}$.

$f''(x)$ changes sign only when $6x^3 - 45x^2 - 24x + 20 = 0$ but we cannot solve this exactly. To solve $f''(x) > 0$ we find the intervals in which the graph of $y = \text{NDER2}(f(x))$ lies above the x-axis. The solution set is $(-0.911, 0.460) \cup (7.950, \infty)$.

65. $\frac{d}{dx}(cf(x)) = c\frac{df}{dx} + \frac{dc}{dx}f(x) = c\frac{df}{dx} + 0(f(x)) = c\frac{df}{dx}$

67. a) $\frac{d}{dx}(uv) = uv' + u'v$. The value of this function at $x = 0$ is $u(0)v'(0) + u'(0)v(0) = 5(2) + (-3)(-1) = 13$.

b) $\frac{d}{dx}(\frac{u}{v}) = \frac{vu' - uv'}{v^2}$. At $x = 0$ this becomes $\frac{(-1)(-3) - 5(2)}{(-1)^2} = -7$.

c) $\frac{d}{dx}(\frac{v}{u}) = \frac{uv' - vu'}{u^2}$. At $x = 0$ this becomes $\frac{5(2) - (-1)(-3)}{5^2} = \frac{7}{25}$.

d) $\frac{d}{dx}(7v - 2u) = 7v' - 2u'$. At $x = 0$ this becomes $7(2) - 2(-3) = 20$.

69. Let $f(x) = x^2 + 5x$. $f'(x) = 2x + 5$ and $f'(3) = 11$. Answer: c)

71. $y' = 3x^2 - 3$. $y'(2) = 9$ is the slope of the tangent. Hence $-1/9$ is the slope of the perpendicular at $x = 2$. Thus it has equation $y - 3 = (-1/9)(x - 2)$ or $x + 9y = 29$.

73. $y' = 6x^2 - 6x - 12 = 6(x^2 - x - 2) = 6(x + 1)(x - 2) = 0$ at $x = -1$ and $x = 2$. Answer: $(-1, 27)$, $(2, 0)$

75. $y' = \frac{(x^2+1)4 - 4x(2x)}{(x^2+1)^2} = \frac{4 - 4x^2}{(x^2+1)^2} = \frac{4(1-x^2)}{(x^2+1)^2}$. At $x = 0$, $y' = 4$ and the tangent has equation $y = 4x$. At $x = 1$, $y' = 0$ and the tangent has equation $y = 2$.

77. $P = nRT(\frac{1}{V-nb}) - an^2V^{-2}$. $\frac{dP}{dV} = nRT\frac{[(V-nb)0 - 1\cdot 1]}{(V-nb)^2} + 2an^2V^{-3} = -\frac{nRT}{(V-nb)^2} + \frac{2an^2}{V^3}$

79. $R = \frac{CM^2}{2} - \frac{M^3}{3}$. $\frac{dR}{dM} = CM - M^2$

81. $f'(-x) = \lim_{h\to 0} \frac{f(-x+h) - f(-x)}{h} = \lim_{h\to 0} \frac{f[-(x-h)] - f(x)}{h} = \lim_{h\to 0} \frac{f(x-h) - f(x)}{h} = \lim_{h\to 0} -\frac{f(x-h) - f(x)}{-h} = -f'(x)$ so $f'(x)$ is odd.

83. a) $x = 1.442695\ldots$ b) $x = 1.442698\ldots$ c) very close

85. a) Graph $y = abs(y_1 - f''(x))$ where $f''(x) = 20x^3 - 36x^2 + 6x - 12$ in $[-10, 10]$ by $[-0.1, 0.1]$. Its maximum value (which is the maximum error) is found to be 0.0424.

b) Let $y_2 = \frac{y_1(x+0.01) - y_1(x-0.01)}{0.02}$.

c) Graph $y = abs(y_2 - f^{(3)}(x))$ where $f^{(3)}(x) = 60x^2 - 72x + 6$. Its maximum (and rather constant) value on $[-10, 10]$ is about 0.01 (with dubious accuracy). NOTE: The answers for a) and c) depend on the h value used in the NDER algorithm. We used $\delta = 0.01$ here. On our calculator we had $y_1(x + 0.01) = \text{NDER}(f(x), x), x, x + 0.01)$.

3.4 VELOCITY, SPEED, AND OTHER RATES OF CHANGE

1. a) Let $x_1(t) = t^2 - 3t + 2,\ y_1(t) = 3,\ 0 \leqq t \leqq 5$. On a parametric graphing utility with Tstep 0.05 in $[-1, 15]$ by $[-1, 15]$, holding down TRACE, we see the particle moves along the line $y = 3$ first (starting with $t = 0$) to the left and then to the right. b) Using TRACE, we find the positions to be: $(2, 3)$ at $t = 0$, $(0, 3)$ at $t = 1$, $(0, 3)$ at $t = 2$, $(2, 3)$ at $t = 3$. c) Using TRACE, we find the particle changes direction at $(-0.25, 3)$ when $t = 1.5$. The velocity $v = s'(t) = 2t - 3$ at this time is 0 while the acceleration $a = v' = 2$. d) The particle first travels left from $(2, 3)$ to $(-0.25, 3)$ and then right to $(12, 3)$ at $t = 5$. Thus the total distance traveled is $2.25 + 12.25 = 14.5$. e) Along with the graph in a), we graph $x_2(t) = t,\ y_2(t) = t^2 - 3t + 2$. Use simultaneous graphing format. f) Along with the graph in a), we simultaneously graph $x_2(t) = t,\ y_2(t) = 2t - 3 = v$ and $x_3(t) = 5,\ y_3(t) = 2 = a$. The particle is at rest when $v = 0$, that is, when $t = 1.5$sec.

3. For this problem we use $x_1(t) = t^3 - 6t^2 + 7t - 3,\ y_1(t) = 3,\ x_2(t) = t,\ y_2(t) = 3t^2 - 12t + 7 (= v = s')$ and $x_3(t) = t,\ y_3(t) = 6t - 12 (= a(t))$ with $0 \leqq t \leqq 5$, Tstep 0.05 in the viewing rectangle $[-10, 10]$ by $[-15, 25]$. a) The particle first ($t = 0$) moves to the right, then to the left and then to the right again. b) $(-3, 3)$ when $t = 0$, $(-1, 3)$ when $t = 1$, $(-5, 3)$ when $t = 2$, $(-9, 3)$ when $t = 3$sec. c) Using TRACE, we get $(-0.70, 3)$ when $t = 0.7$, $v = 0.07$, $a = -7.8$ and $(-9.30, 3)$ when $t = 3.3$sec $v = 0.07$, $a = 7.8$. d) The particle travels right from $(-3, 3)$ to $(-0.70, 3)$ then left from $(-0.70, 3)$ to $(-9.30, 3)$ then right from $(-9.30, 3)$ to $(7, 3)$. Thus the total distance traveled is $2.3 + 8.6 + 16.3 = 27.2$meters. f) The particle is at rest when $v = 0$. With TRACE our approximation of this is at $t = 0.7$ and $t = 3.3$sec.

5. For this problem we use $x_1(t) = t \sin t,\ y_1(t) = 3,\ x_2(t) = t,\ y_2(t) = \text{NDER}(x_1, t, t) (= v)$ and $x_3(t) = t,\ y_3(t) = \text{NDER}(y_2, t, t) (= a(t))$ with $0 \leqq t \leqq 15$, Tstep 0.05 in the window $[-15, 15]$ by $[-15, 15]$. a) The particle first moves right, then left, then right, then left, then right, then left slightly. b) $(0, 3)$, $(0.84, 3)$, $(1.82, 3)$, $(0.42, 3)$ c) Using TRACE, we obtain $(1.82, 3)$ when $t = 2.05$, $v = -0.06$, $a = -2.7$; $(-4.814, 3)$ when $t = 4.90$, $v = -0.07$, $a = 5.19$; $(7.91, 3)$ when $t = 8$, $v = -0.17$, $a = -8.2$; $(-11.04, 3)$ when $t = 11.1$, $v = .16$, $a = 11.2$; $(14.17, 3)$ at $t = 14.2$,

$v = 0.106$, $a = -14.3$ d) The particle travels right from $(0, 3)$ to $(1.82, 3)$, from $(1.82, 3)$ to $(-4.81, 3)$, from $(-4.81, 3)$ to $(7.91, 3)$, from $(7.91, 3)$ to $(-11.04, 3)$, from $(-11.04, 3)$ to $(14.17, 3)$ and from $(14.17, 3)$ to $(9.75, 3)$. Thus the total distance traveled is $1.82 + 6.63 + 12.72 + 18.95 + 25.21 + 4.42 = 69.75$meters. e) We simultaneously graph x_1, y_1 and $x_4 = t$, $y_4 = t \sin t$. f) The particle is at rest when $v = y_2(t) = 0$. With TRACE our approximations of these times are $t = 0$, 2.05, 4.9, 8, 11.1 and 14.2.

7. On Mars $s = 1.86t^2$, $v = 2(1.86)t = 16.6$ so $t = 4.46$sec. On Jupiter $s = 11.44t^2$, $v = 22.88t = 16.6$ so $t = 0.73$sec.

9. One possibility: Graph $x_1(t) = t$, $y_1(t) = 24t - 0.8t^2$, $0 \leqq t \leqq 30$, in $[0, 30]$ by $[0, 180]$. Then use TRACE and zoom-in if more accuracy is desired.

11. $s = 832t - 2.6t^2 = t(832 - 2.6t) = 0$ when $t = 0$ and when $t = 832/2.6 = 320$sec. Thus it will get back down in 320sec. On earth $s = 832t - 16t^2 = t(832 - 16t) = 0$. When $t = 0$ and when $t = 832/16 = 52$sec so it will get back down in 52sec.

13. $b(t) = 10^6 + 10^4 t - 10^3 t^2$. $b'(t) = 10^4 - (2)10^3 t$. a)$b'(0) = 10^4$ per hour b) $b'(5) = 10^4 - (2)10^3(5) = 0$ c) $b'(10) = 10^4 - (2)10^4 = -10^4$ per hour

15. $c(x) = 2000 + 100x - 0.1x^2$. a) The average cost of one washing machine when producing the first 100 washing machines is $c(100)/100 = \$110$. During production of the first 100 machines the average increase in producing one more machine is: average increase $= \frac{c(100)-c(0)}{100-0} = \$90$; the fixed cost $c(0) = \$2000$ is omitted with this method. b) $c'(x) = 100 - 0.2x$, $c'(100) = \$80$ c) $c(101) - c(100) = \$79.90$.

17. $s = t^3 - 6t^2 + 9t$. $v = s' = 3t^2 - 12t + 9 = 3(t^2 - 4t + 3) = 3(t-1)(t-3) = 0$ when $t = 1$ and 3sec. $a = v' = 6t - 12$. $a(1) = -6$, $a(3) = 6$. Thus $a = -6\text{m/sec}^2$ when $t = 1$sec and $a = 6\text{m/sec}^2$ when $t = 3$sec.

19. a) We use $-12 \leqq t \leqq 12$, Tstep 0.05 in the viewing rectangle $[-35, 35]$ by $[-3, 10]$. b) For this graph we can use $0 \leqq t \leqq 6.29 \approx 2\pi$, Tstep 0.05 in $[-3, 8]$ by $[-3, 3]$. c) This line segment can be viewed using $0 \leqq t \leqq 6.29$, Tstep 0.05 in $[-6, 10]$ by $[-6, 2]$.

21. a) All have derivative $3x^2$. c) The result of a) suggests that the family consists of all functions of the form $x^3 + C$ where C can be any constant. b) From c) $f(x)$ has the form $f(x) = x^3 + C$. Since $f(0) = 0$, $f(0) = 0^3 + C = 0$ so $C = 0$ and $f(x) = x^3$ is the unique function with the two properties. e) Yes. $g(x) = x^3 + C$ and $g(0) = 0^3 + C = 3$ so $g(x) = x^3 + 3$.

23. a) 190ft/s b) 2 c) At 8s when $v = 0$ d) At 10.8s when it was falling at 90ft/s e) From $t = 8s$ to 10.8s, i.e., 2.8s f) Just before burnout, i.e., just before $t = 2s$. The acceleration was constant from $t = 2s$ to $t = 10.8s$ during free fall. Note that the graph of v has constant slope during this time.

25. a) 0, 0 b) 1700, 1400 c) Rabbits per day and foxes per day

27. (b) **29.** (d)

31. a)

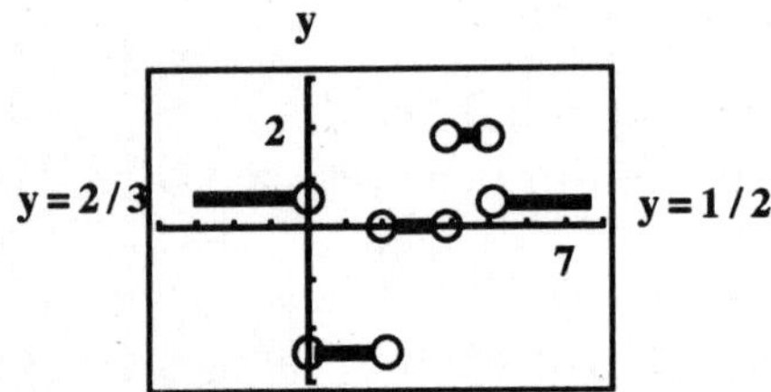

b) $x = 0, 2, 4, 5$

33. We must solve the equation $x_1(t) = 4t^3 - 16t^2 + 15t = 5$ or $4t^3 - 16t^2 + 15t - 5 = 0$. The approximate solution is $t = 2.832$.

35.

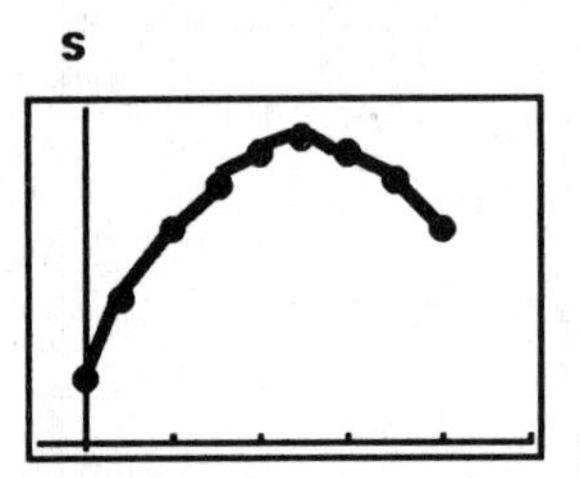

a) $v \approx 18$ft/sec b) $v \approx 0$ft/sec c) $v \approx -12$ft/sec

3.5 DERIVATIVES OF TRIGONOMETRIC FUNCTIONS

1. $y = 1 + x - \cos x,\ y' = 1 + \sin x$ **3.** $y = \frac{1}{x} + 5\sin x,\ y' = -\frac{1}{x^2} + 5\cos x$

5. $y = \csc x - 5x + 7,\ y' = -\csc x \cot x - 5$

7. $y = x\sec x,\ y' = x\sec x\tan x + \sec x = \sec x(x\tan x + 1)$

9. $y = x^2\cot x,\ y' = -x^2\csc^2 x + 2x\cot x = x(2\cot x - x\csc^2 x)$

11. $y = 3x + x\tan x,\ y' = 3 + x\sec^2 x + \tan x$

13. $y = \sin x\sec x = \tan x,\ y' = \sec^2 x$

15. $y = \tan x\cot x = 1,\ y' = 0$

17. $y = \frac{4}{\cos x} = 4\sec x,\ y' = 4\sec x\tan x$

19. $y = \frac{\cos x}{x},\ y' = \frac{x(-\sin x)-(\cos x)(1)}{x^2} = -\frac{x\sin x+\cos x}{x^2}$

21. $y = \frac{x}{1+\cos x},\ y' = \frac{(1+\cos x)(1)-x(-\sin x)}{(1+\cos x)^2} = \frac{1+\cos x+x\sin x}{(1+\cos x)^2}$

23. $y = \frac{\cot x}{1+\cot x},\ y' = \frac{(1+\cot x)(-\csc^2 x)-\cot x(-\csc^2 x)}{(1+\cot x)^2} = -\frac{\csc^2 x}{(1+\cot x)^2}$

25. $y = \csc x,\ y' = -\csc x\cot x,\ y'' = -[\csc x(-\csc^2 x) + (-\csc x\cot x)\cot x] = \csc^3 x + \csc x\cot^2 x = \csc x(\csc^2 x + \cot^2 x)$

27. $y = \sin x,\ x = 0.\ y' = \cos x.\ f(0) = \cos 0 = 1$ is the slope m of the tangent line. Hence the tangent line has equation $y - 0 = 1(x - 0)$ or $y = x$. For the normal line we need the slope $-1/m = -1$. The normal line has equation $y - 0 = -1(x - 0)$ or $y = -x$. We graph the three functions $y = \sin x$, $y = x$ and $y = -x$ in the viewing rectangle $[-3, 3]$ by $[-2, 2]$. Notice that the horizontal length 6 and the vertical height 4 of the screen are in the ratio 3 to 2 to ensure that perpendicular lines appear to intersect in a right angle.

29. $y = 2\sin^2 x,\ x = 2.\ y = 2\sin x\sin x$ so by the product rule $y' = 2(\sin x\cos x + \cos x\sin x) = 4\sin x\cos x$. $f'(2) = 4\sin 2\cos 2$. Tangent: $y - 2\sin^2 2 = 4\sin 2\cos 2(x - 2)$. Normal: $y - 2\sin^2 2 = -(4\sin 2\cos 2)^{-1}(x - 2)$. We may view $y = 2(\sin x)^2$, the tangent and the normal in $[0, 4]$ by $[0, 2.7]$.

31. $D_x\cos x = \lim_{h\to 0}\frac{\cos(x+h)-\cos x}{h} = \lim_{h\to 0}\frac{\cos x\cosh - \sin x\sinh - \cos x}{h} = \lim_{h\to 0}\left[\cos x\frac{(\cosh - 1)}{h} - \sin x\frac{\sin h}{h}\right] = (\cos x)0 - (\sin x)1 = -\sin x$

33. $(\tan x)' = \sec^2 x = \frac{1}{\cos^2 x}$ and $(\cot x)' = -\csc^2 x = -\frac{1}{\sin^2 x}$ cannot be 0 for any value of x.

35. $y = 2x + \sin x,\ y' = 2 + \cos x = 0$ is equivalent to $\cos x = -2$. Since the latter equation has no solution, the graph has no horizontal tangent.

37. $y = x + 2\cos x$, $y' = 1 - 2\sin x = 0$ is equivalent to $\sin x = \frac{1}{2}$ which has two solutions $x = \pi/6$, $5\pi/6$ in the interval. Thus there are horizontal tangents at the points $(\pi/6, (\pi/6) + \sqrt{3})$ and $(5\pi/6, (5\pi/6) - \sqrt{3})$.

39. $y' = -\sqrt{2}\sin x$. $f'(\pi/4) = -\sqrt{2}\sin(\pi/4) = -1 = m$. $y - 1 = (-1)(x - \pi/4)$, $y = -x + 1 + \pi/4$ is an equation of the tangent. $y - 1 = (-1/m)(x - \pi/4) = x - \pi/4$, $y = x + 1 - \pi/4$ is an equation of the normal.

41. $y = \cot x - \sqrt{2}\csc x$, $0 < x < \pi$. $y' = -\csc^2 x + \sqrt{2}\csc x \cot x = \csc x(\sqrt{2}\cot x - \csc x)$. Since $\csc x \neq 0$, $y' = 0$ is equivalent to $\sqrt{2}\cot x - \csc x = 0$ which leads to (by multiplication by $\sin x$) $\sqrt{2}\cos x - 1 = 0$, $\cos x = 1/\sqrt{2}$, $x = \pi/4$. $f(\pi/4) = \cot(\pi/4) - \sqrt{2}\csc(\pi/4) = 1 - 2 = -1$. Thus $(\pi/4, -1)$ is the only point where there is a horizontal tangent. The tangent there has equation $y = -1$. This answer is supported by graphing $f(x)$ and $y = -1$ in the viewing window $[0, 3.14]$ by $[-10, 1]$.

43. The graph of $\tan x$ and its derivative $\sec^2 x$, $-\pi/2 < x < \pi/2$ may be viewed in the rectangle $[-1.57, 1.57]$ by $[-5, 5]$.

45. $\lim_{h\to 0} \frac{1-\cos h}{h^2} = \lim_{h\to 0} \frac{1-\cos h}{h^2}\frac{1+\cos h}{1+\cos h} = \lim_{h\to 0} \frac{1-\cos^2 h}{h^2(1+\cos h)} = \lim_{h\to 0}(\frac{\sin h}{h})^2 \frac{1}{1+\cos h} = (1)^2 \frac{1}{1+\cos 0} = \frac{1}{2}$

47. $(\cot x)' = (\frac{\cos x}{\sin x})' = \frac{\sin x(-\sin x) - \cos x \cos x}{\sin^2 x} = \frac{-(\sin^2 x + \cos^2 x)}{\sin^2 x} = -\frac{1}{\sin^2 x} = -\csc^2 x$

3.6 THE CHAIN RULE

1. $y = \sin(3x + 1)$. $y' = 3\cos(3x + 1)$

3. $y = \cos(-x/3) = \cos(x/3)$. $y' = -\frac{1}{3}\sin(x/3)$

5. $y = \tan(2x - x^3)$. $y' = (2 - 3x^2)\sec^2(2x - x^3)$

7. $y = x\sec(x^2 + \sqrt{2})$. $y' = 1 + 2x\sec(x^2 + \sqrt{2})\tan(x^2 + \sqrt{2})$

9. $y = -\csc(x^2 + 7x)$. $y' = (2x + 7)\csc(x^2 + 7x)\cot(x^2 + 7x)$

11. $y = 5\cot(\frac{2}{x})$. $y' = 5(-\frac{2}{x^2})(-\csc^2(\frac{2}{x})) = \frac{10}{x^2}\csc^2(\frac{2}{x})$

13. $y = \cos(\sin x)$. $y' = [-\sin(\sin x)]\cos x$

15. $y = (2x + 1)^5$. $y' = 5(2x + 1)^4(2) = 10(2x + 1)^4$. We support this result by graphing $10(2x + 1)^4$ and NDER$((2x + 1)^5, x)$ in $[-2, 1]$ by $[0, 810]$ and seeing that the two graphs coincide.

17. $y=(x^2+1)^{-3}$. $y'=(-3)(x^2+1)^{-4}(2x)=-6x(x^2+1)^{-4}$

19. $y=(1-\frac{x}{7})^{-7}$. $y'=-7(1-\frac{x}{7})^{-8}(-\frac{1}{7})=(1-\frac{x}{7})^{-8}$

21. $y=(\frac{x^2}{8}+x-\frac{1}{x})^4$. $y'=4(\frac{x^2}{8}+x-\frac{1}{x})^3(\frac{x}{4}+1+\frac{1}{x^2})$

23. $y=(\csc x+\cot x)^{-1}$. $y'=-(\csc x+\cot x)^{-2}(-\csc x\cot x-\csc^2 x)=\csc x(\csc x+\cot x)^{-2}(\cot x+\csc x)=\csc x(\csc x+\cot x)^{-1}$

25. $y=\sin^4 x+\cos^{-2} x$. $y'=4\sin^3 x\cos x-2\cos^{-3} x(-\sin x)=2\sin x(2\sin^2 x\cos x+\cos^{-3} x)$

27. $y=x^3(2x-5)^4$. $y'=4x^3(2x-5)^3(2)+3x^2(2x-5)^4=x^2(2x-5)^3(8x+3(2x-5))=x^2(2x-5)^3(14x-15)$

29. $y=(4x+3)^4(x+1)^{-3}$. $y'=(4x+3)^4[-3(x+1)^{-4}]+4(4x+3)^3 4(x+1)^{-3}=(4x+3)^3(x+1)^{-4}[-3(4x+3)+16(x+1)]=(4x+3)^3(x+1)^{-4}(4x+7)$

31. $y=(\frac{\sin x}{1+\cos x})^2$. $y'=2(\frac{\sin x}{1+\cos x})\frac{(1+\cos x)\cos x-\sin x(-\sin x)}{(1+\cos x)^2}=\frac{2\sin x}{(1+\cos x)^3}(\cos x+\cos^2 x+\sin^2 x)=\frac{2\sin x}{(1+\cos x)^2}$

33. $y=(\frac{x}{x-1})^{-3}=(\frac{x-1}{x})^3$. $y'=3(\frac{x-1}{x})^2\frac{x-(x-1)}{x^2}=3(\frac{x-1}{x})^2\frac{1}{x^2}$

35. $y=\sin^3 x\tan 4x$. $y'=\sin^3 x(\sec^2 4x)4+3\sin^2 x\cos x\tan 4x=\sin^2 x(4\sec^2 4x+3\cos x\tan 4x)$

37. $y=\sqrt{\sin x}$. $y'=\frac{1}{2\sqrt{\sin x}}(\cos x)=\frac{\cos x}{2\sqrt{\sin x}}$. The result is supported by graphing $\frac{\cos x}{2\sqrt{\sin x}}$ and NDER(y,x) in $[-10,10]$ by $[-3,3]$ and seeing that the two graphs coincide.

39. $y=4\sqrt{\sec x+\tan x}$. $y'=4\frac{1}{2\sqrt{\sec x+\tan x}}(\sec x\tan x+\sec^2 x)=\frac{2\sec x(\tan x+\sec x)}{\sqrt{\sec x+\tan x}}=2\sec x\sqrt{\sec x+\tan x}$. The result is supported by graphing the last function and NDER(y,x) in $[-10,10]$ by $[0,10]$ and seeing that the two graphs coincide.

41. $y=\frac{3}{\sqrt{2x+1}}=3(2x+1)^{-1/2}$. $y'=3(-\frac{1}{2})(2x+1)^{-3/2}(2)=-\frac{3}{(2x+1)^{3/2}}$

43. $y=(2x-6)\sqrt{x+5}$. $y'=(2x-6)\frac{1}{2\sqrt{x+5}}+2\sqrt{x+5}=\frac{2x-6+4(x+5)}{2\sqrt{x+5}}=\frac{6x+14}{2\sqrt{x+5}}=\frac{3x+7}{\sqrt{x+5}}$

45. $s=\cos(\frac{\pi}{2}-3t)=\sin 3t$(complementary angles). $\frac{ds}{dt}=3\cos 3t$ or $3\sin(\frac{\pi}{2}-3t)$

47. $s=\frac{4}{3\pi}\sin 3t+\frac{4}{5\pi}\cos 5t$. $\frac{ds}{dt}=\frac{4}{\pi}\cos 3t-\frac{4}{\pi}\sin 5t=\frac{4}{\pi}(\cos 3t-\sin 5t)$

49. $r = \tan(2-\theta)$. $\frac{dr}{d\theta} = [\sec^2(2-\theta)](-1) = -\sec^2(2-\theta)$

51. $r = \sqrt{\theta}\sin\theta$. $\frac{dr}{d\theta} = \frac{1}{2\sqrt{\theta \sin\theta}}(\theta\cos\theta + \sin\theta) = \frac{\theta\cos\theta+\sin\theta}{2\sqrt{\theta\sin\theta}}$

53. $y = \sin^2(3x-2)$. $y' = 2\sin(3x-2)\cos(3x-2)\cdot 3 = 6\sin(3x-2)\cos(3x-2)$

55. $y = (1+\cos 2x)^2$. $y' = 2(1+\cos 2x)(-2\sin 2x) = -4(\sin 2x)(1+\cos 2x)$

57. $y = \sin(\cos(2x-5))$. $y' = \cos(\cos(2x-5))(-\sin(2x-5))2 = -2[\sin(2x-5)]\cos(\cos(2x-5))$

59. $y = \cot\sqrt{2x}$. $y' = (-\csc^2\sqrt{2x})(\frac{1}{2\sqrt{2x}})2 = -\frac{\csc^2\sqrt{2x}}{\sqrt{2x}}$

61. $y = \tan x$. $y' = \sec^2 x$. $y'' = 2\sec x\sec x\tan x = 2(\sec^2 x)\tan x$

63. $y = \cot(3x-1)$. $y' = -3\csc^2(3x-1)$, $y'' = -3(2)\csc(3x-1)[-\csc(3x-1)\cot(3x-1)(3)] = 18\csc^2(3x-1)\cot(3x-1)$

65. We have $f'(u) = 5u^4$ and $g'(x) = \frac{1}{2\sqrt{x}} \cdot (f\circ g)^1(1) = f'(g(1))g'(1) = 5(g(1))^4(\frac{1}{2}) = \frac{5}{2}$

67. $f(u) = \cot\frac{\pi u}{10}$, $f'(u) = -\frac{\pi}{10}\csc^2(\frac{\pi u}{10})$. $g(x) = 5\sqrt{x}$, $g'(x) = \frac{5}{2\sqrt{x}}$. $(f\circ g)'(1) = f'(g(1))g'(1) = f'(5)(\frac{5}{2}) = -\frac{5}{2}\frac{\pi}{10}\csc^2(\frac{5\pi}{10}) = -\frac{\pi}{4}$

69. $f(u) = \frac{2u}{u^2+1}$. $f'(u) = \frac{2[(u^2+1)\cdot 1 - u(2u)]}{(u^2+1)^2} = \frac{2(1-u^2)}{(u^2+1)^2}\cdot g(x) = 10x^2+x+1$, $g'(x) = 20x+1$. $(f\circ g)'(0) = f'(g(0))g'(0) = f'(1)(1) = 0$

71. a) $\frac{dy}{dx} = \frac{dy}{du}\frac{du}{dx} = -\sin u(6) = -6\sin(6x+2)$

b) $\frac{dy}{dx} = \frac{dy}{du}\frac{du}{dx} = -2\sin 2u(3) = -6\sin[2(3x+1)] = -6\sin(6x+2)$

73. a) $\frac{dy}{dx} = \frac{dy}{du}\frac{du}{dx} = \frac{1}{5}5 = 1$

b) $\frac{dy}{dx} = \frac{dy}{du}\frac{du}{dx} = -\frac{1}{u^2}[-\frac{1}{(x-1)^2}] = \frac{1}{u^2}u^2 = 1$

75. $\frac{ds}{dt} = \frac{ds}{d\theta}\frac{d\theta}{dt} = -\sin\theta(5) = -5\sin\theta$. When $\theta = 3\pi/2$, $\frac{ds}{dt} = -5\sin(3\pi/2) = 5$.

77. Slope $= y' = \frac{1}{2}\cos(x/2)$. Since the largest value of the cosine function is 1, the largest value of the slope is $1/2$.

79. $y = 2\tan(\pi x/4)$, $y' = 2\sec^2(\pi x/4)(\pi/4) = (\pi/2)\sec^2(\pi x/4)$. When $x = 1$, $y = 2$, $y' = (\pi/2)2 = \pi$. Tangent: $y - 2 = \pi(x-1)$. Normal: $y - 2 = -\frac{1}{\pi}(x-1)$.

81. a) $\left[\frac{d}{dx}\{2f(x)\}\right]_{x=2} = [2f'(x)]_{x=2} = 2f'(2) = 2/3$

b) $f'(3) + g'(3) = 2\pi + 5$

c) $[f(x)g'(x) + f'(x)g(x)]_{x=3} = 3.5 + 2\pi(-4) = 15 - 8\pi$

d) $\left[\frac{g(x)f'(x)-f(x)g'(x)}{g^2(x)}\right]_{x=2} = \frac{2(1/3)-8(-3)}{2^2} = \frac{2/3+24}{4} = 1/6 + 6 = 37/6$

e) $f'(g(2))g'(2) = f'(2)(-3) = -1$

f) $\left[\frac{1}{2\sqrt{f(x)}}f'(x)\right]_{x=2} = \frac{1}{2\sqrt{8}}\left(\frac{1}{3}\right) = \frac{1}{12\sqrt{2}} = \frac{\sqrt{2}}{24}$

g) $\left[\frac{d}{dx}\{g^{-2}(x)\}\right]_{x=3} = [-2g^{-3}(x)g'(x)]_{x=3} = -2(-4)^{-3}5 = \frac{2(5)}{64} = \frac{5}{32}$

h) $\left[\frac{1}{2\sqrt{f^2(x)+g^2(x)}}2f(x)f'(x) + 2g(x)g'(x)\right]_{x=2} = \left[\frac{f(x)f'(x)+g(x)g'(x)}{\sqrt{f^2(x)+g^2(x)}}\right]_{x=2}$
$= \frac{8(1/3)+2(-3)}{\sqrt{8^2+2^2}} = \frac{-5\sqrt{17}}{51}$

83. $s = A\cos(2\pi bt)$, $V = s' = -2\pi bA\sin(2\pi bt)$ and $a = v' = -4\pi^2b^2A\cos(2\pi bt)$. Now let $s_1 = A\cos[2\pi(2b)t] = A\cos(4\pi bt)$. Then the new velocity and acceleration are given by $v_1 = -4\pi bA\sin(4\pi bt) = 2(-2\pi bA)\sin(4\pi bt)$ and $a_1 = -16\pi^2b^2A\cos(4\pi bt) = 4(-4\pi^2b^2A)\cos(4\pi bt)$. Thus the amplitude of v is doubled and the amplitude of a is quadrupled.

85. a) Graph $y = 37\sin\left[\frac{2\pi}{365}(t-101)\right] + 25$ in $[0, 365]$ by $[-12, 62]$.

b) $y' = 37(\frac{2\pi}{365})\cos\left[\frac{2\pi}{365}(t-101)\right]$ is largest when $t - 101 = 0$, $t = 101$ or on April 12 (of a non-leap year)

c) When $t = 101$, $y' = \frac{37(2\pi)}{365} = .6369..^\circ F/\text{day}$.

3.7 IMPLICIT DIFFERENTIATION AND FRACTIONAL POWERS

1. $y = x^{9/4}$. $y' = (9/4)x^{(9/4)-1} = (9/4)x^{5/4}$

3. $y = \sqrt[3]{x} = x^{1/3}$. $y' = (1/3)x^{-2/3} = \frac{1}{3\sqrt[3]{x^2}}$

5. $y = (2x+5)^{-1/2}$. $y' = -(1/2)(2x+5)^{-3/2}(2) = -(2x+5)^{-3/2}$

7. $y = x\sqrt{x^2+1} = x(x^2+1)^{1/2}$. $y' = x(1/2)(x^2+1)^{-1/2}(2x) + 1\cdot(x^2+1)^{1/2} = \frac{x^2}{\sqrt{x^2+1}} + \sqrt{x^2+1} = \frac{x^2+x^2+1}{\sqrt{x^2+1}} = \frac{2x^2+1}{\sqrt{x^2+1}}$

9. $x^2y + xy^2 = 6$. $\frac{d}{dx}(x^2y) + \frac{d}{dx}(xy^2) = \frac{d}{dx}(6)$. $(x^2\frac{dy}{dx} + 2xy) + (x2y\frac{dy}{dx} + 1\cdot y^2) = 0$. $x^2y' + 2xyy' + 2xy + y^2 = 0$. $y'(x^2+2xy) = -2xy - y^2$. $y' = -\frac{2xy+y^2}{x^2+2xy}$

11. $2xy + y^2 = x + y$. $2xy' + 2y + 2yy' = 1 + y'$, $y'(2x + 2y - 1) = 1 - 2y$, $y' = (1 - 2y)/(2x + 2y - 1)$

13. $x^2y^2 = x^2 + y^2$. $x^2 2yy' + 2xy^2 = 2x + 2yy'$, $x^2yy' + xy^2 = x + yy'$, $y'(x^2y - y) = x - xy^2$, $y' = x(1 - y^2)/[y(x^2 - 1)]$

15. $y^2 = \frac{x-1}{x+1}$. $2yy' = \frac{(x+1)\cdot 1-(x-1)\cdot 1}{(x+1)^2} = \frac{2}{(x+1)^2}$, $y' = \frac{1}{y(x+1)^2}$

17. $y = \sqrt{1 - \sqrt{x}}$. $y' = \frac{1}{2\sqrt{1-\sqrt{x}}}(-\frac{1}{2\sqrt{x}}) = \frac{-1}{4\sqrt{x}\sqrt{1-\sqrt{x}}}$

19. $y = 3(\csc x)^{3/2}$. $y' = (9/2)(\csc x)^{1/2}(-\csc x \cot x) = -(9/2)\csc^{3/2} x \cot x$

21. $x = \tan y$. $1 = (\sec^2 y)y'$. $y' = 1/\sec^2 y = \cos^2 y$

23. $x + \tan(xy) = 0$. $1 + \sec^2(xy)[xy' + y] = 0$, $xy' \sec^2(xy) + y\sec^2(xy) = -1$, $y' = \frac{-[1+y\sec^2 xy]}{x\sec^2(xy)} = -\frac{[\cos^2(xy)+y]}{x}$

25. $y\sin(\frac{1}{y}) = 1 - xy$. $y\cos(\frac{1}{y})(-\frac{1}{y^2})y' + y'\sin(\frac{1}{y}) = -y - xy'$, $y'[-\frac{1}{y}\cos(\frac{1}{y}) + \sin(\frac{1}{y}) + x] = -y$, $y' = y/[\frac{1}{y}\cos(\frac{1}{y}) - \sin(\frac{1}{y}) - x]$

27. a) $f(x) = \frac{3}{2}x^{2/3} - 3$. $f'(x) = x^{-1/3}$, $f''(x) = -\frac{1}{3}x^{-2/3}$ so a) is not true. b) $f(x) = \frac{9}{10}x^{5/3} - 7$ leads to $f'(x) = \frac{3}{2}x^{2/3}$, $f''(x) = x^{-1/3}$ so b) could be true. c) Since $f''(x) = x^{-1/3}$ leads to $f'''(x) = -\frac{1}{3}x^{-4/3}$, c) is true. d) $f'(x) = \frac{3}{2}x^{2/3} + 6$ leads to $f''(x) = x^{-1/3}$, d) could be true. Answer: b), c), d).

29. $x^2 + y^2 = 1$. $2x + 2yy' = 0$, $y' = -x/y$. $y'' = -\frac{y\cdot 1-xy'}{y^2} = -\frac{y-x(-x/y)}{y^2} = -\frac{y^2+x^2}{y^3} = -\frac{1}{y^3}$

31. $y^2 = x^2 + 2x$. $2yy' = 2x + 2$, $y' = (x+1)/y$. $y'' = \frac{y\cdot 1-(x+1)y'}{y^2} = \frac{y-(x+1)(x+1)/y}{y^2} = \frac{y^2-(x+1)^2}{y^3}$

33. $y + 2\sqrt{y} = x$. $y' + 2(\frac{1}{2\sqrt{y}})y' = 1$, $y'(1 + \frac{1}{\sqrt{y}}) = 1$, $y'(\frac{\sqrt{y}+1}{\sqrt{y}}) = 1$, $y' = \frac{\sqrt{y}}{\sqrt{y}+1}$. $y'' = \frac{(\sqrt{y}+1)(1/2\sqrt{y})y' - \sqrt{y}(1/(2\sqrt{y}))y'}{(\sqrt{y}+1)^2} = \frac{y'/(2\sqrt{y})}{(\sqrt{y}+1)^2} = \frac{(1/2)}{(\sqrt{y}+1)^3} = \frac{1}{2(\sqrt{y}+1)^3}$

35. $x^2 + xy - y^2 = 1$ at $(2,3)$. $2x + xy' + y - 2yy' = 0$, $y'(x - 2y) = -(2x + y)$, $y' = \frac{(2x+y)}{2y-x}$. At $(2,3)$, $y' = \frac{4+3}{6-2} = 7/4$. Tangent: $y - 3 = \frac{7}{4}(x - 2)$ or $7x - 4y = 2$. Normal: $y - 3 = -\frac{4}{7}(x - 2)$ or $4x + 7y = 29$.

37. $x^2y^2 = 9$ at $(-1,3)$. $2xy^2 + x^2(2yy') = 0$, $x^2yy' = -xy^2$, $y' = -\frac{y}{x}$ $(= 3$ at $(-1,3))$. a) Tangent: $y - 3 = 3(x+1)$ or $y - 3x = 6$. b) Normal: $y - 3 = (-1/3)(x+1)$ or $x + 3y = 8$.

39. $6x^2 + 3xy + 2y^2 + 17y - 6 = 0$ at $(-1,0)$. $12x + 3xy' + 3y + 4yy' + 17y' = 0$, $y'(3x + 4y + 17) = -12x - 3y$, $y' = \frac{-3(4x+y)}{3x+4y+17}$ $(= \frac{12}{14} = \frac{6}{7}$ at $(-1,0))$. a) Tangent: $y = \frac{6}{7}(x+1)$ b) Normal: $y = -\frac{7}{6}(x+1)$

41. $2xy + \pi \sin y = 2\pi$ at $(1, \pi/2)$. $2xy' + 2y + \pi(\cos y)y' = 0$, $y'(2x + \pi\cos y) = -2y$, $y' = \frac{-2y}{2x+\pi\cos y}$ $(= \frac{-\pi}{2+0} = -\frac{\pi}{2}$ at $(1, \frac{\pi}{2}))$. a) Tangent: $y - \frac{\pi}{2} = -\frac{\pi}{2}(x-1)$ or $y = -\frac{\pi}{2}x + \pi$ b) Normal: $y - \frac{\pi}{2} = \frac{2}{\pi}(x-1)$ or $y = \frac{2}{\pi}x + \frac{\pi^2-4}{2\pi}$

43. $y = 2\sin(\pi x - y)$ at $(1,0)$. $y' = (2\cos(\pi x - y))(\pi - y')$, $y'[1 + 2\cos(\pi x - y)] = 2\pi\cos(\pi x - y)$, $y' = \frac{2\pi\cos(\pi x - y)}{1+2\cos(\pi x-y)}$ $(= \frac{-2\pi}{-1} = 2\pi$ at $(1,0))$. a) Tangent: $y = 2\pi(x-1)$ b) Normal: $y = -\frac{1}{2\pi}(x-1)$

45. $2xy + \pi\sin y = 2\pi$. $2xy' + 2y + \pi(\cos y)y' = 0$, $y'(2x + \pi\cos y) = -2y$, $y' = -2y/(2x + \pi\cos y)$. At $(1, \pi/2)$, $y' = -\pi/(2 + \pi\cdot 0) = -\pi/2$.

47. a) $y^4 = y^2 - x^2$ leads to $4y^3y' = 2yy' - 2x$, $y'(4y^3 - 2y) = -2x$, $y' = \frac{2x}{2y-4y^3} = \frac{x}{y(1-2y^2)}$. At $(\frac{\sqrt{3}}{4}, \frac{\sqrt{3}}{2})$, the slope is $\frac{\sqrt{3}/4}{(\sqrt{3}/2)(1-3/2)} = -1$ and at $(\frac{\sqrt{3}}{4}, \frac{1}{2})$ the slope is $\frac{\sqrt{3}/4}{(1/2)(1-1/2)} = \sqrt{3}$ b) Graph $x_1 = \sqrt{t^2 - t^4}$, $y_1 = t$ and $x_2 = -x_1$, $y_2 = t$, $-1 \leqq t \leqq 1$ in $[-0.5, 0.5]$ by $[-1, 1]$.

49. $x^3y^2 = \cos(\pi y)$. a) Substituting $x = -1$, $y = 1$, we get $(-1)^3(1)^2 = \cos\pi$, a true statement. b) Differentiating both sides with respect to x, we obtain $x^3(2yy') + 3x^2y^2 = -\sin(\pi y)(\pi y')$. This leads to $y'[2x^3y + \pi\sin(\pi y)] = -3x^2y^2$, $y' = \frac{-3x^2y^2}{2x^3y+\pi\sin(\pi y)}$. At $(-1,1)$, $y' = \frac{-3}{-2+0} = \frac{3}{2}$.

51. Differentiating $x^2 + xy + y^2 = 7$, we get $2x + xy' + y + 2yy' = 0$, $y'(x + 2y) = -(2x + y)$, $y' = -(2x + y)/(x + 2y)$. For the x-intercepts we set $y = 0$: $x^2 + x(0) + 0^2 = 7$ and obtain $x = \pm\sqrt{7}$. For these points $y' = -[2(\pm\sqrt{7}) + 0]/[\pm\sqrt{7} + 2(0)] = -2$. This shows that the two tangents are parallel with -2 as their common slope.

53. Let C_1 be the curve $2x^2 + 3y^2 = 5$ and let C_2 be the curve $y^2 = x^3$. $4x + 6yy' = 0$ leads to $y' = -\frac{2x}{3y}$ for the slope of C_1. $y^2 = x^3$ leads to $2yy' = 3x^2$, $y' = \frac{3x^2}{2y}$ for the slope of C_2. At $(1,1)$ the slopes are $-\frac{2}{3}$ and $\frac{3}{2}$ (negative reciprocals) so C_1 and C_2 are orthogonal at $(1,1)$. At $(1,-1)$ the slopes are $\frac{2}{3}$ and $-\frac{3}{2}$ so we reach the same conclusion at $(1,-1)$. Graph $y_1 = \sqrt{(5 - 2x^2)/3}$, $y_2 =$

$-y_1$, $y_3 = x^{1.5}$, $y_4 = -y_3$, $y_5 = -(2/3)x + 5/3$, $y_6 = (3/2)x - 1/2$, $y_7 = (2/3)x - 5/3$, $y_8 = -(3/2)x + 1/2$ in $[-6, 6]$ by $[-3.5, 3.5]$.

55. $a = \frac{dv}{dt} = \frac{k}{2\sqrt{s}}\frac{ds}{dt} = \frac{k}{2\sqrt{s}}v = \frac{k}{2\sqrt{s}}(k\sqrt{s}) = \frac{k^2}{2}$

57. a) $-1 \leqq -\sin xy \leqq 1$ so $-1 \leqq y^5 \leqq 1$. This implies $-1 \leqq y \leqq 1$. b) $\sin xy = -y^5$ leads to $xy = \sin^{-1}(-y^5) + 2k\pi$ or $xy = \pi - \sin^{-1}(-y^5) + 2k\pi = (2k+1)\pi + \sin^{-1}(y^5)$. c) When we graph $x(t) = t^{-1}\sin^{-1}(-t^5)$, $y(t) = t$, $-1 \leqq t \leqq 1$, we see that $-\frac{\pi}{2} \leqq x < 0$ so the domain is $[-\frac{\pi}{2}, 0)$. The range of the graphed relation is $[-1, 0) \cup (0, 1]$. d) Graph $x(t) = \frac{\sin^{-1}(-t^5) + 2\pi}{t}$, $y(t) = t$, $-1 \leqq t \leqq 1$ in $[-50, 50]$ by $[-2, 2]$. The domain (set of x-values) is $(-\infty, \frac{-5\pi}{2}] \cup [\frac{3\pi}{2}, \infty)$. The range is $[-1, 0) \cup (0, 1]$.

59. $\frac{dy}{dx} = \frac{y - 2x}{2y - x}$ does not exist when $2y - x = 0$ or $x = 2y$. Substituting this into $x^2 - xy + y^2 = 7$, we obtain $4y^2 - 2y^2 + y^2 = 3y^2 = 7$, $y = \pm\sqrt{\frac{7}{3}}$. Since $x = 2y$, the points we seek are $(-2\sqrt{\frac{7}{3}}, -\sqrt{\frac{7}{3}})$ and $(2\sqrt{\frac{7}{3}}, \sqrt{\frac{7}{3}})$.

3.8 LINEAR APPROXIMATIONS AND DIFFERENTIALS

1. $f(x) = x^4$, $a = 1$. $f'(x) = 4x^3$, $f'(1) = 4$. $L(x) = f(a) + f'(a)(x - a) = 1 + 4(x - 1) = 4x - 3$.

3. $f(x) = x^3 - x$, $a = 1$. $f'(x) = 3x^2 - 1$. $L(x) = f(a) + f'(a)(x - a) = 0 + 2(x - 1) = 2(x - 1)$.

5. $f(x) = \sqrt{x}$, $x = 4$. $f(4) = 2$, $f'(x) = \frac{1}{2\sqrt{x}}$, $f'(4) = \frac{1}{4}$. $L(x) = f(a) + f'(a)(x - a) = 2 + \frac{1}{4}(x - 4) = \frac{x}{4} + 1$.

7. $a = 0$. $f(x) = x^2 + 2x$, $f'(x) = 2x + 2$, $f(0) = 0$, $f'(0) = 2$. $L(x) = f(a) + f'(a)(x - a) = 0 + 2(x - 0) = 2x$.

9. $a = -1$. $f(x) = 2x^2 + 4x - 3$, $f'(x) = 4x + 4$, $f(-1) = -5$, $f'(-1) = 0$. $L(x) = f(-1) + f'(-1)(x + 1) = -5$.

11. $f(x) = \sqrt[3]{x}$, $a = 8$. $f'(x) = \frac{1}{3}x^{-2/3} = \frac{1}{3(\sqrt[3]{x})^2}$. $f(8) = 2$, $f'(8) = \frac{1}{3(2)^2} = \frac{1}{12}$. $L(x) = f(a) + f'(a)(x - a) = 2 + \frac{1}{12}(x - 8)$.

13. $f(x) = \sin x$, $a = 0$. $f'(x) = \cos x$. $f(0) = 0$, $f'(0) = 1$. $L(x) = 0 + (1)(x - 0) = x$.

15. $f(x) = \sin x$, $a = \pi$. $f'(x) = \cos x$. $f(\pi) = 0$, $f'(\pi) = -1$. $L(x) = f(a) + f'(a)(x-a) = 0 - (x-\pi) = \pi - x$.

17. $f(x) = \tan x$, $a = \pi/4$. $f'(x) = \sec^2 x$, $f(\pi/4) = 1$, $f'(\pi/4) = 2$. $L(x) = f(a) + f'(a)(x-a) = 1 + 2(x - \pi/4)$.

19. a) $(1+x)^2 \approx 1+2x$ b) $(1+x)^{-5} \approx 1-5x$ c) $2(1+(-x))^{-1} \approx 2[1+(-1)(-x)] = 2(1+x)$ d) $(1+x)^6 \approx 1+6x$ and replacing x by $-x$, $(1-x)^6 \approx 1-6x$ e) $3(1+x)^{1/3} \approx 3[1+(1/3)x] = 3+x$ f) $(1+x)^{-1/2} \approx 1-(1/2)x$

21. $f(x) = \sqrt{x+1} + \sin x$. $f(0) = \sqrt{0+1} + \sin 0 = 1+0 = 1$. $f'(x) = (\sqrt{x+1})' + (\sin x)' = \frac{1}{2\sqrt{x+1}} + \cos x$. $f'(0) = \frac{1}{2} + 1 = \frac{3}{2}$. $L(x) = 1 + \frac{3}{2}x$ is the sum of the linearizations of $\sqrt{x+1}$ and $\sin x$.

23. b) From Example 1, $\sqrt{1+x} \approx 1 + \frac{x}{2}$. Thus when x represents the decimal part of $1+x$ and the square root of $1+x$ is taken, the result is nearly the same as halving the decimal part.

c) $\sqrt{2} = \sqrt{1+1} \approx 1+\frac{1}{2}$ using the linearization. Then $(2^{\frac{1}{2}})^{\frac{1}{2}} \approx (1+\frac{1}{2})^{\frac{1}{2}} \approx 1+\frac{1}{4}$ or $1+\frac{1}{2^2}$ applying the linearization to $\sqrt{1+\frac{1}{2}}$. Repeating the process, we arrive at $((2^{\frac{1}{2}})^{\frac{1}{2}})^{\frac{1}{2}} \approx 1 + \frac{1}{2^3}$ and in general $2^{1/2^n} \approx 1 + \frac{1}{2^n}$ if the process is carried out n times. These numbers are approaching $1+0=1$.

d) Starting with $m > 1$, $\sqrt{m} = \sqrt{1+(m-1)} \approx 1 + \frac{m-1}{2}$ and $(m)^{1/2^n} \approx 1 + \frac{m-1}{2^n}$ approaches 1.

25. $f(x) = x^2 + 2x$, $x_0 = 0$, $dx = 0.1$. a) $\Delta f = f(x_0 + dx) - f(x_0) = f(0.1) - f(0) = 0.01 + 0.2 - 0 = 0.21$. b) $f'(x) = 2x+2$. $df = f'(x_0)dx = 2(0.1) = 0.2$. c) $|\Delta f - df| = |0.21 - 0.2| = 0.01$.

27. $f(x) = x^3 - x$, $x_0 = 1$, $dx = 0.1$. a) $\Delta f = f(x_0 + dx) - f(x_0) = (1.1)^3 - 1.1 - 0 = 0.231$. b) $f'(x) = 3x^2 - 1$. $df = f'(x_0)dx = 2(0.1) = 0.2$. c) $|\Delta f - df| = |0.231 - 0.2| = 0.031$.

29. $f(x) = x^{-1}$, $x_0 = 0.5$, $dx = 0.1$. a) $\Delta f = f(x_0 + dx) - f(x_0) = (0.6)^{-1} - (0.5)^{-1} = (5/3) - 2 = -1/3$. b) $f'(x) = -1/x^2$, $df = f'(x_0)dx = (-1/0.25)(0.1) = -4(0.1) = -0.4 = -2/5$ c) $|\Delta f - df| = |-1/3 - (-2/5)| = 1/15$.

31. $dV = \frac{dV}{dr}dr = 4\pi r^2 dr$ and, at r_0, $dV = 4\pi r_0^2 dr$.

33. $V(x) = x^3$. $V'(x) = 3x^2$. $dV = 3x_0^2 dx$.

35. $V(r) = \pi r^2 h$. $V'(r) = 2\pi r h$. $dV = 2\pi r_0 h dr$.

37. a) $A(r) = \pi r^2$, $A'(r) = 2\pi r$. $dA = A'(2.00)dr = \pi 4(0.02) = 0.08\pi \approx 0.2513m^2$ b) $\frac{0.2513}{4\pi}(100) \approx 2.000\%$.

39. $V = x^3$, $V'(x) = 3x^2$, $dV = V'(x_0)dx = V'(10)(\frac{10}{100}) = 3(10)^2(\frac{1}{10}) = 30$ is the estimated error. The estimated percentage error is $\frac{30}{100}(100) = 3\%$.

41. $V = \frac{4}{3}\pi r^3 = \frac{4}{3}\pi(\frac{d}{2})^3 = \frac{\pi}{6}d^3$. $V'(d) = \frac{\pi d^2}{2}$. The estimated error in the volume calculation is $dV = V'(d_0)d(d) = V'(d_0)$ (differential of d) $= V'(100)(1) = 5000\pi$. The estimated percentage error is $\frac{dV}{V}(100) = \frac{5000\pi(100)}{(\pi/6)100^3} = \frac{(100^2/2)\pi 100}{(\pi/6)100^3} = \frac{6}{2} = 3\%$.

43. We have $V = \pi h^3$, $dV = 3\pi h^2 dh$. We require $dV/V \leqq 0.01$, $3\pi h^2 dh/\pi h^3 = 3dh/h \leqq 0.01$ or $dh/h \leqq 0.01/3$. Thus h should be measured with an error of no more than 1/3 of 1%.

45. Let weight of coin $= W = kV$ when k is the density and V is the volume. $V = \pi r^2 h$. Thus $W = kh\pi r^2$, $dW = 2kh\pi r dr$. It is required that $dW \leqq (1/1000)W$, $2kh\pi r dr \leqq (1/1000)kh\pi r^2$ or $dr \leqq (1/2000)r$. The variation of the radius should not exceed 1/2000 of its ideal value, that is, 0.05% of the ideal value.

47. $s = 16t^2$, $ds = 32tdt$. $ds/s = 32tdt/16t^2 = 2dt/t$. Thus the relative error in s is twice the relative error in measuring t.

49. a) $g(x) = \sqrt{x} + \sqrt{1+x} - 4$. $g(3) = \sqrt{3} + \sqrt{4} - 4 = \sqrt{3} - 2 < 0$. $g(4) = \sqrt{4} + \sqrt{5} - 4 = \sqrt{5} - 2 > 0$.

b) We use the linearization of $g(x)$: $L(x) = g(3) + g'(3)(x-3)$. $g'(x) = 1/(2\sqrt{x}) + 1/(2\sqrt{x+1})$. $g'(3) = 1/(2\sqrt{3}) + 1/2\sqrt{4} = 1/(2\sqrt{3}) + 1/4$. $L(x) = (\sqrt{3} - 2) + (1/(2\sqrt{3}) + 1/4)(x-3) = 0$ leads to $[\frac{1}{2\sqrt{3}} + \frac{1}{4}](x-3) = 2 - \sqrt{3}$, $\frac{2+\sqrt{3}}{4\sqrt{3}}(x-3) = 2 - \sqrt{3}$, $x - 3 = \frac{4\sqrt{3}}{2+\sqrt{3}}(2-\sqrt{3})\frac{2-\sqrt{3}}{2-\sqrt{3}} = \frac{4\sqrt{3}(4-4\sqrt{3}+3)}{4-3} = 28\sqrt{3} - 48$, $x = 28\sqrt{3} - 45 \approx 3.497$.

c) We find $g(28\sqrt{3} - 45) \approx -0.009$. d) $x = 3.516$ with error at most 0.01.

e) $\sqrt{x} + \sqrt{1+x} - 4 = 0$, leads to $\sqrt{1+x} = 4 - \sqrt{x}$, $1 + x = 16 - 8\sqrt{x} + x$, $\sqrt{x} = \frac{15}{8}$, $x = \frac{225}{64}$.

51. $y = x^3 - 3x$. $dy = (3x^2 - 3)dx = 3(x^2 - 1)dx$.

53. $y = \frac{2x}{1+x^2}$. $dy = 2\frac{[(1+x^2)\cdot 1 - x(2x)]dx}{(1+x^2)^2} = \frac{2(1-x^2)dx}{(1+x^2)^2}$.

55. $y + xy - x = 0$ leads to $y(1+x) - x = 0$, $y = \frac{x}{1+x}$, $dy = \frac{[(1+x)\cdot 1 - x(1)]dx}{(1+x)^2} = \frac{dx}{(1+x)^2}$.

57. $y = \sin(5x)$. $dy = [\cos(5x)]5dx = 5\cos(5x)dx$.

59. $y = 4\tan(x/2)$. $dy = 4\sec^2(x/2)(1/2)dx = 2\sec^2(x/2)dx$.

61. $y = 3\csc(1-(x/3))$. $dy = -3\csc(1-(x/3))\cot(1-(x/3))(-1/3)$. $dy = -3\csc(1-(x/3))\cot(1-(x/3))(-1/3)dx = \csc(1-(x/3))\cot(1-(x/3))dx$.

63. a) We graphed $y = \sqrt{1+x}$ in $[-0.17, 0.17]$ by $[0.91, 1.08]$. b) We use TRACE to estimate the coordinates of the extreme points of the graph in a): $m = \frac{y_2-y_1}{x_2-x_1} = \frac{1.08-0.91}{0.17-(-0.17)} = 0.5$. $f'(x) = \frac{1}{2\sqrt{1+x}}$. $f'(0) = 0.5$. c) Also graph $y = 1 + \frac{x}{2}$. The graphs nearly coincide (especially near $x = 0$) in the above window.

65. $\lim_{x\to 0} \frac{\sqrt{1+x}}{1+(x/2)} = \frac{\sqrt{1+0}}{1+0} = 1$.

67. $E(x) = f(x) - g(x) = f(x) - m(x-a) - c$. $E(a) = 0$ implies $f(a) - 0 - c = 0$ and so $c = f(a)$. $E(x) = f(x) - f(a) - m(x-a)$. $0 = \lim_{x\to a} \frac{E(x)}{x-a} = \lim_{x\to a} \frac{f(x)-f(a)}{x-a} - m = f'(a) - m$. Thus $m = f'(a)$ and $g(x) = L(x)$.

PRACTICE EXERCISES, CHAPTER 3

1. $y = x^5 - \frac{1}{8}x^2 + \frac{1}{4}x$. $y' = 5x^4 - \frac{1}{4}x + \frac{1}{4}$.

3. $y = (x+1)^2(x^2+2x)$. $y' = (x+1)^2(2x+2) + 2(x+1)(x^2+2x) = 2(x+1)[(x+1)^2 + x^2 + 2x] = 2(x+1)(2x^2+4x+1)$.

5. $y = 2\sin x \cos x = \sin 2x$. $y' = 2\cos 2x$.

7. $y = \frac{x}{x+1}$. $y' = \frac{(x+1)\cdot 1 - x\cdot 1}{(x+1)^2} = \frac{1}{(x+1)^2}$.

9. $y = (x^3+1)^{-4/3}$. $y' = (-4/3)(x^3+1)^{-7/3}(3x^2) = -4x^2(x^3+1)^{-7/3}$.

11. $y = \cos(1-2x)$. $y' = [-\sin(1-2x)](-2) = 2\sin(1-2x)$.

13. $y = (x^2+x+1)^3$. $y' = 3(x^2+x+1)^2(2x+1)$.

15. $y = \sqrt{2u+u^2}$, $u = 2x+3$. $\frac{dy}{dx} = \frac{dy}{du}\frac{du}{dx} = \left(\frac{2+2u}{2\sqrt{2u+u^2}}\right)(2) = \frac{2(1+u)}{\sqrt{2u+u^2}} = \frac{2(1+2x+3)}{\sqrt{2(2x+3)+(2x+3)^2}} = \frac{4(x+2)}{\sqrt{4x^2+16x+15}}$.

17. $xy + y^2 = 1$. $xy' + y + 2yy' = 0$, $y'(x+2y) = -y$, $y' = \frac{-y}{x+2y}$.

19. $x^2+xy+y^2-5x=2$. $2x+xy'+y+2yy'-5=0$, $y'(x+2y)=5-2x-y$, $y'=\frac{5-2x-y}{x+2y}$.

21. $5x^{4/5}+10y^{6/5}=15$. $4x^{-1/5}+12y^{1/5}y'=0$, $y'=\frac{-4x^{-1/5}}{12y^{1/5}}=\frac{-1}{3(xy)^{1/5}}$.

23. $y^2=\frac{x}{x+1}$. $2yy'=\frac{1}{(x+1)^2}$ by Exercise 7. $y'=\frac{1}{2y(x+1)^2}$.

25. $y^2=\frac{(5x^2+2x)^{3/2}}{3}$. $2yy'=\frac{1}{2}(5x^2+2x)^{1/2}(10x+2)=(5x+1)(5x^2+2x)^{1/2}$. $y'=\frac{(5x+1)(5x^2+2x)^{1/2}}{2y}$.

27. $y=\sqrt{x}+1+\frac{1}{\sqrt{x}}=\sqrt{x}+1+x^{-1/2}$. $y'=\frac{1}{2\sqrt{x}}-\frac{1}{2}x^{-3/2}=\frac{1}{2\sqrt{x}}-\frac{1}{2x\sqrt{x}}=\frac{x-1}{2x\sqrt{x}}$.

29. $y=\sec(1+3x)$. $y'=\sec(1+3x)\tan(1+3x)\cdot 3=3\sec(1+3x)\tan(1+3x)$.

31. $y=\cot x^2$. $y'=(-\csc^2 x^2)2x=-2x(\csc x^2)^2$.

33. $y=\sqrt{\frac{1-x}{1+x^2}}=\sqrt{u}$ where $u=\frac{1-x}{1+x^2}$. $\frac{dy}{dx}=\frac{dy}{du}\frac{du}{dx}=\frac{1}{2\sqrt{u}}\frac{(1+x^2)(-1)-(1-x)2x}{(1+x^2)^2}=\frac{1}{2}\sqrt{\frac{1}{u}}\frac{-1-x^2-2x+2x^2}{(1+x^2)^2}=\frac{1}{2}\sqrt{\frac{1+x^2}{1-x}}\frac{x^2-2x-1}{(1+x^2)^2}=\frac{x^2-2x-1}{(1-x)^{1/2}(1+x^2)^{3/2}}$.

35. a)

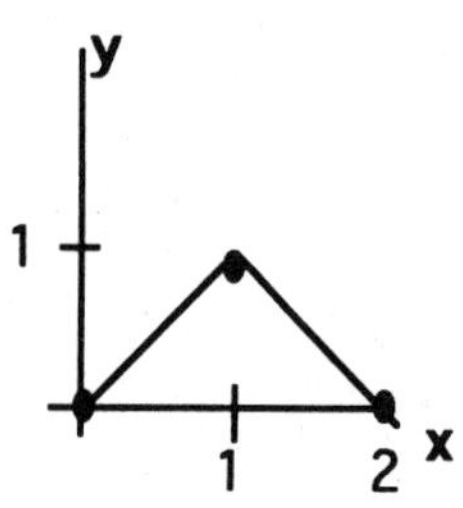

b) Yes. $f(x)$ is continuous at $x=1$ because $\lim_{x\to 1} f(x)=f(1)$.

c) f is not differentiable at $x=1$ because its left-hand derivative (1) is not equal to its right-hand derivative (-1) at $x=1$.

37. $y=2x^3-3x^2-12x+20$. $y'=6x^2-6x-12=6(x^2-x-2)=6(x+1)(x-2)$. The tangent is horizontal if and only if $y'=0$. Thus $x=-1,2$. The points on the curve are $(-1,27)$ and $(2,0)$.

39. $s(t) = 10\cos(t + \pi/4)$. b) $s(0) = 10\cos(0 + \pi/4) = 5\sqrt{2}$. c) The smallest and largest values of the cosine function are -1 and 1 so the largest and smallest values of s are -10 and 10. d) $v = \frac{ds}{dt} = -10\sin(t+\pi/4)$ and $a = \frac{dv}{dt} = -10\cos(t + \pi/4)$. $s(3\pi/4) = -10$ and $s(7\pi/4) = 10$. $v(3\pi/4) = 0$, $a(3\pi/4) = 10$ and $v(7\pi/4) = 0$, $a(7\pi/4) = -10$. e) $s(t)$ will first be 0 when $t + \pi/4 = \pi/2$ or $t = \pi/4$. $v(\pi/4) = -10$, speed $= |v(\pi/4)| = 10$, $a(\pi/4) = 0$.

41. a) $s = 490t^2$. $490t^2 = 160$, $t^2 = 16/49$, $t = (4/7)$ sec. Average velocity $= \frac{s(t_2)-s(t_1)}{t_2-t_1} = \frac{s(4/7)-s(0)}{(4/7)-0} = \frac{490(4/7)^2}{(4/7)} = 280$ cm/sec.

b) $v = \frac{ds}{dt} = 980t$, $a = 980$. $v(4/7) = 560$ cm/sec, $a(4/7) = 980$ cm/sec^2.

43. The steadily increasing distance is shown in (iii). Its rate of increase (velocity) which is always positive is shown in (i). (ii) is the graph of the rate of change of (i): it is positive (above t-axis) between $t = 0$ and $t = 1$ where (i) is increasing, and negative between $t = 1$ and $t = 2$ where (i) is decreasing. Thus (i) shows velocity and (ii) shows acceleration. a) (iii) b) (i) c) (ii)

45. The graph of f looks like this:

Note that for $0 \le x \le 1$, $f' = 0$

and thus f = constant.

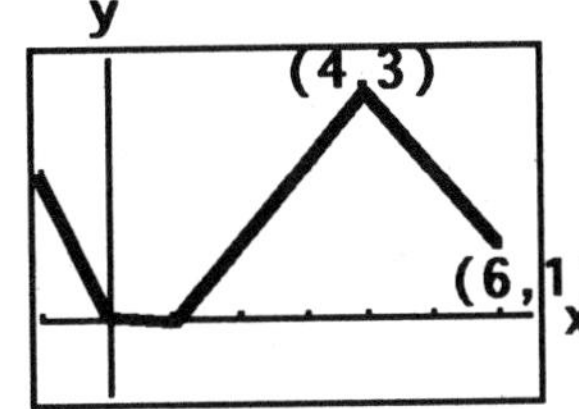

47.

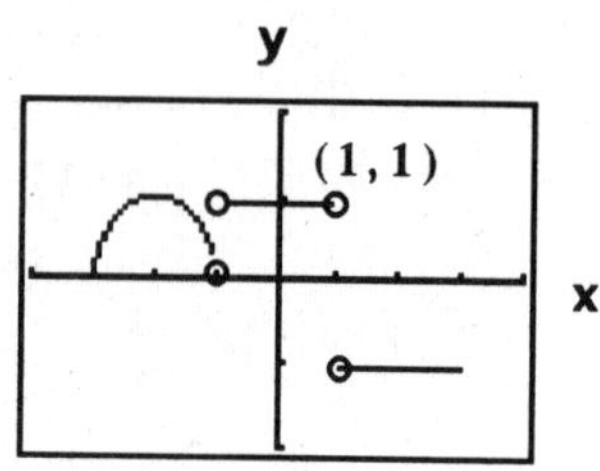

49. $V = \pi[10 - (x/3)]x^2$. $dV/dx = \pi[(10 - (x/3))2x - (1/3)x^2] = \pi(20x - x^2)$.

51. $y = 4 + \cot x - 2\csc x$, $y' = -\csc^2 x + 2\csc x \cot x = \csc x(2\cot x - \csc x)$. a) Since $\csc x \neq 0$, $y' = 0$ when $2\cot x = \csc x$ which leads to $2\cos x = 1$, $\cos x = \frac{1}{2}$, $x = \pm\frac{\pi}{3} + 2k\pi$. Evaluating y at these points, we get the points $(\pm\frac{\pi}{3} + 2k\pi,\ 4 \mp \sqrt{3})$. b) $f'(\frac{\pi}{2}) = -1$ and so the tangent at p has equation $y - 2 = -(x - \frac{\pi}{2})$.

53. $y = f(x) = \sin(x - \sin x)$, $y' = f'(x) = \cos(x - \sin x)[1 - \cos x]$. $y' = 0$ when $\cos(x - \sin x) = 0$ or $1 - \cos x = 0$. From the first equation we get $x - \sin x = (2n + 1)\pi/2$. At a solution x of this equation we see that $f(x) = \sin(x - \sin x) = \sin[(2n+1)\pi/2] = \pm 1$ so $(x, f(x))$ is not on the x-axis. The second equation yields $\cos x = 1$, $x = 2n\pi$. $f(2n\pi) = \sin(2n\pi - 0) = 0$. Thus $(2n\pi, 0)$, $n = 0, \pm 1, \pm 2, \ldots$ are points on the curve where the tangent is horizontal.

55. a) $(5f(x) - g(x))'_{x=1} = (5f'(x) - g'(x))_{x=1} = 5f'(1) - g'(1) = 1$.

b) $(f(x)g^3(x))'_{x=0} = f(0)3g^2(0)g'(0) + f'(0)g^3(0) = 6$.

c) $\left(\frac{f(x)}{g(x)+1}\right)'_{x=1} = \frac{(g(1)+1)f'(1) - f(1)g'(1)}{(g(1)+1)^2} = 1$.

d) $(f(g(x)))'_{x=0} = f'(g(0))g'(0) = f'(1)(1/3) = -1/9$.

e) $(g(f(x)))'_{x=0} = g'(f(0))f'(0) = g'(1)5 = -40/3$.

f) $((x + f(x))^{3/2})'_{x=1} = (3/2)(1 + f(1))^{1/2}(1 + f'(1)) = 2$.

g) $(f(x + g(x)))'_{x=0} = f'(0 + g(0))(1 + g'(0)) = -4/9$.

57. Differentiating both sides of the identity $\sin(x+a) = \sin x \cos a + \cos x \sin a$ with respect to x, we obtain the identity $\cos(x+a) = \cos x \cos a - \sin x \sin a$. We cannot do the same with $x^2 - 2x - 8 = 0$ because this is not an identity between two functions.

59. $s = t^2 + 5t$, $t = (u^2 + 2u)^{1/3}$. $\frac{ds}{du} = \frac{ds}{dt}\frac{dt}{du} = (2t+5)\frac{1}{3}(u^2+2u)^{-2/3}(2u+2)$. If $u = 2$, $t = 2$ and $\frac{ds}{du} = (9)\frac{1}{3}(8)^{-2/3}(6) = \frac{9}{2}$.

61. $y = \sqrt{x}$, $y' = 1/(2\sqrt{x})$. At $x = 4$, $y' = 1/4$ and the tangent has equation $y - 2 = (1/4)(x-4)$ which intersects the coordinates at $(0,1)$ and $(-4,0)$.

63. a) $x^2 + 2y^2 = 9$, $2x + 4yy' = 0$, $y' = -x/(2y)$. At $(1,2)$, $m = y' = -1/4$. Tangent: $y - 2 = (-1/4)(x-1)$ or $x + 4y = 9$. Normal: $y - 2 = 4(x-1)$ or $4x - y = 2$. b) $x^3 + y^2 = 2$, $3x^2 + 2yy' = 0$, $y' = -3x^2/(2y)$. At $(1,1)$, $m = y' = -3/2$. Tangent: $y - 1 = (-3/2)(x-1)$ or $3x + 2y = 5$. Normal: $y - 1 = (2/3)(x-1)$ or $3y - 2x = 1$. c) $xy + 2x - 5y = 2$, $xy' + y + 2 - 5y = 0$, $y'(x-5) = -(y+2)$, $y' = (y+2)/(5-x)$. At $(3,2)$, $m = y' = 2$. Tangent: $y - 2 = 2(x-3)$ or $y = 2x - 4$. Normal: $y - 2 = (-1/2)(x-3)$ or $x + 2y = 7$.

65. From the figure we see that if we find the tangent to the circle at $(12,-9)$ and then find its x-value when $y = -15 - 8 = -23$ we will then have half the desired width. $x^2+y^2 = 225$, $2x+2yy' = 0$, $y' = -x/y$. At $(12,-9)$, $y' = 4/3$ and the tangent has equation $y+9 = (4/3)(x-12)$. When $y = -23$, $x = 3/2$ and so the width is $2(3/2) = 3$ ft.

67. a) $x^3 + y^3 = 1$. $3x^2 + 3y^2y' = 0$, $y' = -x^2/y^2$. $y'' = -\frac{y^2(2x) - x^2(2yy')}{y^4} = -\frac{2xy^2 - 2x^2y(-x^2/y^2)}{y^4} = -\frac{2xy^3 + 2x^4}{y^5} = \frac{-2x(x^3+y^3)}{y^5} = -\frac{2x}{y^5}$.

b) $y^2 = 1 - \frac{2}{x}$. $2yy' = \frac{2}{x^2}$, $y' = \frac{1}{x^2y}$. $y'' = \frac{0-(x^2y'+2xy)}{x^4y^2} = \frac{-((1/y)+2xy)}{x^4y^2} = \frac{-(1+2xy^2)}{x^4y^3} = -\frac{(1+2x(1-2/x))}{x^4y^3} = \frac{3-2x}{x^4y^3}$.

69. a) $y = \sqrt{2x+7}$. $y' = \frac{1}{2\sqrt{2x+7}}(2) = \frac{1}{\sqrt{2x+7}}$. $y'' = \frac{0-1/\sqrt{2x+7}}{2x+7} = -\frac{1}{(2x+7)^{3/2}}$.
b) $x^2 + y^2 = 1$. $2x + 2yy' = 0$, $y' = -x/y$. $y'' = -\frac{y - xy'}{y^2} = -\frac{y + x^2/y}{y^2} = -\frac{x^2+y^2}{y^3} = \frac{-1}{y^3}$.

71. a) Let $f(x) = \tan x$. $f'(x) = \sec^2 x$. $L(x) = f(a) + f'(a)(x-a) = \tan(-\pi/4) + \sec^2(-\pi/4)(x + \pi/4) = -1 + 2(x + \pi/4) = 2x + (\pi/2) - 1$. Graph $y = \tan x$ and $y = 2x + (\pi/2) - 1$ in the viewing rectangle $[-5,5]$ by $[-8,8]$.

b) Let $f(x) = \sec x$. $f'(x) = \sec x \tan x$. $L(x) = f(a) + f'(a)(x - a) = \sec(-\pi/4) + \sec(-\pi/4)\tan(-\pi/4)(x + \pi/4) = \sqrt{2} - \sqrt{2}(x + \pi/4)$. Graph $y = \sec x$ and $y = \sqrt{2} - \sqrt{2}(x + \pi/4)$ in the viewing rectangle $[-8, 8]$ by $[-10, 10]$.

73. $f(x) = \sqrt{1+x} + \sin x - 0.5$. a) $f(-\pi/4) = \sqrt{1 - \pi/4} - \frac{\sqrt{2}}{2} - \frac{1}{2} < 0$. $f(0) = \sqrt{1} + 0 - 0.5 = 0.5 > 0$. b) (The linearization of a sum is the sum of the linearizations.) $f'(x) = \frac{1}{2\sqrt{1+x}} + \cos x$. $L(x) = f(0) + f'(0)(x - 0) = 0.5 + 1.5x$. We solve $0.5 + 1.5x = 0$ and obtain $x = -1/3$. c) $f(-1/3) \approx -0.01$. d) By zoom-in, $x = -0.326$. We cannot find the exact solution with the tools at hand.

75. $V = \frac{1}{3}\pi r^2 h$. $dV = \frac{2}{3}\pi r h dr$. An estimate of the change in question is $dV = \frac{2}{3}\pi r_0 h dr$.

77. a) Surface area $= S = 6x^2$. An estimate of the error in measuring S is $dS = 12x dx$. We require $\frac{dS}{S} \leqq 0.02$, $\frac{12x dx}{6x^2} = \frac{2}{x}dx \leqq 0.02$ or $\frac{dx}{x} \leqq 0.01$. Thus the edge should be measured with an error of no more than 1%.

b) $V = x^3$. $dV = 3x^2 dx$. $\frac{dV}{V} = \frac{3x^2 dx}{x^3} = 3\frac{dx}{x} \leqq 0.03$. Therefore the volume can be measured with an estimated error of no more than 3%.

79. $1° = 1°\frac{\pi}{180°}$ radian $= \frac{\pi}{180}$ radian is the possible error in measuring θ. $h = 100\tan\theta$, $dh = 100\sec^2\theta d\theta$. An estimate of the possible error in calculating h is $dh = 100\sec^2(\pi/6)(\pi/180) = (5/9)\pi(4/3) = (20/27)\pi \approx 2.33$ ft.

81.

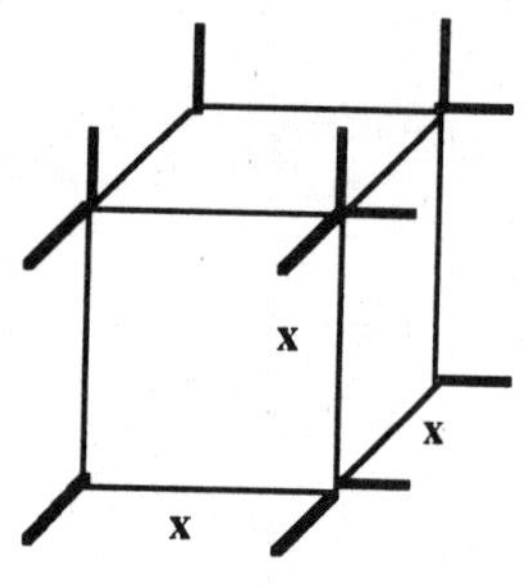

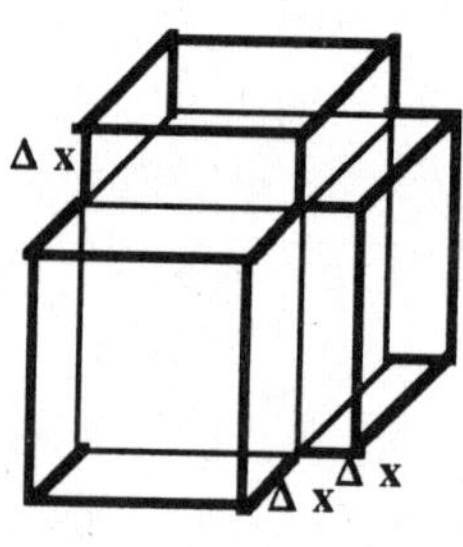

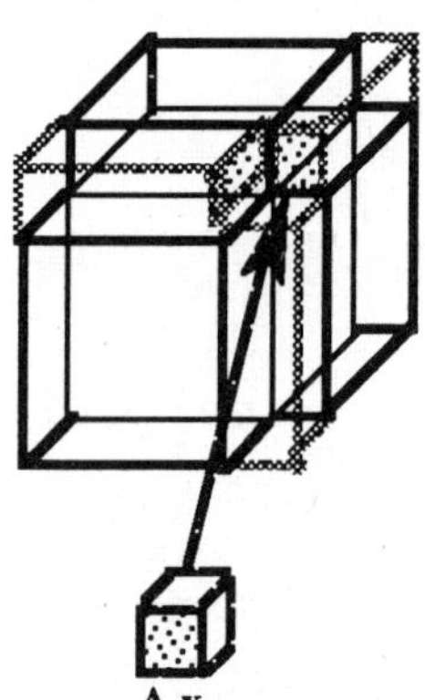

CHAPTER 4

APPLICATIONS OF DERIVATIVES

4.1 MAXIMA, MINIMA, AND THE MEAN VALUE THEOREM

1. f' is defined for all x; $f' = 3x^2 + 2x - 8 = (3x-4)(x+2)$. The critical points are at $x = 4/3$ and $x = -2$.

3. $F(x) = x^{1/3} \Rightarrow F'(x) = \frac{1}{3}x^{-2/3}$; F' does not exist at $x = 0$.

5. g' is not defined at $x = -1$ and $x = 3$. $g' = 0$ at $x = 1$. The critical points are $-1, 1, 3$.

7. h' is not defined at $x = 0$; where defined, $h' \neq 0$. The only critical point of h is at $x = 0$.

9. a) Let F denote the zeros of y, D a zero of its derivative. The screen ranges and orders of zeros are

i $[-6, 6]$ by $[-6, 25]$

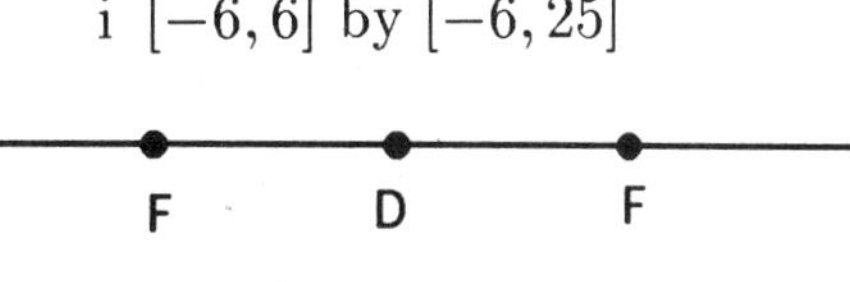

ii $[-6, 6]$ by $[-6, 6]$

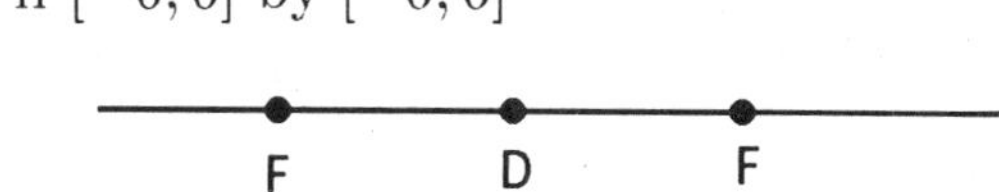

iii $[-2, 40]$ by $[-500, 500]$

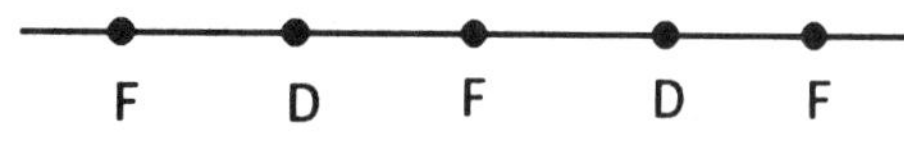

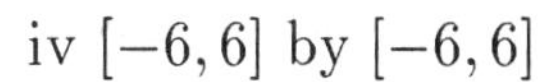
iv $[-6, 6]$ by $[-6, 6]$

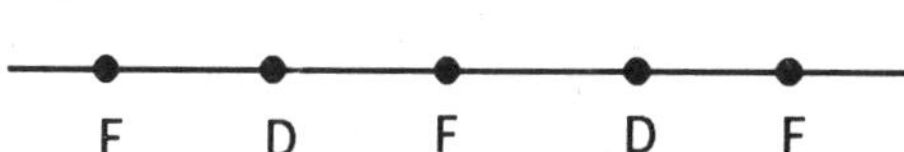

Between consecutive F's there is a D

b) Let $f(x) = x^n + a_{n-1}x^{n-1} \cdots + a_0$; $f'(x) = nx^{n-1} + \cdots + a_1$. Let x_1, x_2 be consecutive zeros of f. By Rolle's Theorem, there exists c, $x_1 < c < x_2$ such that $f'(c) = 0$.

11. (b) Let $f(x) = \sin x$. Between every two zeros of $f(x)$ there is a zero of $f'(x) = \cos x$.

13. a) Use the window $[-5, 5]$ by $[-1, 15]$

b) $f'(0)$ does not exist; the graph does not have a horizontal tangent at $(0, 0)$

c) $f'(3)$ does not exist; the graph does not have a horizontal tangent at $(3, 0)$

d) f has a local minimum at each of $(-3,0),(0,0)$ and $(3,0)$. f has a local maximum at $(\pm1.73, 10.39)$

e) f' is defined at ±1.73; the tangent line at each local maximum is horizontal f' is not defined at $\pm3, 0$

f) f is increasing on $[-3,-1.73] \cup [0,1.73] \cup [3,\infty)$. f is decreasing on $(-\infty,-3] \cup [-1.73,0] \cup [1.73,3]$

15. $f(0) = 0$, $f(6) = -6$. $f'(x) = 5 - 2x$; set $f'(x) = 0$ to find the critical point $x = 5/2$. $f(5/2) = 6.25$. f has a local maximum at $(5/2, 6.25)$. f is increasing on $[0, 5/2]$; f is decreasing on $[5/2, 6]$. The absolute minimum of f on $[0,6]$ is at $(6,-6)$. The absolute maximum of f on $[0,6]$ is at $(5/2, 6.25)$. f has a local minimum at $(0,0)$.

17. $f(4) = 0$, $f(8) = 2$; $f'(x) = \frac{1}{2\sqrt{x-4}}$, not defined at $x = 4$. Since $f' \neq 0$ between 4 and 8 there are no interior local extrema. The absolute minimum occurs at $(4,0)$; the absolute maximum occurs at $(8,2)$; f is increasing on $[4,8]$.

19. $y(-3) = 0$; $y(3) = 0$. $y'(x) = 4x^3 - 20x = 4x(x^2 - 5)$. Critical points are $x = 0, \pm\sqrt{5}$: $y(0) = 9$, $y(\sqrt{5}) = y(-\sqrt{5}) = -16$. There are local minima at $(\pm\sqrt{5}, -16)$; local maximum at $(0,9)$. Absolute minima at $(\pm\sqrt{5}, -16)$; absolute maximum at $(0,9)$. $(-3,0)$ and $(3,0)$ are local maxima. f is increasing on $(-\sqrt{5}, 0) \cup (\sqrt{5}, 3)$. f is decreasing on $(-3, -\sqrt{5}) \cup (0, \sqrt{5})$.

21. Graph h and NDER on $[-5,5]$ by $[-5,5]$. h has absolute maximum at $x = 5$, absolute minimum at $y = -5$; local minimum at $(0.559, -2.639)$, local maximum at $(-1.126, -0.362)$. Increasing on $[-5,-1.126] \cup [0.559, 5]$; decreasing on $[-1.126, 0.559]$.

23. $f'(x) = 2x + 2$; $\frac{f(1)-f(0)}{1-0} = \frac{2-(-1)}{1} = 3$. By the M.V.T., $2c + 2 = 3$. $c = \frac{1}{2}$.

25. $f'(x) = 1 - \frac{1}{x^2}$; $\frac{f(2)-f(1/2)}{2-1/2} = \frac{2.5-2.5}{1.5} = 0$. By the M.V.T., $1 - \frac{1}{c^2} = 0$. $c = \pm1$. The value of c in $(\frac{1}{2}, 2)$ is $c = 1$.

27. Let $s(t)$ = distance traveled in t hours. $s'(t) = v(t)$, the velocity. By the M.V.T., there exists c such that $v(c) = \frac{s(2)-s(0)}{2-0} = \frac{159-0}{2} = 79.5$. At time $t = c$, the trucker was going 79.5mph.

29. Let $s(t)$ = sea miles covered in t hours. By the M.V.T. there exists c, $0 < c < 24$ such that $s'(c) = v(c) = \frac{s(24)-s(0)}{24-0} = \frac{184-0}{24-0} = 7.66$ sea miles/hr $= 7.66$ knot.

31. $[-2, 9]$ by $[-7, 3]$

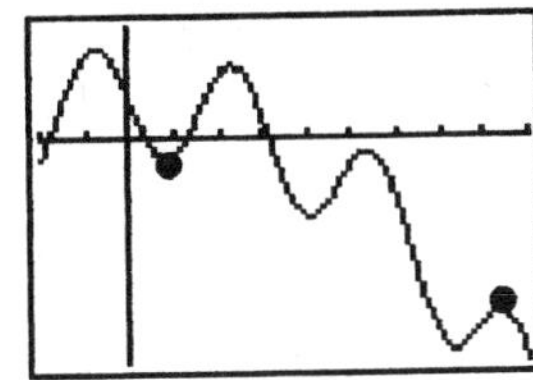

33. $f(-4) = f(4) = 4\sin 4$; $f'(x) = \sin x + x\cos x$. $f'(0) = 0$, $f'(\pm 2.029) = 0$.

35. $f = x^2 - 2x$; $f(-2) = f(4) = 8$; $f' = 2x - 2 = 2(x-1)$; $f'(1) = 0$

37. $y' = \frac{-1}{x^2} < 0$ wherever it is defined. By Corollary 1 to the M.V.T., y is a decreasing function of x on any interval on which it is defined. NOTE: $y(-1) = -1$, $y(1) = 1$, but y is *not* defined on $[-1, 1]$, so this is not a contradiction.

39. If $f(0) = f(1)$, there would exist c such that $f'(c) = \frac{f(1)-f(0)}{1-0} = 0$, which is precluded.

41. $\frac{f(b)-f(a)}{b-a} < 0$. By the M.V.T., there exists c such that $f'(c) = (f(b) - f(a))/(b-a)$.

43. $f(-2) = 11$; $f(-1) = 1$. On $(-2, -1)$, $f'(x) = 4x^3 + 3$. The only zero of $f'(x)$ is at $-(.75)^{1/3} \approx -0.91$. Hence, $f'(x) \neq 0$ on $(-2, 1)$.

45. $f(1) = -1$, $f(3) = \frac{7}{3}$. $f'(x) = 1 + \frac{2}{x^2}$, which is never zero.

47. Let $x = b$, where b is an arbitrary real number, $b \neq 0$. $\frac{f(b)-f(0)}{b-0} = \frac{f(b)-3}{b-0} = f'(c) = 0$. Thus $f(b) - 3 = 0$, or $f(b) = 3$. Since b was an arbitrary real number, $f(x) = 3$ for all x.

49. a) $f' > 0$ where f increases: $(2, 5)$; b) $f' < 0$ where f decreases: $(-2, 2)$; $f'(2) = 0$.

51. Let $f(x) = \sqrt{(x-a)(b-x)}$; at $x = a$ and $x = b$ the tangent lines are vertical.

53.

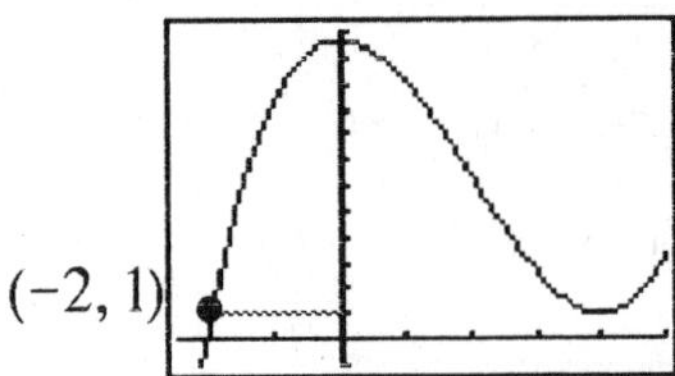

55. Graphing $y = xe^{-x}$ on $[0, 3]$ by $[-.5, .5]$. a) If $a = 1.000$, the rectangle has width 0. b) For each value of $\underline{a}$, use trace to find $\underline{b}$ such that $f(b) = f(a)$.

a	$f(a)$	b	area
0.5	0.303	1.762	0.382
0.8	0.359	1.238	0.157
1.0	0.368	1.0	0
1.2	0.361	0.817	0.138
1.5	0.335	0.627	0.292

c) When $a = .25$, area $\approx .455$, when $a = .30$, area $= .459$, when $a = .35$, area $\approx .450$. The maximum appears to occur near $x = 0.3$, area ≈ 0.459.

4.2 ANALYZING HIDDEN BEHAVIOR

1. $f' > 0$ on $(-\infty, -1)$ and $(1, \infty)$

$f' < 0$ on $(-1, 1)$

3. $f' > 0$, so graph of f is rising on $(-\infty, 2) \cup (0, 2)$; f is falling on $(-2, 0) \cup (2, \infty)$ local maxima at $x = -2, x = 2$; local minimum at $x = 0$.

5. $y' = 2x - 1$; $y'' = 2$; $y' = 0$ at $x = 1/2$; $y'' \neq 0$. The graph is falling for $x < 1/2$, rising for $x > 1/2$. The graph is always concave up. There is a local minimum at $(1/2, -5/4)$.

7. $y' = 3x^2 - 12x + 9 = 3(x-3)(x-1)$; $y'' = 6x - 12$

x	y	y'	y''	concavity	ext/inf	
$(-\infty, 1)$		pos	neg	down		rising
$x = 1$	5	0	neg	down	max	
$(1, 2)$		neg	neg	down		falling
$x = 2$	3	neg	0	–	inf	falling
$(2, 3)$		neg	pos	up		falling
$x = 3$	1	0	pos	up	min	
$(3, \infty)$		pos	pos	up		rising

9. $y' = 8x^3 - 8x = 8x(x^2-1) = 8x(x-1)(x+1)$
$y'' = 24x^2 - 8 = 8(3x^2-1) = 24(x^2 - 1/3) = 24(x - \frac{1}{\sqrt{3}})(x + \frac{1}{\sqrt{3}})$

x	y	y'	y''	concavity	ext/inf	
$(-\infty, -1)$		–	+	up		falling
$x = -1$	-1	0	+	up	min	—
$(-1, -1/\sqrt{3})$		+	+	up		rising
$x = -1/\sqrt{3}$	-1/9	+	0		inf	rising
$(-1/\sqrt{3}, 0)$		+	–	down		rising
$x = 0$	1	0	–	down	max	—
$(0, 1/\sqrt{3})$		–	–	down		falling
$x = 1/\sqrt{3}$	-1/9	–	0		inf	falling
$(1/\sqrt{3}, 1)$		–	+	up		falling
$x = 1$	-1	0	+	up	min	—
$(1, \infty)$		+	+	up		rising

11. The viewing rectangle $[-3, 3]$ by $[-5, 15]$ shows the global behavior. Zoom in to determine the behavior of the function at the "bend". ($[0, 2]$ by $[10.5, 11.3]$ indicates the falling behavior.) $y' = 6x^2 - 10x + 4 = 2(3x^2 - 5x + 2) = 2(x-1)(3x-2)$. $y'' = 12x - 10 = 2(6x-5)$. The graph is concave down on $(-\infty, 5/6)$, concave up on $(5/6, \infty)$. $(5/6, 11.019)$ is an inflection point. By the second derivative test, there is a local maximum at $(1, 11)$ and a local minimum at $(1.5, 11.5)$. Falling on $[2/3, 1]$, rising elsewhere.

13. The viewing rectangle $[-3, 3]$ by $[-12, 4]$ indicates the behavior. $y' = 12x^3 - 2x = 2x(6x^2-1) = 12x(x - 1/\sqrt{6})(x + 1/\sqrt{6})$. $y'' = 36x^2 - 2 = 36(x - 1/\sqrt{18})(x + 1/\sqrt{18})$. There are inflection points at $(\pm 1/\sqrt{18}, -10.046))$; local minima at $(\pm 1/\sqrt{6}, -121/12)$ and a local maximum at $(0, -10)$.

15. The viewing rectangle $[0, 2\pi]$ by $[0, 7]$ indicates the behavior of the function. $y' = 1 + \cos x$; $y' = 0$ at (π, π). For $x \neq \pi$, $y'(x) > 0$ and the graph is rising. $y'' = -\sin x$; $y''(\pi) = 0$, $y''(\pi^-) < 0$, $y''(\pi^+) > 0$, so (π, π) is an inflection point $y(0) = 0$, $y(2\pi) = 2\pi$. Always rising; minimum at $(0, 0)$, maximum at $(2\pi, 2\pi)$. Concave down for $0 < x < \pi$, concave up for $\pi < x < 2\pi$.

17. The viewing rectangle $[-4, 4]$ by $[-25, 25]$ shows the complete graph. To find the local extrema, graph $y' = 4x^3 - 16x + 4 = 4(x^3 - 4x + 1)$. $y'' = 4(3x^2 - 4) = 12(x - \frac{2}{\sqrt{3}})(x + \frac{2}{\sqrt{3}})$. The zeros of y' are, approximately $x = -2.13, 0.24, 1.88$. Using technology to evaluate y:

x	y	
-2.13	-22.23	local minimum
$-2/\sqrt{3}$	-11.50	inflection point
0.24	2.50	local maximum
$2/\sqrt{3}$	-2.27	inflection point
1.88	-6.26	local minimum

19. $y' = -4x^3 + 4x - 3$; $y'' = -12x^2 + 4 = -4(3x^2 - 1)$. The only zero of y' is, approximately -1.263. Using technology, we find.

x	y		
-1.263	2.4350	local maximum	
			concave down
$-\sqrt{1/3}$	0.2876	inflection point	
			concave up
$\sqrt{1/3}$	-3.1760	inflection point	
			concave down

The viewing rectangle $[-3, 3]$ by $[-6, 6]$ shows a complete graph.

21. Graph on $[-4, 4]$ by $[-6, 6]$. Graph is always rising; inflection point at $(0, 3)$; concave up for $x < 0$, concave down for $x > 0$.

23. Defined for $x \geq 0$; always rising and concave down; local minimum at $(0, -1)$.

25. The viewing rectangle $[-3, 3]$ by $[0, 3]$ shows a complete graph. The function is always increasing, always concave up.

27. Graph on $[-4, 4]$ by $[-6, 6]$. $f(0) = 1$ is a local minimum. By definition $(0, 1)$ is also an inflection point.

29. Graph $y = (x\hat{}(1/3))^2(3-x)$ on $[-3,4]$ by $[-6,15]$. Local minimum and inflection point at $(0,0)$; local maximum at $(1.200, 2.033)$, concave up for $x<0$, concave down for $x>0$.

31. Graphing this does not show the concavity: $y' = \frac{4}{3}x^{1/3}(1-\frac{1}{x})$ and $y'' = \frac{4}{9}x^{-2/3}[1+\frac{2}{x}]$ tell us that $x=0$ and $x=-2$ are inflection points; local minimum at $(1,-3)$; concave up for $x>0$ and $x<-2$, concave down for $-2<x<0$.

33. $s(t) = t^2 - 4t + 3 = (t-3)(t-1)$, $\nu(t) = 2t - 4 = 2(t-2)$, $a(t) = 2$.

time	$a(t)$	$\nu(t)$	speed $= \lvert\nu\rvert$	direction
$t<1$	pos	neg, inc	slowing	to the left
$1<t<2$	pos	neg, inc	slowing	to the left
$t=2$	pos	0		stopped
$2<t<3$	pos	pos, inc	gaining	to the right
$3<t$	pos	pos, inc	gaining	to the right

35. $s = t^3 - 3t + 3$, $\nu = 3t^2 - 3 = 3(t-1)(t+1)$, $a = 6t$.

time	$a(t)$	$\nu(t)$	speed	direction
$t<-2.104$	neg	pos, dec	dec	to the right
$-2.104<t<-1$	neg	pos, dec	dec	to the right
$t=-1$	neg	0	0	stopped
$-1<t<0$	neg	neg, dec	inc	to the left
$0<t<1$	pos	neg, inc	dec	to the left
$1<t$	pos	pos	inc	to the right

37. $y' = 0$ at $1, 2$;

$$\begin{aligned} y'' &= 2(x-1)(x-2) + (x-1)^2 \\ &= (x-1)[2x-4+x-1] \\ &= (x-1)[3x-5] \end{aligned}$$

x	y'
1^-	neg
1	0
1^+	neg
2^-	neg
2	0
2^+	pos

x	y''
1^-	pos
1	0
1^+	neg
$5/3^-$	neg
$5/3$	0
$5/3^+$	pos

y has a local minimum at $x=2$; y has inflection points at $x = 1,\ 5/3$.

39. $y' = 1 - \frac{1}{x^2}$; $y'' = \frac{2}{x^3}$; $y'(1) = 0$, $y''(1) > 0$; $y'(-1) = 0$; $y''(-1) < 0$. y has a local maximum at $(-1, -2)$ and a local minimum at $(1, 2)$.

41. No, f might have an inflection point at c. Example: $f(x) = x^3 + 1$ has $f' = 0$ at 0, but $x = 0$ is not a local extremum.

43. $y'' = 2a$ which is never 0. The statement is true.

45. a) The velocity is zero when the tangent to $y = s(t)$ is horizontal: at $t = 2$, $t = 6$, $t = 9.5$.

b) The acceleration is zero when the graph has an inflection point: at $t = 4$, $t = 8$, $t = 11.5$.

47.

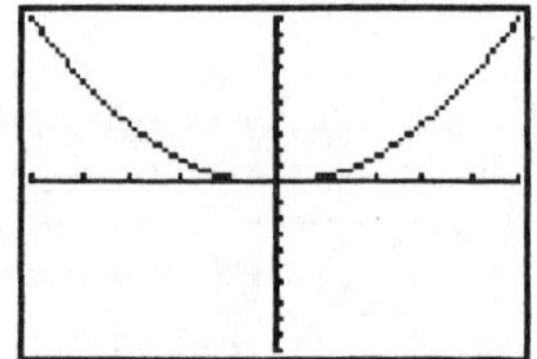

49.

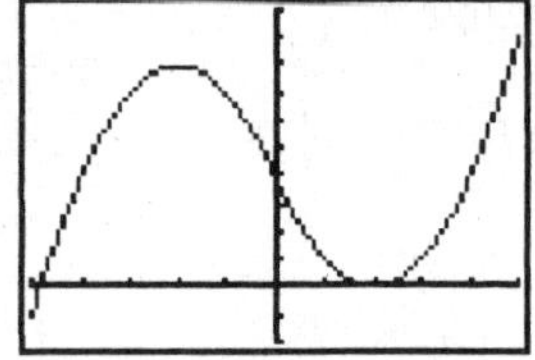

51. Look at both graphs in the viewing rectangle $[-1, 1]$ by $[-1, 1]$.

53. Look at both graphs in $[-0.5, 0.5]$ by $[-5, 5]$.

55. a) $f = x^3 - 9x \Rightarrow f' = 3x^2 - 9 \Rightarrow f'' = 6x$ which changes sign at $x = 0$.

b) $y - f(a) = (x-a)f'(a) \Rightarrow y = a^3 - 9a + (x-a)(3a^2-9) = (-2a^3) + x(3a^2-9)$.

c) Graph on $[-4, 4]$ by $[-12, 12]$; y_1 lies above y_2; $f''(1) > 0 \Rightarrow$ the graph is concave up at $x = 1$.

57. Graph in $[0, 300]$ by $[-20, 150]$ using $0 \le t \le 10$.

59. Setting range = maximum height $\Rightarrow \frac{V_0^2 \sin\alpha\cos\alpha}{16} = \frac{V_0^2 \sin^2\alpha}{64} \Rightarrow \cos\alpha = \frac{1}{4}\sin\alpha \Rightarrow \tan\alpha = 4 \Rightarrow \alpha = 75.964°$.

4.3 POLYNOMIAL FUNCTIONS, NEWTON'S METHOD, AND OPTIMIZATION

1. The procedure: $0 \to$ Ans, followed by repeating the iteration: Ans $-$ (Ans^2 + Ans $- 1) \div (2$ Ans $+ 1)$ leads to the sequence:

1
0.6666666667
0.619047619
0.6180344478
0.6180339888
0.6180339888

and we conclude 0.6180339888 is a solution. Starting with -5, gives the solution -1.618033989.

3. A graph in $[-10, 10]$ by $[-5, 5]$ indicates there are two solutions. The iteration for Newton's method is ANS $-$ (ANS^4 + ANS $- 3)/(4$ ANS$^3 + 1)$. A starting value of $x = 1$ leads to $x = 1.16403514$. A starting value of $x = -1$ leads to $x = -1.452626879$.

If $y1 = x^4 + x + 3$, $y2 = \text{NDER}(y1, x)$, the sequence: $1 \to x$ (enter), followed by $x - y1/y2 \to x$ (enter) (enter) (enter) ... will display the first set of iterates.

5. Graphing the function in $[-10, 10]$ by $[-10, 10]$ indicates there are two solutions. The iteration for Newton's method is $x - (x^4 - 2x^3 - x^2 - 2x + 2)/(4x^3 - 6x^2 - 2x - 2) \to x$. Starting with $x = 1$ leads to 0.6301153962. Starting with $x = 5$ leads to 2.5732719864.

7. A graph in $[-10,10]$ by $[-10,10]$ indicates there are two solutions. The iteration is: $x - (9 - \frac{3}{2}x^2 + x^3 - \frac{1}{4}x^4)/(-3x + 3x^2 - x^3) \rightarrow x$. If $x_0 = 1$, Newton's method converges to 3.216451347. If $x_0 = -1$, Newton's method converges to -1.564587289.

9. The iteration is $x_{n+1} = x_n - (x_n^4 - 2)/4x^3$ $x_0 = 1 \Rightarrow x_1 = 1.25$, $x_2 = 1.1935$.

11. If $f(x_0) = 0$, then $x_1 = x_0 - f(x_0)/f'(x_0) = x_0 - 0/f'(x_0)$. If $f'(x_0) \neq 0$ then x_1 and all subsequent approximations are equal to x_0. If $f'(x_0$, a calculator may show a "math error".

13.

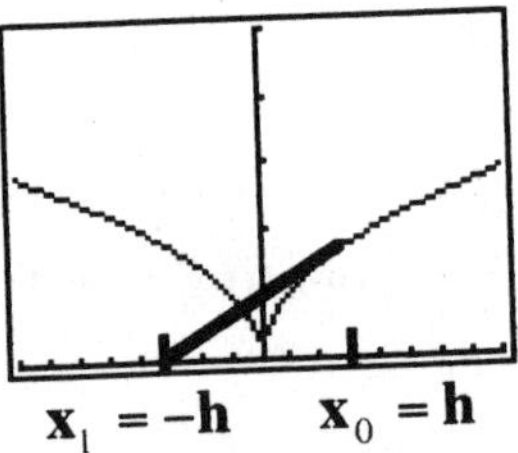

If $x_0 = h > 0$, $f(x) = \sqrt{x}$ and $x_1 = x_0 - \frac{\sqrt{x_0}}{\frac{1}{2\sqrt{x_0}}} = x_0 - 2x_0 = -x_0$. If $x_0 = -h < 0$, $f(x) = \sqrt{-x}$ and $x_1 = x_0 - \frac{\sqrt{-x_0}}{\frac{-1}{2\sqrt{-x_0}}} = x_0 + 2\sqrt{x_0^2} = x_0 + 2|x_0| = -h + 2h = h$.

15. $y' = 3x^2 - 6x + 5$ which is always > 0. The graph is always rising. $y'' = 6x - 6 = 6(x-1)$; the graph is concave down for $x < 1$, concave up for $x > 1$, has an inflection point at $(1, -1)$. The viewing rectangle $[-5,5]$ by $[-10,10]$ shows a complete graph. There is one real root.

17. $y' = 3x^2 - 4x - 3$; $y' = 0$ at $(2 \pm \sqrt{13})/3$, $y'' = 6x - 4$; $y'' < 0$ for $x < 2/3$, $y'' > 0$ for $x > 2/3$; local maximum at $(\frac{2-\sqrt{13}}{3}, 8.879)$; local minimum at $(\frac{2+\sqrt{13}}{3}, 1.9354)$; inflection point at $(\frac{2}{3}, 5.407)$. A viewing rectangle of $[-5,5]$ by $[-10,10]$ shows a complete graph. There is one real root. Rising on $(-\infty, \frac{2-\sqrt{13}}{3}] \cup [\frac{2+\sqrt{13}}{3}, \infty)$; falling on $[\frac{2-\sqrt{13}}{3}, \frac{2+\sqrt{13}}{3}]$; concave down for $x < \frac{2}{3}$, concave up for $x > \frac{2}{3}$.

19. $y' = 1 - 12x^2$; $y' = 0$ at $\pm\frac{1}{\sqrt{12}}$, $y'' = -24x$. Local minimum at $(-1/\sqrt{12}, 11.808)$: local maximum at $(1/\sqrt{12}, 12.192)$ inflection point at $(0, 12)$. The viewing rectangle $[-5,5]$ by $[-10,10]$ shows a complete graph. There is one real root.

21. The viewing rectangle $[-10,10]$ by $[-60,60]$ appears to show a complete graph. Analytically, $y' = -3 - x^2 < 0$. There are no local extremes, the graph is always falling. $y'' = -2x \Rightarrow (0,20)$ is an inflection point. The graph was complete. Concave up for $x < 0$, concave down for $x > 0$. One real root.

23. The viewing rectangle $[-5,5]$ by $[0,20]$ appears to show a complete graph. To be sure that a slight "wiggle" hasn't been missed, look at $y' = 4x^3+2x+1$ and $y'' = 12x^2 + 2 > 0$. The graph is always concave up. There is a local minimum at $(-0.383, 7.785)$. The function is never zero.

25. The rectangle $[-1,6]$ by $[-30,10]$ gives a complete graph. To find the local extremes and inflection points, graph the derivative $y' = 12x^2 - 34x + 8$. The zeros of y' are found, by zooming to be 0.258 and 2.574. $y'' = 24x - 34$, which is zero at $x = 34/24 = 1.417$ y has a local maximum at $(0.258, -0.001)$, a local minimum at $(2.574, -24.825)$ and an inflection point at $(1.417, -12.412)$. There is only one zero.

27. The rectangle $[-5,5]$ by $[-17,17]$ suggests a complete graph, but the behavior of the function between $x = -1$ and $x = 1$ is not clear. A graph of the derivative, $y' = 8x^2 - 4x^3 = 4x^2(2-x)$, shows that $y' \geq 0$ on $[-1,1]$, i.e., there is no hidden local maximum. $(2, 15.33)$ is an local maximum. Tracing along the graph of y' shows that $y'' = 0$ (y' has a local maximum) at $x = 0$ and $x = 1.33$, Hence y has inflection points at $(0,10)$ and $(1.33, 13.14)$.

29. Use $[-3,3]$ by $[0,34]$. There appear to be two local minima and a local maximum. Graph the derivative to find its zeros; y has local minima $x = -1.106$ and $x = 0.835$ and a local maximum at $x = 0.270$. Graphing y'' to find its zeros (use $[-1,1]$ by $[-1,1]$) shows that y has inflection points at $x = \pm 0.578$. Evaluating the function gives: falling to a local minimum at $(-1.106, 17.944)$; rising to a local maximum at $(0.270, 20.130)$; falling to a local minimum at $(0.835, 19.927)$; rising thereafter. Concave down between inflection points $(-0.578, 18.865)$ and $(0.578, 20.021)$; concave up elsewhere.

31. Use $[-8,4]$ by $[-20,35]$; graph both y and its numerical derivative. Use Trace to find the local extremes of y. Use Trace on y' to determine the x-coordinates of its local extrema; these will be the x-coordinates of the inflection points of y. The graph is rising to a local maximum at $(-4,32)$; falling to a local minimum at $(-1.52, -0.18)$; rising to a local maximum at $(0.29, 13.10)$;

falling thereafter. Concave up between inflection points $(-2.86, 15.16)$ and $(-0.50, 7.03)$; concave down elsewhere.

33. Use $[-4, 4]$ by $[-10, 55]$. Falling to local minimum at $(-2.601, -7.580)$; rising to local maximum at $(-1.097, 21.196)$; falling to local minimum at $(0.534, -0.495)$; rising to $(2.364, 53.006)$, falling thereafter. Concavity: up until $(-2.016, 4.530)$, down until $(-0.266, 10.206)$, up until $(1.681, 31.029)$, down thereafter. Five real roots.

35. Use $[-2, 2]$ by $[-50, 50]$; graph y and y'. Since $y' > 0$, there are no local extremes. y' has a local minimum at $x = -0.2857$; hence y has an inflection point at $(-0.286, -2.887)$. The graph is always rising. For $x < -0.286$, the graph is concave down.

37. Let x be one number; $20 - x$ is the other. $S(x) = x^2 + (20 - x)^2$, $0 \leq x \leq 20$. $S'(x) = 2x + 2(20 - x)(-1) = 4x - 40$. $S''(x) = 4$. $S'(10 = 0$. The maximum value of $S(x)$ is the maximum of $S(10) = 20$, $S(0) = 400$ and $S(20) = 400$. The numbers are 0 and 20.

39.

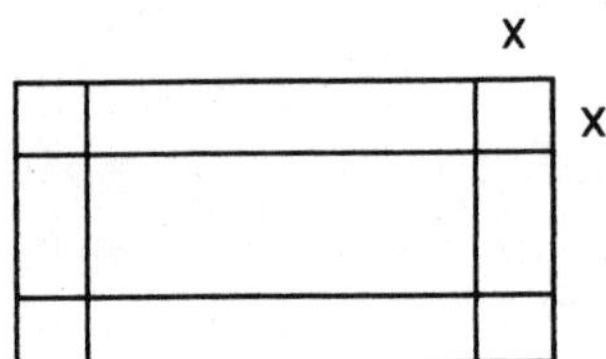

$V(x) = x(8 - 2x)(15 - 2x)$, $0 \leq x \leq 4$. $V'(x) = (8 - 2x)(15 - 2x) - 2x(15 - 2x) - 2x(8 - 2x) = 4x^2 - 46x + 120 + 4x^2 - 30x + 4x^2 - 16x = 12x^2 - 92x + 120$. Solving $V'(x) = 0$ gives $x = \frac{92 \pm 52}{24} = 6$ or $5/3$. 6 is not in the domain of V. Since $V(0) = V(4) = 0$, $V(5/3)$ is the maximum volume. The box has dimensions $5/3 \times 14/3 \times 35/3$.

41. a) Since OAB is an isosceles right triangle, B has coordinates $(0, 1)$ and the equation of the line AB is $y = -(x - 1) = 1 - x$

b) $A(x) = 2x(1 - x), \quad 0 \leq x \leq 1$

c) $A'(x) = 2(1 - x) + 2x(-1) = -4x + 2$. $A'(1/2) = 0$. Since $A(0) = 0$, and $A(1) = 0$, the largest area is $A(1/2) = 1/2$.

43. $v = -32t + 100$

a) $v(0) = 100\text{ft/sec}$.

b) $v = 0$ when $t = 100/32$ seconds; $s(100/32) = 170.122$ feet.

c) $s = 0$ when $t = (-100 - \sqrt{(100^2 + 64 \cdot 200)})/(-32) = 7.844$. $v(7.844) \approx -150.00$ feet/second.

45. Let one side of the square be x. The length of the box is $108 - 4x$ and its volume is $v(x) = x^2(108 - 4x)$, $0 \le x \le 27$. $v'(x) = 2x(108 - 4x) - 4x^2 = 4x(54 - 2x - x) = 12x(18 - x)$. Since $v(0) = 0$ and $v'(18) < 0$, v has a maximum at $x = 18$. The dimensions of the box are $18 \times 18 \times 36$ inches.

47. a) A circle with circumference x has area $\frac{x^2}{2\pi^2}$; in this problem, $2x+2y = 36$, or $y = 18-x$. The volume of the cylinder is $v = \frac{x^2 y}{2\pi^2} = x^2(18-x)/(2\pi^2)$, $0 \le x \le 18$. $v'(x) = \frac{1}{2\pi^2}[2x(18-x) - x^2] = \frac{x}{2\pi^2}[36 - 2x - x]$. Since $v(0) = v(18) = 0$, the maximum value of v occurs when $x = 12$ and $y = 6$. (The maximum value is 4.05cm^3).

b) From the diagram, $v = \pi x^2 y = \pi x^2(18 - x)$. As above, the maximum occurs when $x = 12$, $y = 6$; the maximum value is 251.33cm^3.

49. Exactly 1/4 of the volume of the buoy is submerged, i.e., $\frac{1}{4}(\frac{4}{3}\pi) = \frac{\pi}{3}$ is submerged. The volume of a spherical segment of height x and radius r is $v = \frac{\pi x}{6}(3r^2 + x^2)$. From the diagram, $r^2 + (1-x)^2 = 1^2$. To find x, solve the equation $v = \frac{\pi}{3}$. $\frac{\pi x}{6}(3 - 3(1-x)^2 + x^2) = \frac{\pi}{3}$. $x(3 + x^2 - 3 + 6x - 3x^2) = 2$ or $x^3 - 3x^2 + 1 = 0$. Using technology, the root between 0 and 1 is 0.653.

51. $f'(x) = 3x^2 - 2ax + b$

a) Since $f'(-1) = f'(3) = 0$, $f'(x) = c(x+1)(x-3) = cx^2 - 2cx - 3c$. Equating coefficients gives $c = 3$, $a = 3$, $b = -9$. Hence $f'(x) = 3x^2 - 6x - 9$; $f''(x) = 6x - 6$; $f''(-1) < 0$, hence f has a maximum at $x = -1$; $f''(3) > 0$, hence f has a minimum at 3.

b) $f'' = 6x - 2a$. Since $f''(1) = 0$, $a = 3$. $f'(x) = 3x^2 - 6x + b$. If $f'(4) = 0$, $b = 12$. $f''(12) > 0 \Rightarrow f(12)$ is a minimum.

53. $h = (y + 3)$, radius of cone $= x = \sqrt{9 - y^2}$, $v = \frac{1}{3}\pi(9 - y^2)(y + 3)$; $v' = \frac{\pi}{3}[(-2y)(y+3) + (9 - y^2)]$; $v' = 0 \Rightarrow -2y^2 - 6y + 9 - y^2 = 0$ or $3y^2 + 6y - 9 = 0 \Rightarrow y^2 + 2y - 3 = 0 \Rightarrow y = 1$; $x = \sqrt{8} = 2\sqrt{2}$, $v = \frac{1}{3}\pi(8)4 = 32\pi/3 =$ maximum volume.

55. Let F be the stiffness; $F(d) = kwd^2 = k(\sqrt{144-d^2})d^3$. $[F(d)]^2 = k^2d^6(144-d^2)$. Differentiating gives $2F(d)F'(d) = k^2[6d^5(144-d^2) - 2d^7]$. Setting $F'(d) = 0$ leads to $3(144-d^2) = d^2$. The dimensions of the stillest beam are $d = 6 \times w = 6\sqrt{3}$ inches.

57. To maximize $R(p)$, calculate $R'(p)$: $R'(p) = r(\frac{K-p}{K}) - \frac{rp}{K} = r(K-2p)/K$; $R'(p) = 0$ when $p = K/2$, one half the carrying capacity. $R''(p) = \frac{-2r}{K} < 0$ so rate is maximized at $p = K/2$.

59. $R = M^2(\frac{c}{2} - \frac{M}{3})$; $\frac{dR}{dM} = cM - M^2$, $\frac{d^2R}{dM^2} = c - 2M$; $\frac{d^3R}{dM^3} = -2$; $\frac{dR}{dM}$ is maximized if $\frac{d^2R}{dM^2} = 0$ and $\frac{d^3R}{dM^3} < 0$; this occurs at $M = c/2$.

61. The iteration is $x - \frac{(x-1)^{40}}{40(x-1)^{39}}$. The stopping value depends on the computer/calculator used, but is usually about 1.003.

63. a) $f(x) = 0$ can be written as $x^3 = 3x + 1$ and as $x^3 - 3x = 1$. $f(x) = g'(x)$, hence $f = 0$ when $g' = 0$.

b) Use $[-2, 2]$ by $[-1, 1]$.

c) Using technology, the roots are: $1.879, -0.347, -1.532$.

65. Use the iteration $x - (\tan x) \times (\cos x)^2 \to x$. Start with $x_0 = 3$. After two iterations $x = 3.14159$.

67. b) $v(x) = 2x(24-2x)(18-2x)$ c) $[-10, 10]$ by $[-1500, 1500]$ shows a complete graph; d) $0 < x < 9$, $[0, 9]$ by $[0, 1500]$ is a graph of the problem situation; e) $x = 3.394$; $v = 1309.955$; analytically $v' = 2(24-2x)(18-2x) - 4x(18-2x) - 4x(24-2x) = 24x^2 - 336x + 864 = 24(x^2 - 14x + 36)$. $v' = 0$ at $\frac{14-\sqrt{52}}{2} = 3.3944$; f) $v(x) = 1120$ becomes $x(12-x)(9-x) = 140$; since a factor of 10 is needed, try $x = 2$. Looking at the graph suggests $x = 5$ also works.

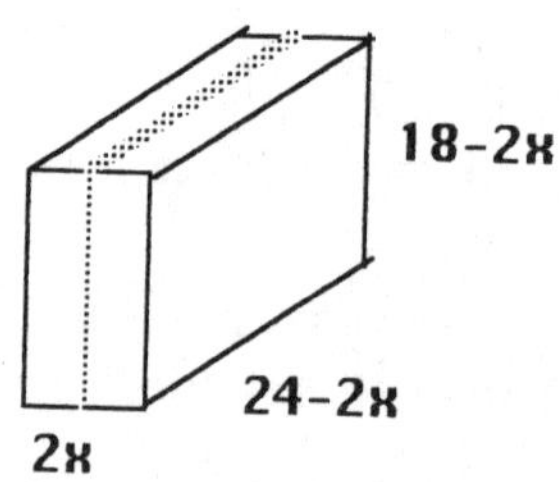

69. $v = x(15-2x)(5-x)$. Graph on $[0,5]$ by $[0,100]$; $v_{\max} = 66.019$ at $x = 1.962$.

71. f and g will have extrema at the same values of x-although their nature may be reversed.

73. $y' = 3ax^2 + 2bx + c$; at $x = 0$, $y' = c$. Since the path is tangential to the runway at the point of contact, $c = 0$.

75. $y(0) = 0 \Rightarrow d = 0$. #73 $\Rightarrow c = 0$, #74 $\Rightarrow 3aL - 2b = 0$ since $L \neq 0$, $y(-L) = H \Rightarrow -aL^3 + bL^2 = H$. Solving for a and $b \Rightarrow a = 2H/L^3$, $b = 3H/L^2$. Hence $y = 2H\frac{x^3}{L^3} + 3H\frac{x^2}{L^2}$.

4.4 RATIONAL FUNCTIONS AND ECONOMIC APPLICATIONS

1. Rewrite as $y = x - \frac{1}{x}$; then $y' = 1 + \frac{1}{x^2}$, $y'' = -\frac{2}{x^3}$. $y' > 0$ for $x \neq 0$, y'' changes sign at $x = 0$. The graph is always rising, vertical asymptote at $x = 0$, concave up for $x < 0$, concave down $x > 0$. The rectangle $[-5,5]$ by $[-5,5]$ shows a complete graph.

3. Rewrite as $y = x^2 + \frac{1}{x^2}$; $y' = 2x - \frac{2}{x^3}$, $y'' = 2 + \frac{6}{x^4}$. Concave up if $x \neq 0$. Local minima at $(\pm 1, 2)$. $[-5,5]$ by $[0,5]$ shows a complete graph. Always concave up. Rising on $[-1,0) \cup [1,\infty)$, falling on $(-\infty,-1] \cup (0,1]$.

5. $y = \frac{x}{x^2-4} \Rightarrow y' = \frac{x^2-4-2x^2}{(x^2-4)^2} = -\frac{(4+x^2)}{(x^2-4)^2} < 0$ when defined so the graph is always falling. $y'' = \left[\frac{-2x(x^2-4)^2+(4+x^2)2(x^2-4)\cdot 2x}{(x^2-4)^4}\right]$; clearly $y''(0) = 0$. $[-5,5] \times [-5,5]$ appears to be a complete graph. Concavity is: down $(-\infty,-2)$, up $(-2,0)$, down $(0,2)$, up $(2,\infty)$. Always falling. Concavity: down $(-\infty,-2)$, up $(-2,0)$, down $(0,2)$, up $(2,\infty)$.

7. $[-5,5]$ by $[-5,5]$ shows a complete graph of $y = \frac{1}{x^2-1}$; $y' = \frac{-2x}{(x^2-1)^2}$. Graph is rising for $x < 0$, falling for $x > 0$. Local maximum at $(0,-1)$. A graph of y' shows that y' has no local extremes; hence y has no inflection points.

9. Rewrite as $y = \frac{1}{x^2-1} - 1$. The graph will be the graph in #7, decreased by 1. Local maximum at $(0,-2)$.

11. Rewrite as $y = x+1-\frac{3}{x-1}$; $y' = 1+\frac{3}{(x-1)^2}$, $y'' = \frac{-6}{(x-1)^3}$, $[-10,10] \times [-10,10]$ shows a complete graph with end behavior $y = x+1$. No local extrema, no inflection points. Always rising; concave up for $x < 1$, concave down for $x > 1$.

13. $[-5, 4]$ by $[10, 10]$ gives a complete graph of both y and y'; local minimum at $(0.575, 0.144)$. The graph of y' has no local extrema, hence y has no inflection points.

15. Graph y and y' in $[-4, 4]$ by $[-0.5, 1.5]$. Find the zeros and local extrema of y' to locate the local extrema and inflection points of y. $(-2.414, -0.207)$ is a local minimum; $(-0.268, 0.683)$ is an inflection point; $(0.414, 1.207)$ is a local maximum; $(1, 1)$ is an inflection point. Rising on $[-2.414, 0.414]$, falling on $(-\infty, -2.414]$ and $[0.414, \infty)$; concave down on $(-0.268, 1)$, concave up elsewhere.

17. Graph y and y' in $[-4, 4]$ by $[-10, 10]$. $(-0.475, -3.331)$ local maximum, $(0.490, 0.800)$ local minimum. Rising on $(-\infty, -2) \cup (-2, -0.475] \cup [0.490, \infty)$; falling on $[-0.475, 0) \cup (0, 0.490]$. Concave up on $(-\infty, -2) \cup (0, \infty)$; concave down on $(-2, 0)$.

19. Two views are needed for a complete graph: $[-2, 6]$ by $[-10, 30]$ shows the end behavior, while $[-2, 2]$ by $[-2, 2]$ shows the local behavior of the left-hand branch. Local minima at $(0.243, -1.589)$ and $(2.543, 18.459)$. Local maximum at $(1.214, -0.869)$; inflection point at $(0.855, -1.158)$. Rising on $[0.243, 1.214] \cup [2.543, \infty)$; falling on $(-\infty, 0.243] \cup [1.214, 2) \cup (2, 2.543]$. Concave up on $(-\infty, 0.855) \cup (2, \infty)$, concave down on $(0.855, 2)$.

21. Graph both y and y' in $[-5, 5]$ by $[-15, 15]$. The graph of y' shows that y has two inflection points, at $(0, -0.5)$ and $(-3.005, 10.792)$, and a local minimum at $(1.666, 2.884)$. Falling on $(-\infty, -2) \cup (-2, 1) \cup (1, 1.666]$; rising on $[1.666, \infty)$; concave down on $(-3.005, -2)$ and $(0, 1)$, concave up on $(-\infty, -3.005)$, $(-2, 0)$ and $(1, \infty)$.

23. Three views are necessary to show the complete graph: $[-5, 5]$ by $[-100, 100]$ shows the end behavior, $[-3.2, -2.8]$ by $[-100, 100]$ and $[2.8, 3.2]$ by $[-100, 100]$ show a local maximum at $(2.919, 45.572)$ and a local minimum at $(3.077, 62.540)$. Inflection points at $(-3.257, -67.885)$, $(0.004, -0.222)$ and $(2.727, 39.274)$. Rising on $(-\infty, -3) \cup (-3, 2.919] \cup [3.077, \infty)$, falling on $[2.919, 3) \cup (3, 3.077)$; concave up for $-3.257 < x < -3$, $0.004 < x < 2.727$, and $x > 3$; concave down on $(-\infty, -3.257) \cup (-3, 0.004) \cup (2.919, 3)$.

25. Graph y and y' on $[-4, 4]$ by $[-1, 2]$. Local maximum at $(0, 2)$; inflection points at $(\pm 1.155, 1.5)$. Rising on $(-\infty, 0]$, falling on $[0, \infty)$; concave down for $|x| < 1.155$, concave up for $|x| > 1.55$.

27. $P = 2x + 2y$, where $xy = 16$. $P(x) = 2x + \frac{32}{x}$, $x > 0$; $P'(x) = 2 - \frac{32}{x^2}$: $P''(x) = \frac{64}{x^3}$. Set $P'(x) = 0$ to obtain $x = 4$. Since $P''(4) > 0$, $P(4)$ is a minimum. $P(4) = 16$.

29.

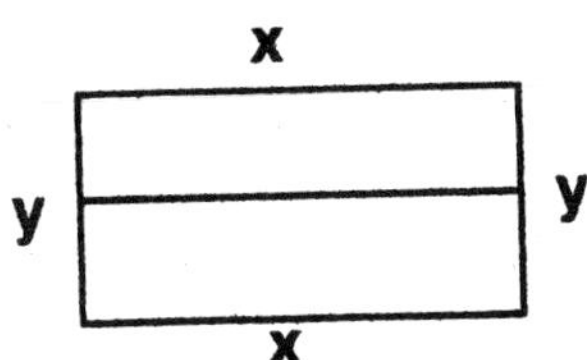

$F = 3x + 2y$ where $xy = 216$.

$F(x) = 3x + \frac{432}{x}$. $F'(x) = 3 - \frac{432}{x^2}$.

Set $F'(x) = 0$ to obtain $x^2 = 432/3$, i.e., $x = 12$. Since $F''(12) > 0$, $F(12)$ is a minimum. $y = 18$. $F(12) = 72$ m (the "equal parts" condition does not affect the answer).

31. From the problem, $x^2y = 1,125$. Thus $C = 5(x^2 + 4x \cdot \frac{1,125}{x^2}) + 10x \cdot \frac{1,125}{x^2} = 5(x^2 + \frac{5,400}{x}) + \frac{11,250}{x} = 5x^2 + \frac{33,750}{x}$. $C'(x) = 10x - \frac{33,750}{x^2}$. Set $C'(x) = 0$: $x^3 = 3,375 \Rightarrow x = 15$, $x = 15 \Rightarrow y = 5$.

33. For this can, $A = 2\pi rh + \pi r^2$ where $\pi r^2 h = 1000$, $A = 2\pi r(\frac{1000}{\pi r^2}) + \pi r^2 = \frac{2000}{r} + \pi r^2$, $\frac{dA}{dr} = \frac{-2000}{r^2} + 2\pi r$. Set $\frac{dA}{dr} = 0$, giving $2000 = 2\pi r^3$ or $r = \frac{10}{\sqrt[3]{\pi}}$, $h = \frac{1000}{\pi r^2} = \frac{1000}{100 \cdot \pi}\pi^{2/3} = \frac{10}{\sqrt[3]{\pi}} = r$.

35. Draw $\overline{RT}$. $\overline{PB} = 8.5 - x$, $\overline{QB}$ $= \sqrt{x^2 - (8.5-x)^2} = s$, $\overline{TQ} = \sqrt{L^2 - x^2} - s$. Use triangle RQT to obtain: $L^2 - x^2 = (8.5)^2$ $+(\sqrt{L^2-x^2} - s)^2 \Rightarrow L^2 = x^2 + \frac{17x^2}{4x-17} = \frac{4x^3}{4x-17}$.

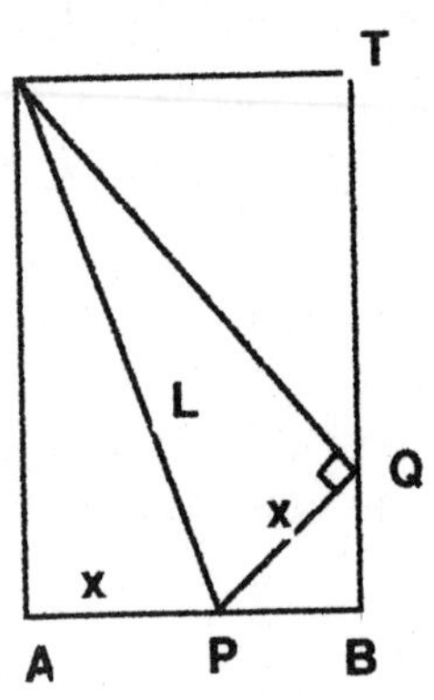

c. Differentiating gives $2LL' = \frac{12x^2(4x-17) - 4x^3(4)}{(4x-17)^2} =$ $\frac{4x^2[12x-51-4x]}{(4x-17)^2}$, $L' = \frac{1}{2L}\frac{4x^2}{(4x-17)^2}[8x-51]$.
$L' = 0$ at $x = 51/8$, $L'(51/8^-) < 0$, $L'(51/8^+) > 0$, so there is a local minimum of L at $51/8$.

d. $L(51/8) = 11.04$ inches.

37. $f'(x) = 2x - \frac{a}{x^2}$; $f''(x) = 2 + \frac{2a}{x^3}$.

a. For f to have a local minimum at $x = 2$, $f'(2) = 0$ and $f''(2) > 0$. Set $f'(2) = 0 \Rightarrow 0 = 4 - \frac{a}{4} \Rightarrow a = 16$. Check that $f''(16) > 0$.

b. Set $f''(1) = 0 \Rightarrow 0 = 2 + 2a \Rightarrow a = -1$. If $f''(x) = 2 - 2/x^3$ then $f''(1^-) < 0$, $f''(1^+) > 0$ and $x = 1$ is an inflection point.

39. Method 1 – using algebra:

$$\begin{aligned} 6x^2 + 8x + 19 &= 6(x^2 + \frac{8}{6}x + \frac{4}{9}) + 19 - \frac{4}{9}\cdot 6 \\ &= 6(x + \frac{2}{3})^2 + \frac{147}{9} > 0. \end{aligned}$$

Method 2 – using calculus: To show $y > 0$ for all x, find the minimum value of y and show that $y \geq \min y > 0$. $y' = 12x + 8$, $y'' = 12$, $y' = 0$ at $x = -2/3$; since $y'' > 0$, this is a minimum $y(-2/3) = \frac{147}{9} > 0$.

41. The profit is $P = Nx - Nc = (x-c)[\frac{a}{(x-c)} + b(100-x)] = a + b(x-c)(100-x)$. $P'(x) = b[100 - x - (x-c)] = b[100 + c - 2x]$; $P''(x) = -2b < 0$. P is a minimum when $x = 50 + c/2$.

43. $A'(q) = \frac{-km}{q^2} + \frac{h}{2} = \frac{-2km + hq^2}{2q^2}$; $A''(q) = \frac{2km}{q^3} > 0$, $A'(q) = 0$ when $q = \sqrt{(2km/h)}$. $A'' > 0 \Rightarrow$ this is a local minimum.

45. The profit is $p(x) = r(x) - c(x) = 6x - x^3 + 6x^2 - 15x$. $p'(x) = 6 - 3x^2 + 12x - 15 = -3(x^2 - 4x + 3)$. $p''(x) = -6x + 12$. The critical points of p are $x = 1$ and $x = 3$. At $x = 1$, since $p''(1) > 0$, p has a local minimum.

$p''(3) < 0$; the maximum value of p is $p(3) = 0$. p is never positive, i.e., there is never a profit.

47. If $f(x) = \frac{p(x)}{q(x)} = \frac{p}{q}$, then $f' = \frac{p'q-q'p}{q^2}$. The zeros of $q^2(x)$ are the zeros of $q(x)$. If $q(x) \neq 0$, x is in the domain of f; if $q(\bar{x}) = 0$, then $x = \bar{x}$ is a vertical asymptote of f. Moreover, if $q(x) \neq 0$, x is in the domain of f'. The tricky part is to show that $q(\bar{x}) = 0 \Rightarrow \bar{x}$ not in the domain of f' (i.e., that the numerator does not cancel the denominator). This requires writing $q(x) = (x-\bar{x})^m r(x)$, where $r(\bar{x}) \neq 0$. Then

$$f'(x) = \frac{p'(x-\bar{x})^m r - p(m(x-\bar{x})^{m-1} r + (x-\bar{x})^m r')}{(x-\bar{x})^{2m} r^2}$$

Divide by $(x-\bar{x})^{m-1}$,

$$f'(x) = \frac{(p'r - pr')(x-\bar{x}) - pmr}{(x-\bar{x})^{m+1} r^2}$$

at $x = \bar{x}$, the denominator is zero but the numerator is $p(\bar{x})mr(\bar{x}) \neq 0$. Hence a rational function and its derivative have the same domain and same vertical asymptotes. Since f'' is the derivative of f', the assertion extends to f''.

49. Let the specified volume be V; let the cylinder have radius r and height h. Then $V = \pi r^2 h$; the material M is given by $M = 2\pi rh + 2\pi r^2 = \frac{2\pi rV}{\pi r^2} + 2\pi r^2 = 2[\frac{V}{r} + \pi r^2]$. $M' = 0$ when $\frac{V}{r^2} = 2\pi r$, i.e., $2r = \frac{V}{\pi r^2} = h$. The height will equal the diameter (twice the radius).

4.5 RADICAL AND TRANSCENDENTAL FUNCTIONS

1. $y = x^{1/3} \Rightarrow y' = \frac{1}{3}x^{-2/3}$ and $y'' = \frac{-2}{9}x^{-5/3}$. The derivative is not defined at $x = 0$; elsewhere $y' > 0$ and y is increasing. For $x < 0$, $y'' > 0$ and y is concave up, for $x > 0$, $y'' < 0$ and y is concave down. $[-5,5]$ by $[-5,5]$ confirms this analysis.

3. $y = x^{3/2}$, $y' = \frac{3}{2}x^{1/2}$, $y'' = \frac{3}{4}x^{-1/2}$; y defined for $x \geq 0$, $y' > 0 \Rightarrow y$ is always increasiang, $y'' > 0 \Rightarrow y$ is concave up; local minimum $(0,0)$.

5. $y = (2x+3)^{1/3} \Rightarrow y' = \frac{1}{2}(2x+3)^{-1/2}2$ and $y'' = -(2x+3)^{-3/2}$. For $x < -\frac{3}{2}$, y is not defined. For $x > -\frac{3}{2}$, y is rising and concave down. $[-5,5]$ by $[-5,5]$ shows the complete graph; $(-1.5,0)$ is a local minimum.

7. $[-1, 3]$ by $[0, 6]$ gives a good view of slightly more than two complete periods; $[-0.5, 2]$ by $[0, 6]$ shows one period. The "bottoms" correspond to $3x + 5 = 3\pi/2$ and $3x + 5 = 7\pi/2$, i.e., $-0.144 \le x \le 1.999$ describes a complete period. The graph has a local maximum at $(.951, 5)$. To find the inflection points analytically, look for the extrema of $y' = 6\cos(3x + 5)$. These occur at $3x + 5 = 2\pi$ and 3π. The inflection points are $(.432, 3.01)$, $(1.147, 3.01)$.

9. $y = \sin 3x + \cos 3x \Rightarrow y' = 3(\cos 3x - \sin 3x)$ three successive zeros of y' will determine a complete period. $\cos 3x = \sin 3x$ at $3x = \pi/4$, $5\pi/4$, $9\pi/4$; the function has a local maximum of $\sqrt{2}/2 + \sqrt{2}/2 = \sqrt{2}$ at $x = \pi/12$ and $x = 9\pi/12$. It has a local minimum of $-\sqrt{2}$ at $x = 5\pi/12$. Inflection points are found by locating the extremes of y'; they are $(3\pi/12, 0)$ and $(7\pi/12, 0)$.

11. Graph $y = (x\verb|^|(1/3))\verb|^|5$ on $[-5, 5]$ by $[-5, 5]$. The graph appears to be always increasing, concave down for $x < 0$, inflection point at $x = 0$, concave up for $x > 0$. Analytically, $y' = \frac{5}{3}x^{2/3}$, $y'' = \frac{10}{9}x^{-1/3}$; $y' \ge 0$ for all x, y'' changes sign at $x = 0$.

13. $y = \sqrt{2x - 3}$; graph on $[0, 5]$ by $[0, 5]$. The graph appears to be concave down and increasing from a minimum at $(\frac{3}{2}, 0)$; $y' = \frac{2}{2\sqrt{2x-3}}$; $y'' = (-\frac{1}{2})\frac{(2)1}{(2x-3)^{3/2}}$ which is < 0 for $x > \frac{3}{2}$.

15. Graphing y' on $[3, 5]$ by $[-10, 10]$ shows the interval $3 \le x \le 5$ includes all possible extrema and inflection points (the domain is $[4, \infty)$) $[3, 6]$ by $[-3, 1]$ shows the function.

17. $y = \frac{3}{\sin(2x+\pi)} - 5$. A graph in $[-\pi, \pi]$ by $[-10, 6]$ shows that $(0, \pi)$ is a complete period. y has a minimum of -2 when $2x + \pi = 5\pi/2$, i.e. at $(3\pi/4, -2)$ and a maximum of -8 when $2x + \pi = 3\pi/2$, i.e. at $x = \pi/4$. For this function, a knowledge of trigonometry is more helpful than calculus. Concave down on $(0, \pi/2)$, concave up on $(\pi/2, \pi)$.

19. Graph y and y' (use NDER) on $[-1, 1]$ by $[-3, 3]$. There will be an inflection point at $x = 0$ and where y' has a local minimum, i.e. at $(0, 0)$ and also at $(0.27, 1.34)$. y has a local minimum at $(-0.35, -0.58)$.

21. Look at y and y' (use NDER) in $[-5, 5]$ by $[-1, 2]$ $y' > 0$ and decreasing for $x > 2$ hence y is increasing and concave down. (Graph $y = ((x)\verb|^|(1/5))\verb|^|3$.

23. Graph on $[-\pi, \pi]$ by $[-2, 2]$. Concave up on $(-\pi/4, \pi/4)$ with minimum at $(0, 1)$; concave down on $(\pi/4, 3\pi/4)$ with maximum at $(\pi/2, -1)$.

25. $[-5,5]$ by $[0,100]$ shows a complete graph. The graph is always rising, always concave up.

27. Graph $y = 3(\ell n(x+1))/\ell n\, 2$ in $[-5,5]$ by $[-10,10]$ the graph is always rising, always concave down.

29. Use $[-5,5]$ by $[0,4]$. Concave down $x \leq 0$ and $x \geq 2$.

31. $[-\pi/2, 7\pi/2]$ by $[-8,8]$ shows two periods. Inflection points at $(0,0)$ and $(\pi,0)$.

33. Graph y and y' in $[-\pi/2, 3\pi/2]$ by $[-3,3]$, using a horizontal scale of $\pi/16$.

Local minima at	**Local maxima at**	**Inflection points at**
(−0.968, −1.906)	(0.216, 1.216)	(−0.413, −0.408)
(1.228, −0.223)	(1.914, 0.223)	(0.673, 0.542)
(2.925, −1.216)	(4.109, 1.906)	(1.571, 0)
		(2.469, −0.542)
		(3.554, 0.408)
		(4.712, 0)

35. Looking at the graph in $[-2\pi, 2\pi]$ by $[-10,10]$ indicates that $[-\pi,\pi]$ by $[-6,6]$ will show a complete period. Graph y to find that y has: a local minimum at $(-1.298, -4.132)$, an inflection point at $(-0.858, -3.890)$, a local maximum at $(-0.578, -3.718)$, a local minimum at $(0.578, 3.718)$, an inflection point at $(0.858, 3.890)$, and a maximum at $(1.298, -4.132)$.

37. Graph y and y' on $[0,5]$ by $[-2,5]$. Using trace, y has a minimum at $(0.36, -0.37)$. Analytically, $y' = \ell n\, x + 1$; $y' = 0$ at $x = 1/e$. $y(1/e) = -1/e$.

39. Graph y and y' on $[-3,3]$ by $[-1,2]$. Using trace, y has a minimum at $(\pm.60, -0.18)$. Analytically, for $x > 0$. $y' = 2x\, \ell n\, x - x = x(\ell n\, x^2 - 1)$; $y' = 0$ at $x = \sqrt{(1/e)}$. y has a local minimum at $(\pm\sqrt{(1/e)}, -0.18)$. y is not defined at 0, but, if we define $y(0) = 0$, then y has a local maximum at $(0,0)$.

41. Use $[-3,3]$ by $[-1,6]$. The graph of y' shows that there is no hidden behavior near $x = 0$. $(0, 1/2)$ is a minimum. Analytically, $y' = 2^{x^2-1}(\ell n\, 2)(2x)$.

43. Graph y and y' on $[-3,3]$ by $[-1,1]$. Local extrema at $(\mp 0.86, \mp 0.52)$; inflection points at $(\mp 1.48, \mp 0.33)$ and $(0,0)$.

45. Graphing the function on $[-4\pi, 4\pi]$ by $[-20, 20]$ shows the symmetry with respect to $x = 0$. Graph y and y' on $[0, 4\pi]$ by $[-15, 12]$ and tracing shows local maxima at $(\pm 4.493, -4.603)$ and $(\pm 10.904, -10.950)$, local minima at $(\pm 7.725, 7.790)$. The function is not defined at $x = 0$, but if we define $y(0) = 1$, y is continuous at 0 and has a minimum at $(0, 1)$. Concavity changes at $x = k\pi$, $k = \pm 1, \pm 2, \ldots$.

47. Graph y and y' on $[-2, 6]$ by $[-1, 1]$. There are inflection points at $(-2.626, -0.500)$, $(1.090, 0.089)$ and $(5.333, 0.182)$. If $y(0)$ is defined to be 0, y has a minimum at $(0, 0)$. Concave down for $-2.626 < x < -1$ and $1.090 < x < 5.333$, concave up on $(-\infty, -2.626) \cup (-1, 1.090) \cup (5.333, \infty)$ (assuming $y(0) = 0$).

49. Graph $y = (x$^3)^(1/5) on $[-5, 5]$ by $[-3, 3]$ in order to see the entire function. $\min y = -2.63$, $\max y = 2.63$.

51. Graph y on $[-6, 6]$ by $[-200, 20]$. To find the maximum and minimum, graph $y' = e^x \sin x + e^x \cos x = e^x(\sin x + \cos x)$ on $[0, 6]$ by $[-10, 10]$. $y' = 0$ at $3\pi/4$ (2.356), where $y = 7.460$; and at $7\pi/4$ (5.498) where $y = -172.64$.

53-57

problem	function	window for complete graph	inflection point at	local maximum	local minimum
53.	$\sinh x$	[-3,3] by [-3,3]	(0,0)		
55.	$\coth x$	[-3,3] by [-3,3]	none		concavity changes at 0
57.	$\operatorname{csch} x$	[-3,3] by [-3,3]	none		

59. If x and y are the legs, then $x^2 + y^2 = 25$ and $A = \frac{1}{2}xy = \frac{1}{2}x\sqrt{25 - x^2}$. Graphing A on $[0, 5]$ by $[0, 7]$ shows maximum area is 6.25 cm^2.

61. $x^2 + y^2 = 5 \Rightarrow s = 2x + y = 2x + \sqrt{5 - x^2}$. Graphing s on $[0, 3]$ by $[0, 6]$ shows $\max s = 5$ (when $x = 2$).

63. From the diagram, the height of the trough is $h = 1 \cdot \cos\theta$; the triangles form a rectangle of width $\sin\theta$. The area of a cross section is $A(\theta) = \cos\theta \sin\theta + (\cos\theta) \cdot 1$, $0 \le \theta < \pi/2$; volume is maximized when $A(\theta)$ is a maximum. $A'(\theta) = -\sin^2\theta + \cos^2\theta - \sin\theta = 1 - 2\sin^2\theta - \sin\theta = -(2\sin\theta - 1)(\sin\theta + 1)$: $A'(\pi/6) = 0$, $A' > 0$ for $0 < \theta < \pi/6$, $A' < 0$ for $\pi/6 < \theta < \pi/2$. The volumeis maximized at $\pi/6$.

65. Sketch the semicircles $y = \sqrt{16 - x^2}$ and $y = \sqrt{4 - x^2}$; they are 2 units apart. The point $(1, \sqrt{3})$ lies on the second semicircle.

67. $f' = (3^x \ln 3 - 3^{-x}\ln 3)/2 = \frac{\ln 3}{2}[3^x - 3^{-x}]$; a graph of f' shows f has a minimum at $x = 0$ where f' changes from negative to positive). Hence $f(x) \geq f(0) = 0$. f is never negative.

69. Let P represent the closest point on the shore, C the city, and x the distance from P where the boat lands. The necessary time is $T(x) = \frac{\sqrt{4+x^2}}{2} + \frac{6-x}{5}$, $0 \leq x \leq 6$. Graph $T(x)$ on $[0,6]$ by $[0,4]$; finding its minimum value shows $x = 0.873$ miles.

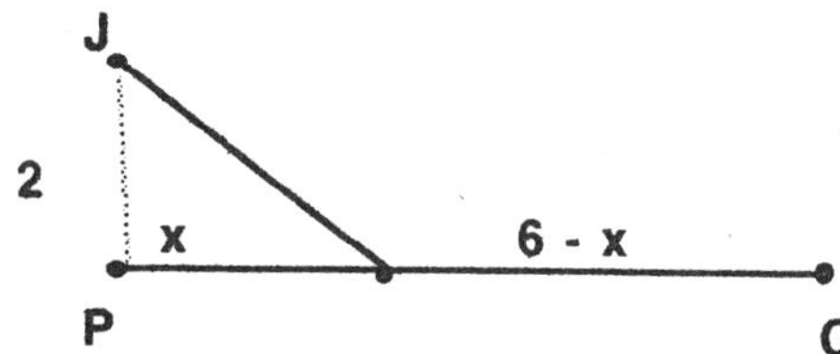

71. Height of rectangle is $2\sin\theta$, width is $2(2\cos\theta)$. Hence $A(\theta) = 8\sin\theta\cos\theta = 4\sin 2\theta$; $A'(\theta) = -8\cos 2\theta$, $A' = 0$ when $2\theta = \pi/2$ or $\theta = \pi/4$; $A(\pi/4) = 8(\frac{1}{\sqrt{2}})(\frac{1}{\sqrt{2}}) = 4$.

73. $A(x) =$ base $\times$ height $= 2x(8\cos 0.3x)$. Graphing on $[0, \frac{\pi}{2(.3)}]$ by $[0,30]$ shows a maximum area of 29.925.

4.6 RELATED RATES OF CHANGE

1. $A = \pi r^2 \Rightarrow dA/dt = 2\pi r\ dr/dt$

3. $dV/dt = (2/3)\pi rh\ dr/dt$

5. a) $\frac{dP}{dt} = I^2\frac{dR}{dt} + 2IR\frac{dI}{dt}$

b) $\frac{dP}{dt} = 0 \Rightarrow \frac{dR}{dt} = -\frac{2}{I}R\frac{dI}{dt}$ or $\frac{1}{R}\frac{dR}{dt} = -2\frac{1}{I}\frac{dI}{dt}$

7. $\frac{ds}{dt} = \frac{1}{2\sqrt{x^2+y^2+z^2}}(2x\frac{dx}{dt} + 2y\frac{dy}{dt} + 2z\frac{dz}{dt}) = \frac{x}{s}\frac{dx}{dt} + \frac{y}{s}\frac{dy}{dt} + \frac{z}{s}\frac{dz}{dt}$

9. $A = \pi r^2$; $dA/dt = 2\pi r\ dr/dt$. When $r = 50$ cm and $dr/dt = 0.01$ cm/min, $dA/dt = 2\pi 50(0.01) = \pi$ cm^2/min.

11. $A = \ell w \Rightarrow dA/dt = \ell dw/dt + w d\ell/dt = \ell(+2) + w(-2)$ cm^2/sec. $P = 2(\ell + w) \Rightarrow dP/dt = 2(d\ell/dt + dw/dt) = 2(-2+2) = 0$ cm/sec. A diagonal has length $D = \sqrt{\ell^2 + w^2}$; hence

$$\begin{aligned} dD/dt &= \frac{1}{2\sqrt{\ell^2 + w^2}}(2\ell\ d\ell/dt + 2w\ dw/dt) \\ &= \frac{\ell\ d\ell/dt + w\ dw/dt}{\sqrt{\ell^2 + w^2}} = \frac{\ell(-2) + w(2)}{\sqrt{\ell^2 + w^2}} \text{ cm/sec} \end{aligned}$$

Evaulating these rates at $\ell = 12$ and $w = 15$ gives $dA/dt = 12(+2)+5(-2) = +14$ cm^2/sec, increasing; $dP/dt = 0$ cm/sec, constant; $dD/dt = \frac{12(-2)+5(2)}{\sqrt{12^2+5^2}} = \frac{-14}{\sqrt{169}} = -14/13$ cm/sec., decreasing.

13. $D^2 = x^2 + y^2$ where plane A is x miles from the intersection point of the courses. $2D\frac{dD}{dt} = 2x\frac{dx}{dt} + 2y\frac{dy}{dt} = 2[x(-520) + y(-520)]$; when $x = 5$, $y = 12$, $D = 13$, so $13\frac{dD}{dt} = -520[5 + 12]$ or $\frac{dD}{dt} = -680$ miles/hr.

15. $V = \frac{4}{3}\pi r^3 \Rightarrow \frac{dV}{dt} = 4\pi r^2 \frac{dr}{dt} \Rightarrow \frac{dr}{dt} = \frac{dV/dt}{4\pi r^2}$; $\frac{dV}{dt} = -0.08$ mL $= -0.08$ cm^3; when $r = 10$ mm $= 1$cm, $\frac{dr}{dt} = \frac{-0.08}{4\pi \cdot 1^2} = -.006366$ cm/min $= -0.06366$ mm/min.

17. Since $h = \frac{3}{8}(2r)$, $V = \frac{\pi}{3}r^2 h = \frac{\pi}{3}r^2(\frac{3}{8}2r) = \frac{\pi}{4}r^3$; $\frac{dV}{dt} = \frac{3\pi}{4}r^2\frac{dr}{dt}$. When $h = 4$ m, $r = \frac{4}{3} \cdot 4 = \frac{16}{3}$ m; a) $\frac{dh}{dt} = \frac{3}{4}\frac{dr}{dt} = \frac{\frac{3}{4}\frac{dV}{dt}}{\frac{3}{4}\pi r^2} = \frac{10}{\pi(\frac{16}{3})^2}$ m/min $= \frac{90 \cdot 100}{256\pi}$ cm/min $= 11.191$ cm/min.

b) $\frac{dr}{dt} = \frac{4}{3}\frac{dh}{dt} = \frac{4}{3} \cdot \frac{1125}{32\pi}$ cm/min $= 14,921$ cm/min.

19. $V = (\frac{\pi}{3})y^2(3 \cdot 13 - y) \Rightarrow \frac{dV}{dt} = \frac{\pi}{3}[78y - 3y^2]\frac{dy}{dt}$.

a) $\frac{dV}{dt} = -6$, $y = 8 \Rightarrow \frac{dy}{dt} = -13.263$ cm/min

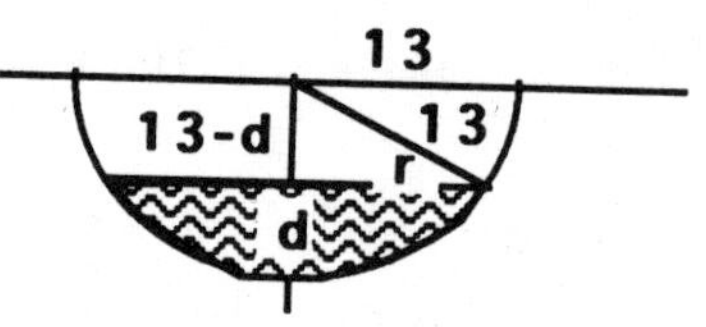

b) From the diagram, for depth d,

$x^2 = 13^2 - (13 - d)^2 = 26d - d^2$.

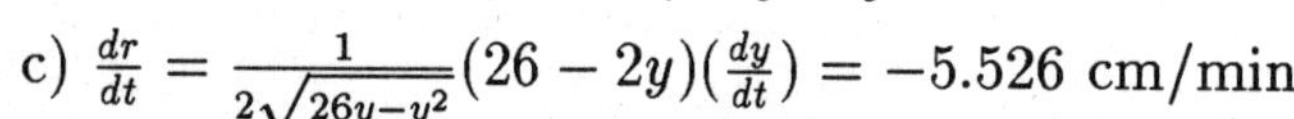

Hence, for depth $y, r = \sqrt{26y - y^2}$.

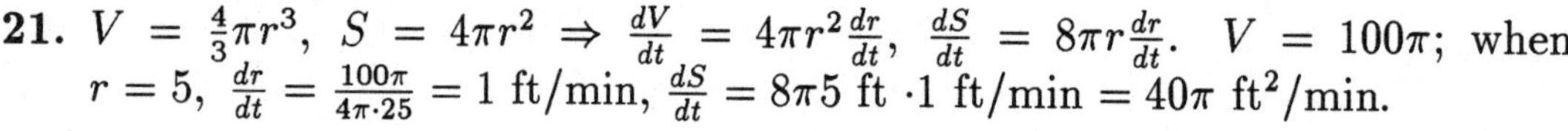

c) $\frac{dr}{dt} = \frac{1}{2\sqrt{26y - y^2}}(26 - 2y)(\frac{dy}{dt}) = -5.526$ cm/min.

21. $V = \frac{4}{3}\pi r^3$, $S = 4\pi r^2 \Rightarrow \frac{dV}{dt} = 4\pi r^2 \frac{dr}{dt}$, $\frac{dS}{dt} = 8\pi r \frac{dr}{dt}$. $V = 100\pi$; when $r = 5$, $\frac{dr}{dt} = \frac{100\pi}{4\pi \cdot 25} = 1$ ft/min, $\frac{dS}{dt} = 8\pi 5$ ft $\cdot 1$ ft/min $= 40\pi$ ft^2/min.

23. $[s(t)]^2 = x^2(t) + y^2(t)$. $2s(t)\frac{ds}{dt} = 2x(t)\frac{dx}{dt} + 2y(t)\frac{dy}{dt} = 2x \cdot 17 + 26 \cdot 1$. When $t = 3$, $x = 17 \cdot 3 = 51$, $y = 65 + 3 = 68$, $s = \sqrt{51^2 + 68^2} = 85$. Hence $\frac{ds}{dt} = \frac{1}{85}[51 \cdot 17 + 68 \cdot 1] = 11$ ft/sec.

25. $\frac{dr}{dt} = -0.2$, $\frac{dV}{dt} = 4Kr^3\frac{dr}{dt}$; $\frac{\frac{dV}{dt}}{V} = \frac{4Kr^3\frac{dr}{dt}}{Kr^4} = 4\frac{\frac{dr}{dt}}{r} = -0.8$. The volume decreases 80% per minute.

27. Let $P(x, y)$ be a point on the curve. Then $y/x = \tan\theta$, where θ is the angle of inclination of the line OP. Since $y = x^2$, we have $x = \tan\theta$; $dx/dt = \sec^2\theta \; d\theta/dt$ or

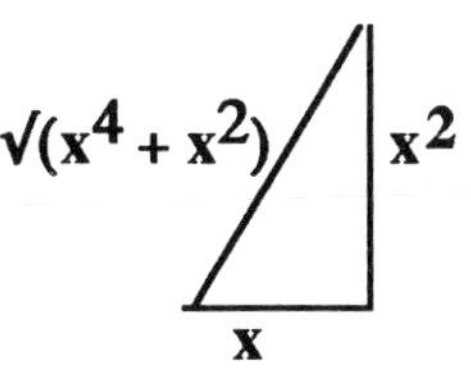

$d\theta/dt = (\cos^2\theta)dx/dt = 10\cos^2\theta$.

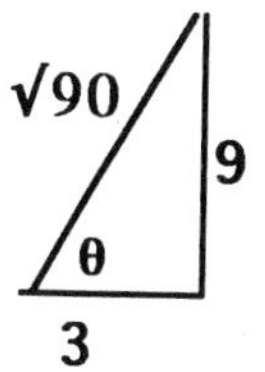

When $x = 3$, $y = 9$, $\cos\theta = 3/\sqrt{9 + 81} = 1/\sqrt{10}$ and $d\theta/dt = 10/10 = 1$ rad/sec.

As $x \to \infty$, $\frac{d\theta}{dt} = 10\frac{x^2}{x^4 + x^2}$.

$\text{Lim}_{x\to\infty} d\theta/dt = 0$.

29. a) $\frac{dr}{dt} = 9\frac{dx}{dt}$, $\frac{dc}{dt} = 3x^2\frac{dx}{dt} - 12x\frac{dx}{dt} + 15\frac{dx}{dt} = (3x^2 - 12x + 15)\frac{dx}{dt}$; when $x = 2$, $\frac{dx}{dt} = 0.1$, $\frac{dr}{dt} = 0.9$, $\frac{dc}{dt} = (3)(0.1) = 0.3$; $\frac{dp}{dt} = 0.9 - 0.3 = 0.6$.

b) $\frac{dr}{dt} = 70\frac{dx}{dt}$, $\frac{dc}{dt} = (3x^2 - 12x - \frac{45}{x^2})\frac{dx}{dt}$; when $\frac{dx}{dt} = 0.05$ and $x = 1.5$, $\frac{dr}{dt} = 3.5$, $\frac{dc}{dt} = -1.5625$; $\frac{dp}{dt} = 3.5 - (-1.5625) = 5.0625$.

31. Let ℓ = distance from pole to the ball's shadow, h the height of the ball.

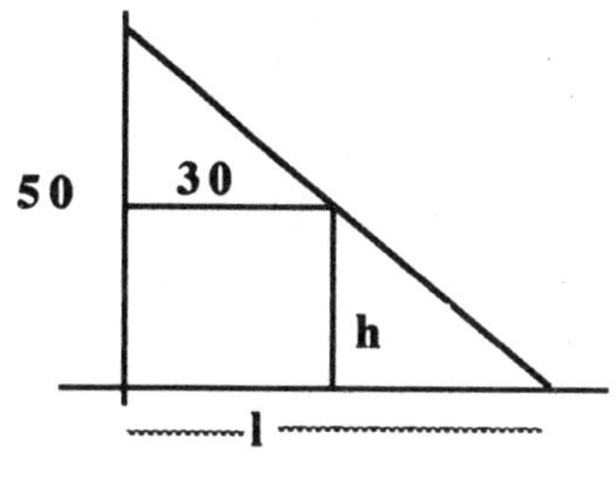

$h = 50 - 16t^2$, $\frac{50-h}{30} = \frac{50}{\ell}$, or $\ell(50 - h) = 1500$

$\frac{d\ell}{dt}(50 - h) + \ell(-\frac{dh}{dt}) = 0$; $\frac{dh}{dt} = -32t$,

$t = \frac{1}{2} \Rightarrow h = 46 \Rightarrow \ell = 375$,

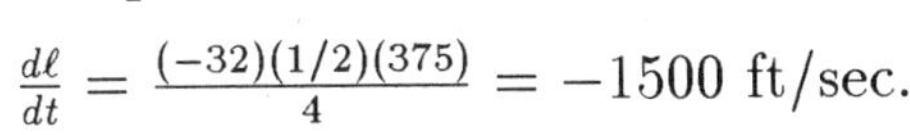

$\frac{d\ell}{dt} = \frac{(-32)(1/2)(375)}{4} = -1500$ ft/sec.

33. When the ice is 2 in thick, the radius of the ice ball is $(4 + 2) = 6$ in. $V = \frac{4}{3}\pi r^3$, $S = 4\pi r^2$. $\frac{dV}{dt} = 4\pi r^2\frac{dr}{dt}$, $\frac{dS}{dt} = 8\pi r\frac{dr}{dt}$. When $\frac{dV}{dt} = -10$ and $r = 6$, $\frac{dr}{dt} = \frac{-10}{4\pi \cdot 36} = \frac{-5}{72\pi}$ in/min. $\frac{dS}{dt} = 8\pi \cdot 6(\frac{-5}{72\pi}) = \frac{-10}{3}$ in^2/min.

35. $0.27° = 0.27\pi/180$ radians. Let s be the length of the shadow; $\frac{80}{s} = \tan\theta \Rightarrow \frac{-80}{s^2}\frac{ds}{dt} = \sec^2\theta\frac{d\theta}{dt}$; $\frac{ds}{dt} = \frac{-s^2\sec^2\theta}{80}\frac{d\theta}{dt}$. When $s = 60$, $\frac{d\theta}{dt} = 0.27\pi/180$, $\sec^2\theta = \tan^2\theta + 1 = (\frac{80}{60})^2 + 1$. Hence $\frac{ds}{dt} = \frac{-60^2(\frac{80^2}{60^2}+1)}{80}\cdot\frac{0.27\pi}{180} = 0.589$ ft/min ≈ 7.1 in/min.

37. By the law of cosines: $AB^2 = OA^2 + OB^2 - 2\cdot OA\cdot OB\cdot\cos 120$ or $AB^2 = OA^2 + OB^2 + OA\cdot OB$. Hence

$2\cdot AB\cdot dAB/dt = 2OA\cdot dOA/dt + 2OB\cdot dOB/dt + OA\cdot dOB/dt + OB\cdot dOA/dt.$

When $OA = 5$ and $OB = 3$, $AB = \sqrt{(25+9+15)} = \sqrt{(49)} = 7$ and

$$2\cdot 7\cdot dAB/dt = 2\cdot 5\cdot 14 + 2\cdot 3\cdot 21 + 5\cdot 21 + 3\cdot 14$$

or $dAB/dt = 29.5$ knots.

4.7 ANTIDERIVATIVES, INITIAL VALUE PROBLEMS, AND MATHEMATICAL MODELING

1. a) $x^2 + C$, b) $\frac{x^3}{3} + C$ c) $\frac{x^3}{3} - x^2 + x + C$

3. a) $y' = -3x^{-4} \Rightarrow y = -3x^{-3}/(-3) + C = x^{-3} + C$; b) $y' = x^{-4} \Rightarrow y = x^{-3}/(-3) + C$ c) $y' = x^{-4} + 2x + 3 \Rightarrow y = -x^{-3}/3 + x^2 + 3x + C$

5. a) $y' = x^{-2} \Rightarrow y = -x^{-1} + C$, b) $y = \frac{-5}{x} + C$ c) $y = 2x + \frac{5}{x} + C$

7. a) $y' = \frac{3}{2}x^{1/2} \Rightarrow y = \frac{(\frac{3}{2})x^{1/2+1}}{1/2+1} + C = \frac{(\frac{3}{2})}{(\frac{3}{2})}x^{3/2} + C = x^{3/2} + C$ b) $y' = 4x^{1/2} \Rightarrow y = \frac{4}{(\frac{3}{2})}x^{3/2} + C = \frac{8}{3}x^{3/2} + C$ c) $y' = x^2 - 4x^{1/2} \Rightarrow y = x^3/3 - (8/3)x^{3/2} + C$

9. a) $y' = \frac{2}{3}x^{-1/3} \Rightarrow y = \frac{\frac{2}{3}}{1-\frac{1}{3}}x^{-1/3+1} + C = x^{2/3} + C$ b) $y' = \frac{1}{3}x^{-2/3} \Rightarrow y = \frac{\frac{1}{3}}{\frac{1}{3}}x^{1/3} + C = x^{1/3} + C$ c) $y' = -\frac{1}{3}x^{-4/3} \Rightarrow y = \frac{-\frac{1}{3}}{-\frac{4}{3}+1}x^{-1/3} + C = x^{-1/3} + C$

11. a) $y' = \sin(3x) \Rightarrow y = \frac{\cos(3x)}{3} + C$ b) $y' = 3\sin x \Rightarrow y = -3\cos x + C$ c) $y' = 3\sin x - \sin(3x) \Rightarrow y' = -3\cos x + \frac{\cos(3x)}{3} + C$

13. a) $y' = \sec^2 x \Rightarrow y = \tan x + C$ b) $y' = 5\sec^2(5x) \Rightarrow y = 5\frac{\tan(5x)}{5} + C = \tan(5x) + C$ c) $y' = \sec^2(5x) \Rightarrow y = \frac{\tan(5x)}{5} + C$

15. a) $y' = \sec x\tan x \Rightarrow y = \sec x + C$ b) $y' = 2\sec 2x\tan 2x \Rightarrow y = \sec 2x + C$ c) $y' = 4\sec 2x\tan 2x \Rightarrow y = \frac{4\sec 2x}{2} + C = 2\sec 2x + C$

17. $y' = (\sin x - \cos x)^2 = \sin^2 x - 2\sin x\cos x + \cos^2 x = 1 - 2\sin x\cos x = 1 - \sin 2x \Rightarrow y = x + \frac{\cos 2x}{2} + C$

19. a) $1 - \sqrt{x} + C$ or, since $1 + C$ is a constant, the general antiderivative can be written $-\sqrt{x} + C$ b) $x + C$ c) $\sqrt{x} + C$ d) $-x + C$ e) $x - \sqrt{x} + c$ f) $-3\sqrt{x} - 2x + C$ g) $\frac{x^2}{2} - \sqrt{x} + C$ h) $x - 4x + C = -3x + C$

21. $y' = 2x \Rightarrow y = x^2 + C$; $y(1) = 4 \Rightarrow y = x^2 + 3$; graph b.

23. $y' = 2x - 7 \Rightarrow y(-x) = x^2 - 7x + C$. Evaluating at $x = 2$ gives $0 = 2^2 - 7\cdot 2 + C$, $C = 10$, $y = x^2 - 7x + 10$.

25. $y' = x^2 + 1 \Rightarrow y = x^3/3 + x + C$; $y(0) = 1 \Rightarrow 1 = 0 + 0 + C$; $y = x^3/3 + x + 1$

27. $y' = -5x^{-2} \Rightarrow y = \frac{-5}{-1}x^{-1} + C = 5x^{-1} + C$; $3 = 5(5^{-1}) + C \Rightarrow C = 2$. $y = 5/x + 2$

29. $y' = 3x^2 + 2x + 1 \Rightarrow y = x^3 + x^2 + x + C$; $0 = 1 + 1 + 1 + C \Rightarrow C = -3$, $y = x^3 + x^2 + x - 3$

31. $y' = 1 + \cos x \Rightarrow y = x + \sin x + C$; $4 = 0 + \sin 0 + C \Rightarrow C = 4$. $y = x + \sin x + 4$

33. $y'' = 2 - 6x \Rightarrow y' = 2x - 3x^2 + C$; since $y' = 4$ when $x = 0$, $4 = 0 - 0 + C$. Hence $y' = 2x - 3x^2 + 4$ giving $y = x^2 - x^3 + 4x + C$; since $y = 1$ when $V = 0$, $1 = C$. Hence $y = x^2 - x^3 + 4x + 1$.

35. $v = s' = 9.8t \Rightarrow s = \frac{9.8}{2}t^2 + C = 4.9t^2 + C$. $s(0) = 10 \Rightarrow 10 = C$. Hence $s = 4.9t^2 + 10$.

37. $a = v' = 32 \Rightarrow v = 32t + C$; $v(0) = 20 \Rightarrow C = 20$. Hence $v = 32t + 20$. $s' = v = 32t + 20 \Rightarrow s = 16t^2 + 20t + C$; $s(0) = 0 \Rightarrow C = 0$. Hence $s = 16t^2 + 20t$.

39. The slope, y', satisfies the equation $y' = 3x^{1/2}$. Hence, $y = \frac{3}{3/2}x^{3/2} + C = 2x^{3/2} + C$. Since $y = 4$ when $x = 9$, $4 = 2\cdot 27 + C$, or $C = -50$. The curve is given by $y = 2x^{3/2} - 50$.

41. $r' = 3x^2 - 6x + 12 \Rightarrow r = x^3 - 3x^2 + 12x + C$. $r(0) = 0 \Rightarrow C = 0 \Rightarrow r = x^3 - 3x^2 + 12x$

43. $a = -1.6 \Rightarrow v = -1.6t + C$. If the rock is dropped (not thrown), $v(0) = 0$. Hence $v = -1.6t$m/sec. At $t = 30$sec, $v = -48$m/sec. The velocity is 48m/sec in a downwards direction.

45. The problem assumes that the diver does not jump or take a running start, i.e. $v(0) = 0$. $a = v' = -9.8 \Rightarrow v = -9.8t + C$. By assumption, $C = 0$, $s' = v = -9.8t \Rightarrow s = -4.9t^2 + C$. The height of the board is 10m, hence $s(0) = 10$, $s = -4.9t^2 + 10$. You enter the water when $s = 0$. Solving $0 = -4.9t^2 + 10$ gives $t = \sqrt{10/4.9}$. $v(\sqrt{10/4.9}) = -9.8\sqrt{10/4.9} = -14$m/sec.

47. The general antiderivative of $y^{-1/2}\frac{dy}{dt}$ is $\frac{y^{1/2}}{\frac{1}{2}} + C_1 = 2y^{1/2} + C_1$. The general antiderivative of $-k$ is $-kt + C_2$. Hence $2y^{1/2} + C_1 = -kt + C_2$. $2y^{1/2} = -kt + (C_2 - C_1) = -kt + C$, $y^{1/2} = -\frac{kt}{2} + \frac{C}{2} = \frac{C-kt}{2}$, $y = \frac{(C-kt)^2}{4}$.

49. The least common multiple of 8 and 12 is 24; every 24 seconds Renée and Sherrie will return to points R and S. Graph (x_3, y_3) using $x\,\mathrm{Min} = t\,\mathrm{Min} = 0$, $x\,\mathrm{Max} = t\,\mathrm{Max} = 48$, $y\,\mathrm{Min} = 0$, $y\,\mathrm{Max} = 60$. The screen shows two periods of $D(t)$. c) Using zoom, the maximum distance is 45.821 at $t = 15.925$, the minimum is 4.109 at $t = 21.5$ d) Try to find the critical points of $D(t)$: $D^2(t) = (x_1 - x_2)^2 + (y_1 - y_2)^2$; $2DD'(t) = 2(x_1 - x_2)(x_1' - x_2') + 2(y_1 - y_2)(y_1' - y_2')$: Set $D'(t) = 0$ and substitute for the derivatives of x_1, x_2, y_1 and y_2. the algebra is almost impossible.

51. The diagram shows that $y(1) = 1$ and $y'(1) = 2$. The only initial value problem listed satisfying both these conditions is (d).

53. By the sign of y', y decreases for $x < 0$ and increases for $y > 0$; $y'' = 2 > 0 \Rightarrow y$ is always concave up. Actual solutions are of the form $y = x^2 + C$.

55. $y' = 1 - 3x^2 \Rightarrow y$ decreasing for $x < -\frac{1}{\sqrt{3}}$ increasing on $(-\frac{1}{\sqrt{3}}, \frac{1}{\sqrt{3}})$, decreasing thereafter. $y'' = -6x \Rightarrow y$ concave up for $x < 0$, concave down for $x > 0$. The solution curves are $y = x - x^3 + C$.

57. $y' > 0 \Rightarrow y$ is always increasing. $y'' = \frac{4x^3}{2\sqrt{1+x^4}}$ shows y has an inflection point at $x = 0$. To see the shape of the graph of an antiderivative, put your calculator in Dif Eq Mode, select GRAPH and set the ranges $t : [-2, 2]$, $x[-2, 2]$, $y[-4, 4]$. The calculator graphs an antiderivative of $Q'(t)$. Enter, for this problem, $Q'(t) = \sqrt{(1 + t\hat{}4)}$, then select INITC to specify the value of y when $x = x\,\mathrm{min} = t\,\mathrm{min}$. For this problem $y(-2) = -4$ will show the graph of one antiderivative of $\sqrt{(1 + t\hat{}4)}$.

59. $y' \to 0$ as $x \to \pm\infty \Rightarrow y$ approaches a constant. $y'(0) = 0$ and changes from negative to positive there $\Rightarrow$ minimum at $(0, 0)$. To see an antiderivative, use Dif Eq. setting the ranges: $t[-8, 8]$, $x[-8, 8]$, $y[-6, 6]$. Try $y(-8) = 3$ to see the shape of the curve.

61.

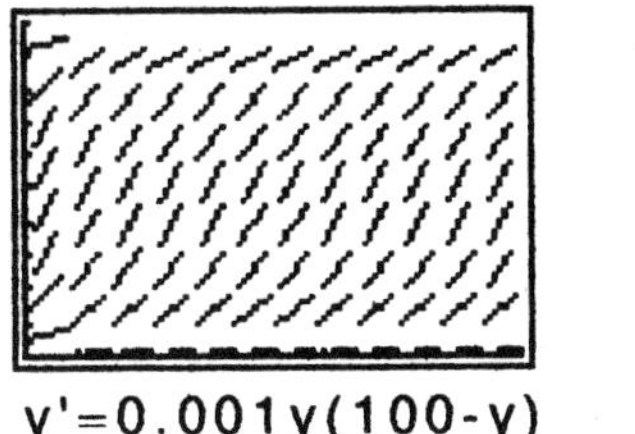

y'=0.001y(100-y)

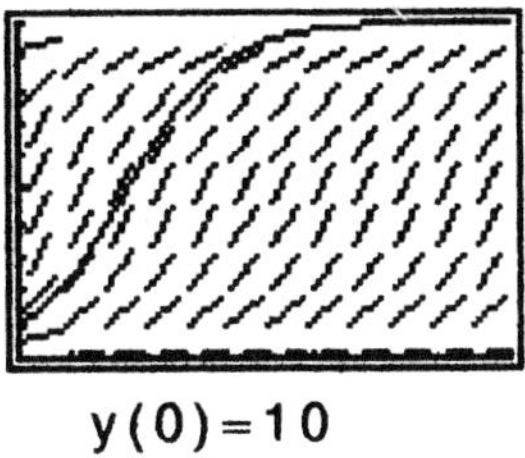

y(0)=10

PRACTICE EXERCISES, CHAPTER 4

1. $y' = \frac{(x+1)-(x)}{(x+1)^2} = \frac{1}{(x+1)^2} > 0$. Hence y is always increasing.

3. $y = x^3 + 2x \Rightarrow y' = 3x^2 + 2$ which is defined for all x but is never 0. y' is always > 0 and y is always increasing.

5. By the M.V.T. the increase in f can be written $f(6) - f(0) = (6-0)f'(c) \leq 6 \cdot 2 = 12$, the maximum increase.

7. Graph y' in $[-3, 3]$ by $[-5, 30]$. $y'(-1^{-1}) < 0$, $y'(-1) = 0$ $y'(-1^{+}) > 0$ means y changes from a decreasing to an increasing function, i.e. y has a local minimum at $x = -1$. The inflection points of y are at the local extremes of y', i.e. at $x = 0$ and $x = 2$.

9. a) For $y' < 0$ and $y'' < 0$, y must be decreasing and concave down. These conditions are met at T. b) $y' < 0$ and $y'' > 0$ means y is decreasing and concave up: P.

11. $y' = -3x^2 - 6x - 4$; $y'' = -6x - 6 = -6(x+1)$. y has an inflection point at $x = -1$, $y = 0$. Set $y' = 0$: $x = (6 \pm \sqrt{36 - 48})/(-6)$; there are no local extrema. $[-6, 6]$ by $[-4, 4]$ shows a complete graph.

13. $y = (x-2)^{1/3} \Rightarrow y' = \frac{1}{3}(x-2)^{-2/3}$, $y'' = \frac{-2}{9}(x-2)^{-5/3}$, y'' changes sign at $x = 2$. y' is never 0; y' is not defined at 2, but $(2, 0)$ is an inflection point. $[-6, 6]$ by $[-4, 4]$ shows the graph.

15. $[-6, 6]$ by $[-4, 4]$ appears to show a complete graph. $y' = 1 - 2x - 4x^3$; $y'' = -2 - 12x^2 = -2(1 + 6x^2)$. There are no inflection points. The unique zero of y', between 0 and 1, corresponds to the plotted maximum.

17. Use a grapher to graph y and y' in $[-6,6]$ by $[-50,50]$. To determine whether y is ever increasing, look at $y' = -8x^2 + 8x - 2$. If y' is positive, the discriminant of the quadratic will be positive; however $64 - 4 \cdot 2 \cdot 8 = 0$. We have not missed any hidden behavior of y.

19. Use $[-6,6]$ by $[-8,3]$; graph both y and y'. This time, zooming in on y' near its local maximum shows that y' does turn positive, and hence, that y is increasing between $x = -0.12$ and $x = 0$. To see the increasing behavior of y, use $[-0.5, 0.5]$ by $[0.99, 1.01]$.

21. Graphing y and y' in $[-5,5]$ by $[-15,25]$ shows the end behavior of y. On the interval $-1.5 \leq x \leq 0$, the graph of y' is steadily increasing. This means there is no concealed behavior in the graph of y. Local minimum at $(-0.578, 0.972)$, inflection point at $(1.079, 13.601)$, local maximum at $(1.692, 20.517)$.

23. Graph y and y' in $[-5,5]$ by $[-15,20]$. Tracing along y' shows that y has an inflection point at $(3.71, -3.41)$.

25. If $y = \log_3 |x|$, then $3^y = |x|$ and $y = \frac{\ell n|x|}{\ell n\, 3}$. The graph will be symmetric about the y-axis. Graphing y and y' in $[-1,4]$ by $[-5,5]$ shows the complete behavior of y. The graph of y' gets closer and closer to zero as $x \to \infty \Rightarrow$ the graph of y grows more and more slowly as $x \to \infty$.

27. Use $[1.9, 4]$ by $[-5,5]$. Defined for $x > 2$, always rising, concave down on $(2,4)$. Inflection point at $(4, 2.773)$.

29. $y = [x(1-x)]^{1/4}$. The domain of y is $0 \leq x \leq 1$; $y \geq 0$. Use $[-1,2]$ by $[-1,2]$. Local maximum at $(0.500, 0.707)$; local minima at $(0,0)$ and $(1,0)$.

31. The question asks for the solution to $-x^{-3} + 3x + 4 = 0$. Let $f(x) = -x^{-3} + 3x + 4$. $f(1) = 6$, $f(+\frac{1}{2}) = -8 + \frac{3}{2} + 4 < 0$. For Newton's method, the iteration is $x_{n+1} = x_n - \frac{(-x_n^{-3} + 3x_n + 4)}{(3x^{-4} + 3)}$. Starting with $x_0 = 1$, leads to division by 0, a "math error". Starting with $x_0 = 0.5$ leads quickly to $x = 0.5604$.

33. Let's blindly start with $x_0 = 0$ and use Newton's method: $x_{n+1} = x_n - (2\cos x_n - \sqrt{1 + x_n})/(-2\sin x_n - \frac{1}{2\sqrt{1+x_n}})$. The iterates are:

$$\begin{aligned} x_1 &= 2 \\ x_2 &= .783\ldots \\ x_3 &= .829\ldots \\ x_4 &= .828\ldots \\ x_5 &= .828\ldots \end{aligned}$$

35. b) $x - f(x)/f'(x) = x - \frac{\frac{1}{x}-3}{-\frac{1}{x^2}} = x + x^2(\frac{1}{x} - 3) = x + x - 3x^2 = x(2-3x)$

37. Graph $f(x)$ on $[-6,1]$ by $[-80,30]$, $f(-6) = -74$ is the minimum value. The maximum value of 16.25 is taken on at -4.550.

39. $A = \frac{1}{2}rs = \frac{1}{2}r(100-2r) = 50r - r^2$, $0 \le r \le 50$. $A' = 50 - 2r \Rightarrow A' = 0$ when $r = 25$. $A'' = -2 < 0$, so this gives a maximum for A ($A(0) = A(50) = 0$). When $r = 25$, $s = 50$.

41. Let the sides of the base be x, let the height be h. Then $V = s^2h$, $108 =$ (area of base) $+4$(area of one side) $= s^2 + 4sh$. Solving for h gives $h = \frac{108-s^2}{4s}$. $V = \frac{s^2(108-s^2)}{4s} = \frac{1}{4}s(108-s^2) = \frac{1}{4}[108s - s^3]$. $V' = \frac{1}{4}[108 - 3s^2]$; $V' = 0$ when $s^2 = \frac{108}{3} = 36$. The maximum volume occurs when $s = 6$ ft, $h = \frac{108-36}{24} = 3$ ft.

43. From the diagram $(\frac{h}{2})^2 + r^2 = (\sqrt{3})^2$, or $r^2 = 3 - h^2/4$. $V = \pi r^2 h = \pi(3 - \frac{h^2}{4})h$, $0 \le h \le 2\sqrt{3}$, $V' = \pi[-\frac{-2h}{4} \cdot h + (3 - \frac{h^2}{4})] = \frac{\pi}{4}[12 - 3h^2]$. $V' = 0$ for $h = 2$; for $h > 0$, $V'' < 0$ so $h = 2$ gives the maximum volume. $r^2 = 3 - 1 = 2 \Rightarrow r = \sqrt{2}$.

45. The cost, $C = x \cdot 40000 + (20 - y)30000$, where $y^2 + 12^2 = x^2$. $C = 10000[4\sqrt{y^2+12^2} + 3(20-y)]$. $C' = 10000[\frac{4 \cdot y}{\sqrt{y^2+12^2}} - 3]$; setting $C' = 0$ gives $4y = 3\sqrt{y^2+12^2} \Rightarrow 16y^2 = 9(y^2+12^2)$, or $y = 36/\sqrt{7}$; $x^2 = \frac{9 \cdot 12^2}{7} + 12^2 = \frac{16 \cdot 12^2}{7} \Rightarrow x = 48/\sqrt{7}$. Alternatively, graphing $C/10000$ on $[0,40]$ by $[-5,150]$ gives $y = 13.607$, $x = \sqrt{y^2+12^2} = 18.143$.

47. The total profit is $P = K(2x + y)$, where K is a constant; $P = K(2x + \frac{40-10x}{5-x}) = K[\frac{40-2x^2}{5-x}]$; $P' = 0$ when $-4x(5-x) - (40-2x^2)(-1) = 0$, i.e. at $x = 2.764$. Since $P(2.764) = 11.048K$, $P(0) = 8K$, $P(4) = 8K$, for maximum profit you should make 276 Grade A tires, 553 Grade B tires.

49. If the squares have side x, the volume of the box is $V = x(16-2x)(10-2x)$. $V' = (16-2x)(10-2x) - 2x(16-2x) - 2x(10-2x) = 12x^2 - 104x + 160 = (x-2)(12x-80)$. $x = 80/12 > 6$ is impossible for this problem. $V(2) = 144$ is the maximum volume. The dimensions are $12 \times 6 \times 2$.

51. $s'(t) = 3t^2 + 2t - 6$. $s' = 0$ at $t = \frac{-2 \pm \sqrt{4+72}}{6} = \frac{-2 \pm \sqrt{76}}{6}$. $s'(0) < 0$, so for $0 < t < (-2 + \sqrt{76})/6$ the particle moves to the left. It then moves to the right. At $t = (-2 + \sqrt{76})/6$, the particle is 0.94 units to the right of the origin.

53. $A = \pi r^2$ and $dr/dt = -2/\pi$. $dA/dt = 2\pi r\, dr/dt = -4r$. $\frac{dA}{dt}|_{r=10} = -40\text{m}^2/\text{sec}$.

55. Let s be the side of the cube. $V = s^3$ and $\frac{dV}{dt} = 1200\text{cm}^3/\text{min}$. $\frac{dV}{dt} = 3s^2\frac{ds}{dt}$. Hence $\frac{ds}{dt} = \frac{1}{3s^2}\frac{dV}{dt} = \frac{1200}{3s^2}|_{s=20} = \frac{1200}{3\cdot 20\cdot 20} = 1\text{cm}/\text{min}$.

57. a) From the figure, $h/r = 10/4$ or $r = 2h/5$ b) $V = \frac{1}{3}\pi r^2 h = \frac{1}{3}\pi\frac{4h^3}{25}$; $dV/dt = \frac{4}{25}\pi h^2 dh/dt = -5$ when $h = 6$, $dh/dt = \frac{-25\cdot 5}{4\pi\cdot 36} = \frac{-125}{144\pi}\text{ft}/\text{min}$.

59. Let x = distance from the car to the point on the track which is right in front of you. $\tan\theta = x/132 \Rightarrow \sec^2\theta\frac{d\theta}{dt} = \frac{1}{132}\frac{dx}{dt} = -\frac{264}{132}$ or $\frac{d\theta}{dt} = -2\cos^2\theta$. When the car is in front of you, $\theta = 0$, $\cos\theta = 1$ and $\frac{d\theta}{dt} = -2\text{rad}/\text{sec}$.

61. Let $H(x) = 3x$; then $H'(x) = 3$. If $g'(x) = 3$, for some function $g(x)$, then $g(x) - H(x) = K$, a constant. Thus, $g(x) = H(x) + K = 3x + K$ and $g(x)$ is one of the functions $F(x) = 3x + C$. All functions with derivative 3 are of the form $F(x) = 3x + C$.

63. a) C b) $x + C$, c) $\frac{x^2}{2} + C$ d) $\frac{x^3}{3} + C$ e) $x^{11}/11 + C$ f) $-x^{-1} + C$ g) $-x^{-4}/4 + C$ h) $\frac{2}{7}x^{7/2} + C$ i) $\frac{3}{7}x^{7/3} + C$ j) $\frac{4}{7}x^{7/4} + C$ k) $\frac{2}{3}x^{3/2} + C$ l) $2x^{1/2} + C$ m) $\frac{7}{4}x^{4/7} + C$ n) $\frac{-3}{4}x^{-4/3} + C$

65. $y' = 3x^2 + 5x - 7 \Rightarrow y = x^3 + (5/2)x^2 - 7x + C$

67. $y' = x^{1/2} + x^{-1/2} \Rightarrow y = \frac{x^{3/2}}{3/2} + \frac{x^{1/2}}{1/2} + C = (2/3)x^{3/2} + 2x^{1/2} + C$

69. $y' = 3\cos 5x \Rightarrow y = 3(\frac{1}{5})\sin 5x + C$

71. $y' = 3\sec^2 3x \Rightarrow y' = 3(\frac{1}{3})\tan 3x + C = \tan 3x + C$

73. $y' = \frac{1}{2} - \cos x \Rightarrow y = (1/2)x - \sin x + C$

75. $y' = \sec\frac{x}{3}\tan\frac{x}{3} + 5 \Rightarrow y = 3\sec(x/3) + 5x + C$

77. $y' = \tan^2 x = \sec^2 x - 1 \Rightarrow y = \tan x - x + C$

79. $y' = 2\sin^2 x = 1 - \cos 2x \Rightarrow y = x - \frac{1}{2}\sin 2x + C$

81. $y' = 1 + x + \frac{x^2}{2} \Rightarrow y = x + \frac{x^2}{2} + \frac{x^3}{6} + C$. Since $y(0) = 1$, $1 = C$. $y = x + \frac{x^2}{2} + \frac{x^3}{6} + 1$. Or in the powers of x, $y = 1 + x + \frac{x^2}{2} + \frac{x^3}{6}$.

83. $y' = \frac{x^2+1}{x^2} = 1 + x^{-2} \Rightarrow y = x - x^{-1} + C$. Since $y(1) = -1$, $-1 = 1 - 1/1 + C$, or $C = -1$; $y = x - 1/x - 1$

85. $y'' = -\sin x \Rightarrow y' = \cos x + C$. Since $dy/dx = 1$ when $x = 0$, $C = 0$. Hence $y' = \cos x$, which gives $y = \sin x + C$. Since $y(0) = 0$, this $C = 0$ also and $y = \sin x$ is the solution.

87. a) $y'' = 0$ for all $x \Rightarrow y' = C$ b) $\frac{dy}{dx} = 1$ when $x = 0 \Rightarrow C = 1$, i.e., $y' = 1$ c) $y' = 1 \Rightarrow y = x + C$. Since $y(0) = 0$, $C = 0$. $y = x$ satisfies all the conditions.

89. Let $s(t)$ be the height of the shovelful of dirt at time t and let the bottom of the hole correspond to $s(0) = 0$. The problem states that $s'(0) = v(0) = 32$. By the law of gravity, $s''(t) = -32$ for all t. Hence $s'(t) = -32t + C$; $32 = -32 \cdot 0 + C$, so $s'(t) = -32t + 32$. $s(t) = -16t^2 + 32t + C$. $s(0) = 0 \Rightarrow C = 0$, so the height of the dirt is $s(t) = -16t^2 + 32t$. Set $s'(t) = 0$ to find the value of t at which s has a maximum: $t = 1$. $s(1) = -16 + 32 = 16$. Duck!

91. For $-\sqrt{13} < x < \sqrt{13}$, $y' > 0$ and the function is increasing. There is a local minimum at $-\sqrt{13} \approx -3.6$ and a local maximum at $\sqrt{13} \approx 3.6$. $y'' = \frac{-x}{\sqrt{x^2+3}}$ is 0 at $x = 0$; there is an inflection point at 0. Graph the slope field on $[-6, 0]$ by $[-4, 4]$ and $[0, 6]$ by $[-4, 4]$; the solution behaves like

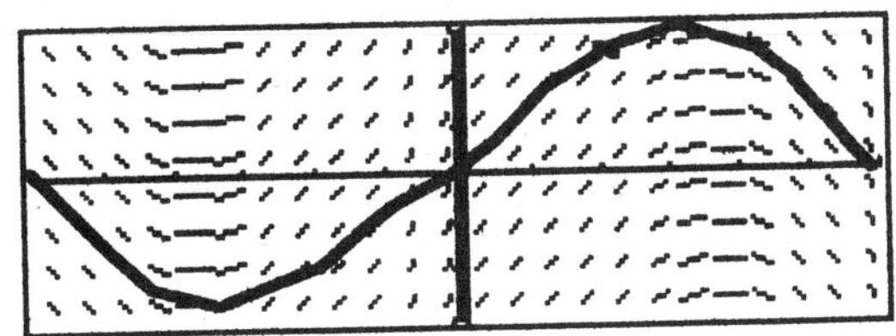

CHAPTER 5
INTEGRATION

5.1 CALCULUS AND AREA

1. a)

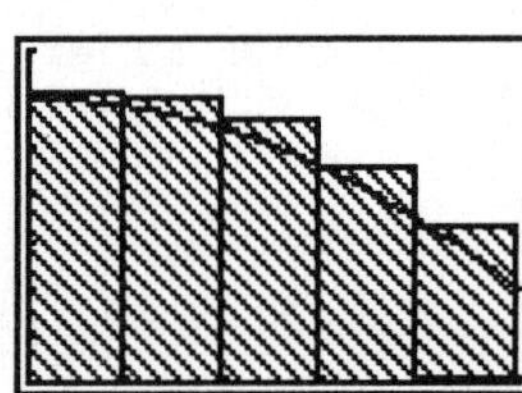  

b) $\mathrm{LRAM}_5(6-x^2) = (6-0^2)\Delta x + (6-(0.4)^2)\Delta x + (6-(0.8)^2)\Delta x + (6-(1.2)^2)\Delta x + (6-(1.6)^2)\Delta x = 10.08$(where $\Delta x = 0.4$).

$\mathrm{RRAM}_5(6-x^2) = (6-0.4^2)\Delta x+(6-0.8)^2)\Delta x+(6-1.2^2)\Delta x+(6-1.6^2)\Delta x+(6-2^2)\Delta x = 8.48$.

$\mathrm{MRAM}_5(6-x^2) = (6-0.2^2)\Delta x+(6-0.6^2)\Delta x+(6-1^2)\Delta x+(6-1.4^2)\Delta x+(6-1.8^2)\Delta x = 9.36$

3. a)

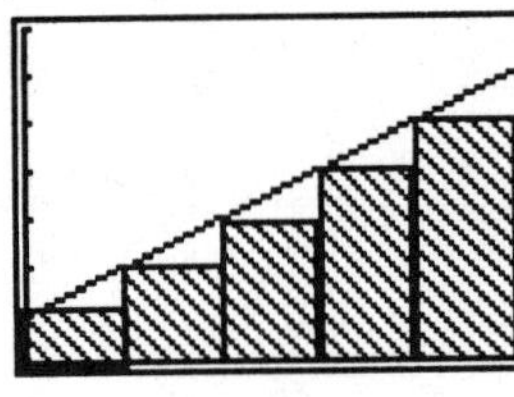  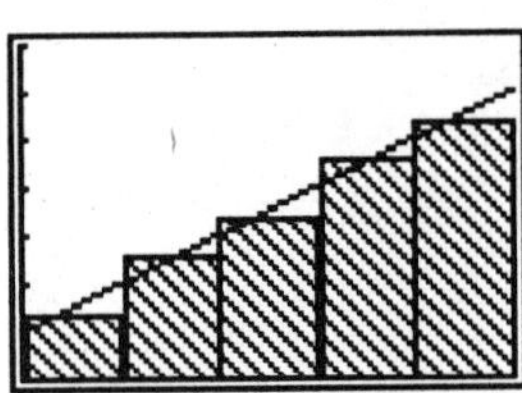

b) $\mathrm{LRAM}_5(x+1) = [(0+1)+(1+1)+(2+1)+(3+1)+(4+1)]\Delta x = 15$(where $\Delta x = 1$).

$\mathrm{RRAM}_5(x+1) = [(1+1)+(2+1)+(3+1)+(4+1)+(5+1)]\Delta x = 20$.

$\mathrm{MRAM}_5(x+1) = [(0.5+1)+(1.5+1)+(2.5+1)+(3.5+1)+(4.5+1)]\Delta x = 17.5$.

5. a)

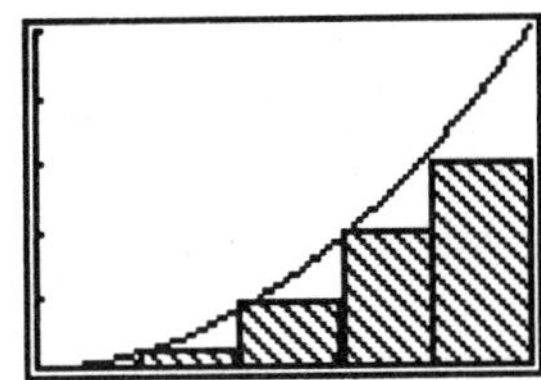 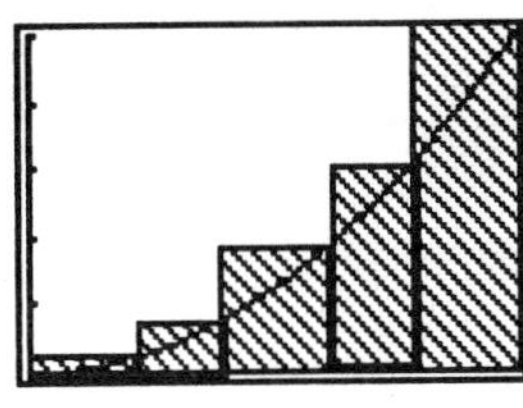 

b) $\text{LRAM}_5(2x^2) = [2(0)^2 + 2(1)^2 + 2(2^2) + 2(3^2) + 2(4^2)]\Delta x = 60$ (where $\Delta x = 1$).

$\text{RRAM}_5(2x^2) = [2(1^2) + 2(2^2) + 2(3^2) + 2(4^2) + 2(5^2)]\Delta x = 110.$

$\text{MRAM}_5(2x^2) = [2(0.5^2) + 2(1.5^2) + 2(2.5^2) + 2(3.5^2) + 2(4.5^2)]\Delta x = 82.5.$

7. We may verify $f(x) = x^2 - x + 3 \geqq 0$ on $[0, 3]$ by seeing that its graph does not fall below the x-axis on the interval.

n	$\text{LRAM}_n f$	$\text{RRAM}_n f$	$\text{MRAM}_n f$
10	12.645	14.445	13.4775
100	13.41045	13.59045	13.499775
1000	13.4910045	13.5090045	13.49999775

9.

n	$\text{LRAM}_n f$	$\text{RRAM}_n f$	$\text{MRAM}_n f$
10	268.125	393.125	325.9375
100	321.28125	333.78125	327.48438..
1000	326.87531..	328.12531..	327.49984..

11.

n	$\text{LRAM}_n f$	$\text{RRAM}_n f$	$\text{MRAM}_n f$
10	1.98352..	1.98352..	2.00825..
100	1.99984..	1.99984	2.00008..
1000	1.99999..	1.99999..	2.00000..

13.

n	$\text{LRAM}_n f$	$\text{RRAM}_n f$	$\text{MRAM}_n f$
10	1.77264..	1.77264..	1.77227..
100	1.77245..	1.77245..	1.77245..
1000	1.77245..	1.77245..	1.77245..

15. 17.5, 83, 13.5, 327.5, 2, respectively.

19. $\sum_{k=1}^{4} \frac{1}{k} = \frac{1}{1} + \frac{1}{2} + \frac{1}{3} + \frac{1}{4} = \frac{25}{12}$

21. $\sum_{k=1}^{3}(k+2) = (1+2) + (2+2) + (3+2) = 12$

23. $\sum_{k=0}^{4} \frac{k}{4} = \frac{0}{4} + \frac{1}{4} + \frac{2}{4} + \frac{3}{4} + \frac{4}{4} = \frac{5}{2}$

25. $\sum_{k=1}^{4} \cos k\pi = \cos(1 \cdot \pi) + \cos 2\pi + \cos 3\pi + \cos 4\pi = 0$

27. $\sum_{k=1}^{4}(-1)^k = (-1)^1 + (-1)^2 + (-1)^3 + (-1)^4 = 0$

29. All

31. $1+2+3+4+5+6 = \sum_{k=1}^{6} k$

33. $\sum_{k=1}^{4} \frac{1}{2^k}$

35. $\sum_{k=1}^{5}(-1)^{k+1}\frac{k}{5}$

37. $\sum_{k=1}^{10} k = \frac{10(11)}{2} = 55$

39. $\sum_{k=1}^{6} -k^2 = -\sum_{k=1}^{6} k^2 = -\frac{6(6+1)(2\cdot 6+1)}{6} = -91$

41. $\sum_{k=1}^{5} k(k-5) = \sum_{k=1}^{5}(k^2-5k) = \sum_{k=1}^{5} k^2 - 5\sum_{k=1}^{5} k = \frac{5(6)11}{6} - \frac{5(5)(6)}{2} = -20$

43. $\sum_{k=1}^{100} k^3 - \sum_{k=1}^{99} k^3 = 100^3 + \sum_{k=1}^{99} k^3 - \sum_{k=1}^{99} k^3 = 1{,}000{,}000$

45. a) $\sum_{k=1}^{n} 3a_k = 3\sum_{k=1}^{n} a_k = 3(-5) = -15$

b) $\sum_{k=1}^{n} \frac{b_k}{6} = \frac{1}{6}\sum_{k=1}^{n} b_k = 1$

c) $\sum_{k=1}^{n}(a_k + b_k) = \sum_{k=1}^{n} a_k + \sum_{k=1}^{n} b_k = -5 + 6 = 1$

d) $\sum_{k=1}^{n}(a_k - b_k) = \sum_{k=1}^{n} a_k - \sum_{k=1}^{n} b_k = -11$

e) $\sum_{k=1}^{n}(b_k - 2a_k) = \sum_{k=1}^{n} b_k - 2\sum_{k=1}^{n} a_k = 16$

47. $\frac{6\cdot 1}{1+1} + \frac{6\cdot 2}{2+1} + \frac{6\cdot 3}{3+1} + \frac{6\cdot 4}{4+1} + \frac{6\cdot 5}{5+1}$. $\sum_{k=1}^{100} \frac{6k}{k+1} = 574.816..$

49. $1(1-1)(1-2) + 2(2-1)(2-2) + 3(3-1)(3-2) + 4(4-1)(4-2) + 5(5-1)(5-2)$. $\sum_{k=1}^{500} k(k-1)(k-2) = 15{,}562{,}437{,}750$

51. $\sum_{k=1}^{12} k = 12(13)/2 = 78$

53. $\sum_{k=1}^{n} k + \sum_{k=1}^{n-1} k = \frac{n(n+1)}{2} + \frac{(n-1)n}{2} = \frac{n}{2}(n+1+n-1) = n^2$

55. Because $0 < x_{k-1} < x_k$ in the partition of $[a, b]$, it follows that $\frac{1}{x_{k-1}} > \frac{1}{x_k}$. Therefore $\text{LRAM}_n(1/x) = (\frac{1}{x_0} + \frac{1}{x_1} + \frac{1}{x_2} + \cdots + \frac{1}{x_{n-1}})\Delta x > (\frac{1}{x_1} + \frac{1}{x_2} + \cdots + \frac{1}{x_n})\Delta x = \text{RRAM}_n(1/x)$.

57. Let A_k be the area of the region under $y = f(x)$ over the x-axis from $x = x_{k-1}$ to $x = x_k$. Since $f(x)$ is decreasing and nonnegative, $f(x_{k-1})\Delta x > A_k > f(x_k)\Delta x$ and therefore $\sum_{k=1}^n f(x_{k-1})\Delta x > \sum_{k=1}^n A_k > \sum_{k=1}^n f(x_k)\Delta x$ which is to say $\text{LRAM}_n f > A_a^b f > \text{RRAM}_n f$ or $\text{RRAM}_n f < A_a^b f < \text{LRAM}_n f$.

59. The results of Example 2 disprove the statement.

61. Since $x = \frac{a+b}{2}$ is a line of symmetry, $f(a) = f(b)$ or $f(x_0) = f(x_n)$. The result now follows from #58 b).

63. $\Delta x = \frac{5}{n}$, $x_k = \frac{5k}{n} \cdot \text{RRAM}_n(2x^2) = 2\sum_{k=1}^n (\frac{5k}{n}^2 \frac{5}{n} = (\frac{250}{n^3})\frac{n(n+1)(2n+1)}{6} = \frac{125}{3}(\frac{n+1}{n})(\frac{2n+1}{n}) \to \frac{125}{3}(1)(2) = \frac{250}{3}$.

65. $\Delta x = \frac{3}{n}$, $x_k = \frac{3k}{n} \cdot \text{RRAM}_n(x^2 - x + 3) = \sum_{k=1}^n \left[\left(\frac{3k}{n}\right)^2 \frac{3}{n} - \frac{3k}{n}\frac{3}{n} + 3\left(\frac{3}{n}\right)\right] = \frac{27}{n^3}\sum_{k=1}^n k^2 - \frac{9}{n^2}\sum_{k=1}^n k + \frac{9}{n}(n) = \frac{27}{n^3}\frac{n(n+1)(2n+1)}{6} - \frac{9}{n^2}\frac{n(n+1)}{2} + 9 = \frac{9}{2}\left(\frac{n+1}{n}\right)\left(\frac{2n+1}{n}\right) - \frac{9}{2}\left(\frac{n+1}{n}\right) + 9 \to \frac{9}{2}(1)(2) - \frac{9}{2}(1) + 9 = \frac{27}{2}$.

67. $\Delta x = \frac{4}{n}$, $x_k = -1 + \frac{4k}{n}$. $\text{RRAM}_n f = \sum_{k=1}^n[(-1 + \frac{4k}{n})^3 + (-1 + \frac{4k}{n})^2$ $2(-1 + \frac{4k}{n}) + 3]\frac{4}{n} = \frac{4}{n}\sum_{k=1}^n[1 + \frac{12k}{n} - \frac{32k^2}{n^2} + \frac{64k^3}{n^3}] = \frac{4}{n}[n + \frac{12}{n}\frac{n(n+1)}{2} - \frac{32}{n^2}\frac{n(n+1)(2n+1)}{6} + \frac{64}{n^3}\frac{n^2(n+1)^2}{4}] = 4 + 24(\frac{n+1}{n}) - \frac{64}{3}(\frac{n+1}{n})(\frac{2n+1}{n}) + 64(\frac{n+1}{n})^2 \to 4 + 24 - \frac{128}{3} + 64 = \frac{148}{3}$.

69. a) $\sum_{k=1}^n(2k-1)^2 = 1^2 + 2^2 + \cdots + (2n)^2 - [2^2 + 4^2 + 6^2 + \cdots + (2n)^2] = \sum_{k=1}^{2n} k^2 - \sum_{k=1}^n (2k)^2 = \sum_{k=1}^{2n} k^2 - 4\sum_{k=1}^n k^2 = \frac{2n(2n+1)(4n+1)}{6} - \frac{4n(n+1)(2n+1)}{6} = \frac{2n(2n+1)}{6}[4n + 1 - 2(n+1)] = \frac{(2n-1)n(2n+1)}{3}$.

b) $\sum_{k=1}^n(2k-1)^3 = \sum_{k=1}^{2n} k^3 - \sum_{k=1}^n (2k)^3 = \sum_{k=1}^{2n} k^3 - 2^3\sum_{k=1}^n k^3 = [\frac{2n(2n+1)}{2}]^2 - 8[\frac{n(n+1)}{2}]^2 = n^2(2n+1)^2 - 2n^2(n+1)^2 = n^2(2n^2-1)$.

71. $\text{LRAM}_n x^3 = \sum_{k=0}^{n-1} x_k^3\, \Delta x = \sum_{k=0}^{n-1}(\frac{5k}{n})^3\frac{5}{n} = \frac{5^4}{n^4}\sum_{k=0}^{n-1} k^3 = \frac{5^4}{n^4}\sum_{k=1}^{n-1} k^3 = \frac{5^4}{n^4}[\frac{(n-1)n}{2}]^2 = \frac{625}{4}(\frac{n-1}{n})^2 \to \frac{625}{4} = 156.25$.

73. In the solution to Exercise 71, we replace the 5 by x. $\text{LRAM}_n t^3 = \frac{x^4}{4}(\frac{n-1}{n})^2$. $A_0^x(t^3) = \lim_{n\to\infty} \frac{x^4}{4}(\frac{n-1}{n})^2 = \frac{x^4}{4}$.

75. The explanation is below the picture.

5.2 DEFINITE INTEGRALS

1. a) b) c)

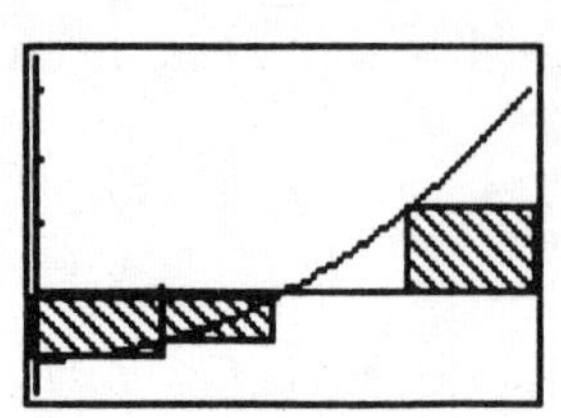
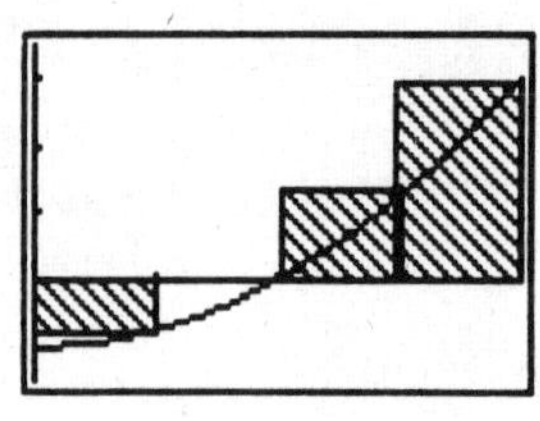
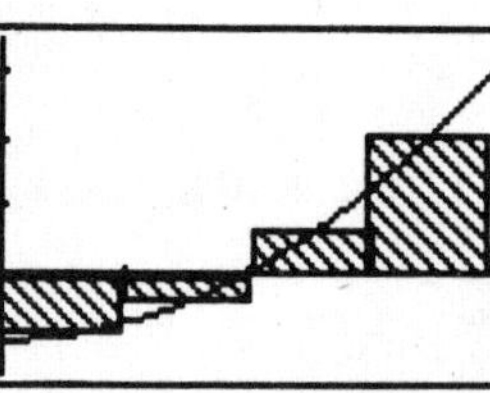

3. a) b) c)

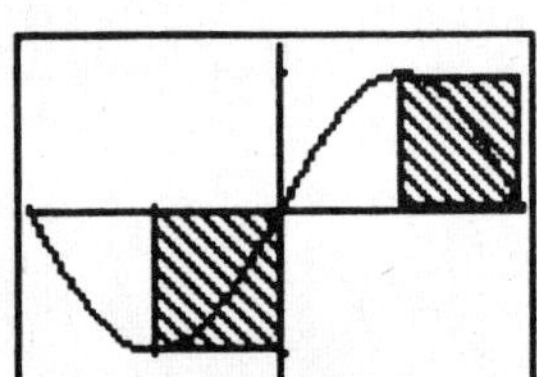
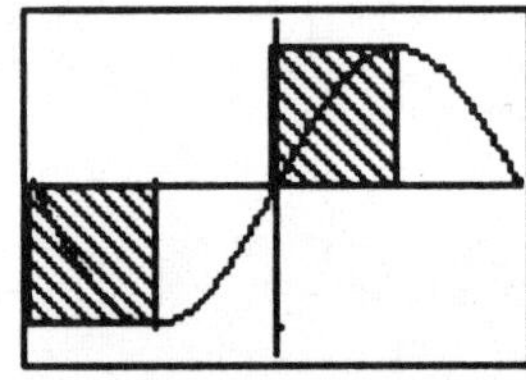
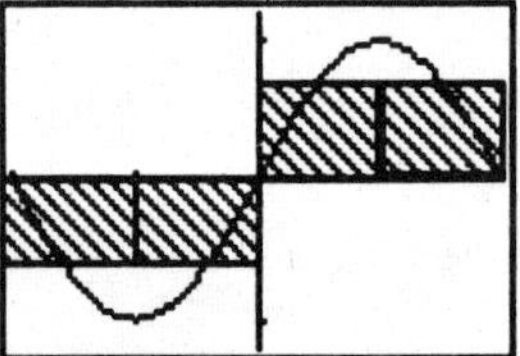

5. $\Delta x_1 = 1.2$, $\Delta x_2 = 1.5-1.2 = 0.3$, $\Delta x_3 = 2.3-1.5 = 0.8$, $\Delta x_4 = 0.3$, $\Delta x_5 = 0.4$. The largest of these is 0.8 so $\|P\| = 0.8$.

7. $\int_0^2 x^2 dx$ **9.** $\int_{-7}^5 (x^2 - 3x)dx$ **11.** $\int_2^3 \frac{1}{1-x} dx$ **13.** $\int_0^4 \cos x \, dx$

15. Graph $y = x^2 - 4$ in $[0, 2]$ by $[-5, 1]$. We see that $y = f(x) \leqq 0$ on the interval. $A = -\int_0^2 (x^2 - 4)dx = 5.333$.

17. $f(x) = \sqrt{25 - x^2} \geqq 0$. $A = \int_0^5 \sqrt{25 - x^2} dx = 19.634$.

19. $f(x) = \tan x \geqq 0$ for $0 \leqq x \leqq \frac{\pi}{4}$. $A = \int_0^{\pi/4} \tan x \, dx = 0.346$.

21. $f(x) = x^2 e^{-x^2}$, $[0, 3]$.

	LRAM$_n$f	RRAM$_n$f	MRAM$_n$f
n = 100	0.44290145	0.44293477	0.44291878
n = 1000	0.44291688	0.44292022	0.44291856
NINT(f)		0.442918559	

23. $f(x) = \frac{\sin x}{x}$, $[1, 10]$.

	LRAM$_n$f	RRAM$_n$f	MRAM$_n$f
n = 100	0.75272914	0.67210056	0.71218935
n = 1000	0.71629745	0.70823459	0.71226377
NINT(f)		0.71226452	

25. $f(x) = x \sin x$, $[-1, 2]$

	LRAM$_n$f	RRAM$_n$f	MRAM$_n$f
n = 100	2.0282123	2.0575260	2.0427050
n = 1000	2.0412951	2.0442265	2.0427592
NINT(f)		2.0427597	

27. NINT$((2 - x - 5x^2, x, -1, 3) = -42.666$.

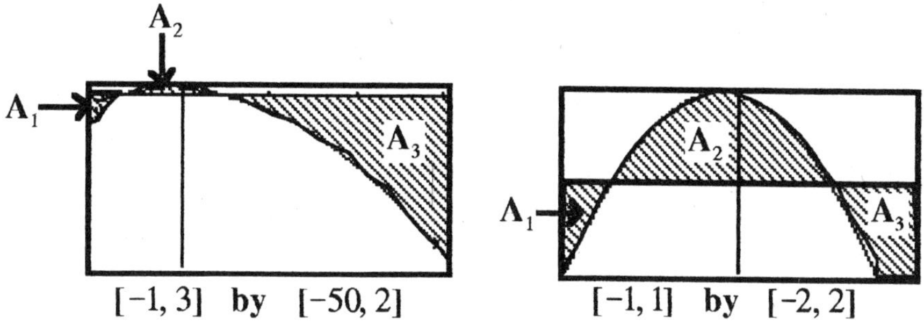

The definite integral has the value $A_2 - A_1 - A_3$. Since A_1 and A_3 are below the x-axis, they each contribute a negative value to the integral.

29. NINT$(\sin(x^2), x, 0, 2\pi) = 0.642$. Graph $y = \sin(x^2)$ in $[0, 2\pi]$ by $[-1, 1]$. The integral is the sum of the signed areas (positive if above the x-axis, negative if below the x-axis) between the x-axis and the curve.

31. $\int_{-1}^{1} \sqrt{1-x^2}dx = \frac{\pi}{2}$ (area of upper half of unit circle). NINT$(\sqrt{1-x^2}, x, -1, 1) = 1.571$.

33. $\int_{-1}^{1}(1-|x|)dx =$ area of rectangle with vertices $(-1, 0)$, $(-1, 1)$, $(1, 1)$, $(1, 0)$ minus the area of the triangle with vertices $(0, 0)$, $(-1, 1)$, $(1, 1)$. $A = 2(1) - \frac{1}{2}(2)1 = 1$. NINT$(1 - |x|, x, -1, 1) = 1$.

35. For $[0, 5] : \Delta x = 5/n$. $x_k = k\Delta x - \frac{\Delta x}{2} = (2k-1)\frac{\Delta x}{2}$. MRAM$_n$f $= \frac{1}{8}\sum_{k=1}^{n}(2k-1)^3(\Delta x)^4 = \frac{5^4}{8n^4}n^2(2n^2-1) = \frac{625}{8}(2 - \frac{1}{n^2})$. Lim$_{n\to\infty}$MRAM$_n$f $= \frac{625}{4}$. For $[0, a] : \Delta x = a/n$, $x_k = k\Delta x - \frac{\Delta x}{2} = (2k-1)\frac{\Delta x}{2}$. MRAM$_n$f $= \frac{1}{8}\sum_{k=1}^{n}(2k-1)^3\Delta x^4 = \frac{a^4}{8n^4}n^2(2n^2-1) = \frac{a^4}{8}(2 - \frac{1}{n^2})$. Lim$_{n\to\infty}$MRAM$_n$f $= \frac{a^4}{4}$.

37. $|x| = x$ if $x \geqq 0$ and $|x| = -x$ if $x < 0$. Thus $\frac{|x|}{x} = 1$ if $x > 0$ and $\frac{|x|}{x} = -1$ if $x < 0$. $x = 0$ is the only discontinuity. Graph $y_1 = 1 + 0\sqrt{x}$ and $y_2 = -1 + 0\sqrt{-x}$ in $[-2, 3]$ by $[-2, 2]$. The value of the integral is the net area between the graph and the x-axis from $x = -2$ to $x = 3$ counting any area below the x-axis as a negative contribution. Thus integral $= -2+3 = 1$.

39. Graph $y = \frac{x^2-1}{x+1}$ in $[-3, 4]$ by $[-5, 4]$. Since $y = \frac{(x-1)(x+1)}{x+1} = x - 1$ when $x \neq -1$, the graph is a straight line with a hole at $x = -1$, the only discontinuity. On $[-3, 4]$ the area between the curve and the x-axis consists of two triangles, one below and one above the x-axis separated at the point $(1, 0)$. Thus the net area $= I = -[\frac{1}{2}(4)(4)] + \frac{1}{2}(3)(3) = -\frac{7}{2}$.

41. a) Since $\lim_{x\to 0} \frac{\sin x}{x} = 1$, $g(x) = \frac{\sin x}{x}$, $x \neq 0$, $g(0) = 1$ is the continuous extension of y. b) 1.848 c) 3.697

43. Let $f(x) = \frac{\sin x}{x}$. $f(-x) = \frac{\sin(-x)}{-x} = \frac{-\sin(x)}{-x} = \frac{\sin x}{x} = f(x)$. Thus $f(-x) = f(x)$, $f(x)$ is an even function, its graph is symmetric about the y-axis. Therefore the net area between the curve and the x-axis from $x = -3$ to $x = 0$ equals that from $x = 0$ to $x = 3$, and is $\frac{1}{2}$ that from $x = -3$ to $x = 3$.

45. They are equal for each of the three functions. Conjecture: If $a \leqq c \leqq b$, then $\int_a^b f(x)dx = \int_a^c f(x)dx + \int_c^b f(x)dx$ when these integrals exist.

47. From the graph of $f(x)$ we see that $\int_{-1}^0 f(x)dx$ and $\int_2^3 f(x)dx$ are negative numbers because they are related to areas below the x-axis while $\int_0^2 f(x)dx$ is positive. We therefore take $a = c = -1$, $b = 1$.

49. In work with $\text{RRAM}_n(x^2)$ on $[0, 1]$, we have $\Delta x = \frac{1-0}{n} = \frac{1}{n}$ and $x_k = \frac{k}{n}$. Thus $\text{RRAM}_n(x^2) = (\frac{1}{n})^2\frac{1}{n} + (\frac{2}{n})^2\frac{1}{n} + \cdots + (\frac{n}{n})^2\frac{1}{n} = S_n$. So $\lim_{n\to\infty} S_n = \lim_{n\to\infty} \text{RRAM}_n(x^2) = \int_0^1 x^2dx = \text{NINT}(x^2, 0, 1) = 0.333$.

51. $S_n = \sum_{k=1}^n \frac{n+k}{n^2} = \sum_{k=1}^n (\frac{n+k}{n})\frac{1}{n} = \sum_{k=1}^n (1 + \frac{k}{n})\frac{1}{n} = \text{RRAM}_n(1 + x)$ on $[0, 1]$. $\lim_{n\to\infty} S_n = \int_0^1 (1 + x)dx = \text{NINT}(1 + x, 0, 1) = 1.5$.

53. $S_n = \sum_{k=1}^n \frac{(2n+k)^2}{n^3} = \sum_{k=1}^n (\frac{2n+k}{n})^2\frac{1}{n} = \sum_{k=1}^n (2 + \frac{k}{n})^2\frac{1}{n} = \text{RRAM}_n((2 + x)^2)$ on $[0, 1]$. $\text{Lim}_{n\to\infty} S_n = \int_0^1 (2 + x)^2dx = \text{NINT}((2 + x)^2, 0, 1) = 6.333$.

5.3 ANTIDERIVATVES AND DEFINITE INTEGRALS

1. $F(x) = \frac{x^3}{3}$. $F(2) - F(0) = \frac{8}{3}$.

3. $F(x) = \frac{2}{3}x^{3/2}$. $F(4) - F(0) = \frac{16}{3}$.

5. $F(x) = 2x - \frac{2}{3}x^{3/2}$. $F(4) - F(0) = 8 - \frac{2}{3}(4)^{3/2} = 8 - \frac{16}{3} = \frac{8}{3}$.

7. $F(x) = \frac{x^4}{4} - x^3 + 4x$. $F(2) - F(-1) = 4 - 8 + 8 - [\frac{1}{4} + 1 - 4] = \frac{27}{4}$.

9. $F(x) = \sin x$. $F(\frac{\pi}{2}) - F(0) = \sin\frac{\pi}{2} - \sin 0 = 1$.

11. $F(x) = -\frac{1}{2}\cos 2x$. $F(\frac{\pi}{2}) - F(0) = -\frac{1}{2}[\cos\pi - \cos 0] = 1$.

13. $F(x) = \frac{\sin\pi x}{\pi}$. $F(\frac{1}{2}) - F(-\frac{1}{2}) = \frac{1}{\pi}[\sin\frac{\pi}{2} - \sin(-\frac{\pi}{2})] = \frac{2}{\pi}$.

15. $F(x) = \tan x$. $F(\frac{\pi}{3}) - F(-\frac{\pi}{4}) = \tan\frac{\pi}{3} - \tan(-\frac{\pi}{4}) = \sqrt{3} + 1$.

17. $F(x) = \frac{x^{n+1}}{n+1}$ is antiderivative of x^n. $A = F(b) - F(0) = \frac{b^{n+1}}{n+1}$.

19. a) 0 b) $\int_5^1 g(x)dx = -\int_1^5 g(x)dx = -8$ c) $\int_1^2 3f(x)dx = 3\int_1^2 f(x)dx = -12$ d) $\int_2^5 f(x)dx = \int_2^1 f(x)dx + \int_1^5 f(x)dx = -\int_1^2 f(x)dx + 6 = 10$ e) $\int_1^5[f(x) - g(x)]dx = \int_1^5 f(x)dx - \int_1^5 g(x)dx = 6 - 8 = -2$ f) $\int_1^5[4f(x) - g(x)]dx = \int_1^5 4f(x)dx - \int_1^5 g(x)dx = 4\int_1^5 f(x)dx - \int_1^5 g(x)dx = 4(6) - 8 = 16$

21. a) $\int_1^2 f(u)du = 5$ b) $\int_2^1 f(t)dt = -\int_1^2 f(t)dt = -5$ c) $\int_2^3 f(y)dy = \int_1^3 f(y)dy - \int_1^2 f(y)dy = 2 - 5 = -3$

23. An antiderivative of x^3 is $\frac{x^3}{3}$. Hence $\int_{-3}^5 x^2dx = \frac{5^3}{3} - \frac{(-3)^3}{3} = \frac{125+27}{3} = \frac{152}{3}$. $\int_{-3}^5(2 - x^2)dx = \int_{-3}^5 2\,dx - \int_{-3}^5 x^2dx = 2[5 - (-3)] - \frac{152}{3} = 16 - \frac{152}{3} = -\frac{104}{3}$. The average value is $\frac{1}{5-(-3)}\int_{-3}^5(2 - x^2)dx = \frac{1}{8}(-\frac{104}{3}) = -\frac{13}{3}$.

27. We use $\frac{1}{5-0}$ NINT$(x\sin x, x, 0, 5) = -0.475...$

29. Min$(\frac{1}{1+x^2}) = \frac{1}{1+1^2} = \frac{1}{2}$ and max $\frac{1}{1+x^2} = \frac{1}{1+0^2} = 1$ for $0 \leqq x \leqq 1$. By Rule 7, $\frac{1}{2}(1 - 0) \leqq \int_0^1 \frac{1}{1+x^2}dx \leqq 1(1 - 0)$ or $\frac{1}{2} \leqq \int_0^1 \frac{1}{1+x^2}dx \leqq 1$.

31. Since $f(x) = x^3 + 1$ is continuous on $[0, 1]$, the Mean Value Theorem tells us that there is at least one such point c. $(c^3 + 1)(1 - 0) = \frac{5}{4}$, $c^3 = \frac{1}{4}$, $c = \frac{1}{\sqrt[3]{4}}$. Graph $y_1 = x^3 + 1$ and $y_2 = f(c) = \frac{5}{4}$ in $[0, 1]$ by $[-1, 2]$.

33. $f(c)(b - a) = \frac{\pi}{4}$, $\frac{1}{c^2+1}(1 - 0) = \frac{\pi}{4}$, $c^2 + 1 = \frac{4}{\pi}$, $c = \sqrt{\frac{4}{\pi} - 1} \approx 0.52$. Graph $y_1 = \frac{1}{x^2+1}$, $y_2 = f(c) = \frac{\pi}{4}$ in $[0, 1]$ by $[-0.5, 1]$.

35. We are given that $\int_{-1}^{2} f(t)dt = (2-(-1))5 = 15$ and $\int_2^7 f(t)dt = (7-2)3 = 15$. $\int_{-1}^{7} f(t)dt = \int_{-1}^{2} f(t)dt + \int_2^7 f(t)dt = 15 + 15 = 30$. Therefore, av. value of f on $[-1,7] = \frac{1}{7-(-1)} \int_{-1}^{7} f(t)dt = \frac{30}{8} = \frac{15}{4}$.

37. This is an immediate consequence of the Mean Value Theorem for Definite Integrals.

5.4 THE FUNDAMENTAL THEOREM OF CALCULUS

1. $\int_0^3 (4 - x^2)dx = \left[4x - \frac{x^3}{3}\right]_0^3 = 4 \cdot 3 - \frac{3^3}{3} - (4 \cdot 0 - \frac{0^3}{3}) = 3.$

3. $\int_0^1 (x^2 + \sqrt{x})dx = \frac{x^3}{3} + \frac{2}{3}x^{3/2} \,|_0^1 = \frac{1}{3} + \frac{2}{3} = 1.$

5. $\int_1^{32} x^{-6/5}dx = -5x^{-1/5} \,|_1^{32} = -5\left(\frac{1}{32^{1/5}} - 1\right) = -5\left(\frac{1}{2} - 1\right) = \frac{5}{2}.$

7. $\int_0^\pi \sin x dx = -\cos x \,|_0^\pi = -(\cos\pi - \cos 0) = -(-1-1) = 2.$

9. $\int_0^{\pi/3} 2\sec^2 x dx = 2\tan x \Big|_0^{\pi/3} = 2(\tan\pi/3 - \tan 0) = 2\left(\sqrt{3} - 0\right) = 2\sqrt{3}.$

11. $\int_{\pi/4}^{3\pi/4} \csc x \cot x dx = -\csc x \Big|_{\pi/4}^{3\pi/4} = -\left(\csc\frac{3\pi}{4} - \csc\frac{\pi}{4}\right) = -\left(\sqrt{2} - \sqrt{2}\right) = 0.$

13. $\int_{-1}^{1}(r+1)^2 dr = \frac{1}{3}(r+1)^3]_{-1}^{1} = \frac{1}{3}[(1+1)^3 - (-1+1)^3] = \frac{8}{3}$. Alternatively, $\int_{-1}^{1}(r^2+2r+1) = \frac{r^3}{3} + r^2 + r]_{-1}^{1} = (\frac{1}{3}+1+1) - (-\frac{1}{3}+1-1) = \frac{1}{3}+2+\frac{1}{3} = \frac{8}{3}.$

15. $A = \int_0^2 (2-x)dx - \int_2^3 (2-x)dx = \left[2x - \frac{x^2}{2}\right]_0^2 - \left[2x - \frac{x^2}{2}\right]_2^3 = (4-2) - \left[6 - \frac{9}{2} - (4-2)\right] = 2 - \left[-\frac{1}{2}\right] = \frac{5}{2}.$

17. $y = x^3 - 3x^2 + 2x = x(x^2 - 3x + 2) = x(x-1)(x-2)$, $0 \leqq x \leqq 2$. Graphically or algebraically, we see that $y \geqq 0$ for $0 \leqq x \leqq 1$ and $y \leqq 0$ for $1 \leqq x \leqq 2$. Thus $A = \int_0^1 f(x)dx - \int_1^2 f(x)dx = \left[\frac{x^4}{4} - x^3 + x^2\right]_0^1 - \left[\frac{x^4}{4} - x^3 + x^2\right]_1^2 = \left[\frac{1}{4} - 1 + 1\right] - \left[\left(\frac{16}{4} - 8 + 4\right) - \left(\frac{1}{4} - 1 + 1\right)\right] = \frac{1}{4} - \left[\frac{15}{4} - 4\right] = 4 - \frac{14}{4} = \frac{1}{2}.$

19. Let $f(x) = \frac{x^2-1}{x+1}$. If $x \neq -1$, $f(x) = \frac{(x-1)(x+1)}{x+1} = x - 1$. Thus $x - 1$ is a continuous extension of $f(x)$ and the discontinuity at $x = -1$ is removable. In this case $\int_{-2}^{3} f(x)dx = \int_{-2}^{3}(x-1)dx = \left[\frac{x^2}{2} - x\right]_{-2}^{3} = \frac{3}{2} - \frac{8}{2} = -\frac{5}{2}.$

21. Tan x has a discontinuity at $x = \frac{\pi}{2}$ so Theorem 4 does not apply. Tan x is unbounded on $[0, \frac{\pi}{2})$. The integral does not exist.

23. $f(x) = \frac{\sin x}{x}$ has a removable discontinuity at $x = 0$ since $\frac{\sin x}{x} \to 1$ as $x \to 0$. NINT$\left(\frac{\sin x}{x}, x, -1, 2\right) = 2.551..$

25. $\int_0^1 x^2 dx + \int_1^2 (2 - x)dx = \frac{x^3}{3}\Big]_0^1 + \left[2x - \frac{x^2}{2}\right]_1^2 = \frac{1}{3} + (4-2) - (2 - \frac{1}{2}) = \frac{1}{3} + \frac{1}{2} = \frac{5}{6}$.

27. Area of rectangle − area under curve $= 2\pi - \int_0^\pi (1 + \cos x)dx = 2\pi - [x + \sin x]_0^\pi = 2\pi - [\pi + 0 - (0 + 0)] = \pi$.

29. $F(x) = \int_0^x (t - 2)dt = [\frac{t^2}{2} - 2t]_0^x = \frac{x^2}{2} - 2x$. The two graphs are the same in the standard viewing rectangle $[-10, 10]$ by $[-10, 10]$. $F(0.5) = -0.875$ and we get the same value after zooming in. $F(1) = -1.5$ compared to -1.516 as one approximation. $F(1.5) = -1.875$ compared to -1.879 as one approximation. $F(2) = -2$ compared to -2. $F(5) = 2.5$ compared to 2.53 as one approximation.

31. $F(x) = \int_0^x (t^2 - 3t + 6)dt = [\frac{t^3}{3} - \frac{3}{2}t^2 + 6t]_0^x = \frac{x^3}{3} - \frac{3}{2}x^2 + 6x$. The two graphs are indistinguishable in the viewing rectangle $[-15, 15]$ by $[-1,000, 1,000]$. The values of $F(x)$ and NINT$(f(t), t, 0, x)$ agree when accurately calculated.

33. Graph $y =$ NINT$(t^2 \sin t, t, 0, x)$ in the viewing window $[-3, 3]$ by $[0, 9]$.

35. Graph $y =$ NINT$(5e^{-0.3t^2}, t, 0, x)$ in the viewing window $[0, 5]$ by $[0, 10]$.

37. We graph $y =$ NDER(NINT$(4 - t^2, t, 0, x), x, x)$ and $y = 4 - x^2$ in $[-5, 5]$ by $[-21, 4]$ and see that the graphs are identical. This visually supports $D_x(\int_0^x (4 - t^2)dt) = 4 - x^2$.

39. We know the same K works for all x because the two integrals differ by a constant. Let $x = b = 2$. Then the equation becomes $\int_{-1}^2 f(t)dt + K = \int_2^2 f(t)dt$. Evaluating, we get $\frac{3}{2} + K = 0$, $K = -\frac{3}{2}$.

41. We solve $\int_0^x e^{-t^2} dt = 0.6$. Graph $y =$ NINT(e^$(-t^2), t, 0, x)$ and $y = 0.6$ in $[0, 20]$ by $[0, 2]$. Zooming in to the point of intersection, we find $x = 0.70$. Remark: On the TI-85 one may enter the functions and range and then use the ISECT function.

43. $\sqrt{1 + x^2}$

45. $y = \int_0^{\sqrt{x}} \sin(t^2)dt$. Let $u = \sqrt{x}$, $y = \int_0^u \sin(t^2)dt$. $\frac{dy}{dx} = \frac{dy}{du}\frac{du}{dx} = \sin(u^2)\frac{1}{2\sqrt{x}} = \frac{\sin x}{2\sqrt{x}}$.

47. d) **49.** b) **51.** $x = a$

53. Differentiating both sides of the given equation, we obtain $f(x) = \cos \pi x - \pi x \sin \pi x$. $f(4) = \cos 4\pi - 4\pi \sin 4\pi = 1$.

55. If we start with $x = 0$ on the graph of $y = \sin kx$, the first arch will be completed when $kx = \pi$ or $x = \pi/k$. $A = \int_0^{\pi/k} \sin(xk)dx = -\frac{\cos(kx)}{k}]_0^{\pi/k} = -\frac{1}{k}[\cos(k(\pi/k)) - \cos 0] = -\frac{1}{k}[-1-1] = \frac{2}{k}$.

57. a) $c(100) - c(1) = \int_1^{100} \frac{dx}{2\sqrt{x}} = \sqrt{x}]_1^{100} = 9$ dollars. b) $c(400) - c(100) = \int_{100}^{400} \frac{dx}{2\sqrt{x}} = \sqrt{x}]_{100}^{400} = 20 - 10 = 10$ dollars.

59. $I_{av} = \frac{1}{30-0}\int_0^{30}(450 - \frac{x^2}{2})dx = \frac{1}{30}[450x - \frac{x^3}{6}]_0^{30} = \frac{1}{30}[450(30) - \frac{30^3}{6}] = 300$. Average daily holding cost $= 0.02(300) = 6$ dollars per day.

61. a) Compare your drawing with the result of graphing $y = (\cos x)/x$ in $[-15, 15]$ by $[-1, 1]$. The $x-$ and $y-$axes are asymptotes. b) Graph $y = \text{NINT}((\cos t)/t, t, 1, x)$ in $[0, 15]$ by $[-1, 1]$. c) Because $f(0)$ is undefined. d) For $x > 0$, $g(x)$ and $h(x)$ have the same derivative $f(x)$ and so they differ by an additive constant. This is confirmed if one graph can be obtained from the other by a vertical shift. Along with the function in b), graph $y = \text{NINT}((\cos t)/t, t, 0.5, x)$ in $[0.01, 3]$ by $[-3, 3]$ to see that this is the case. Alternatively, $\int_{0.5}^x f(t)dt = \int_{0.5}^1 f(t)dt + \int_1^x f(t)dt \approx 0.5 + \int_1^x f(t)dt$.

63. $F(x) = \int_1^{x^2}\sqrt{1-t^2}dt$. a) The integrand has domain $[-1, 1]$. x^2 is in $[-1, 1]$ if x is. Hence $F(x)$ has domain $[-1, 1]$. b) $F'(x) = \sqrt{1-(x^2)^2}(2x) = 2x\sqrt{1-x^4}$. $F'(x) = 0$ when $x = 0, \pm 1$. $F''(x) = 2x\frac{(-4x^3)}{2\sqrt{1-x^4}} + 2\sqrt{1-x^4} = \frac{-4x^4+2(1-x^4)}{\sqrt{1-x^4}} = \frac{2(1-3x^4)}{\sqrt{1-x^4}}$. We see that $F'' < 0$ for $-1 < x < -\frac{1}{\sqrt[4]{3}}$ and for $\frac{1}{\sqrt[4]{3}} < x < 1$ and $F''(x) > 0$ for $-\frac{1}{\sqrt[4]{3}} < x < \frac{1}{\sqrt[4]{3}}$. Thus F' decreases on $(-1, -\frac{1}{\sqrt[4]{3}})$ and $(\frac{1}{\sqrt[4]{3}}, 1)$ and increases on $(-\frac{1}{\sqrt[4]{3}}, \frac{1}{\sqrt[4]{3}})$. F' has a local maximum at $x = 1$, local minimum at $x = -\frac{1}{\sqrt[4]{3}}$, local maximum at $x = \frac{1}{\sqrt[4]{3}}$ and local minimum at $x = 1$. d) The zeros of $F'(x)$ tell us where the graph has a horizontal tangent and possible local extremes. The graph of F is concave up where F' is increasing and concave down where F' is decreasing. The local extrema of F' at $x = \pm\frac{1}{\sqrt[4]{3}}$ correspond to inflection points of F. Graph $\text{NINT}(\sqrt{1-t^2}, t, 1, x^2)$ in $[-1, 1]$ by $[-2, 1]$.

65. $F(x) = \int_1^{2x}\frac{dt}{\sqrt{1-t^2}}$. a) Since we must have $|t| < 1$, we get $|2x| < 1$, and so F has domain $(-\frac{1}{2}, \frac{1}{2})$. Later work will show that F has a continuous extension to a function with domain $[-\frac{1}{2}, \frac{1}{2}]$. b) By Theorem 4 (extended form), $F'(x) = \frac{1}{\sqrt{1-(2x)^2}}(2) = \frac{2}{\sqrt{1-4x^2}}$. c) $F'(x) > 0$ for all x in the domain

(F is increasing). By inspection, F' is decreasing on $(-\frac{1}{2}, 0)$ and increasing on $(0, \frac{1}{2})$. Hence $x = 0$ is a local minimum of F', the only local extreme in $(-\frac{1}{2}, \frac{1}{2})$. d) c) tells us that the graph of F is concave down on $(-\frac{1}{2}, 0)$, concave up on $(0, \frac{1}{2})$ and that F has an inflection point at $x = 0$. In parametric mode (to save time) graph $x = t$, $y = \text{NINT}(\frac{1}{\sqrt{1-s^2}}, s, 1, 2t)$, $-0.499 \leqq t \leqq 0.499$, t step $= 0.05$ in $[-0.499, 0.499]$ by $[-\pi, 0]$.

67. $F(x) = \int_{x^2}^{x^3} \cos(2t)dt = \frac{1}{2}\sin(2t)]_{x^2}^{x^3} = \frac{1}{2}[\sin(2x^3) - \sin(2x^2)]$. $F'(x) = \frac{1}{2}[6x^2\cos(2x^3) - 4x\cos(2x^2)] = 3x^2\cos(2x^3) - 2x\cos(2x^2)$. This is supported by graphing, in parametric mode, $x_1 = t$, $y_1 = \text{NDER}(\text{NINT}(\cos(2s), s, t^2, t^3), t, t)$ and $x_2 = t$, $y_2 = 3t^2\cos(2t^3) - 2t\cos(2t^2)$, $-1.5 \leqq t \leqq 1.5$, t Step $= 0.05$, in $[-1.5, 1.5]$ by $[-12, 12]$.

5.5 INDEFINITE INTEGRALS

1. $\int x^3 dx = \frac{x^4}{4} + C$. $(\frac{x^4}{4} + C)' = x^3$

3. $\int (x+1)dx = \frac{x^2}{2} + x + C$. $(\frac{x^2}{2} + x + C)' = x + 1$

5. $\int 3\sqrt{x}dx = 3\int \sqrt{x}dx = 3(\frac{2}{3})x^{3/2} + C = 2x^{3/2} + C$

7. $\int x^{-1/3}dx = \frac{3}{2}x^{2/3} + C$

9. $\int (5x^2 + 2x)dx = \frac{5}{3}x^3 + x^2 + C$

11. $\int (2x^3 - 5x + 7)dx = \frac{1}{2}x^4 - \frac{5}{2}x^2 + 7x + C$. We graph the integrand and $\text{NDER}(\frac{1}{2}x^4 - \frac{5}{2}x^2 + 7x, x)$ in $[-3, 3]$ by $[-50, 50]$ and see that two graphs are identical.

13. $\int 2\cos x dx = 2\sin x + C$

15. $\int \sin\frac{x}{3}dx = -3\cos\frac{x}{3} + C$

17. $\int 3\csc^2 x\, dx = -3\cot x + C$

19. $\int \frac{\csc x \cot x}{2}dx = -\frac{1}{2}\csc x + C$

21. $\int (4\sec x\tan x - 2\sec^2 x)dx = 4\sec x - 2\tan x + C$. With $C = 0$, for example, we may graph the result and also $y = \text{NINT}(4\sec t\tan t - 2\sec^2 t, t, 0, x)$ in the relatively small window $[-1.5, 1.5]$ by $[-1, 20]$. Use *tol* $= 1$ to save time. We see that one graph can be obtained from the other by a vertical shift, supporting that the functions differ by a constant.

23. $\int(\sin 2x - \csc^2 x)dx = -\frac{1}{2}\cos 2x + \cot x + C$

25. $\int 4\sin^2 y\, dy = 2\int(1-\cos 2y)dy = 2(y - \frac{1}{2}\sin 2y) + C = 2y - \sin 2y + C$

27. $\int \sin x \cos x dx = \frac{1}{2}\int 2\sin x\cos x dx = \frac{1}{2}\int \sin 2x dx = -\frac{1}{4}\cos 2x + C$

29. $\int(1+\tan^2\theta)d\theta = \int \sec^2\theta\, d\theta = \tan\theta + C$

31. $\lim_{x\to 0}\frac{1-\cos x}{x^2} = \lim_{x\to 0}\frac{1-\cos x}{x^2}(\frac{1+\cos x}{1+\cos x}) = \lim_{x\to 0}\frac{1-\cos^2 x}{x^2(1+\cos x)} = \lim_{x\to 0}(\frac{\sin x}{x})^2\frac{1}{1+\cos x} = (1)^2\frac{1}{1+1} = \frac{1}{2}$. Thus the continuous extension g of f may be written as $g(x) = \frac{1-\cos x}{x^2}$, $-10 \leqq x \leqq 10$, $x \neq 0$, $g(0) = \frac{1}{2}$. Graph $y = \int_0^x f(t)dt = \text{NINT}(f(t), t, 0, x)$ in $[-10, 10]$ by $[-2, 2]$. For quicker results, use parametric mode, $x = t$, $y = \text{NINT}((1-\cos s)/s^2, s, 0, t)$, t step= 0.3, $-10 \leqq t \leqq 10$, in the above window.

33. $[\frac{(7x-2)^4}{28} + C]' = \frac{4(7x-2)^3(7)}{28} = (7x-2)^3$

35. $[-\frac{1}{x+1} + C]' = -[(x+1)^{-1}]' = -[(-1)(x+1)^{-2}] = \frac{1}{(x+1)^2}$

37. a) $(\frac{x^2}{2}\sin x)' = \frac{x^2}{2}\cos x + x\sin x$. Wrong b) $(-x\cos x)' = x\sin x - \cos x$. Wrong c) $(-x\cos x + \sin x)' = x\sin x - \cos x + \cos x = \sin x$. Right

39. $\frac{dy}{dx} = 3\sqrt{x}$, $y = \int 3\sqrt{x}dx = 3(\frac{2}{3})x^{3/2} + C = 2x^{3/2} + C$. When $x = 9$, $y = 2(9)^{3/2} + C = 54 + C = 4$, $C = -50$. $y = 2x^{3/2} - 50$

41. $y = \int_0^x 2^t dt + C$. When $x = 0$, $y = \int_0^0 2^t dt + C = C = 2$. $y = \int_0^x 2^t dt + 2$. Graph $y = 2 + \text{NINT}(2^t, t, 0, x)$ in $[-7.5, 9.5]$ by $[0, 10]$.

43. $\frac{d^2y}{dx^2} = 0$ leads to $\frac{dy}{dx} = C$ and so $\frac{dy}{dx} = 2$. $y = \int 2\, dx = 2x + C_1$. When $x = 0$, $y = 0$ so $C_1 = 0$ and $y = 2x$.

45. $\frac{dy}{dx} = \int\frac{3x}{8}dx = \frac{3}{16}x^2 + C_1$. When $x = 4$, $\frac{dy}{dx} = \frac{3}{16}(4^2) + C_1 = 3$ so $C_1 = 0$ and $\frac{dy}{dx} = \frac{3x^2}{16}$. $y = \int\frac{3}{16}x^2dx = \frac{1}{16}x^3 + C_2$. When $x = 4$, $4 = y = \frac{1}{16}(4^3) + C_2 = 4 + C_2$ so $C_2 = 0$ and $y = \frac{x^3}{16}$.

47. Step 1. $v = \frac{ds}{dt} = \int -k\, dt = -kt + C_1 = -kt + 88$ since $v = 88$ when $t = 0$. $s = \int(-kt + 88)dt = -k\frac{t^2}{2} + 88t + C_2 = -k\frac{t^2}{2} + 88t$ since $s = 0$ when $t = 0$. Step 2. $\frac{ds}{dt} = 0$ when $t = 88/k$. Step 3. $s = -\frac{k}{2}(\frac{88}{k})^2 + 88(\frac{88}{k}) = 242$ leads to $\frac{88^2}{2k} = 242$, $k = \frac{88^2}{484} = 16\text{ft/sec}^2$.

49. $\frac{ds}{dt} = \int 5.2dt = 5.2t + C_1 = 5.2t$ since $\frac{ds}{dt} = 0$ when $t = 0$. $s = \int 5.2t\, dt = \frac{5.2}{2}t^2 + C_2 = 2.6t^2$ since $s = 0$ when $t = 0$. $2.6t^2 = 4$ leads to $\sqrt{2.6}t = 2$, $t = 2/\sqrt{2.6} = 1.24\text{sec}$ approximately.

51. $\int \cos^2 x\,dx = \frac{1}{2}\int(1+\cos 2x)dx = \frac{1}{2}(x + \frac{\sin 2x}{2}) + C = \frac{x}{2} + \frac{\sin 2x}{4} + C$

55. A) From the preceding exercise $f(t) = \frac{\sqrt{\pi^2+4}}{2}t^{-1/2}$. $\int_x^1 f(t)dt = \frac{\sqrt{\pi^2+4}}{2}[2t^{1/2}]_x^1 = \sqrt{\pi^2+4}(1-\sqrt{x})$. $T = \lim_{x\to 0^+}\frac{1}{\sqrt{g}}\sqrt{\pi^2+4}(1-\sqrt{x}) = \frac{\sqrt{\pi^2+4}}{\sqrt{g}}$. C) $T = \lim_{x\to 0^+}\frac{1}{\sqrt{g}}\int_x^1 \pi\,dt = \lim_{x\to 0^+}\frac{\pi(1-x)}{\sqrt{g}} = \frac{\pi}{\sqrt{g}}$

57. In Exercise 55 we found that T for track C is $\pi/\sqrt{g}$. The times T found in Exercise 56 are all less than this.

5.6 INTEGRATION BY SUBSTITUTION - RUNNING THE CHAIN RULE BACKWARD

1. $u = 3x$, $du = 3\,dx$, $dx = \frac{1}{3}du$. $\int \sin 3x\,dx = \int \sin u \frac{1}{3}\,du = \frac{1}{3}\int \sin u\,du = \frac{1}{3}(-\cos u) + C = -\frac{1}{3}\cos 3x + C$

3. $u = 2x$, $du = 2\,dx$, $dx = du/2$. $\int \sec 2x \tan 2x\,dx = \frac{1}{2}\int \sec u \tan u\,du = \frac{1}{2}\sec u + C = \frac{1}{2}\sec 2x + C$

5. $u = 7x-2$, $du = 7dx$, $dx = du/7$. $\int 28(7x-2)^3 dx = \frac{28}{7}\int u^3 du = 4\frac{u^4}{4} + C = (7x-2)^4 + C$

7. $u = 1-r^3$, $du = -3r^2dr$, $r^2dr = -\frac{1}{3}du$. $\int \frac{9r^2\,dr}{\sqrt{1-r^3}} = (-\frac{1}{3})9\int \frac{du}{u^{1/2}} = -3\int u^{-1/2}du = -3\frac{u^{1/2}}{1/2} + C = -6\sqrt{1-r^3} + C$

9. a) $u = \cot 2\theta$, $du = -2\csc^2 2\theta\,d\theta$. $\int \csc^2 2\theta \cot 2\theta\,d\theta = -\frac{1}{2}\int u\,du = -\frac{1}{2}\frac{u^2}{2} + C = -\frac{\cot^2 2\theta}{4} + C$ b) $u = \csc 2\theta$, $du = -2\csc 2\theta \cot 2\theta\,d\theta$. $\int \csc^2 2\theta \cot 2\theta\,d\theta = -\frac{1}{2}\int u\,du = -\frac{1}{2}\frac{u^2}{2} + C = -\frac{\csc^2 2\theta}{4} + C$. The two answers can be seen to be equivalent by using the identity $\csc^2 x = 1 + \cot^2 x$.

11. $u = 2x+1$, $du = 2\,dx$, $dx = \frac{1}{2}du$. Note that $u = 1$ when $x = 0$ and $u = 2$ when $x = \frac{1}{2}$ so the u-limits are 1 to 2. $\int_0^{1/2}\frac{dx}{(2x+1)^3} = \frac{1}{2}\int_1^2 u^{-3}du = \frac{1}{2}\frac{u^{-2}}{-2}]_1^2 = -\frac{1}{4}[\frac{1}{4} - 1] = \frac{1}{4}\frac{3}{4} = \frac{3}{16}$

13. $u = \cos 2x$, $du = -2\sin 2x\,dx$, $\sin 2x\,dx = -\frac{1}{2}du$. $\int_0^{\pi/6}\frac{\sin 2x}{\cos^2 2x}dx = -\frac{1}{2}\int_1^{1/2}\frac{du}{u^2} = -\frac{1}{2}\int_1^{1/2}u^{-2}du = -\frac{1}{2}\frac{u^{-1}}{-1}]_1^{1/2} = \frac{1}{2}[2-1] = \frac{1}{2}$

15. $u = 1-x^2$, $du = -2x\,dx$, $x\,dx = -\frac{1}{2}du$. $\int_{-1}^1 x\sqrt{1-x^2}dx = -\frac{1}{2}\int_0^0 \sqrt{u}\,du = 0$

17. $u = 2+\sin x$, $du = \cos x\,dx$. $\int_{-\pi/2}^{\pi/2}\frac{\cos x}{(2+\sin x)^2}dx = \int_1^3 u^{-2}du = -u^{-1}|_1^3 = -(\frac{1}{3}-1) = \frac{2}{3}$

19. Let $u = 1 - x$, $du = -dx$. $\int \frac{dx}{(1-x)^2} = -\int u^{-2} du = u^{-1} + C = \frac{1}{1-x} + C$

21. Let $u = x + 2$, $du = dx$. $\int \sec^2(x+2)dx = \int \sec^2 u\, du = \tan u + C = \tan(x+2) + C$

23. $u = r^2 - 1$, $du = 2r\, dr$, $r\, dr = \frac{1}{2}du$. $\int 8r(r^2-1)^{1/3}dr = \frac{8}{2}\int u^{1/3}du = 4(\frac{3}{4})u^{4/3} + C = 3(r^2-1)^{4/3} + C$

25. $u = \theta + \frac{\pi}{2}$, $du = d\theta$. $\int \sec(\theta+\frac{\pi}{2})\tan(\theta+\frac{\pi}{2})d\theta = \int \sec u \tan u\, du = \sec u + C = \sec(\theta + \frac{\pi}{2}) + C$

27. $u = 1 + x^4$, $du = 4x^3 dx$, $x^3 dx = \frac{1}{4}du$. $\int \frac{6x^3}{\sqrt[4]{1+x^4}}dx = \frac{6}{4}\int u^{-1/4}du = \frac{3}{2}(\frac{4}{3})u^{3/4} + C = 2(1+x^4)^{3/4} + C$

29. $u = y + 1$, $du = dy$. a) $\int_0^3 \sqrt{y+1}dy = \int_1^4 u^{1/2}du = \frac{2}{3}u^{3/2}]_1^4 = \frac{2}{3}[4^{3/2} - 1^{3/2}] = \frac{2}{3}[8-1] = \frac{14}{3}$ b) $\int_{-1}^0 \sqrt{y+1}dy = \int_0^1 u^{1/2}du = \frac{2}{3}u^{3/2}]_0^1 = \frac{2}{3}(1-0) = \frac{2}{3}$

31. $u = \tan x$, $du = \sec^2 x\, dx$. a) $\int_0^{\pi/4} \tan x \sec^2 x\, dx = \int_0^1 u\, du = \frac{u^2}{2}]_0^1 = \frac{1}{2}$. b) $\int_{-\pi/4}^0 \tan x \sec^2 x\, dx = \int_{-1}^0 u\, du = \frac{1}{2}u^2]_{-1}^0 = \frac{1}{2}[0^2 - (-1)^2] = -\frac{1}{2}$.

33. $u = x^4 + 9$, $du = 4x^3 dx$, $x^3 dx = \frac{1}{4}du$. a) $\int_0^1 \frac{x^3}{\sqrt{x^4+9}}dx = \frac{1}{4}\int_9^{10} u^{-1/2}du = \frac{1}{4}(2)u^{1/2}]_9^{10} = \frac{1}{2}(10^{1/2} - 9^{1/2}) = \frac{1}{2}(\sqrt{10} - 3)$ b) $\int_{-1}^0 \frac{x^3}{\sqrt{x^4+9}}dx = \frac{1}{4}\int_{10}^9 u^{-1/2}du = -\frac{1}{4}\int_9^{10} u^{-1/2}du =$ (from a)) $-\frac{1}{2}(\sqrt{10} - 3) = \frac{1}{2}(3 - \sqrt{10})$

35. $u = x^2 + 1$, $du = 2x\, dx$, $x\, dx = \frac{1}{2}du$. a) $\int_0^{\sqrt{7}} x(x^2+1)^{1/3}dx = \frac{1}{2}\int_1^8 u^{1/3}du = \frac{1}{2}\frac{3}{4}u^{4/3}]_1^8 = \frac{3}{8}(16-1) = \frac{45}{8}$. b) $\int_{-\sqrt{7}}^0 x(x^2+1)^{1/3}dx = \frac{1}{2}\int_8^1 u^{1/3}du = -\frac{1}{2}\int_1^8 u^{1/3}du = -\frac{45}{8}$ from a)

37. $u = 1 - \cos 3x$, $du = 3\sin 3x\, dx$. a) $\int_0^{\pi/6}(1-\cos 3x)\sin 3x\, dx = \frac{1}{3}\int_0^1 u\, du = \frac{1}{3}\frac{u^2}{2}]_0^1 = \frac{1}{6}$ b) $\int_{\pi/6}^{\pi/3}(1-\cos 3x)\sin 3x\, dx = \frac{1}{3}\int_1^2 u\, du = \frac{1}{3}\frac{u^2}{2}]_1^2 = \frac{1}{6}(4-1) = \frac{1}{2}$

39. $u = 2 + \sin x$, $du = \cos x\, dx$. a) $\int_0^{2\pi} \frac{\cos x}{\sqrt{2+\sin x}}dx = \int_2^2 u^{-1/2}du = 0$. b) $\int_{-\pi}^{\pi} \frac{\cos x}{\sqrt{2+\sin x}}dx = \int_2^2 u^{-1/2}du = 0$

41. $u = t^5 + 2t$, $du = (5t^4 + 2)dt$. $\int_0^1 \sqrt{t^5+2t}(5t^4+2)dt = \int_0^3 u^{1/2}du = \frac{2}{3}u^{3/2}]_0^3 = \frac{2}{3}3^{3/2} = 2\sqrt{3}$

43. $u = \cos 2x$, $du = -2\sin 2x\, dx$. $\int_0^{\pi/2} \cos^3 2x \sin 2x\, dx = -\frac{1}{2}\int_1^{-1} u^3 du = -\frac{1}{2}\frac{u^4}{4}]_1^{-1} = -\frac{1}{8}((-1)^4 - (1)^4) = 0$

45. $u = 5 - 4\cos t$, $du = 4\sin 5\,dt$. $\int_0^\pi \frac{8\sin t}{\sqrt{5-4\cos t}}dt = \frac{8}{4}\int_1^9 u^{-1/2}du = 2(2u^{1/2})]_1^9 = 4(3-1) = 8$

47. $u = 5x^3+4$, $du = 15x^2dx$. $\int_0^1 15x^2\sqrt{5x^3+4}dx = \int_4^9 u^{1/2}du = \frac{2}{3}u^{3/2}]_4^9 = \frac{2}{3}(9^{3/2}-4^{3/2}) = \frac{2}{3}(27-8) = \frac{38}{3}$

49. $u = 4-x^2$, $du = -2x\,dx$, $x\,dx = -\frac{1}{2}du$. $A = -\int_{-2}^0 x\sqrt{4-x^2}dx + \int_0^2 x\sqrt{4-x^2}dx = -(-\frac{1}{2})\int_0^4 u^{1/2}du - \frac{1}{2}\int_4^0 u^{1/2}du = \frac{1}{2}\int_0^4 u^{1/2}du + \frac{1}{2}\int_0^4 u^{1/2}du = \int_0^4 u^{1/2}du = \frac{2}{3}u^{3/2}]_0^4 = \frac{16}{3}$

51. Let $u = 3t^2-1$, $du = 6t\,dt$. $s = \int 24t(3t^2-1)^3dt = \frac{24}{6}\int u^3du = 4\frac{u^4}{4} + C = (3t^2-1)^4 + C$. If $t = 0$, $s = (0-1)^4 + C = 1 + C = 0$ so $C = -1$. Thus $s = (3t^2-1)^4 - 1$.

53. Let $u = t+\pi$, $du = dt$. $s = \int 6\sin(t+\pi)dt = 6\int \sin u\,du = -6\cos u + C = -6\cos(t+\pi) + C$. When $t = 0$, $s = -6\cos\pi + C = 6 + C = 0$ so $C = -6$. Thus $s = -6\cos(t+\pi) - 6$.

55. Let $I = \int_0^{\pi/4} \frac{18\tan^2 x\sec^2 x}{(2+\tan^3 x)^2}dx$. a) $u = \tan x$, $du = \sec^2 x\,dx$. $I = 18\int_0^1 \frac{u^2du}{(2+u^3)^2}$. $v = u^3$, $dv = 3u^2du$, $u^2du = \frac{1}{3}dv$. $I = \frac{18}{3}\int_0^1 \frac{dv}{(2+v)^2} = 6\int_0^1 \frac{dv}{(2+v)^2}$. $w = 2+v$, $dw = dv$. $I = 6\int_2^3 w^{-2}dw = -6\frac{1}{w}]_2^3 = -6(\frac{1}{3}-\frac{1}{2}) = 1$ b) $u = \tan^3 x$, $du = 3\tan^2 x\sec^2 x\,dx$. $I = \frac{18}{3}\int_0^1 \frac{du}{(2+u)^2} = 6\int_2^3 v^{-2}dv = 1$ (from end of a)). c) $u = 2+\tan^3 x$, $du = 3\tan^2 x\sec^2 x\,dx$. $I = \frac{18}{3}\int_2^3 u^{-2}du = 1$ as before.

57. The answers can be seen to be equivalent: $\sin^2 x + C_1 = (1-\cos^2 x) + C_1 = -\cos^2 x + (1+C_1) = -\cos^2 x + C_2 = -(\frac{1+\cos 2x}{2}) + C_2 = -\frac{\cos 2x}{2} + (-\frac{1}{2}+C_2) = -\frac{\cos 2x}{2} + C_3$. The graph of any one of the antiderivatives can be obtained from the graph of any other antiderivative by a vertical shift verifying that they differ by an additive constant.

5.7 NUMERICAL INTEGRATION: TRAPEZOIDAL RULE & SIMPSON'S METHOD

1. Let $f(x) = x$. $h = \frac{2-0}{4} = \frac{1}{2}$. a) $T = \frac{1}{2}(\frac{1}{2})[f(0)+2f(\frac{1}{2})+2f(1)+2f(\frac{3}{2})+f(2)] = \frac{1}{4}[0+1+2+3+2] = 2$ b) $S = \frac{1}{3}(\frac{1}{2})[f(0)+4f(\frac{1}{2})+2f(1)+4f(1.5)+f(2)] = 2$ c) $\int_0^2 x\,dx = \frac{x^2}{2}]_0^2 = 2$

3. Let $f(x) = x^3$, $h = \frac{2-0}{4} = \frac{1}{2}$. a) $T = \frac{1}{4}[f(0) + 2f(\frac{1}{2}) + 2f(1) + 2f(\frac{3}{2}) + f(2)] = 4.25$ b) $S = \frac{1}{3}(\frac{1}{2})[f(0) + 4f(\frac{1}{2}) + 2f(1) + 4f(\frac{3}{2}) + f(2)] = 4$ c) $\int_0^2 x^3\,dx = \frac{x^4}{4}]_0^2 = \frac{16}{4} - \frac{0}{4} = 4$.

5. a) $T = \frac{1}{2}(1)[\sqrt{0} + 2\sqrt{1} + 2\sqrt{2} + 2\sqrt{3} + \sqrt{4}] = \frac{1}{2}[2 + 2\sqrt{2} + 2\sqrt{3} + 2] = 2 + \sqrt{2} + \sqrt{3} = 5.146...$ b) $S = \frac{1}{3}(1)[\sqrt{0} + 4\sqrt{1} + 2\sqrt{2} + 4\sqrt{3} + \sqrt{4}] = \frac{1}{3}(6 + 2\sqrt{2} + 4\sqrt{3}) = 5.252...$ c) $\int_0^4 \sqrt{x}\, dx = \frac{2}{3}x^{3/2}\Big|_0^4 = \frac{16}{3} = 5.333...$

7. $\int_{-1}^{3} e^{-x^2}\, dx$

n	TRAP	SIMP	LRAM	RRAM	MRAM
10	1.62316	1.63322	1.69671	1.54961	1.63799
100	1.63293	1.63303150	1.64029	1.62558	1.63308
1000	1.6330305	1.63303148	1.63377	1.63229	1.63303

NINT yields 1.63303148105.

9. $\int_{-5}^{5} x \sin x\, dx$

n	TRAP	SIMP	LRAM	RRAM	MRAM
10	-4.682	-4.73	-4.68	-4.68	-4.79
100	-4.7537	-4.754469	-4.7537	-4.7537	-4.7549
1000	-4.75446	-4.7544704038	-4.75446	-4.75446	-4.754474

NINT yields -4.75447040396.

11. $\int_{3\pi/4}^{4.5} \frac{\tan x}{x} dx$

n	TRAP	SIMP	LRAM	RRAM	MRAM
10	0.257	0.246	0.101	0.413	0.238
100	0.244	0.243771	0.228	0.259	0.244
1000	0.2437718	0.2437703542	0.242	0.245	0.2437696

NINT yields 0.243770354155.

13. $f(x) = x^{-1}$, $f'(x) = -x^{-2}$, $f''(x) = \frac{2}{x^3} \leqq \frac{2}{1^3}$ on $[1, 2]$ so we may take $M = 2$. $|E_T| \leqq \frac{b-a}{12}h^2 M = \frac{1}{12}(\frac{1}{10})^2 2 = \frac{1}{600} = 0.0016666\ldots$.

15. With $f(x) = x$, we have $f''(x) = f^{(4)}(x) = 0$ and so we may take $M = 0$ in both Eq. (6) and Eq. (7). Hence $E_T = E_S = 0$ for all possible n. a) $n = 1$ b) $n = 2$ (n is always even in S)

17. Let $f(x) = x^3$, $f'(x) = 3x^2$, $f''(x) = 6x$. For $0 \leqq x \leqq 2$, $|f''(x)| \leqq 6 \cdot 2 = 12$ so we may take $M = 12$ in Eq. (6). $h = (b/a)/n = (2-0)/n = 2/n$. By Eq. (4), $|E_T| \leqq \frac{2}{12}(\frac{2}{n})^2 12 = \frac{8}{n^2}$. In a) we wish to have $\frac{8}{n^2} < \frac{1}{10^4}$ or $\frac{n^2}{8} > 10^4$, $n^2 > (8)10^4$, $n > 2\sqrt{2}10^2 = 200\sqrt{2} = 282.84\ldots$. Thus, we may take $n = 283$ in a). Since $f^{(4)}(x) = 0$, we may take $M = 0$ in Eq. (7) and S is the exact value of the integral for all even $n > 0$. b) $n = 2$

19. $f(x) = x^{1/2}$, $f'(x) = \frac{1}{2}x^{-1/2}$, $f''(x) = -\frac{1}{4}x^{-3/2}$, $f'''(x) = \frac{3}{8}x^{-5/2}$, $f^{(4)}(x) = -\frac{15}{16}x^{-7/2}$. $|f''(x)| = |\frac{1}{4x^{3/2}}| \leqq \frac{1}{4}$ for $1 \leqq x \leqq 4$ and so we may take $M = \frac{1}{4}$ in Eq. (6). For a) we wish to have $|E_T| \leqq \frac{b-a}{12}h^2 M = \frac{3}{12}(\frac{3}{n})^2 \frac{1}{4} = \frac{9}{16n^2} < \frac{1}{10^4}$ or $\frac{16n^2}{9} > 10^4$, $n^2 > \frac{9}{16}10^4$, $n > \frac{3}{4}(10^2) = 75$. Thus in a) we take $n = 76$.

$|f^{(4)}(x)| = |\frac{15}{16x^{7/2}}| \leqq \frac{15}{16}$ on $[1,4]$ so we may take $M = \frac{15}{16}$ in Eq. (7). $|E_S| \leqq \frac{b-a}{180}h^4 M = \frac{3}{180}(\frac{3}{n})^4 \frac{15}{16} = \frac{81}{64n^4} < \frac{1}{10^4}$ leads to $\frac{64n^4}{81} > 10^4$, $n > (\frac{81}{64}(10^4))^{1/4} = 10.6\ldots$. Since n is even in Simpson's Rule, we take $n = 12$ in b).

21. 3.1379..., 3.14029...

23. 1.3669..., 1.3688...

25. a) 0.057... and 0.0472... b) Let $y_1 = \sin x/x$, $y_2 = \text{der2}(y_1, 2, 2)$, $y_3 = (\frac{1.5\pi}{12})(\frac{1.5\pi}{10})^2 abs y_2$, $y_4 = \text{NDER}(\text{NDER}(y_2, x, x), x, x)$, $y_5 = \frac{1.5\pi}{180}(\frac{1.5\pi}{10})^4 abs y_4$. We graph y_3 in $[\frac{\pi}{2}, 2\pi]$ by $[0, 0.03]$ and y_5 in $[\frac{\pi}{2}, 2\pi]$ by $[0, 3\times 10^{-4}]$. Max $y_3 = 2.168\ldots \times 10^{-2}$, $\max y_5 = 2.26\ldots \times 10^{-4}$ c) $\max E_T(x) = 5.42\ldots \times 10^{-3}$, $\max E_S(x) = 1.41\ldots \times 10^{-5}$ d) $\max E_T(x) = 8.67 \times 10^{-4}$, $\max E_S(x) = 3.62\ldots \times 10^{-7}$ e) We cannot find the exact value, but we can approximate the integral as closely as we like by increasing n. With $n = 50$, Simpson's Rule gives the value 0.0473894... . By d) the error is at most 3.62×10^{-7}.

27. Refer to the method of Exercise 25. a) 3.6664... and 3.65348218... b) $\max E_T = 0.2466\ldots$, $\max E_S = 2.55\ldots \times 10^{-4}$ c) $\max E_T = 6.16\ldots \times 10^{-2}$, $\max E_S = 1.59\ldots \times 10^{-5}$ d) $\max E_T = 9.86\ldots \times 10^{-3}$, $\max E_S = 4.08\ldots \times 10^{-7}$ e) Simpson's Rule with $n = 50$ yields 3.6534844... with error at most $4.08\ldots \times 10^{-7}$.

29. $f(x) = x^2 + \sin x$. On $[1,2]$, $\text{LRAM}_{50}f = 3.2591$, $\text{RRAM}_{50}f = 3.3205$ and $T_{50}f = 3.2898$. We find $2T_{50}f = \text{LRAM}_{50}f + \text{RRAM}_{50}f$ (see Exercise 31).

31. $\frac{\text{LRAM}_{10}f + \text{RRAM}_{10}f}{2} = [(y_0 + y_1 + \cdots + y_9)h + (y_1 + y_2 + \cdots + y_{10})h]/2 = \frac{h}{2}(y_0 + 2y_1 + 2y_2 + \cdots + 2y_9 + y_{10}) = T_{10}f$.

33. We estimate the surface area of the pond using Simpson's Rule. $S = \frac{200}{3}[0+4(520)+2(800)+400(1000)+2(1140)+4(1160)+2(1110)+4(860)+0] =$

1350666.667 ft^2. The volume of the point is $20S$ ft^3. The number of fish initially is $\frac{20S}{1000}$ and the maximum number to be caught is $(\frac{3}{4})\frac{20S}{1000}$. Thus the maximum number of licenses to be sold is $(\frac{1}{20})(\frac{3}{4})\frac{20S}{1000} = 1013$.

35. $S = \frac{1}{3}(\frac{24}{6})[0+4(18.75)+2(24)+4(26)+2(24)+4(18.75)+0] = 466.66\ldots$ in^2.

37. We use the odd-numbered hours. $n = 12$ and $h = 2$. $T = \frac{h}{2}[1.88 + 2(2.02) + 2(2.25) + 2(3.60) + 2(3.05) + 2(2.38) + 2(2.02) + 2(1.72) + 2(1.97) + 2(2.68) + 2(2.65) + 2(2.21) + 1.88] = 56.86$ kwh per customer.

39. In parametric mode graph $x = \pi t - \sin(\pi t)$, $y = 1 + \cos(\pi t)$, $0 \leqq t \leqq 1$, t step $= 0.05$ in $[0, \pi]$ by $[0, 2]$. The fact that the graph appears to pass the vertical line test, supports that the relation is a function. $\frac{dx}{dt} = \pi - \pi\cos(\pi t) = \pi(1 - \cos\pi t) > 0$ for $0 < t \leqq 1$. This shows that x is an increasing function of t on the interval. Hence no value of x is repeated and so the relation is a function. Since x is an increasing function of t, we see that as t goes from 0 to 1, x goes from 0 to π, so the domain is $[0, \pi]$. Similarly, we see that y is a decreasing function of t and that the range is $[0, 2]$.

41. Let $t_k = \frac{k}{n} = \frac{k}{10} = (0.1)k$, $k = 0, 1, 2, \ldots, 10$. Each t_k determines a point (x_k, y_k) on the curve. $n = 10$ trapezoids are determined, the kth one having area $A_k = \frac{1}{2}(y_{k-1} + y_k)(x_k - x_{k-1}) = \frac{1}{2}[1 + \cos((k-1)(0.1)\pi) + 1 + \cos(k(0.1)\pi)] \times [\pi k(0.1) - \sin(k(0.1)\pi) - ((k-1)(0.1)\pi - \sin((k-1)(0.1)\pi)] = \frac{1}{2}[2 + \cos((k-1)(0.1)\pi) + \cos((0.1)k\pi)] \times [(0.1)\pi - \sin((0.1)k\pi) + \sin((k-1)(0.1)\pi)]$. An approximation for the area is $T = \sum_{k=1}^{10} A_k = 1.5965087\ldots$. Your calculator may have a "sum seq" feature that can be used.

43. $g(x) = \frac{\pi}{2} - \sqrt{2x - x^2} - \sin^{-1}(x-1)$. NINT$(g(x), x, 0, 2) = 1.57079603391$. The region under the graph of g from 0 to 2 is the reflection of the region under the graph of f from 0 to π across the line $y = x$.

45. LRAM$_n f$ + RRAM$_n f = \sum_{k=0}^{n-1} y_k h + \sum_{k=1}^{n} y_k h = h[y_0 + 2\sum_{k=1}^{n-1} y_k + y_n] = 2T_n f$ which is equivalent to the first formula. Let $y_0, y_1, \ldots, y_{2n}$ be the f-values corresponding to the regular partition of the interval into $2n$ subintervals. Then $y_1, y_3, y_5, \ldots, y_{2n-1}$ are the f-values needed for MRAM$_n$ for a partition of the interval into n subintervals. With $h = \frac{b-a}{2n}$, $(\text{MRAM}_n f + 2T_{2n} f)/3 = \frac{1}{3}[(y_1 + y_3 + \cdots + y_{2n-1})(2h) + (y_0 + 2y_1 + 2y_2 + \cdots + 2y_{2n-1} + y_{2n})h] = \frac{h}{3}[y_0 + 4y_1 + 2y_2 + 4y_3 + \cdots + 2y_{2n-2} + 4y_{2n-1} + y_{2n}] = S_{2n} f$.

PRACTICE EXERCISES, CHAPTER 5

1. a)

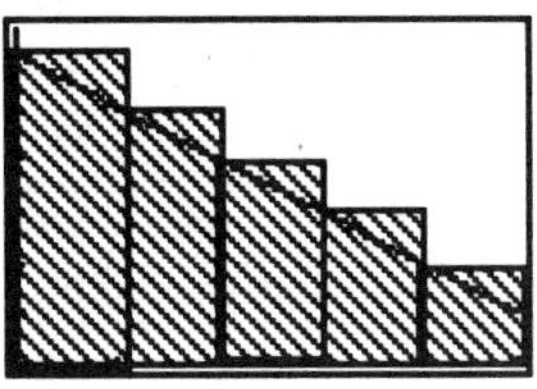 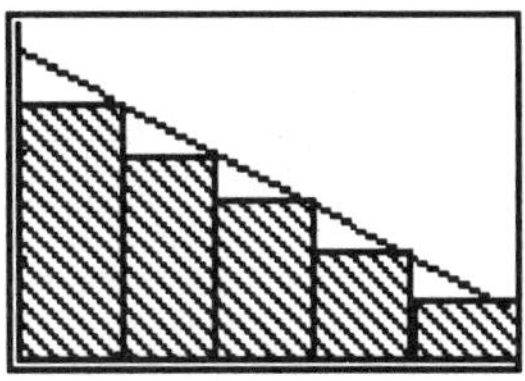 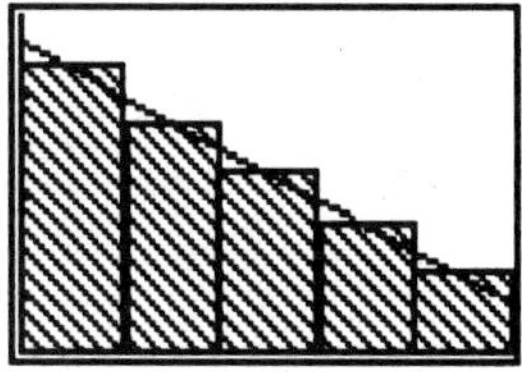

b) $\text{LRAM}_5 f = f(0)\Delta x + f(1)\Delta x + f(2)\Delta x + f(3)\Delta x + f(4)\Delta x = 6\cdot1+5\cdot1+4\cdot1+3\cdot1+2\cdot1 = 20$. $\text{RRAM}_5 f = [f(1)+f(2)+f(3)+f(4)+f(5)]\Delta x = 15$. $\text{MRAM}_5 f = [f(\frac{1}{2}) + f(\frac{3}{2}) + f(\frac{5}{2}) + f(\frac{7}{2}) + f(\frac{9}{2})]\Delta x = 17.5$

3.

	n=10	n=100	n=1000
LRAM_n	22.695	23.86545	23.9865045
RRAM_n	25.395	24.13545	24.0135045
MRAM_5	23.9775	23.999775	23.99999775

5.

	n=10	n=100	n=1000
LRAM_n	3.9670	3.99967	3.9999967
RRAM_n	3.9670	3.99967	3.9999967
MRAM_n	4.0165	4.00016	4.0000016

7. a) $\sum_{k=1}^{10}(k+2) = \sum_{k=1}^{10} k + \sum_{k=1}^{10} 2 = \frac{10(11)}{2} + (10)2 = 75$ b) $\sum_{k=1}^{10}(2k-12) = 2\sum_{k=1}^{10} k - \sum_{k=1}^{10} 12 = 2\frac{(10)(11)}{2} - 12(10) = -10$

9. a) $\sum_{k=1}^{5}(k^3-45) = \sum_{k=1}^{5} k^3 - \sum_{k=1}^{5} 45 = [\frac{5(5+1)}{2}]^2 - 5(45) = 0$ b) $\sum_{k=1}^{6}(\frac{k^3}{7} - \frac{k}{7}) = \frac{1}{7}) = \frac{1}{7}[\sum_{k=1}^{6} k^3 - \sum_{k=1}^{6} k] = \frac{1}{7}[(\frac{6\cdot7}{2})^2 - \frac{6\cdot7}{2}] = \frac{21}{7}[21-1] = 60$

11. a) $\sum_{k=0}^{3} 2^k$ b) $\sum_{k=0}^{4} \frac{1}{3^k}$ c) $\sum_{k=1}^{5}(-1)^{k+1}k$ d) $\sum_{k=1}^{3} \frac{5}{2^k}$

13. $\int_0^1 e^x dx$

15. $x_k = 0 + k\Delta x = k(b-a)/n = 5k/n$. $\text{RRAM}_n f = \sum_{k=1}^{n} f(x_k)\Delta x = \Delta x \sum_{k=1}^{n}(2x_k^3 + 3x_k) = \frac{5}{n}[2\sum_{k=1}^{n} \frac{5^3k^3}{n^3} + 3\sum_{k=1}^{n} \frac{5k}{n}] = \frac{10}{n}\frac{5^3}{n^3}[\frac{n(n+1)}{2}]^2 + 3(\frac{5}{n})^2\frac{n(n+1)}{2} = \frac{1250}{4}\frac{n^2(n+1)^2}{n^4} + \frac{75}{2}\frac{n(n+1)}{n^2} = \frac{625}{2}(\frac{n+1}{n})^2 + \frac{75}{2}\frac{n+1}{n} = \frac{625}{2}(1+\frac{1}{n})^2 + \frac{75}{2}(1+\frac{1}{n})$. $\lim_{n\to\infty} \text{RRAM}_n f = \frac{625}{2} + \frac{75}{2} = 350$

17. a) $\int_0^1 f(t)dt = \pi$ b) $\int_1^0 f(y)dy = -\int_0^1 f(y)dy = -\pi$ c) $\int_0^1 -3f(z)dz = -3\int_0^1 f(z)dz = -3\pi$

19. Area=area of rectangle − area under curve$=\pi - \int_0^\pi \sin x\, dx = \pi - [-\cos x]_0^\pi = \pi + [-1-1] = \pi - 2 \approx 1.14159$. $\pi - \text{NINT}(\sin x, x, 0, \pi) \approx 1.14159$

21. Total area $= \int_0^6 |4-x|dx = \int_0^4 (4-x)dx + \int_4^6 (x-4)dx = [4x - \frac{x^2}{2}]_0^4 + [\frac{x^2}{2} - 4x]_4^6 = 8 + (18-24) - (8-16) = 10$

23. $\int_{-1}^1 (3x^2-4x+7)dx = [x^3-2x^2+7x]_{-1}^1 = (1-2+7)-(-1-2-7) = 6+10 = 16$

25. $\int_1^2 \frac{4}{x^2}dx = [-\frac{4}{x}]_1^2 = -[\frac{4}{2} - \frac{4}{1}] = 2$

27. $\int_1^4 \frac{dt}{t\sqrt{t}} = \int_1^4 t^{-3/2}dt = [-2t^{-1/2}]_1^4 = -2[4^{-1/2} - 1] = 1$

29. $\int_0^\pi \sin 5\theta\, d\theta = [-\frac{\cos 5\theta}{5}]_0^\pi = -\frac{1}{5}[\cos 5\pi - \cos 0] = \frac{2}{5}$

31. $\int_0^{\pi/3} \sec^2\theta\, d\theta = [\tan\theta]_0^{\pi/3} = \sqrt{3}$

33. $I = \int_\pi^{3\pi} \cot^2 \frac{x}{6}dx = \int_\pi^{3\pi}(\csc^2 \frac{x}{6} - 1)dx$. Let $u = \frac{x}{6}$, $du = \frac{1}{6}dx$. $I = 6\int_{\pi/6}^{\pi/2}(\csc^2 u - 1)du = -6[\cot u + u]_{\pi/6}^{\pi/2} = -6[\cot\frac{\pi}{2} + \frac{\pi}{2} - \cot\frac{\pi}{6} - \frac{\pi}{6}] = -6[0 + \frac{\pi}{2} - \sqrt{3} - \frac{\pi}{6}] = 6(\sqrt{3} - \frac{\pi}{3}) = 6\sqrt{3} - 2\pi$

35. Let $\int_0^1 \frac{36dx}{(2x+1)^3} = I$. Let $u = 2x+1$, $du = 2\,dx$. $I = \frac{36}{2}\int_1^3 u^{-3}du = \frac{18}{-2}[u^{-2}]_1^3 = -9[\frac{1}{9} - 1] = -1 + 9 = 8$

37. $\int_{-\pi/3}^0 \sec x \tan x\, dx = [\sec x]_{-\pi/3}^0 = 1 - 2 = -1$

39. $I = \int_0^{\pi/2} 5(\sin x)^{3/2}\cos x\, dx$. Let $u = \sin x$, $du = \cos x\, dx$. $I = 5\int_0^1 u^{3/2}du = 5(\frac{2}{5})[u^{5/2}]_0^1 = 2$

41. $\int_4^8 \frac{1}{5}dt = \text{NINT}(\frac{1}{t}, 4, 8) = 0.6931\ldots$

43. $\int_0^2 \frac{x\,dx}{x^2+5} = \text{NINT}(\frac{x}{x^2+5}, 0, 2) = 0.2938\ldots$

45. $\int_2^3 (t - \frac{2}{t})(t + \frac{2}{t})dt = \int_2^3 (t^2 - \frac{4}{t^2})dt = \int_2^3 (t^2 - 4t^{-2})dt = \frac{t^3}{3} + 4t^{-1}]_2^3 = 9 + \frac{4}{3} - (\frac{8}{3} + 2) = 7 - \frac{4}{3} = \frac{17}{3}$. $\text{NINT}((x - 2/x)(x + 2/x), x, 2, 3) = 5.6666\ldots$

47. $\int_{-4}^0 |x|dx = -\int_4^0 x\, dx = -[\frac{x^2}{2}]_{-4}^0 = -[0 - \frac{16}{2}] = 8$

49. $I = \int_{-\pi/2}^{\pi/2} 15\sin^4 3x \cos 3x\, dx$, $u = \sin 3x$, $du = 3\cos 3x\, dx$. $I = \frac{15}{3}\int_1^{-1} u^4 du = [u^5]_1^{-1} = -2$

51. We graph $y = \text{NINT}((\ell n(5t))/t, t, 1, x)$ and $y = \frac{1}{2}(\ell n\, 5x)^2$ in $[0.01, 7]$ by $[-1.3, 6.4]$. This suggests that the two functions differ by a constant because their graphs appear to be obtainable, one from the other, by a vertical shift. We may also graph $\frac{\ell n\, 5x}{x}$ and $\text{NDER}(\frac{1}{2}(\ell n\, 5x)^2, x)$ in the above rectangle and see that the graphs coincide.

53. Graph $e^x \sin x$ and $\text{NDER}(\frac{e^x}{2}(\sin x - \cos x), x)$ in the viewing window $[-3, 3]$ by $[-1, 8]$ and see that the two graphs coincide.

55. Graph $\text{NINT}((\sin(3t))/t, t, 1, x)$ in $[0.01, 8]$ by $[-2, 0.004]$. It was mentioned earlier that $\frac{\sin x}{x}$ has no explicit antiderivative in terms of elementary functions. Let $I = \int \frac{\sin 3x}{x} dx$. Let $u = 3x$, $du = 3dx$, $dx = \frac{1}{3} du$. Then $I = \frac{1}{3} \int \frac{\sin u}{u/3} du = \int \frac{\sin u}{u} du$. Thus there is no explicit antiderivative of $\frac{\sin 3x}{x}$ otherwise $\sin u/u$ and so $\sin x/x$ would have an explicit antiderivative.

57. Solve $I = \int_0^x (t^3 - 2t + 3)dt = 4$. $I = [\frac{t^4}{4} - t^2 + 3t]_0^x = \frac{x^4}{4} - x^2 + 3x = 4$, $x^4 - 4x^2 + 12x - 16 = 0$. We graph the left-hand side of the last equation and zoom in to its x-intercepts obtaining $x = -3.091, 1.631$. Alternatively, we may graph $y_1 = \text{NINT}(t^3 - 2t + 3, t, 0, x)$ and $y_2 = 4$ and find the points of intersection.

59. We graph $y = \text{NINT}(\cos t/t, t, 1, x)$ in $[0.01, 20]$ by $[-1, 0.5]$. We see that $y < 1$ and so there is no solution.

61. a) and c)

63. By the Fundamental Theorem of Calculus the function $F(x)$ defined by $F(x) = \int_0^x f(t)dt$, $0 \leqq x \leqq 1$, is differentiable and $F'(x) = f(x)$. On the other hand we are given $F(x) = \sin x$ and so $F'(x) = \cos x$. Therefore $f(x) = \cos x$, $0 \leqq x \leqq 1$.

65. $\int_0^1 \sqrt{1 + x^4} dx = F(1) - F(0)$

67. $\frac{d^2 s}{dt^2} = \pi^2 \cos \pi t$. $\frac{ds}{dt} = \pi \sin \pi t + v_0 = \pi \sin \pi t + 8$. $s = -\cos \pi t + 8t + C$, $0 = s(0) = -1 + C$, $C = 1$. $s = -\cos \pi t + 8t + 1$. $s(1) = -\cos \pi + 8 + 1 = 10\text{m}$.

69. $a = \frac{d^2 s}{dt^2} = -k$. $v = \frac{ds}{dt} = -kt + v_0 = -kt + 44$. $s = -\frac{k}{2}t^2 + 44t + s_0 = -\frac{k}{2}t^2 + 44t$. $v(t^*) = -kt^* + 44 = 0$, $t^* = \frac{44}{k}$. $s(t^*) = -\frac{k}{2}(\frac{44}{k})^2 + 44(\frac{44}{k}) = -\frac{1}{2}\frac{44^2}{k} + \frac{44^2}{k} = \frac{1}{2}\frac{44^2}{k} = 45$, $k = \frac{44^2}{90} = 21.511\text{ft/sec}^2$

71. a) Because of periodicity we may use any interval of length 1 and we choose $[0,1]$. $(V^2)_{av} = (\int_0^1 V^2 dt)/1 = (V_{\max})^2 \int_0^1 \sin^2(120\pi t)dt = \frac{(V_{\max})^2}{2}\int_0^1 (1 - \cos(240\pi t))dt = \frac{(V_{\max})^2}{2}[t - \frac{\sin(240\pi t)}{240\pi}] = \frac{V_{\max}^2}{2}$. $V_{rms} = \sqrt{(V^2)_{av}} = \sqrt{\frac{(V_{\max})^2}{2}} = \frac{V_{\max}}{\sqrt{2}}$ verifying (1). b) $V_{\max} = \sqrt{2}V_{rms} = \sqrt{2}(240) = 339.411$ volts.

73. Average value of f' on $[a,b] = \frac{1}{b-a}\int_a^b f'(x)dx = \frac{1}{b-a}[f(b) - f(a)] = \frac{f(b)-f(a)}{b-a}$.

75. Let $f(x) = \frac{1}{x}$. Then we may find that $f^{(4)}(x) = \frac{24}{x^5} \leqq 24 = M$ on $[1,3]$. $|E_S| \leqq \frac{b-a}{180}h^4 M = \frac{3-1}{180}h^4(24) = \frac{4}{15}h^4 \leqq 10^{-4}$. $h^4 \leqq \frac{15}{4}\frac{1}{10^4}$, $h \leqq \frac{1}{10}\sqrt[4]{3.75} = 0.13915788\ldots$ But $h = \frac{b-a}{n} = \frac{2}{n} \leqq 0.13915788\ldots$ which implies $n \geq \frac{2}{0.13915788} = 14.372$. Since n must be even, we have $n \geqq 16$, $h = \frac{b-a}{n} \leqq \frac{2}{16} = \frac{1}{8}$.

77. $T = ((\pi/6)/2)(2\sin^2 0 + 4\sin^2\frac{\pi}{6} + 4\sin^2\frac{\pi}{3} + 4\sin^2\frac{\pi}{2} + 4\sin^2(\frac{2\pi}{3}) + 4\sin^2(\frac{5\pi}{6}) + 2\sin^2\pi) = 3.14159265359\ldots$ $|E_T| = |\pi - T| \leq 10^{-11}$.

$S = \frac{\pi}{18}(2\sin^2 0 + 8\sin^2\frac{\pi}{6} + 4\sin^2\frac{\pi}{3} + 8\sin^2\frac{\pi}{2} + 4\sin^2(\frac{2\pi}{3}) + 8\sin^2(\frac{5\pi}{6}) + 2\sin^2\pi) = 3.14159265359\ldots$ $|E_S| < 10^{-11}$. T and S both agree with π up to the limits of calculator accuracy.

79. By Simpson's Rule the approximate area of the lot is $\frac{15}{3}[0 + 4(36) + 2(54) + 4(51) + 2(49.5) + 4(54) + 2(64.4) + 4(67.5) + 42] = 6059\text{ft}^2$. The cost for the job is $0.10(6059) + (2.00)6059 = \$12,723.90$.

CHAPTER 6

APPLICATIONS OF DEFINITE INTEGRALS

6.1 AREAS BETWEEN CURVES

1. $A = \int_{-2}^{3}[(3x+7)-(x^2+2x+1)]dx = [3\frac{x^2}{2}+7x-\frac{x^3}{3}-x^2-x]_{-2}^{3} = \frac{125}{6}$.

3. $A = \int_{-3}^{2}[(2-y)-(y^2-4)]dy = [6y-\frac{y^2}{2}-\frac{y^3}{3}]_{-3}^{2} = \frac{125}{6}$.

5. $\int_0^\pi (1-\cos^2 x)dx = \int_0^\pi \sin^2 x\, dx = \frac{1}{2}\int_0^\pi (1-\cos 2x)dx = \frac{1}{2}[x-\frac{1}{2}\sin 2x]_0^\pi = \frac{1}{2}[\pi-0]-\frac{1}{2}[0] = \frac{\pi}{2}$.

7. The curves intersect at $(-2,2)$ and $(2,2)$, $\int_{-2}^{2}(2-(x^2-2))dx = \int_{-2}^{2}(4-x^2)dx = 2\int_0^2(r-x^2)dx$ by symmetry $= 2[4x-\frac{x^3}{3}]_0^2 = 2[8-\frac{8}{3}] = \frac{32}{3}$.

9.

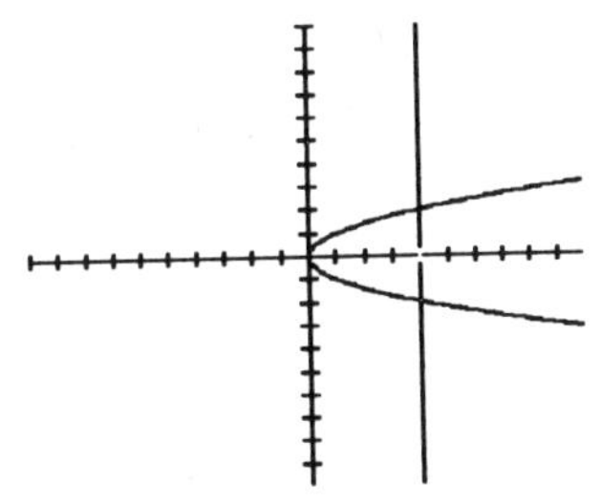

The curves intersect at $(4,-2)$ and $(4,2)$.
Using symmetry, $A = 2\int_0^2(4-y^2)dy$
$= 2[4y-\frac{y^3}{3}]_0^2 = 2[8-\frac{8}{3}] = \frac{32}{3}$.

11. $\int_0^1(x-x^2)dx = \frac{1}{6}$.

13.

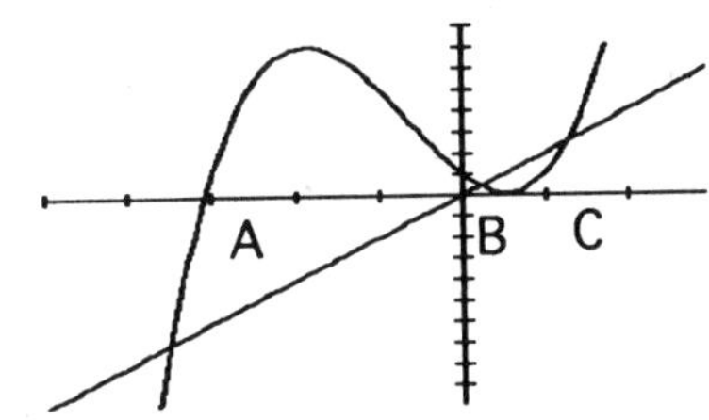

Two regions are enclosed by the curves.
Use technology to find the x-coordinates:
$A: -3.50700,\ B: 0.22187,\ C: 1.28514$
$\int_A^B(x^2+2x^2-3x+1-2x)dx$
$+\int_B^C(2x-(x^3+2x^2-3x+1))dx$
$= 25.299838+0.853568 = 26.15341$.

15. $\int_0^3(x-(x^2-2x))dx = 9/2$.

17. Use technology to find the right-hand value of $B = 0.41936\ldots$. Use technology to compute NINT(e^$(-x^2) - 2x, x, 0, B) = 0.22016$.

19. $\int_1^2 (2-(x-2)^2 - x)dx = 1/6$. (With the change of variable $t = x-2$, the integral becomes $\int_{-1}^{0}(-t-t^2)dt$.)

21. Graph using $[-2,5]$ by $[-10,10]$. Use technology to find the x-coordinates of the points of intersection $A = -1.39138$, $B = 0.22713$, $C = 3.16425$. $\int_A^B (x^3 - 2x^2 - 3x + 1 - x)dx + \int_B^C (x - (x^3 - 2x^2 - 3x + 1))dx = 2.64735 + 13.03641 = 15.68376$.

23. This area can be found by both methods. $A = \int_0^1 (y^2 - y^3)dy = [\frac{y^3}{3} - \frac{y^4}{4}]_0^1 = \frac{1}{12}$ and $A = \int_0^1 (x^{1/3} - x^{1/2})dx = [\frac{x^{4/3}}{4/3} - \frac{x^{3/2}}{3/2}]_0^1 = \frac{3}{4} - \frac{2}{3} = \frac{1}{12}$.

25. $\int_0^1 12(y^2 - y^3)dy = 1$.

27. The curves meet at $x = \pi/4$. $\int_0^{\pi/4}(\cos x - \sin x)dx = [\sin x + \cos x]_0^{\pi/4} = [\frac{2\sqrt{2}}{2}] - [1] = \sqrt{2} - 1$.

29.

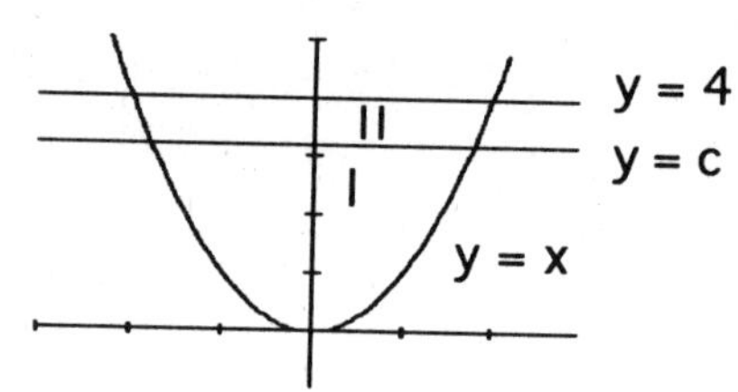

By symmetry, regions I and II will have equal areas

a) $\int_0^c y^{1/2}dy = \int_c^4 y^{1/2}dy$

b) $\int_0^{\sqrt{c}}(c - x^2)dx = (4-c)\sqrt{c} + \int_{\sqrt{c}}^2 (4 - x^2)dx$

Integrating equation a) leads to $c^{3/2} = 4^{3/2} - c^{3/2}$ or $c = 2^{4/3}$.

31. a) $A(t) = \int_0^{t/\sqrt{2}}(2x^2 - x^2)dx + \int_{t/\sqrt{2}}^{t}(t^2 - x^2)dx = \frac{x^3}{3}]_0^{t/\sqrt{2}} + [t^2 x - \frac{x^3}{3}]_{t/\sqrt{2}}^{t} = \frac{2}{3}t^3(1 - \frac{1}{\sqrt{2}})$. b) $A(t) = \int_0^{t^2}(\sqrt{y} - \sqrt{\frac{y}{2}})dy = (1 - \frac{1}{\sqrt{2}})\int_0^{t^2} y^{1/2}dy = (1 - \frac{1}{\sqrt{2}})\frac{2}{3}\cdot t^3$. c) $R(t) = t \cdot t^2 \Rightarrow A(t)/R(t) = (1 - \frac{1}{\sqrt{2}})\frac{2}{3} = 0.195\ldots$.

33. We are given: $\int_a^b f(x)dx = 4$, therefore $\int_a^b (2f(x) - f(x))dx = \int_a^b f(x)dx = 4$

6.2 VOLUMES OF SOLIDS OF REVOLUTION - DISKS AND WASHERS

1. $V = \pi\int_0^2 (2-x)^2 dx = \pi\int_0^2 (x-2)^2 dx = \pi[\frac{(x-2)^3}{3}]_0^2 = \pi[0 - \frac{(-2)^3}{3}] = \frac{8\pi}{3}$

3. $V = 2\pi \int_0^3 (\sqrt{9-x^2})^2 dx = 2\pi \int_0^3 (9-x^2)dx = 2\pi[9x - \frac{x^3}{3}]_0^3 = 36\pi$

5. $V = \pi \int_0^2 (x^3)^2 dx = \pi[\frac{x^7}{7}]_0^2 = 128\pi/7$

7. $V = \pi \int_0^{\pi/2} (\sqrt{\cos x})^2 dx = \pi[\sin x]_0^{\pi/2} = \pi$

9. $V = \pi \int_0^2 (2y)^2 dy = 4\pi[\frac{y^3}{3}]_0^2 = 32\pi/3$

11.

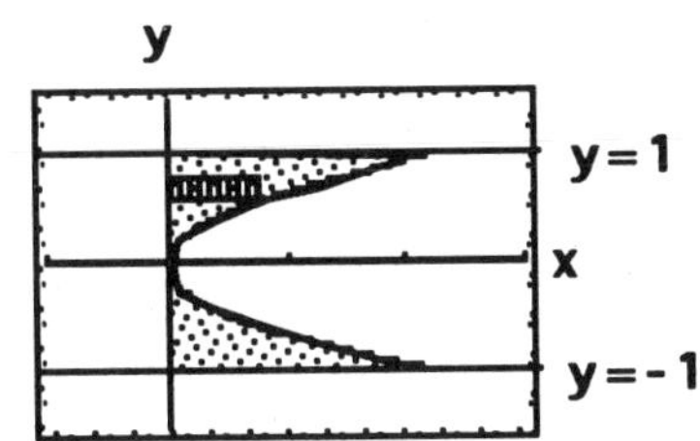

Using symmetry,

$V = 2\pi \int_0^1 (\sqrt{5}y^2)^2 dy$

$= 10\pi[\frac{y^5}{5}]_0^1 = 2\pi$

13. $V = \pi \int_0^2 (y^{3/2})^2 dy = \frac{\pi \cdot 2^4}{4} = 4\pi$

15. $V = \pi \int_0^3 [\frac{2}{\sqrt{y+1}}]^2 dy = 4\pi \int_0^3 \frac{1}{y+1} dy = 4\pi[\ln(y+1)]_0^3 = 4\pi \ln 4$

17. $V = \pi \int_0^1 (1^2 - x^2)dx = 2\pi/3$

19. $V = \pi \int_0^2 (4^2 - (x^2)^2)dx = 128\pi/5$

21. $V = \pi \int_{-1}^2 ((x+3)^2 - (x^2+1)^2)dx = \pi \int_{-1}^2 (x^2 + 6x + 9 - x^4 - 2x^2 - 1)dx =$
$\pi \int_{-1}^2 (8 + 6x - x^2 - x^4)dx = \pi[8x + 3x^2 - x^3/3 - x^5/5]_{-1}^2 = \frac{117\pi}{5}$

23. $V = 2\pi \int_0^{\pi/4} ((\sqrt{2})^2 - \sec^2 x)dx = 2\pi[2x - \tan x]_0^{\pi/4} = 2\pi(\frac{\pi}{2} - 1) = \pi^2 - 2\pi$

25.

$V = \pi \int_0^1 ((y+1)^2 - 1^2)dy = 4\pi/3$

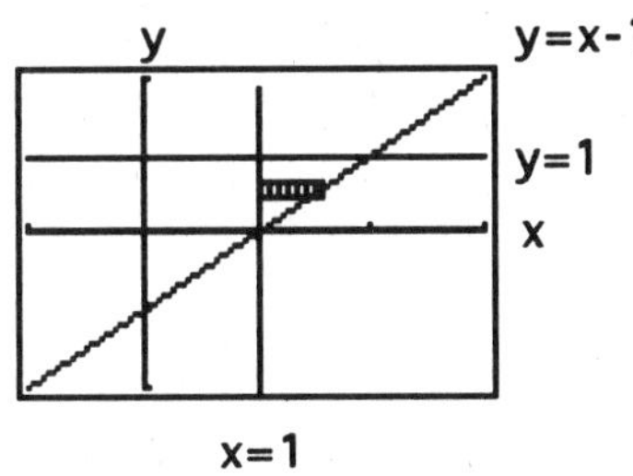

27. $V = \pi \int_0^4 (2^2 - (\sqrt{y})^2)dy = \pi[4y - \frac{y^2}{2}]_0^4 = 8\pi$

29. $V = 2\pi \int_0^5 (\sqrt{25 - y^2})^2 dy = 2 \cdot 250\pi/3 = 500\pi/3$

31. $V = 2\pi \int_0^{\pi/2} (1^2 - (\sqrt{\cos x})^2) dx = 2x[x - \sin x]_0^{\pi/2} = 2\pi(\frac{\pi}{2} - 1) = \pi^2 - 2\pi$

33. $V = \pi \int_0^1 (1^2 - (\sqrt{1 - y^2})^2) dy = \pi/3$

35. a) $V = \pi \int_0^4 (2^2 - (\sqrt{x})^2) dx = 8\pi$ b) $V = \pi \int_0^2 (y^2)^2 dy = 32\pi/5$ c) $V = \pi \int_0^4 (2 - \sqrt{x})^2 dx = 8\pi/3$ d) $V = \pi \int_0^2 (4^2 - (4 - y^2)^2) dy = 224\pi/15$

37. a) $V = 2\pi \int_0^1 (1 - x^2)^2 dx = 2\pi \int_0^1 (1 - 2x^2 + x^4) dx = 2\pi[x - \frac{2x^3}{3} + \frac{x^5}{5}]_0^1 = 16\pi/15$ b) $V = 2\pi \int_0^1 ((2 - x^2)^2 - 1^2) dx = 2\pi \int_0^1 (3 - 4x^2 + x^4) dx = 2\pi[3x - \frac{4x^3}{3} + \frac{x^5}{5}]_0^1 = 56\pi/15$ c) $V = 2\pi \int_0^1 (2^2 - (x^2 + 1)^2) dx = 2\pi \int_0^1 (3 - x^4 - 2x^2) dx = 2\pi[3x - \frac{x^5}{5} - \frac{2x^3}{3}]_0^1 = 64\pi/15$

39. The line has equation $y = -\frac{h}{r}x + h$, or $x = -\frac{r}{h}y + r$. $V = \pi \int_0^h = (-\frac{r}{h}y + r)^2 dy = \pi \int_0^h (\frac{r^2}{h^2}y^2 - \frac{2r^2}{h}y + r^2) dy = \pi[\frac{r^2}{h^2}\frac{y^3}{3} - \frac{r^2}{h}y^2 + r^2 y]_0^h = \pi r^2(\frac{h}{3} - h + h) = \frac{\pi r^2 h}{3}$

41. a) $x = -2$, $x = 0.59375\ldots$ Store the value of x in A.

b) $V = \pi \int_{-2}^{0.59375\ldots} [5\cos(0.5x + 1)]^2 - [x^2 + 1]^2 dx = \pi$ NINT$((5\cos(0.5x + 1))^2 - (x^2 + 1)^2, x, -2, A) = 76.815307\ldots$

43. Revolve the area bounded by $y = -7$, $x = 0$, $x^2 + y^2 = 16^2$ about the y-axis $V = \pi \int_{-16}^{-7} (16^2 - y^2) dy = 1053\pi \approx 3.3L$

6.3 CYLINDRICAL SHELLS – AN ALTERNATIVE TO WASHERS

1. $V = 2\pi \int_0^2$ (radius)(height) $dx = 2\pi \int_0^2 x(x - (-x/2)) dx = 8\pi$.

3. $V = 2\pi \int_0^1$ (radius)(height) $dx = 2\pi \int_0^1 x(x^2 + 1) dx = 3\pi/2$.

5.

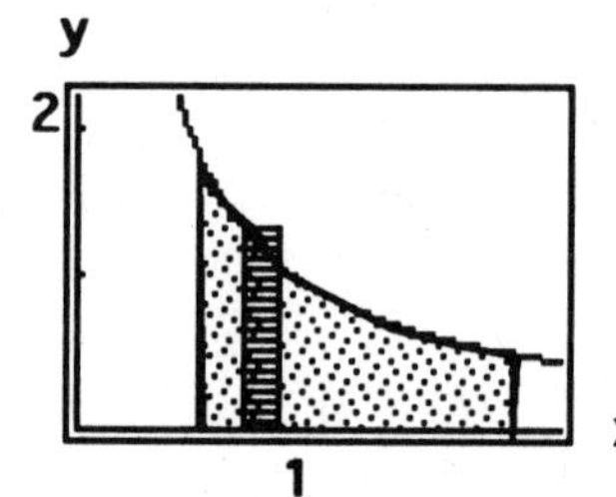

$V = 2\pi \int_{1/2}^2 x(\frac{1}{x}) dx = 2\pi \cdot \frac{3}{2} = 3\pi$.

7.

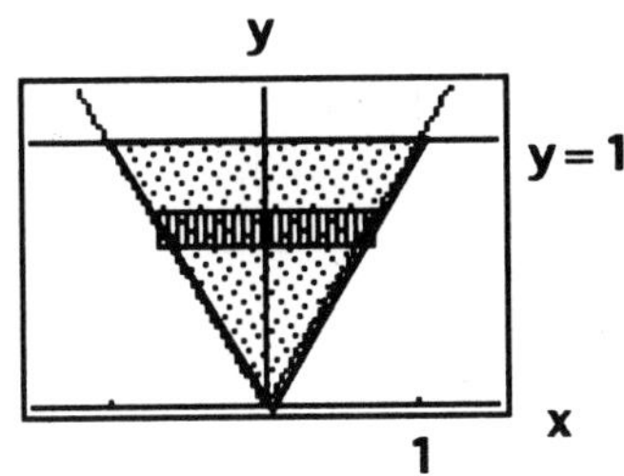

Using symmetry,

$V = 2(2\pi)\int_0^1 y(y)dy = 4\pi/3.$

9.

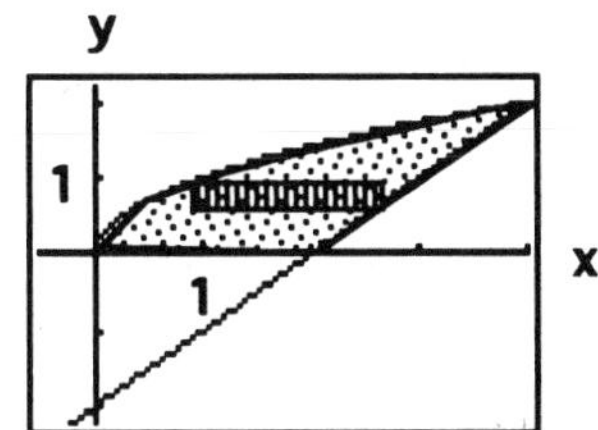

$V = 2\pi\int_0^2 y(y+2-y^2)dy = 16\pi/3.$

11.

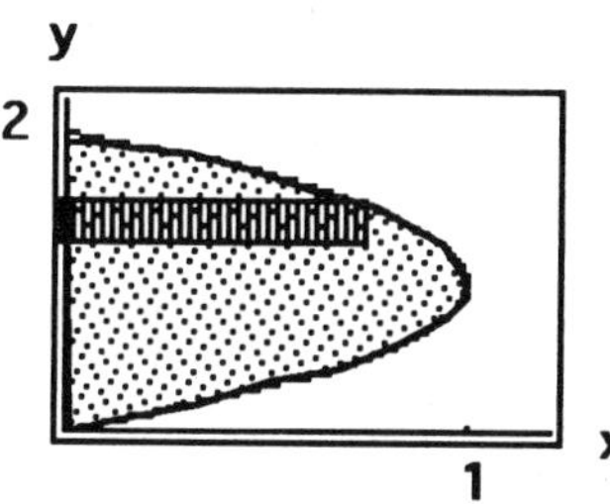

$V = 2\pi\int_0^2 y(2y-y^2)dy = 8\pi/3.$

13. Use shells. $V = 2\pi\int_0^{\sqrt{3}} x\sqrt{x^2+1}dx$. Let $u = x^2+1$, then $du = 2xdx$. $V = \pi\int_1^4 u^{1/2}du = \frac{\pi u^{3/2}}{3/2}\Big|_1^4 = \frac{2\pi}{3}[8-1] = 14\pi/3.$

15. Use shells. $V = 2\pi\int_0^1 y(12(y^2-y^3))dy = 6\pi/5.$

17.

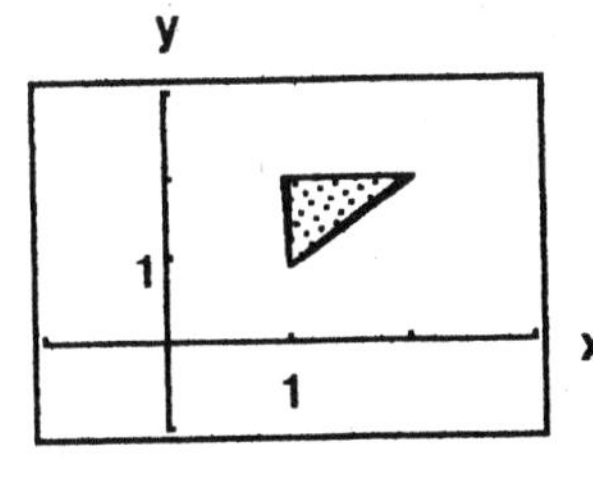

The region is bounded by $y = x$, $x = 1$ and $y = 2$

a) Use shells.

$V = 2\pi\int_1^2 y(y-1)dy = 2\pi[\frac{y^3}{3} - \frac{y^2}{2}]_1^2 = \frac{5\pi}{3}.$

b) Use shells.

$V = 2\pi\int_1^2 x(2-x)dx = 4\pi/3.$

19. 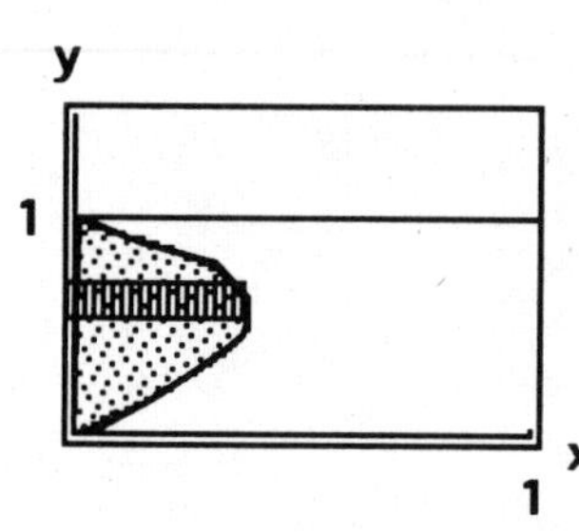

a) Use shells.

$V = 2\pi \int_0^1 y(1-(y-y^3))dy = 11\pi/15.$

b) Use washers.

$V = \pi \int_0^1 (1^2 - (y-y^3)^2)dy = 97\pi/105.$

c) Use washers.

$V = \pi \int_0^1 (1-(y-y^3))^2 dy = 121\pi/210.$

d) Use shells. $V = 2\pi \int_0^1 (1-y)(1-(y-y^3))dy = 23\pi/30.$

21. The curves intersect at (2,8). a) Using shells, $V = 2\pi \int_0^8 y(y^{1/3} - \frac{y}{4})dy$. Using washers, $V = \pi \int_0^2 ((4x)^2 - (x^3)^2)dx = 512\pi/21.$

b) Using washers, $V = \pi \int_0^2 ((8-x^3)^2 - (8-4x)^2)dx = 832\pi/21.$

23. a) Use shells. $V = 2\pi \int_0^1 x(2x - x^2 - x)dx = \pi/6.$

b) Use shells. $V = 2\pi \int_0^1 (1-x)(2x - x^2 - x)dx = \pi/6.$

25. Following the approach of Exploration 2,

a) The solid is generated by revolving the regions OBD, ODCE, and ECA about the x-axis

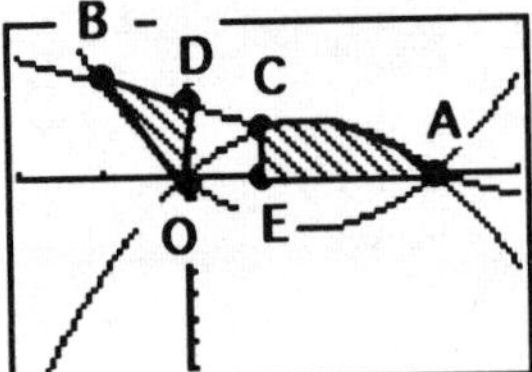

$V = \pi \int_{-1}^{0} [(-x+3)^2 - (x^2-3x)^2]dx + \pi \int_0^1 (-x+3)^2 dx + \pi \int_1^3 (3x - x^2)^2 dx = \pi[7.6333 + 6.3333 + 6.4] = 20.367\pi.$

b) 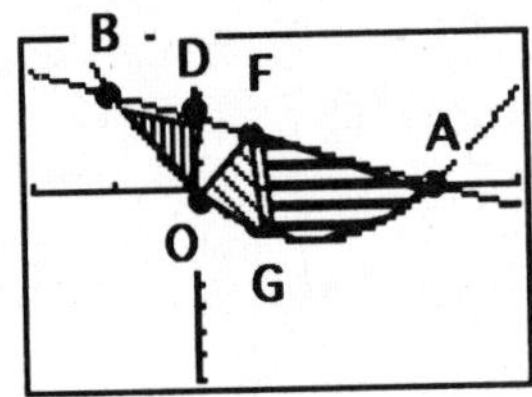

The solid is generated by revolving the regions OBD, OFG, and FGA about the y-axis

$V = 2\pi \int_{-1}^{0} -x(-x+3-(x^2-3x))dx + 2\pi \int_0^{0.6458} x(x^2+3x-(x^2-3x))dx + 2\pi \int_{0.6458}^{3} x(-x+3-(x^2-3x))dx = 2\pi[0.58333 + 0.53867 + 10.4883] = 2\pi(11.610)$

27. $V = 2 \cdot 2\pi \int_1^3 x\sqrt{9-x^2}dx$

$= \frac{4\pi \cdot 16\sqrt{2}}{3} = 94.782.$

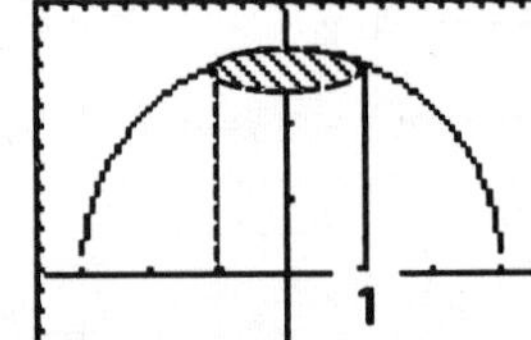

29. Use shells. Graph on $[0,3]$ by $[-1,2]$.

a) $V = 2\pi\int_0^b$ radius $\cdot$ height $dx = 2\pi\int_0^b (2-x)(1-(1-\frac{\sin 4x}{4x}))dx = 2\pi\int_0^b (2-x)\frac{(\sin(4x))}{4x}dx$.

b) The curves intersect when $1 = \frac{\sin 4x}{4x} = 1$, i.e., at $x = \pi/4$.

c) $V = 2\pi\int_0^{\pi/4}\frac{(2-x)\sin(4x)}{4x}dx = 5.033$, using NINT. Because the integrand is not defined at $x=0$, it may be necessary to use a lower limit of 0.000001.

6.4 LENGTHS OF CURVES IN THE PLANE

1. $\frac{dy}{dx} = \frac{1}{3}\cdot\frac{3}{2}(x^2+2)^{1/2}2x = x(x^2+2)^{1/2}$; $L = \int_0^3\sqrt{1+x^2(x^2+2)}\,dx = \int_0^3\sqrt{x^4+2x^2+1}\,dx = \int_0^3(x^2+1)dx = [\frac{x^3}{3}+x]_0^3 = 12$.

3. $y = (\frac{9x^2}{4})^{1/3}$ is not differentiable at $x=0$. Use $x = \frac{2}{3}y^{3/2}$. Then $\frac{dx}{dy} = y^{1/2}$ and $L = \int_0^3\sqrt{1+y}\,dy = \frac{2}{3}[(1+y)^{3/2}]_0^3 = \frac{2}{3}[8-1] = \frac{14}{3}$.

5. $1+(\frac{dy}{dx})^2 = 1+(x^2-\frac{1}{4x^2})^2 = (x^2+\frac{1}{4x^2})^2$; $L = \int_1^3(x^2+\frac{1}{4x^2})dx = [x^3/3 - 1/(4x)]_1^3 = \frac{53}{6}$.

7. $1+(\frac{dx}{dy})^2 = 1+(y^3-\frac{1}{4y^3})^2 = (y^3+\frac{1}{4y^3})^2$; $L = \int_1^2(y^3+\frac{1}{4}y^{-3})dy = [\frac{y^4}{4}-\frac{1}{8}y^{-2}]_1^2 = \frac{123}{32}$.

9. $1+(\frac{dy}{dx})^2 = 1+[\frac{e^x-e^{-x}}{2}]^2 = \frac{1}{4}(e^x+e^{-x})^2$; $L = \frac{1}{2}\int_{-\ln 2}^{\ln 2}(e^x+e^{-x})dx = \frac{1}{2}[e^x-e^{-x}]_{-\ln 2}^{\ln 2} = \frac{1}{2}[(2-\frac{1}{2})-(\frac{1}{2}-2)] = \frac{3}{2}$.

11. $1+(\frac{dy}{dx})^2 = 1+\sec^2 x\tan^2 x$; $L = \int_{-\pi/3}^{\pi/3}\sqrt{1+\sec^2 x\tan^2 x}\,dx = 3.1385$, using NINT.

13. $y' = (e^2-e^{-x})/2$; $1+(y')^2 = \frac{4+(e^{2x}-2+e^{-2x})}{4} = \frac{(e^x+e^{-x})^2}{4}$; $L = \int_{-3}^3\frac{e^x+e^{-x}}{2}dx = \frac{1}{2}[e^x-e^{-x}]_{-3}^3 = e^3-e^{-3} = 20.036$.

15. $y = (1-x^{2/3})^{3/2} \Rightarrow \frac{dy}{dx} = \frac{3}{2}(1-x^{2/3})^{1/2}(-\frac{2}{3}x^{-1/3}) = -x^{-1/3}(1-x^{2/3})^{1/2}$; $1+(\frac{dy}{dx})^2 = 1+x^{-2/3}(1-x^{2/3}) = x^{-2/3}$. This function is not continuous on $[0,1]$. It is, however, continuous on $[\sqrt{2}/4, 1]$, which is where the curve and the line $y=x$ intersect. $L = 8\int_{\sqrt{2}/4}^1 x^{-1/3}dx = \frac{8x^{2/3}}{2/3}\Big|_{\sqrt{2}/4}^1 = \frac{3\cdot 8}{2}[1-(2^{-3/2})^{2/3}] = 4\cdot 3(\frac{1}{2}) = 6$.

17. $1+(\frac{dy}{dx})^2 = 1+1/(4x)$, $\frac{dy}{dx} = \frac{1}{2\sqrt{x}}$, $y = \int\frac{1}{x}x^{-1/2}\,dx = x^{1/2}+C$. Since $y(0)=0$, $C=0$ and the curve is $y = x^{1/2}$, $0\le x\le 4$.

19. $\frac{dy}{dx} = 1/x$. $y = \ln x + C$. Since $y(1) = 3$, $C = 3$ and the curve is $y = \ln x + 3$.

21. $\frac{dy}{dx} = \frac{-2x}{1-x^2}$. $L = \int_0^{1/2} \sqrt{1 + \frac{4x^2}{(1-x^2)^2}}dx = \int_0^{1/2} \frac{\sqrt{1+2x^2+x^4}}{(1-x^2)}dx = \int_0^{1/2} \frac{1+x^2}{1-x^2}dx = 0.598612\ldots$.

23. The width of the material equals the length of the curve which is found by evaluating $\int_0^{20}[1 + \frac{9\pi^2}{400}\cos^2 \frac{3\pi}{20}x]^{1/2}dx = 21.068$ inches, using NINT.

25. The path is a hyperbola; the asteroid will be close to earth at the vertex $(\sqrt{5}, 0)$. The distance travelled is the arc length $L = \int_{\sqrt{5}}^{10}[1 + (\frac{dy}{dx})^2]^{1/2}dx$. Either solve explicitly for y and differentiate, or use the following method: Since $y^2 = 0.2x^2 - 1$, $2yy' = 0.4x$ and $y' = \frac{0.4x}{2y} = \frac{0.2x}{y}$; $(y')^2 = \frac{0.04x^2}{y^2} = \frac{0.04x^2}{0.2x^2-1}$. $L = \text{NINT}(\sqrt{(1 + 0.04x^2/(0.2x^2 - 1))}, x, \sqrt{5}, 10) = 9.0333173\ldots$.

27. The pool is shaped like a teardrop. Use $[0, 20]$ by $[-30, 30]$. The limits are $x = 0$ and $x = 20$, $y = 2x(2-0.1x)^{1/2}$; $\frac{dy}{dx} = 2(2-0.1x)^{1/2} - \frac{2x(0.1)}{2(2-0.1x)^{1/2}}$; $\frac{dy}{dx} = \frac{2(2-0.1x)-x(0.1)}{(2-0.1x)^{1/2}} = \frac{4-0.3x}{(2-0.1x)^{1/2}}$; $ds^2 = 1+(\frac{dy}{dx})^2 = 1+\frac{(4-0.3x)^2}{(2-0.1x)}$; $L = 2\int_0^{19.999999} ds = 100.89\ldots$.

6.5 AREAS OF SURFACES OF REVOLUTION

1. $S = 2\pi \int_0^4 y\,ds = 2\pi \int_0^4 \frac{x}{2}\sqrt{1 + (\frac{1}{2})^2}dx = \frac{\pi\sqrt{5}}{2}[\frac{x^2}{2}]_0^4 = 4\pi\sqrt{5}$

[(base circumference)/2] $\times$ slant height $= \frac{\pi \cdot 4}{2}\sqrt{4^2 + 2^2} = 4\pi\sqrt{5}$.

3. $S = 2\pi \int y\,ds = 2\pi \int_1^3(\frac{x}{2} + \frac{1}{2})\sqrt{1 + (\frac{1}{2})^2}dx = \frac{\pi\sqrt{5}}{2}\int_1^3(x+1)dx = \frac{\pi\sqrt{5}}{2}[\frac{x^2}{2} + x]_1^3 = 3\pi\sqrt{5}$.

5. $S = 2\pi \int_0^2 \frac{x^3}{9}\sqrt{1 + (\frac{x^2}{3})^2}dx = (2\pi/27)\int_0^2 x^3\sqrt{9 + x^4}dx$. Let $u = x^4 + 9$, then $du = 4x^3dx$ and $S = \frac{(2\pi/27)}{4}\int_{x=0}^{x=2} u^{1/2}du = \frac{\pi}{54}\int_9^{25} u^{1/2}du = \frac{\pi}{54} \cdot \frac{2}{3}[u^{3/2}]_9^{25} = \frac{\pi}{27 \cdot 3}[125 - 27] = 98\pi/81$.

7. $S = 2\pi \int_{3/4}^{15/4} \sqrt{x}\sqrt{1 + (\frac{1}{2\sqrt{x}})^2}dx = \pi \int_{3/4}^{15/4} \sqrt{4x + 1}dx = 28\pi/3$.

9. $S = 2\pi \int_1^5 \sqrt{x+1}\,\sqrt{1 + (\frac{1}{2\sqrt{x+1}})^2}dx = \pi \int_1^5 \sqrt{4x + 5}dx = \frac{\pi}{4} \cdot \frac{2}{3}[(4x + 5)^{3/2}]_1^5 = (\pi/6)[125 - 27] = 49\pi/3$.

11. $S = 2\pi \int_0^1 x\sqrt{1 + x^2}dx = \pi \cdot \frac{2}{3}[(1 + x^2)^{3/2}]_0^1 = 2\pi(2\sqrt{2} - 1)/3$.

13. $S = 2\pi \int_1^2 y\sqrt{1 + (y^3 - \frac{1}{4y^3})^2}dy = 2\pi \int_1^2 y\sqrt{(y^3 + \frac{1}{4y^3})^2}dy = 2\pi \int_1^2(y^4 + \frac{1}{4y^2})dy = 253\pi/20$.

15. $S = \int_0^{1/2} 2\pi x^3(1+9x^4)^{1/2}dx$; let $\omega = (1+9x^4)$, $d\omega = 36x^3dx$. $S = \frac{2\pi}{36}\int_{x=0}^{x=1/2}\omega^{1/2}d\omega = \frac{2\pi}{36}\frac{(1+9x^4)^{3/2}}{3/2}\Big|_0^{1/2} = \frac{\pi}{27}[(\frac{5}{4})^3 - 1]$.

17. $S = 2\pi\int y\,ds = 2\pi\int_{-\pi/2}^{\pi/2}\cos x(1+\sin^2 x)^{1/2}dx = 4.591\pi$.

19. For one wok, the area of the outside is $S = 2\pi\int_{y=-16}^{y=-7} x\,ds$. Since $dx = \frac{-y}{x}dy$, $ds = \frac{16}{x}dy$. $S = 2\pi\int_{-16}^{-7} x\cdot\frac{16}{x}dy = 32\pi\cdot 9 = 904.779$ cm^2; the amount of paint is $904.779(.1)$ cm^3 $= 0.0904779$ liters per side per wok. For 5000 woks, 452.390 liters of each color will be needed.

21. $S = 2\pi\int_{x=0}^{x=40} y\,ds = 2\pi\int_0^{40} 10x^{13/30}\sqrt{1+(\frac{13}{3}x^{-17/30})^2}\,dx = 11,899.571$. 11,900 tiles are needed.

6.6 WORK

1. $W = \int_0^{20} 40(\frac{20-x}{20})dx = 2\int_0^{20}(20-x)dx = 400$ ft $\cdot$ lb.

3. $W = \int_0^{50} 0.74(50-x)dx = 0.74[50x - \frac{x^2}{2}]_0^{50} = 925\ N\cdot m$.

5. $W = \int_0^{180} 4(180-x)dx = 64,800$ ft $\cdot$ lb.

7. $6 = K(0.4)$, so $K = 15N/m$. $W = \int_0^{0.4} 15\,x\,dx = 1.2\ N\cdot m$.

9. $10,000 = K(1)$. a) $W = \int_0^{1/2} 10,000\,x\,dx = 1250$ in $\cdot$ lb $= 104\frac{1}{6}$ ft $\cdot$ lb.

b) $W = \int_{1/2}^{1} 10,000\,x\,dx = 3750$ in $\cdot$ lb $= 312.5$ ft $\cdot$ lb.

11. $W = \int_0^{10} 25\pi w y\,dy = 62.5\times 25\times\pi\cdot 10^2/2 = 245,436.926$ ft $\cdot$ lb.

13. $W = 51.2\int_0^{30}\pi(10^2)(30-y)dy = 7,238,229.473$ ft $\cdot$ lb.

15. a) $W = \int_0^{20} 62.5(20-y)120dy = 120(62.5)[20y - \frac{y^2}{2}]_0^{20} = 1,500,000$ ft $\cdot$ lb.

b) $1,500,000/250 = 6000$ sec $= 100$ minutes.

c) $W = \int_{10}^{20} 62.5(20-y)120dy = 375,000$ ft $\cdot$ lb are required to lower the level 10 feet. The pump will require $375,000/250 = 1500$ sec $= 25$ minutes.

17. $W = \omega\int_0^{16}(16-y)\pi(\sqrt{y})^2\,dy = \omega\pi\int_0^{16}(16y-y^2)dy = \frac{2048}{3}\pi\omega = 21,446,605.85$ $N\cdot m$.

19. The water weighs $\pi \cdot 2^2 \cdot 6 \cdot 62.5 = 4712.389$ pounds. If it is pumped to the top, $W = 4712.389 \cdot (15+6) = 98,960.17$ ft · lb. If it is pumped to the level of the valve and then pumped in, $W = 4712.389 \cdot 15 + 62.5(2^2)\pi \int_0^6 y\,dy = 84,823$ ft·lb through the valve is faster.

21. $W = 56 \int_0^{10} \pi(\sqrt{100-y^2})^2(12-y)dy = 56\pi \int_0^{10}(1200 - 12y^2 - 100y + y^3)dy = 56\pi(5500) = 967,610.537$ ft · lb. Cost is \$4838.05.

23. $\int_a^b \frac{mMG}{r^2}dr = -mMG[\frac{1}{r}]_1^b = -1000 \times 5.975 \times 10^{24} \times 6.6720 \times 10^{-11}[\frac{1}{3578} \times 10^{-4} - \frac{1}{637} \times 10^{-4}] = -1000 \times 5.975 \times 6.6720[\frac{1}{3578} - \frac{1}{637}] \times 10^9 = 0.051441 \times 10^3 \times 10^9 = 5.1441 \times 10^{-2+3+9} = 5.1441 \times 10^{10}\ N \cdot m.$

6.7 FLUID PRESSURES AND FLUID FORCES

1. $F = \int_0^3 (62.5)2(7-y)y\,dy = 125 \int_0^3 (7y - y^2)dy = 2812.5$ lb.

3. $F = 62.5 \int_0^3 2x(3-y)dy = 125 \int_0^3 \frac{2y}{3}(3-y)dy = 375$ lb. The length does not affect the force.

5. The side of the plate lies along the line through $B(2,-1)$ and $(0,-5)$: $y = 2x - 5$. $F = 62.5 \int_{-5}^{-1} 2(\frac{y+5}{2})(-y)dy = 62.5 \int_{-5}^{-1}(5+y)(-y)dy = 62.5 \cdot \frac{56}{3} = 1166.67$ lb.

7. $F = 62.5 \int_{-1}^0 2\sqrt{1-y^2}(-y)dy = +62.5 \int_0^1 u^{1/2}du = 62.5[\frac{2}{3}] = 41.67$ lb. Let $u = 1 - y^2$.

9. $F = \omega \int_{-33.5}^{-0.5} 63(-y)dy = -\omega \cdot 63[\frac{y^2}{2}]_{-33.5}^{-0.5} = \omega \cdot 63[561]$, $\omega = 64$ lb/ft^3 $= 0.03073$ lb/in^3, $F = 1309$ lb/in^3.

11. $F = 64.5 \int_{-3}^0 2\sqrt{9-y^2}(-y)dy = 64.5 \int_0^9 u^{1/2}du = 64.5(\frac{2}{3})9^{3/2} = 1161$ lb. Let $u = 9 - y^2$.

13. Assume the depth of the water is h. Then $V = 30 \cdot h(\frac{2h}{5}) = 12h^2$. $F = 62.5 \int_0^h 2(2y/5)(h-y)dy = 50 \int_0^h (hy - y^2)dy = 50[\frac{hy^2}{2} - \frac{y^3}{3}]_0^h = 50 \cdot \frac{h^3}{6}$. $F = 6667 \Rightarrow h = 9.23$, $V = 1034.16$ ft^3.

15. Let the plate have width w, height h. Assume the top of the plate is d units below the surface of the fluid which is on the x-axis. The force on one side of the plate is $F = \omega \int_{-h-d}^{-3} w(-y)dy = \omega w[\frac{h^2+2dh}{2}]$. The average of the pressure at the top and bottom is $\frac{\omega d + \omega(d+h)}{2}$. Multiplied by the area, this is $\frac{\omega}{2}(h+2d)hw)$.

6.8 CENTERS OF MASS

1. $80 \cdot 5 = 100 \cdot x$. $x = 4$ ft.

3. $M_0 = \int_0^2 \delta(x) x dx = 4\int_0^2 x dx = 4[\frac{x^2}{2}]_0^2 = 8$. $M = \int_0^2 \delta dx = 4\int_0^2 dx = 8$. $\bar{x} = \frac{8}{8} = 1$.

5. $M_0 = \int_0^4 x\delta(x) dx = \int_0^4 x(1+\frac{x}{4})^2 dx = \int_0^4 x(1+\frac{x}{2}+\frac{x^2}{16}) dx = [\frac{x^2}{2}+\frac{x^3}{6}+\frac{x^4}{64}]_0^4 = \frac{68}{3}$. $M = \int_0^4 (1+\frac{x}{4})^2 dx = \frac{4}{3}[(1+\frac{x}{4})^3]_0^4 = \frac{4}{3}(8-1) = \frac{28}{3}$. $\bar{x} = M_0/M = 68/28 = 17/7$.

7.

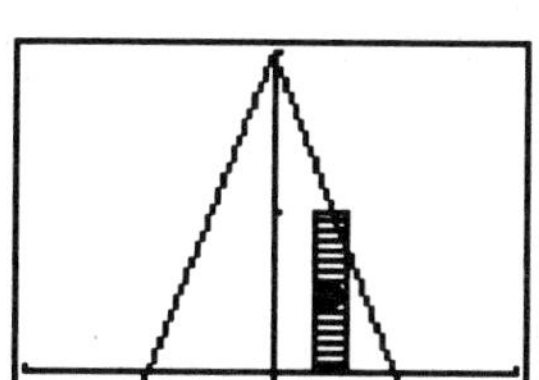

$\bar{x} = 0$. Using vertical strips and symmetry,

$M_x = 2\delta \int_{x=0}^{x=1} \tilde{y} dA =$

$2\delta \int_0^1 \frac{(-2x+2)}{2}(-2x+2)dx =$

$4\delta \int_0^1 (1-x)^2 dx = 4\delta[\frac{(1-x)^3}{-3}]_0^1 = 4\delta/3$.

$M = 2\delta \int_0^1 dA = 2\delta \int_0^1 (-2x+2)dx =$

$4\delta \int_0^1 (1-x)dx = 4\delta[\frac{(1-x)^2}{-2}]_0^1 = 2\delta$. $\bar{y} = 2/3$.

9.

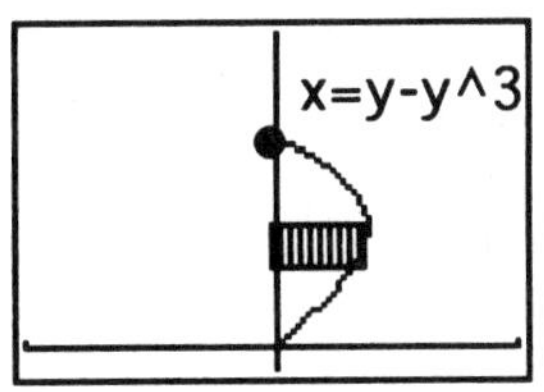

$M_x = \delta \int_0^1 y(y-y^3)dy$

$= \delta[\frac{y^3}{3} - \frac{y^5}{5}]_0^1 = 2\delta/15$.

$M_y = \delta \int_0^1 (\frac{y-y^3}{2})(y-y^3)dy = 4\delta/105$.

$M = \delta \int_0^1 (y-y^3)dy = \delta[\frac{y^2}{2} - \frac{y^4}{4}]_0^1 = \delta/4$.

$\bar{x} = M_y/M = 16/105$; $\bar{y} = M_x/M = 8/15$.

11.

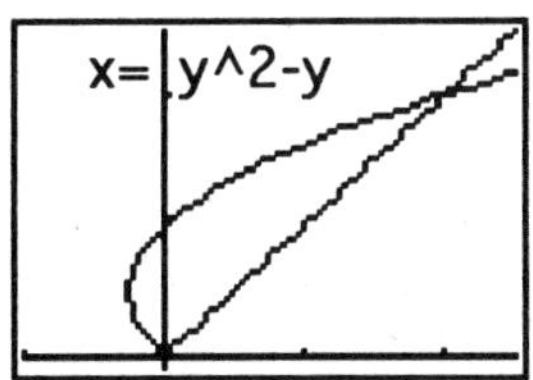

$M_x = \delta \int_0^2 y dA = \delta \int_0^2 y[y-(y^2-y)]dy$

$= \delta \int_0^2 (2y^2 - y^3)dy = \delta[\frac{2y^3}{3} - \frac{y^4}{4}]_0^2 = 4\delta/3$.

$M_y = \delta \int_0^2 \tilde{x} dA = \delta \int_0^2 \frac{[y+(y^2-y)]}{2}[y-(y^2-y)]dy$

$= \frac{\delta}{2}\int_0^2 [y^2-(y^2-y)^2]dy = \frac{\delta}{2}\int_0^2 (2y^3-y^4)dy$

$= 4\delta/5$.

$M = \delta \int_0^2 [y-(y^2-y)]dy = \delta[y^2 - \frac{y^3}{3}]_0^2 = 4\delta/3$,

$\bar{x} = M_y/M = 3/5$, $\bar{y} = M_x/M = 1$.

13. By symmetry $\bar{x} = 0$. $M_x = \delta \int_{-\pi/2}^{\pi/2} \frac{\cos x}{2} \cos x dx = \frac{\delta}{2}\frac{1}{2}[x + \frac{1}{2}\sin 2x]_{-\pi/2}^{\pi/2} = \frac{\delta}{4}(\pi)$. $M = \delta \int_{-\pi/2}^{\pi/2} \cos x dx = \delta[\sin x]_{-\pi/2}^{\pi/2} = 2\delta$, $\bar{y} = M_x/M = \pi/8$.

15. $M = \delta \int_0^2[(2x - x^2) - (2x^2 - 4x)]dx = \delta \int_0^2(6x - 3x^2)dx = 4\delta$; $M_x = \delta \int_0^2 \frac{(2x-x^2)+(2x^2-4x)}{2}(6x - 3x^2)dx = \frac{\delta}{2}\int_0^2(x^2 - 2x)(6x - 3x^2)dx = \frac{\delta}{2}\int_0^2(6x^3 - 3x^4 - 12x^2 + 6x^3)dx = \frac{-\delta}{2}\cdot\frac{16}{5} = \frac{-8\delta}{5}$. $M_y = \delta \int_0^2 x(6x - 3x^2)dx = \delta[2x^3 - 3\frac{x^4}{4}]_0^2 = 4\delta$. $\bar{x} = M_y/M = 1$; $\bar{y} = M_x/M = -2/5$.

17. Area $= \frac{1}{4}$ (area of square of side 6 - area of circle) $= \frac{1}{4}(36 - \pi \cdot 3^2) = \frac{9}{4}(4 - \pi)$; $M = \frac{9}{4}(4 - \pi)\delta$; $M_x = \delta \int_0^3 \frac{3+\sqrt{9-x^2}}{2} \cdot (3 - \sqrt{9 - x^2})\, dx = \frac{\delta}{2}\int_0^3(9 - (9 - x^2))dx = \frac{\delta}{2}[\frac{x^3}{3}]_0^3 = \frac{9\delta}{2}$; $\bar{y} = M_x/M = \frac{2}{4-\pi}$. The center of mass of a symmetric region lies on its axis of symmetry. Hence $\bar{x} = \bar{y} = (\frac{2}{4-\pi})$.

19. $\bar{y} = \frac{1}{3}(3 - 0)$. $\bar{x} = 0$.

21. By symmetry, the centroid lies on the line $y = x$. One median coincides with the line joining $(0, a)$ and $(\frac{a}{2}, 0) : y = -2x + a$. The lines intersect at $(\frac{a}{3}, \frac{a}{3})$.

23. $M = \int_1^2 \delta dA = \int_1^2 x^3 \cdot \frac{2}{x^2}dx = [x^2]_1^2 = 3$. $M_x = \int_1^2 \frac{1}{2}(\frac{2}{x^2})x^3 \cdot \frac{2}{x^2}dx = 2\int_1^2 \frac{dx}{x} = 2\ln 2 = \ln 4$. $M_y = \int_1^2 x \cdot x^3 \cdot \frac{2}{x^2}dx = 2\int_1^2 x^2 dx = 2[\frac{x^3}{3}]_1^2 = \frac{14}{3}$. $\bar{x} = 14/9$, $\bar{y} = (\ln 4)/3$.

25. By symmetry, the centroid is at $(2, 2)$. The area of the square is $(\sqrt{8})^2$. $V = 2\pi \cdot 2 \cdot 8 = 32\pi$. The perimeter of the square is $4\sqrt{8}$. $S = 2\pi \cdot 2 \cdot 4\sqrt{8} = 32\sqrt{2}\,\pi$.

27. The centroid is at $(2, 0)$. The area of the circle is π. Hence $V = 2\pi \cdot 2 \cdot \pi = 4\pi^2$.

29. By symmetry $\bar{x} = 0$. The length of the semicircle is πa. By Theorem 2, $4\pi a^2 = 2\pi\bar{y}(\pi a)$. $\bar{y} = 2a/\pi$.

31. The area of the semicircle is $\pi a^2/2$. Hence $2\pi\bar{y} \cdot \pi a^2/2 = (4/3)\pi a^3$ implies $\bar{y} = 4a/(3\pi)$. By symmetry, $\bar{x} = 0$.

33.

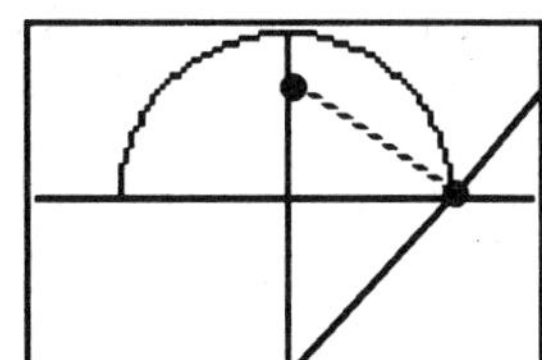

By Problem 31, the centroid is located at $(0, 4a/3\pi)$. To find the distance the centroid travels, first find the line through $(0, 4a/3\pi)$ with slope $-1 : y = -x + 4a/3\pi$. This line meets $y = x - a$ at $(\frac{a(4+3\pi)}{6\pi}, \frac{a(4-3\pi)}{6\pi})$. The distance from the centroid to the axis of rotation:

$\sqrt{[\frac{a(4+3\pi)}{6\pi}]^2 + [\frac{a(4-3\pi)}{6\pi} - \frac{8a}{6\pi}]^2} = \sqrt{2[\frac{a(4+3\pi)}{6\pi}]^2} = \frac{a(4+3\pi)}{6\pi}\sqrt{2}$; this is the radius of the circle swept out by the centroid. By Pappus, $V = \text{area} \cdot \text{distance} = \frac{\pi a^2}{2} \cdot 2\pi \frac{a(4+3\pi)}{6\pi}\sqrt{2} = \frac{\pi a^3}{6}(4 + 3\pi)\sqrt{2}$.

6.9 THE BASIC IDEA. OTHER MODELING APPLICATIONS

1. Cross-sections are squares with diagonal of length $2\sqrt{x}$. The area of the square is $\frac{(2\sqrt{x})^2}{2} = 2x$ Hence $V = \int_0^4 2x dx = [x^2]_0^4 = 16$.

3. The side of each square is $2\sqrt{1-x^2}$. $V = \int_{-1}^{1}(2\sqrt{1-x^2})^2 dx = 4\int_{-1}^{1}(1-x^2)dx = 4[x - \frac{x^3}{3}]_{-1}^{1} = \frac{16}{3}$.

5. The radius of each disk is $\frac{1}{\sqrt{x}}$. $V = \int_1^2 \pi(\frac{1}{x})dx = [\pi \ln x]_1^2 = \pi \ln 2$.

7. The base and height of each triangle are $2\sqrt{1-x^2}$. Hence $A(x) = \frac{1}{2}2\sqrt{1-x^2} \cdot 2\sqrt{1-x^2} = 2 - 2x^2$. $V = \int_{-1}^{1}(2-2x^2)dx = 2\int_0^1(2-2x^2)dx = 2[2x - \frac{2x^3}{3}]_0^1 = \frac{8}{3}$.

9.

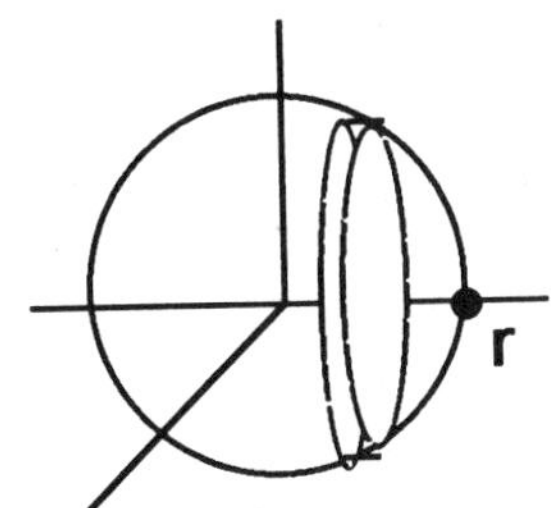

$$V = 2\int_0^r \text{area of slice } dx = 2\int_0^r \pi(\sqrt{r^2 - x^2})^2 dx = 2\pi\int_0^r (r^2 - x^2)dx = 2\pi(r^3 - \frac{r^3}{3}) = \frac{4\pi r^3}{3}$$

11. Twisted or not, the area of each square is s^2. $V = \int_0^h s^2 dx = s^2 h$. V is independent of the number of revolutions of the square.

13. b) Total distance $= \int_0^{2\pi} |5\cos t| dt = 5\int_0^{\pi/2} \cos t dt - 5\int_{\pi/2}^{3\pi/2} \cos t\, dt + 5\int_{3\pi/2}^{2\pi} \cos t = 5[\sin t]_0^{\pi/2} - 5[\sin t]_{\pi/2}^{3\pi/2} + 5[\sin t]_{3\pi/2}^{2\pi} = 5[1-0] - 5[-1-(+1)] + 5(0-(-1)) = 20$ meters. c) $\int_0^{2\pi} 5\cos t\, dt = 5[\sin t]_0^{2\pi} = 0$ m

15. b) Total distance $= \int_0^{\pi/2} |6\sin 3t| dt = 6\int_0^{\pi/3} \sin 3t\, dt - 6\int_{\pi/3}^{\pi/2} \sin 3t\, dt = -\frac{6}{3}[\cos 3t]_0^{\pi/3} + \frac{6}{3}[\cos 3t]_0^{\pi/2} = -2[-1-(1)] + 2[0-(-1)] = 6$ meters

c) $\int_0^{\pi/2}(6\sin 3t)dt = -\frac{6}{3}[\cos 3t]_0^{\pi/2} = -2[0-1] = 2$ meters.

17. b) $\int_0^{10} |49 - 9.8t| dt = 2\int_0^5 (49 - 9.8t)dt = 2[49t - \frac{9.8t^2}{2}]_0^5 = 245$ meters.

c) By symmetry, 0 meters.

19. b) $\int_0^2 |6(t-1)(t-2)| dt = \int_0^1 (6t^2 - 18t + 12)dt - \int_1^2 (6t^2 - 18t + 12)dt = [2t^3 - 9t^2 + 12t]_0^1 - [2t^3 - 9t^2 + 12t]_1^2 = [5] - [16 - 36 + 24 - (2 - 9 + 12)] = [5] - [4 - (5)] = 5 - (-1) = 6$ meters. c) $\int_0^2 (6t^2 - 18t + 12)dt = 5 - 1 = 4$ meters.

21. a) distance travelled $= 1\cdot 1 + 0\cdot 1 + 1\cdot 1 + 0\cdot 1 = 2$; position shift $= 1\cdot 1 + 0\cdot 1 + 1\cdot 1 + 0\cdot 1 = 2$. The velocity is always non-negative; hence position shift = distance travelled. b) distance travelled $= 1\cdot 1 + 1\cdot 1 + 1\cdot 1 + 1\cdot 1 = 4$; position shift $= 1\cdot 1 + (-1)\cdot 1 + 1\cdot + (-1)\cdot 1 = 0$ c) distance travelled $= 4\cdot\frac{1}{2}(1)(2) = 4$; position shift = 4 d) distance travelled $= \frac{1}{2}(2)(1) + 0 + \frac{1}{2}(2)(1) = 2$; position shift = 2

23. $\int_0^{\sqrt{3}} \frac{2\pi x}{\sqrt{3}} dx = \frac{\pi \cdot x^2}{\sqrt{3}}\Big|_0^{\sqrt{3}} = \frac{3\pi}{\sqrt{3}} \approx 1.73\pi$ instead of 2π

PRACTICE EXERCISES, CHAPTER 6

1. $A = \int_1^2 (x - \frac{1}{x^2})dx = [\frac{x^2}{2} + \frac{1}{x}]_1^2 = 1$

3. $A = \int_{-2}^1 (3 - x^2 - (x+1))dx = \int_{-2}^1 (2 - x - x^2)dx = [2x - \frac{x^2}{2} - \frac{x^3}{3}]_{-2}^1 = \frac{9}{2}$

5. $A = \int_0^3 2y^2 dy = [\frac{2y^3}{3}]_0^3 = 18$

7. $A = \int_{-1}^2 (\frac{y+2}{4} - \frac{y^2}{4})dy = \frac{1}{4}\int_{-1}^2 (y + 2 - y^2)dy = \frac{1}{4}[\frac{y^2}{2} + 2y - \frac{y^3}{3}]_{-1}^2 = \frac{9}{8}$

9. $A = \int_0^{\pi/4} (x - \sin x)dx = [\frac{x^2}{2} + \cos x]_0^{\pi/4} = \frac{\pi^2}{32} + \frac{\sqrt{2}}{2} - 1 = 0.0155$

11. $A = \int_0^{\pi} (2\sin x - \sin 2x)dx = [-2\cos x + \frac{1}{2}\cos 2x]_0^{\pi} = 4$

13. $A = \int_1^2 (\sqrt{y} - (2 - y))dy = [\frac{2y^{3/2}}{3} - 2y + \frac{y^2}{2}]_1^2 = (8\sqrt{2} - 7)/6$

15. $A = \int_{-\ln 2}^{\ln 2} 2e^{-x}dx = -2[e^{-x}]_{-\ln 2}^{\ln 2} = -2[e^{-\ln 2} - e^{\ln 2}] = -2[\frac{1}{2} - 2] = 3$

17. a) $V = \pi \int_{-1}^{1}(3x^4)^2 dx = 2\pi \cdot 9 \int_0^1 x^8 dx = 2\pi \cdot 9[\frac{x^9}{9}]_0^1 = 2\pi$ b) The solid is generated by revolving the region between $x = 0$ and $x = 1$. $V = 2\pi \int_0^1 x \cdot 3x^4 dx = 6\pi[\frac{x^6}{6}]_0^1 = \pi$

19. a) $V = \pi \int_1^5 y^2 dx = \pi \int_1^5 (x-1)dx = \pi[\frac{x^2}{2} - x]_1^5 = 8\pi$ b) $V = 4\pi \int_1^5 x\sqrt{x-1}dx = 4\pi \int_0^4 (u+1)u^{1/2}du = 4\pi[\frac{2u^{5/2}}{5} + \frac{2u^{3/2}}{3}]_0^4 = 4\pi[\frac{64}{5} + \frac{16}{3}] = \frac{1088\pi}{15}$, where $u = x - 1$. c) $V = 4\pi \int_1^5 (5-x)\sqrt{x-1}dx = 4\pi \int_0^4 (4-u)u^{1/2}du = 4\pi[\frac{4u^{3/2}\cdot 2}{3} - \frac{2u^{5/2}}{5}]_0^4 = 512\pi/15$, where $u = x - 1$.

21. $V = \pi \int_0^{\pi/3} (\tan x)^2 dx = \pi \int_0^{\pi/3} (\sec^2 x - 1)dx = \pi[\sqrt{3} - \frac{\pi}{3}]$

23. $V = \pi \int_1^{16} (\frac{1}{\sqrt{x}})^2 dx = \pi[\ln x]_1^{16} = \pi \ln 16$

25. Revolve the strip bounded by $x = 0, x = \sqrt{3}, x^2 + y^2 = 4$ about the y-axis. $V = 4\pi \int_0^{\sqrt{3}} x\sqrt{4 - x^2}dx = -\frac{4\pi}{3}[(4 - x^2)^{3/2}]_0^{\sqrt{3}} = \frac{28\pi}{3}$

27. $L = \int_0^3 \sqrt{1 + [\frac{1}{2}x^{-1/2} - \frac{1}{2}x^{1/2}]^2}dx = \int_0^3 [\frac{1}{2}x^{-1/2} + \frac{1}{2}x^{1/2}]dx = \frac{1}{2}[2x^{1/2} + \frac{2x^{3/2}}{3}]_0^3 = 2\sqrt{3}$

29. $y' = \frac{x^3}{2} - \frac{1}{2x^3}$; $1 + (y')^2 = 1 + \frac{x^6}{4} - \frac{1}{2} + \frac{1}{4x^6} = (\frac{x^3}{2} + \frac{1}{2x^3})^2$; $s = \frac{1}{2}\int_1^3 (x^3 + x^{-3})dx = \frac{1}{2}[\frac{x^4}{4} - \frac{1}{2x^2}]_1^3 = \frac{92}{9}$

31. $1 + (y')^2 = 1 + [\frac{2}{2\sqrt{2x+1}}]^2 = \frac{2x+2}{2x+1}$; $S = 2\pi \int_0^{12} \sqrt{2x+1}\frac{\sqrt{2x+2}}{\sqrt{2x+1}}dx = 2\pi \int_0^{12} (2x + 2)^{1/2}dx = \pi[\frac{2(2x+2)^{3/2}}{3}]_0^{12} = \frac{2\pi}{3}[26^{3/2} - 2^{3/2}] \approx 86.5\pi$

33. $1 + (y')^2 = 1 + [\frac{x^{1/2}}{2} - \frac{1}{2}x^{-1/2}]^2 = [\frac{x^{1/2}}{2} + \frac{x^{-1/2}}{2}]^2$. The graph lies below the x axis. Hence $S = 2\pi \int_0^3 -(\frac{x^{3/2}}{3} - x^{1/2})\frac{1}{2}(x^{1/2} + x^{-1/2})dx = -\pi \int_0^3 (\frac{1}{3}x^2 + \frac{1}{3}x - x - 1)dx = -\pi[\frac{x^3}{9} - \frac{x^2}{3} - x]_0^3 = -\pi[-3] = 3\pi$

35. $1 + (y')^2 = 1 + x^2(x^2 + 2) = (x^2 + 1)^2$. $S = 2\pi \int_0^1 x(x^2 + 1)dx = 2\pi[\frac{x^4}{4} + \frac{x^2}{2}]_0^1 = 3\pi/2$

37. The work needed to haul up the equipment is $10(9.8)(40) = 3920N \cdot m$. For the rope, $W = \int_0^{40} 0.8(40 - x)dx = [32x - 0.4x^2]_0^{40} = 640N \cdot m$. Total work $= 3920 + 640 = 4560N \cdot m$.

39. $K = 20$ lb/ft. $W = \int_0^1 20x\, dx = [10x^2]_0^1 = 10$ ft · lb. To stretch an additional foot requires $W = \int_1^2 20x dx = [10x^2]_1^2 = 30$ ft · lb.

41. $W = \omega\int_0^8 (14-y)\pi r^2 dy = \omega\int_0^8 (14-y)\pi(\frac{10y}{8})^2 dy = \omega\pi\, 6400/3$ ft · lb, where $\omega = 62.5$

43. $F = 2\omega\int_0^2 (2-y)2y\,dy = 4\cdot\omega\int_0^2 (2y-y^2)dy = 4\omega[y^2 - \frac{y^3}{3}]_0^2 = \frac{16\omega}{3} = \frac{16(62.5)}{3} =$ 333.3 lb

45. $F = 2\int_0^4 (9-y)62.5\frac{\sqrt{y}}{2}dy = 62.5\int_0^4 (9y^{1/2} - y^{3/2})dy = 62.5[\frac{2}{3}\cdot 9y^{3/2} - \frac{2}{5}y^{5/2}]_0^4 =$ $62.5 \times \frac{176}{5} = 2200$ lb

47. $M_x = \int_{-1}^1 \tilde{y}\,dm = \delta\int_{-1}^1 [\frac{3-x^2+2x^2}{2}](3-x^2-2x^2)dx = \frac{\delta}{2}\int_{-1}^1 (3+x^2)(3-3x^2)dx =$ $\frac{3\delta}{2}\int_{-1}^1 [3-2x^2-x^4]dx = \frac{3\delta}{2}\cdot 2[3x - \frac{2x^3}{3} - \frac{x^5}{5}]_0^1 = \frac{32}{5}\delta$. $M = \int_{-1}^1 \delta\,dA =$ $2\delta\int_0^1 (3-3x^2)dx = 6\delta[x - \frac{x^3}{3}]_0^1 = 48$. $\bar{y} = M_x/M = 8/5$. By symmetry, $\bar{x} = 0$.

49. $M_x = \int \tilde{y}\,dm = \delta\int_0^4 y(2\sqrt{y})dy = 2\delta\frac{2}{5}[y^{5/2}]_0^4 = \frac{128}{5}\delta$; $M_y = \int \tilde{x}\,dm = \delta\int_0^4 x(4 - \frac{x^2}{4})dx = \delta[2x^2 - \frac{x^4}{16}]_0^4 = 16\delta$; $M = \delta\int_0^4 (4 - \frac{x^2}{4})dx = \delta[4x - \frac{x^3}{12}]_0^4 = \frac{32}{3}\delta$; $\bar{x} = M_y/M = 3/2$; $\bar{y} = M_x/M = 12/5$

51. $M_x = \int \tilde{y}\,dm = \delta\int_0^1 y\cdot 2(y+2)dy = 2\delta[\frac{y^3}{3} + y^2]_0^1 = 8\delta/3$; $M = 2\delta\int_0^1 (y+2)dy =$ $2\delta[\frac{y^2}{2} + 2y]_0^1 = 5\delta$; $\bar{y} = M_x/M = 8/15$. By symmetry, $\bar{x} = 0$.

53. $M_x = \int \tilde{y}\,dm = \delta\int_1^{16} \frac{1}{2\sqrt{x}}\cdot\frac{1}{\sqrt{x}}dx = \frac{1}{2}\delta[\ln x]_1^{16} = \delta\ln 4 = 2\delta\ln 2$; $M_y = \int \tilde{x}\,dm =$ $\delta\int_1^{16} x\frac{1}{\sqrt{x}}dx = \delta\cdot\frac{2}{3}[x^{3/2}]_1^{16} = 42\delta$; $M = \int dm = \delta\int_1^{16}\frac{1}{\sqrt{x}}dx = 2\delta[x^{1/2}]_1^{16} = 6\delta$. $\bar{x} = M_y/M = 7$; $\bar{y} = M_x/M = \frac{1}{3}\ln 2$

55. The radius of each cross-section is $\frac{1}{2}[x^2 - \sqrt{x}]$. $V = \int_0^1 \pi[\frac{1}{2}(x^2 - \sqrt{x})]^2 dx =$ $\frac{\pi}{4}\int_0^1 (x^4 - 2x^{5/2} + x)dx = \frac{\pi}{4}[\frac{x^5}{5} - 2\cdot\frac{2}{7}x^{7/2} + \frac{x^2}{x}]_0^1 = \frac{9\pi}{280}$.

57. The radius of each cross-section is $\frac{1}{2}[2\sin x - 2\cos x]$. $V = \pi\int_{\pi/4}^{5\pi/4}[\sin x -$ $\cos x]^2 dx = \pi\int_{\pi/4}^{5\pi/4}[\sin^2 x - 2\sin x\cos x + \cos^2 x]dx = \pi\int_{\pi/4}^{5\pi/4}(1 - 2\sin x\cos x)dx =$ $\pi[x + \frac{\cos^2 x}{2}]_{\pi/4}^{5\pi/4} = \pi^2$.

59. Each cross-section is an isosceles triangle with acute angles 45°. $A(y) = \frac{1}{2}$ base · height $= \frac{1}{2}(\text{base})^2 = \frac{1}{2}x^2 = \frac{1}{2}(9-y^2)$; $V = \int_{-3}^3 \frac{1}{2}(9-y^2)dy = 2\cdot\frac{1}{2}\int_0^3 (9-$ $y^2)dy = [9y - \frac{y^3}{3}]_0^3 = 18$.

61. b) $\int_0^6 |v(t)|dt = \int_0^2 (1-t/2)dt + \int_2^6 (t/2-1)dt = [t - t^2/4]_0^2 + [t^2/4 - t]_2^6 =$ 5 ft c) $\int_0^6 v(t)dt = \int_0^6 (t/2-1)dt = [t^2/4 - t]_0^6 = 3$ ft.

63. b) $\int_0^{3\pi/2} |v(t)|dt = \int_0^{\pi/2} 5\cos t\,dt - \int_{\pi/2}^{3\pi/2} 5\cos t\,dt = [5\sin t]_0^{\pi/2} - [5\sin t]_{\pi/2}^{3\pi/2} =$ 15ft. c) $\int_0^{3\pi/2} 5\cos t\,dt = [5\sin t]_0^{3\pi/2} = -5$ ft.

CHAPTER 7

THE CALCULUS OF TRANSCENDENTAL FUNCTIONS

7.1 THE NATURAL LOGARITHM FUNCTION

1. $\ln 4/9 = \ln 4 - \ln 9 = \ln 2^2 - \ln 3^2 = 2\ln 2 - 2\ln 3 = 2(\ln 2 - \ln 3)$.

3. $\ln(1/2) = \ln 1 - \ln 2 = -\ln 2$.

5. $\ln 4.5 = \ln(9/2) = \ln 3^2 - \ln 2 = 2\ln 3 - \ln 2$.

7. $\ln 3\sqrt{2} = \ln 3 + \ln 2^{1/2} = \ln 3 + \frac{1}{2}\ln 2$.

9. $y = \ln(x^2) = 2\ln x$. $\frac{dy}{dx} = \frac{2}{x}$. The result is supported by graphing $2/x$ and NDER$(\ln(x^2), x, x)$ in $[-5, 5]$ by $[-5, 5]$.

11. $y = \ln(1/x) = \ln x^{-1} = -\ln x$. $\frac{dy}{dx} = -\frac{1}{x}$.

13. $y = \ln(x+2)$. $\frac{dy}{dx} = \frac{1}{x+2}(x+2)' = \frac{1}{x+2}$.

15. $y = \ln(2 - \cos x)$. $\frac{dy}{dx} = \frac{\sin x}{2-\cos x}$.

17. $y = \ln(\ln x)$. $\frac{dy}{dx} = \frac{\frac{1}{x}}{\ln x} = \frac{1}{x\ln x}$.

19. $y = \sqrt{x(x+1)} = [x(x+1)]^{1/2}$. $\ln y = \ln[x(x+1)]^{1/2} = \frac{1}{2}\ln[x(x+1)] = \frac{1}{2}[\ln x + \ln(x+1)]$. $\frac{y'}{y} = \frac{1}{2}[\frac{1}{x} + \frac{1}{x+1}]$, $y' = \frac{y}{2}[\frac{x+1+x}{x(x+1)}] = \sqrt{x(x+1)}\,\frac{2x+1}{2x(x+1)} = \frac{2x+1}{2\sqrt{x(x+1)}}$. See the remark in the solution of Exercise 21.

21. $y = (x+3)^{1/2}\sin x$. $\ln y = \ln(x+3)^{1/2} + \ln\sin x$, $\ln y = \frac{1}{2}\ln(x+3) + \ln\sin x$. $\frac{y'}{y} = \frac{1}{2}\frac{1}{x+3} + \frac{\cos x}{\sin x}$, $y' = (x+3)^{1/2}\sin x[\frac{1}{2(x+3)} + \frac{\cos x}{\sin x}] = \frac{\sin x}{2\sqrt{x+3}} + \sqrt{x+3}\,\cos x = \frac{\sin x + 2(x+3)\cos x}{2\sqrt{x+3}}$. Remark: To be sure we are "taking log's" of positive numbers only, we could proceed as follows. Since $y > 0$, $y = |y| = |x+3|^{1/2}|\sin x|$ in the domain of y. $\ln|y| = \frac{1}{2}\ln|x+3| + \ln|\sin x|$. Differentiating the last equation yields $\frac{y'}{y} = \frac{1}{2(x+3)} + \frac{\cos x}{\sin x}$ and we may proceed as above. Other exercises in this section may be treated in a similar fashion.

23. $y = x(x+1)(x+2)$, $x > 0$. $\ln y = \ln x + \ln(x+1) + \ln(x+2)$, $\frac{y'}{y} = \frac{1}{x} + \frac{1}{x+1} + \frac{1}{x+2}$, $y' = x(x+1)(x+2)(\frac{1}{x} + \frac{1}{x+1} + \frac{1}{x+2}) = (x+1)(x+2) + x(x+2) + x(x+1) = 3x^2 + 6x + 2$.

25. $y = \frac{x+5}{x\cos x}$, $x > 0$. $\ln y = \ln(x+5) - \ln x - \ln\cos x$. $\frac{y'}{y} = \frac{1}{x+5} - \frac{1}{x} + \frac{\sin x}{\cos x}$, $y' = \frac{x+5}{x\cos x}\left[\tan x - \frac{5}{x(x+5)}\right]$.

27. $y = \frac{x(x^2+1)^{1/2}}{(x+1)^{2/3}}$. $\ln y = \ln x + \frac{1}{2}\ln(x^2+1) - \frac{2}{3}\ln(x+1)$. $\frac{y'}{y} = \frac{1}{x} + \frac{1}{2}\frac{2x}{x^2+1} - \frac{2}{3}\frac{1}{x+1}$, $y' = \frac{x\sqrt{x^2+1}}{(x+1)^{2/3}}\left[\frac{1}{x} + \frac{x}{x^2+1} - \frac{2}{3(x+1)}\right]$.

29. $y = (\frac{x(x-2)}{x^2+1})^{1/3}$. $\ln y = \frac{1}{3}[\ln x + \ln(x-2) - \ln(x^2+1)]$, $\frac{y'}{y} = \frac{1}{3}[\frac{1}{x} + \frac{1}{x-2} - \frac{2x}{x^2+1}]$. $y' = \frac{2}{3}\sqrt[3]{\frac{x(x-2)}{x^2+1}}\,\frac{(x^2+x-1)}{x(x-2)(x^2+1)}$.

31. $\int_{-3}^{-2}\frac{dx}{x} = \ln|x|\,]_{-3}^{-2} = \ln 2 - \ln 3$.

33. $I = \int_{-1}^{0}\frac{3dx}{3x-2}$. Let $u = 3x-2$, $du = 3dx$ and note that $u = -5$ when $x = -1$ and $u = -2$ when $x = 0$. $I = \int_{-5}^{-2}\frac{du}{u} = \ln|u|\,]_{-5}^{-2} = \ln 2 - \ln 5$.

35. $\int_3^4 \frac{dx}{x-5} = \ln|x-5|\,]_3^4$ (by inspection) $= \ln 1 - \ln 2 = -\ln 2$.

37. Let $I = \int_0^3 \frac{2xdx}{x^2-25}$, $u = x^2 - 25$, $du = 2xdx$. Then $I = \int_{-25}^{-16}\frac{du}{u} = \ln|u|\,]_{-25}^{-16} = \ln 16 - \ln 25 = \ln(16/25) = \ln(4/5)^2 = 2\ln(4/5) = 2\ln(0.8)$.

39. $\int_0^3 \frac{1}{x+1}dx$, $u = x+1$, $du = dx$. $\int_1^4 \frac{du}{u} = \ln u]_1^4 = \ln 4 - \ln 1 = \ln 4$.

41. Let $I = \int_0^\pi \frac{\sin x}{2-\cos x}dx$. Let $u = 2 - \cos x$. Then $du = \sin x dx$, $u = 1$ when $x = 0$ and $u = 3$ when $x = \pi$, $I = \int_1^3 \frac{du}{u} = \ln|u|\,]_1^3 = \ln 3 - \ln 1 = \ln 3$.

43. Let $I = \int_1^2 \frac{2\ln x}{x}dx$, $u = \ln x$, $du = \frac{1}{x}$. Then $I = 2\int_0^{\ln 2} udu = u^2]_0^{\ln 2} = (\ln 2)^2$.

45. $I = \int_0^{\pi/2}\tan\frac{x}{2}dx = \int_0^{\pi/2}\frac{\sin\frac{x}{2}}{\cos\frac{x}{2}}dx$. Let $u = \cos\frac{x}{2}$. Then $du = -\frac{1}{2}(\sin\frac{x}{2})dx$, $-2du = (\sin\frac{x}{2})dx$. $I = -2\int_1^{\sqrt{2}/2}\frac{du}{u} = -2\ln|u|\,]_1^{\sqrt{2}/2} = -2[\ln(\sqrt{2})^{-1} - \ln 1] = 2\ln 2^{1/2} = \ln 2$.

47. Let $I = \int_{\pi/2}^{\pi} 2\cot\frac{x}{3}dx = 2\int_{\pi/2}^{\pi}\frac{\cos\frac{x}{3}}{\sin\frac{x}{3}}dx$. Let $u = \sin\frac{x}{3}$, $du = \frac{1}{3}(\cos\frac{x}{3})dx$. Then $I = 6\int_{1/2}^{\sqrt{3}/2}\frac{du}{u} = 6\ln|u|\,]_{1/2}^{\sqrt{3}/2} = 6[(\ln\sqrt{3} - \ln 2) - (\ln 1 - \ln 2)] = 6(\frac{1}{2})\ln 3 = 3\ln 3$.

49. $\int\frac{2dx}{x} = 2\ln|x| + C = \ln x^2 + C$.

51. Let $I = \int \frac{xdx}{x^2+4}$, $u = x^2 + 4$, $du = 2xdx$, $\frac{du}{2} = xdx$. Then $I = \frac{1}{2}\int \frac{du}{u} = \frac{1}{2}\ln|u| + C = \frac{1}{2}\ln(x^2+4) + C$.

53. Let $I = \int \tan\frac{x}{3}dx = \int \frac{\sin(x/3)dx}{\cos(x/3)}$, $u = \cos\frac{x}{3}$, $du = -\frac{1}{3}\sin\frac{x}{3}dx$, $-3du = \sin\frac{x}{3}dx$. Then $I = -3\int\frac{du}{u} = -3\ln|u| + C = -3\ln|\cos\frac{x}{3}| + C$.

55. $\lim_{x\to\infty}\ln\frac{1}{x} = \lim_{u\to 0^+}\ln u = -\infty$.

57. $\lim_{x\to 0}\ln|x| = -\infty$.

59. $\int_0^{\pi/3}\tan x\,dx = -\ln|\cos x|\Big|_0^{\pi/3} = -[\ln|\cos(\pi/3)| - \ln|\cos 0|\,] = -[\ln(1/2) - \ln 1] = -\ln 2^{-1} = \ln 2$.

61. Suppose $f(x)$ is an increasing function on its domain. Let $x_1 < x_2$ be distinct points of the domain. Then $f(x_1) < f(x_2)$ and so $f(x_1) \neq f(x_2)$. Therefore $f(x)$ is one-to-one. The proof for decreasing functions is similar. Since $\frac{d}{dx}(\ln x) = \frac{1}{x} > 0$ for $x > 0$, $\ln x$ is an increasing function. Therefore $\ln x$ is one-to-one.

63. a) Graph $y_1 = \sqrt{x+3}\,\sin x$ in $[-4, 20]$ by $[-5, 5]$. b) In the same viewing rectangle graph $y_2 = \text{NDER}(y_1, x)$ and $y_3 = \frac{dy}{dx}$ (found in Exercise 21). The fact that y_2 and y_3 coincide supports that y_3 is valid where $y < 0$.

65. Graph $y_1 = \frac{x+5}{x\cos x}$ in $[-20, 20]$ by $[-20, 20]$. Proceed now as in Exercise 63.

67. Graph $y_1 = (x(x-2)/(x^2+1))^{1/3}$ in $[-20, 20]$ by $[-1.1, 1.2]$. Proceed now as in Exercise 63.

69. $y = \frac{f}{g}$. $|y| = \frac{|f|}{|g|}$. $\ln|y| = \ln|f| - \ln|g|$. $\frac{y'}{y} = \frac{f'}{f} - \frac{g'}{g}$, $y' = \frac{f}{g}(\frac{f'}{f} - \frac{g'}{g}) = \frac{f'}{g} - \frac{fg'}{g^2} = \frac{gf' - fg'}{g^2}$ or $\frac{dy}{dx} = \frac{\frac{df}{dx}g - f\frac{dg}{dx}}{g^2}$.

71. a) Graph $y = f(x) = \text{NINT}(\cot t, t, \pi/2, x)$ in $[0, \pi]$ by $[-5, 5]$. (Use $tol = 1$.) Conjecture: $\lim_{x\to 0^+} f(x) = \lim_{x\to\pi^-} f(x) = -\infty$. b) You may see "*tol* not met" due to the discontinuity of the integrand at $x = \pi$. c) Except for the vertical asymptote at $x = \pi$ the graph of the antiderivative $\ln|\sin x|$ is smooth. The two graphs are identical on $(0, \pi)$.

7.2 THE EXPONENTIAL FUNCTION

1. $e^{\ln 7} = 7$ **3.** $\ln e^2 = 2$ **5.** $e^{2+\ln 3} = e^2 e^{\ln 3} = 3e^2$

7. $e^{2k} = 4$. $2k = \ln e^{2k} = \ln 4$. $k = \frac{1}{2}\ln 4 = \ln 2$

9. $100e^{10k} = 200$. $e^{10k} = 2$, $10k = \ln 2$, $k = \frac{\ln 2}{10}$

11. $2^{k+1} = 3^k$. $(k+1)\ln 2 = k\ln 3$, $k[\ln 3 - \ln 2] = \ln 2$, $k = \frac{\ln 2}{\ln \frac{3}{2}}$

13. $e^t = 1$. $t = \ln e^t = \ln 1 = 0$.

15. $e^{-0.3t} = 27$. $-0.3t = \ln 27$, $t = -\frac{\ln 27}{0.3} = -\frac{3\ln 3}{0.3}\frac{10}{10} = -10\ln 3$

17. $2^{e^t} = 2 - t$. $y = 2^{e^t}$ is an increasing function while $y = 2 - t$ is a linear decreasing function. There is therefore at most one point of intersection. By inspection $(0, 2)$ is one such point. Therefore $t = 0$ is the only solution.

19. $\ln y = 2t = 4$. $y = e^{2t+4}$ **21.** $\ln(y - 40) = 5t$. $y - 40 = e^{5t}$, $y = 40 + e^{5t}$

23. $5 + \ln y = 2^{x^2+1}$. $\ln y = 2^{x^2+1} - 5$, $y = e^{(2^{x^2+1}-5)}$

25. $y = 2e^x$. $dy/dx = 2e^x$ **27.** $y = e^{-x}$. $dy/dx = e^{-x}(-1) = -e^{-x}$

29. $y = e^{2x/3}$. $\frac{dy}{dx} = \frac{2}{3}e^{2x/3}$ **31.** $y = xe^2 - e^x$. $\frac{dy}{dx} = e^2 - e^x$

33. $y = e^{\sqrt{x}}$. $\frac{dy}{dx} = e^{\sqrt{x}}\frac{1}{2\sqrt{x}} = \frac{e^{\sqrt{x}}}{2\sqrt{x}}$

35. $\int_1^{e^2} \frac{dx}{x} = \ln x]_1^{e^2} = \ln e^2 - \ln 1 = 2 - 0 = 2$

37. $\int_{\ln 2}^{\ln 3} e^x dx = e^x]_{\ln 2}^{\ln 3} = e^{\ln 3} - e^{\ln 2} = 3 - 2 = 1$

39. $\int_{\ln 3}^{\ln 5} e^{2x}dx$, $u = 2x$, $du = 2\,dx$. $\frac{1}{2}\int_{2\ln 3}^{2\ln 5} e^u du = \frac{1}{2}e^u]_{\ln 9}^{\ln 25} = \frac{1}{2}(e^{\ln 25} - e^{\ln 9}) = \frac{1}{2}(25 - 9) = 8$.

41. $\int_0^1 (1+e^x)e^x dx$, $u = 1+e^x$, $du = e^x dx$. $\int_2^{1+e} u\,du = \frac{u^2}{2}]_2^{1+e} = \frac{1}{2}[(1+e)^2 - 2^2] = \frac{1}{2}(e^2 + 2e - 3)$

43. $\int_2^4 \frac{dx}{x+2}$, $u = x + 2$, $du = dx$. $\int_4^6 \frac{du}{u} = \ln u]_4^6 = \ln 6 - \ln 4 = \ln(6/4) = \ln(3/2)$

45. $\int_{-1}^1 2xe^{-x^2}dx$, $u = -x^2$, $du = -2xdx$. $-\int_{-1}^{-1} e^u du = 0$

47. $\int_1^4 \frac{e^{\sqrt{x}}dx}{2\sqrt{x}}$, $u = \sqrt{x}$, $du = \frac{dx}{2\sqrt{x}}$. $\int_1^2 e^u du = e^u]_1^2 = e^2 - e$

49. Let $I = \int 2e^x\cos(e^x)dx$, $u = e^x$, $du = e^x dx$. Then $I = 2\int \cos u du = 2\sin u + C = 2\sin(e^x) + C$.

51. Let $I = \int \frac{e^x dx}{1+e^x}$, $u = 1 + e^x$, $du = e^x dx$. Then $I = \int \frac{du}{u} = \ln|u| + C = \ln(1 + e^x) + C$.

53. Let $I = \int \frac{\tan(\sqrt{x})dx}{\sqrt{x}}$, $u = \sqrt{x}$, $du = \frac{dx}{2\sqrt{x}}$, $\frac{dx}{\sqrt{x}} = 2du$. Then $I = 2\int \tan u\, du = -2\ln|\cos u| + C = -2\ln|\cos(\sqrt{x})| + C$

55. a) Let $y = f(x) = 2x + 3$. The inverse relation is determined by $x = 2y + 3$, $2y = x - 3$, $y = \frac{x-3}{2}$, $f^{-1}(x) = \frac{x-3}{2}$. b) We may graph $y = 2x + 3$ and $y = \frac{x-3}{2}$ together in $[-10, 10]$ by $[-10, 10]$. c) $\frac{df}{dx}|_{x=a} = 2$, $\frac{df^{-1}}{dx}|_{x=f(a)} = \frac{1}{2}$ which verifies the Equation.

57. $y = f(x) = \frac{1-2x}{x+2}$. $x = \frac{1-2y}{y+2}$, $xy + 2x = 1 - 2y$, $xy + 2y = 1 - 2x$, $y(x + 2) = 1 - 2x$, $y = \frac{1-2x}{x+2}$, $f^{-1}(x) = \frac{1-2}{x+2}$. Thus $f = f^{-1}$, f is self-inverse. b) We see that the graph of f is symmetric about the line $y = x$ confirming that it is self-inverse. c) $f' = (f^{-1})' = \frac{(x+2)(-2)-(1-2x)}{(x+2)^2} = \frac{-5}{(x+2)^2}$. $f'(a) = \frac{5}{(a+2)^2}$. $f(a) = \frac{1-2a}{a+2}$, $\frac{df^1}{dx}|_{f(a)} = \frac{-5}{(\frac{1-2a}{a+2}+2)^2} = \frac{-5(a+2)^2}{(1-2a+2a+4)^2} = -\frac{(a+2)^2}{5}$, verifying the Equation.

59. $f'(x) = 2x - 4$. $\frac{df^{-1}}{dx}|_{x=f(4)=-3} = \frac{1}{f'(4)} = \frac{1}{4}$

61. $f(x) = \frac{2x+3}{x+3} = 2 - \frac{3}{x+3}$ whence $f'(x) = \frac{3}{(x+3)^2}$. $f^{-1}(x) = \frac{-3x+3}{x-2} = -3 - \frac{3}{x-2}$ whence $\frac{df^{-1}(x)}{dx} = \frac{3}{(x-2)^2}$. $f'(-5) = \frac{3}{(-2)^2} = \frac{3}{4}$. $(f^{-1})'(3.5) = \frac{3}{(3/2)^2} = \frac{4}{3}$.

63. $\lim_{x\to-\infty} e^{-x} = \infty$

65. $\lim_{x\to\infty} \int_x^{2x} \frac{1}{t}dt = \lim_{x\to\infty} \ln|t|]_x^{2x} = \lim_{x\to\infty}[\ln|2x| - \ln|x|] = \lim_{x\to\infty} \frac{\ln 2|x|}{|x|} = \ln 2$

67. We graph $y = \mathrm{NINT}(2^t \ln t, t, 1, x) - 0.1$ in $[0.001, 3]$ by $[-1, 10]$ and zoom in to the x-intercepts getting $\{0.679\ldots, 1.3086\ldots\}$.

69. $f(x) = x + e^{4x}$, $f'(x) = 1 + 4e^{4x}$. $L(x) = f(0) + f'(0)(x - 0) = 1 + 5x$

71. Since $f(x)$ is periodic of period 2π, we need only investigate the extreme values of f on $[0, 2\pi]$. Since e^x is an increasing function, we can use the extreme values of $\sin x$. Thus the maximum and minimum values of f are, respectively, e and e^{-1}.

73. $\frac{dy}{dx} = 1 + \frac{1}{x}$, $y(1) = 3$. $y(x) = 3 + \int_1^x (1 + \frac{1}{t})dt = 3 + [t + \ln|t|]_1^x = 3 + [x + \ln|x|] - 1 = 2 + x + \ln|x|$

75. The first integral is the area under the curve $y = \ln x$ from 1 to a. The second integral is the area in the rectangle to the left of the curve. The area, $a \ln a$, of the rectangle is the sum of these two areas.

77. $\frac{1}{e^x} = \frac{e^0}{e^x} = e^{0-x} = e^{-x}$

79. Using $y = y_0 e^{kt}$, when $t = 3$, we get $10000 = y_0 e^{3k}$ and when $t = 5$, $40000 = y_0 e^{5k}$. Dividing the second relation by the first, we have $4 = e^{2k}$, $\ln 4 = 2k$, $k = \frac{1}{2}\ln 4 = \frac{1}{2}\ln 2^2 = \ln 2$. $10000 = y_0 e^{3k}$ leads to $y_0 = \frac{10000}{e^{\ln 2^3}} = \frac{10000}{8} = 1250$ for the initial population.

81. a) $A(t) = A_0 e^{rt}$ and here $r = 1$ so the amount in the account after t years is $A(t) = A_0 e^t$. b) $A(t) = 3A_0 = A_0 e^t$, $3 = e^t$, $t = \ln 3 = 1.0986$ years (rounded). c) $A(1) = A_0 e^1 = eA_0$, e times the original amount.

83. $A(t) = A_0 e^{rt}$ and we wish to find r so that $131A_0 = A(100) = A_0 e^{100r}$, $131 = e^{100r}$, $\ln|3| = 100r$, $r = \frac{\ln 131}{100} = 0.04875\ldots$. The interest rate should be $4.875\ldots\%$ per year.

85. a) $y = \frac{70}{i} = \frac{70}{5} = 14$ years. $y = \frac{70}{7} = 10$ years b) $y = \frac{70}{i}$. $5 = \frac{70}{i}$ leads to $i = 14\%$. $20 = \frac{70}{i}$ leads to $i = 3.5\%$.

87. $p(t) = 144.4e^{0.031t}$. $2(144.4) = 144.4e^{0.031t}$, $2 = e^{0.031t}$, $\ln 2 = 0.031t$, $t \approx 22.36$ years. $22\frac{1}{3}$ years from May, 1993 would be August, 2015.

89. $1.04 = \frac{p(t+1)}{p(t)} = \frac{p_0 e^{k(t+1)}}{p_0 e^{kt}} = e^{kt+k-kt} = e^k$. $k = \ln 1.04 \approx 0.039$. The inflation rate was about 3.9%.

91. The required number of years t satisfies $\frac{100}{p(t)} = \frac{1}{2}\frac{100}{p_0}$ or $p(t) = 2p_0$, $p_0 e^{(r/100)t} = 2p_0$. This leads to $e^{(r/100)t} = \frac{r}{100}t = \ln 2$, $t = \frac{100 \ln 2}{r}$ years. .

93. a) $p(x) = p_0 e^{-0.01x}$. $p(100) = 20.09 = p_0 e^{-1}$ yielding $p_0 = 20.09e$ and $p(x) = 20.09e^{1-0.01x}$. b) $p(10) = 20.09e^{1-0.1} = 20.09e^{0.9} = \$49.41$ (rounded). $p(90) = 20.09e^{1-0.9} = 20.09e^{0.1} = \$22.20$ (rounded). c) $r(x) = xp(x) = 20.09xe^{1-0.01x}$. $r'(x) = 20.09[e^{1-0.01x} - 0.01xe^{1-0.01x}] = 20.09e^{1-0.01x}(1 - 0.01x)$. We see from this that r increases ($r' > 0$) until x reaches 100 and from then on decreases. Thus $r(100)$ is the maximum value of r. d) Graph $y = r(x) = 20.09xe^{1-0.01x}$ in $[0, 200]$ by $[0, 2100]$.

95. a) $0.04\int_0^9 \frac{d\tau}{1+\tau} = 0.04\ln(1+\tau)]_0^9 = 0.04[\ln 10 - \ln 1] = 0.04\ln 10$ b) $p(9) = 100e^{0.04\ln 10} = 109.65$ (rounded) c) The exponent $\int_0^t k(\tau)d\tau = (\frac{1}{t-0}\int_0^t k(\tau)d\tau)t = kt$ where k is the average value of $k(\tau)$ from $\tau = 0$ to $\tau = t$. Hence we will always get the same result in b) and c).

97. a) $P(0) = \frac{1000}{1+e^{4.8}} \approx 8$ b) $\lim_{t\to\infty} e^{4.8-0.7t} = e^{4.8}\lim_{t\to\infty}\frac{1}{e^{0.7t}} = 0$. Hence $\lim_{t\to\infty} P(t) = 1000$ is the limiting maximum of the population. c) By b) $P(t) = 1200$ is impossible. $P(t) = 700$ leads to $1000 = 700(1+e^{4.8-0.7t})$, $10 = 7 + 7e^{4.8}e^{-0.7t}$, $e^{-0.7t} = \frac{3}{7e^{4.8}}$, $e^{0.7t} = \frac{7e^{4.8}}{3}$, $t = \frac{1}{0.7}\ln\frac{7e^{4.8}}{3} \approx 8.03$ months d) Determining this analytically would be quite messy. Instead we graph NDER$(P(x), x)$ in $[0, 15]$ by $[0, 200]$ and zoom in to the maximum. We find $\max P' = 173.919$rabbits per month which occurs when $t = 7.082$ months.

99. a) We know $D_x F^{-1}(1) = \frac{1}{F'(a)} = \frac{1}{f(a)}$ where a is the number with $F(a) = 1$. So we must find an estimate for a. That is, we must estimate the x-coordinate of the point of intersection of the graphs of $y_1 = F(x) = \text{NINT}(e^{-0.7t^2}, t, 0, x)$ and $y_2 = 1$. Using ZOOM-IN after graphing in $[0, 2]$ by $[0, 2]$, we find $a = 1.616$. Hence $D_x F^{-1}(1) = \frac{1}{f(a)} \approx e^{0.7(1.616)^2} = 6.222$. b) In parametric mode, graph $x_1 = t$, $y_1 = F(t) = \text{NINT}(e^{-0.7s^2}, s, 0, t)$ and the inverse, $x_2 = y_1(t), y_2 = t$, in $[0, 2]$ by $[0, 2]$. Using TRACE to find the coordinates of two points on $y = F^{-1}(x)$, one to the left and one to the right of $x = 1$, we estimate the slope, $D_x F^{-1}(1)$, and obtain 6.348. This estimate supports the estimate found in a).

7.3 OTHER EXPONENTIAL AND LOGARITHMIC FUNCTIONS

1. $y = x^\pi$. $\frac{dy}{dx} = \pi x^{\pi-1}$

3. $y = x^{-\sqrt{2}}$. $\frac{dy}{dx} = -\sqrt{2}\,x^{-\sqrt{2}-1}$

5. $y = 8^x$. $\frac{dy}{dx} = 8^x \ln 8$

7. $y = 3^{\csc x}$. $\frac{dy}{dx} = (\ln 3)3^{\csc x}(-\csc x\cot x) = -\csc x\cot x(3^{\csc x})\ln 3$

9. $y = x^{\ln x}$, $x > 0$. $\ln y = (\ln x)\ln x = (\ln x)^2$. $\frac{y'}{y} = 2(\ln x)\frac{1}{x}$, $y' = y(2\frac{\ln x}{x}) = 2(\frac{\ln x}{x})x^{\ln x}$.

11. $y = (x+1)^x$. $\ln y = x\ln(x+1)$, $\frac{y'}{y} = \frac{x}{x+1} + \ln(x+1)$, $y' = y[\frac{x}{x+1} + \ln(x+1)] = (x+1)^x[\frac{x}{x+1} + \ln(x+1)]$.

13. $y = x^{\sin x}$. $\ln y = \sin x(\ln x)$, $\frac{y'}{y} = (\sin x)\frac{1}{x} + (\cos x)\ln x$, $y' = x^{\sin x}[\frac{\sin x}{x} + (\cos x)\ln x]$.

15. $y = \log_4 x^2 = 2\log_4 x = 2\frac{\ln x}{\ln 4}$. $\frac{dy}{dx} = \frac{2}{(\ln 4)x} = \frac{2}{2(\ln 2)x} = \frac{1}{x\ln 2}$.

17. $y = \log_2(3x+1) = \frac{\ln(3x+1)}{\ln 3}$. $\frac{dy}{dx} = \frac{3}{(\ln 3)(3x+1)}$.

19. $y = \log_2(1/x) = -\log_2 x = -\frac{\ln x}{\ln 2}$. $\frac{dy}{dx} = -\frac{1}{x\ln 2}$, $x > 0$.

21. $y = (\ln 2)\log_2 x = \ln 2\frac{\ln x}{\ln 2} = \ln x$. $\frac{dy}{dx} = \frac{1}{x}$.

23. $y = \log_{10} e^x = \frac{\ln e^x}{\ln 10} = \frac{x}{\ln 10}$. $\frac{dy}{dx} = \frac{1}{\ln 10}$.

25. $\int_0^1 3x^{\sqrt{3}}dx = 3\frac{x^{\sqrt{3}+1}}{\sqrt{3}+1}\Big]_0^1 = \frac{3}{\sqrt{3}+1} = \frac{3}{\sqrt{3}+1}\frac{\sqrt{3}-1}{\sqrt{3}-1} = \frac{3(\sqrt{3}-1)}{3-1} = \frac{3}{2}(\sqrt{3}-1)$.

27. $\int_0^1 5^x dx = \frac{5^x}{\ln 5}\Big]_0^1 = \frac{5-5^0}{\ln 5} = \frac{4}{\ln 5}$.

29. Let $I = \int_0^1 2^{-x}dx$. Let $u = -x$, $du = -dx$. Then $I = -\int_0^{-1} 2^u du = \frac{-2^u}{\ln 2}\Big]_0^{-1} = \frac{-(2^{-1}-2^0)}{\ln 2} = \frac{1}{2\ln 2} = \frac{1}{\ln 4}$.

31. Let $I = \int_{-1}^0 4^{-x}\ln 2\, dx$. Let $u = -x$, $du = -dx$. Then $I = -\ln 2\int_1^0 4^u du = -\ln 2\frac{4^u}{\ln 4}\Big|_1^0 = -\frac{\ln 2}{2\ln 2}(1-4) = \frac{3}{2}$.

33. Let $I = \int_1^{\sqrt{2}} x2^{x^2}dx$. Let $u = x^2$, $du = 2xdx$, $\frac{1}{2}du = xdx$. Then $I = \frac{1}{2}\int_1^2 2^u du = \frac{1}{2}\frac{2^u}{\ln 2}\Big]_1^2 = \frac{2^2-2^1}{2\ln 2} = \frac{1}{\ln 2}$.

35. Let $I = \int_1^{10}\frac{\log_{10} x}{x}dx = \frac{1}{\ln 10}\int_1^{10}\frac{\ln x}{x}dx$. Let $u = \ln x$. Then $du = \frac{dx}{x}$, $I = \frac{1}{\ln 10}\int_0^{\ln 10} udu = \frac{1}{\ln 10}\frac{u^2}{2}\Big]_0^{\ln 10} = \frac{1}{2}\frac{(\ln 10)^2}{\ln 10} = \frac{\ln 10}{2}$.

37. Let $I = \int_0^2\frac{\log_2(x+2)}{x+2}dx = \frac{1}{\ln 2}\int_0^2\frac{\ln(x+2)}{(x+2)}dx$. Let $u = \ln(x+2)$. Then $du = \frac{dx}{x+2}$ and $I = \frac{1}{\ln 2}\int_{\ln 2}^{\ln 4} udu = \frac{1}{\ln 2}\frac{u^2}{2}\Big|_{\ln 2}^{\ln 4} = \frac{(\ln 4)^2-(\ln 2)^2}{2\ln 2} = \frac{(2\ln 2)^2-(\ln 2)^2}{2\ln 2} = \frac{3(\ln 2)^2}{2\ln 2} = \frac{3}{2}\ln 2$.

39. $\int_0^9\frac{2\log_{10}(x+1)}{x+1}dx = \frac{2}{\ln 10}\int_0^9\frac{\ln(x+1)}{(x+1)}dx = \frac{2}{\ln 10}\frac{[\ln(x+1)]^2}{2}\Big]_0^9 = \frac{1}{\ln 10}[(\ln 10)^2 - (\ln 1)^2] = \ln 10$.

41. Let $I = \int 2^{\sin x}\cos x dx$, $u = \sin x$, $du = \cos x dx$. Then $I = \int 2^u du = \frac{2^u}{\ln 2} + C = \frac{2^{\sin x}}{\ln 2} + C$.

43. Let $I = \int\frac{\log_3(x-2)dx}{x-2} = \frac{1}{\ln 3}\int\frac{\ln(x-2)dx}{x-2}$, $u = \ln(x-2)$, $du = \frac{dx}{x-2}$. Then $I = \frac{1}{\ln 3}\int udu = \frac{u^2}{2\ln 3} + C = \frac{(\ln(x-2))^2}{2\ln 3} + C$.

45. $y = x^{-\sqrt{3}} = \frac{1}{x^{\sqrt{3}}}$. As $x \to 0^+$, $y \to \infty$ and as $x \to \infty$, $y \to 0$. $y' = -\sqrt{3}\, x^{-\sqrt{3}-1} = \frac{-\sqrt{3}}{x^{\sqrt{3}+1}} < 0$ since the domain of the function is $x > 0$. Hence the graph is falling on $(0, \infty)$. $y'' = \sqrt{3}(\sqrt{3}+1)x^{-\sqrt{3}-2} = \frac{\sqrt{3}(\sqrt{3}+1)}{x^{\sqrt{3}+2}} > 0$ on $(0, \infty)$. The curve is concave up for $x > 0$. There are no local extrema or inflection points. Check your graph by graphing y in $[0, 10]$ by $[0, 3]$.

47. Graph $y_1 = x^{\sqrt{x}}$ in $[-1, 3]$ by $[-2, 8]$. The domain is $(0, \infty)$. $\ln y = \sqrt{x}\,\ln x$, $\frac{y'}{y} = \frac{\sqrt{x}}{x} + \frac{\ln x}{2\sqrt{x}}$, $y' = y(\frac{1}{\sqrt{x}} + \frac{\ln x}{2\sqrt{x}}) = x^{\sqrt{x}}\frac{(2+\ln x)}{2\sqrt{x}}$. $y' = 0$ when $\ln x = -2$ or when $x = e^{-2}$. $y' < 0$ on $(0, e^{-2})$ and $y' > 0$ on (e^{-2}, ∞). Therefore y is decreasing on $(0, e^{-2}]$, has a relative minimum at $x = e^{-2}$, and is increasing on $[e^{-2}, \infty)$. It is more practical to deal with y'' graphically. Also graph separately $y' = y_2 =$ NDER(y_1, x) and $y'' = y_3 =$ NDER(y_2, x) in the above window. We see that $y'' > 0$ for all $x > 0$ (y' is increasing steadily) and so the graph of y is concave up for all such x.

49. $y = 2^{\sec x}$ has period 2π and the graph is symmetric with respect to the vertical lines $x = n\pi$. In order to include the relative extrema in the interior we work in the interval $[-\frac{\pi}{2}, \frac{3\pi}{2})$. Graph y in $[-\frac{\pi}{2}, \frac{3\pi}{2}]$ by $[0, 4]$. $y' = 2^{\sec x}(\ln 2)\sec x \tan x$ and we use $y'' =$ NDER(y', x, x). Rel. min. at $(0, 2)$, rel. max. at $(\pi, \frac{1}{2})$. A root of $y'' = 0$ is $v = 1.90392136$. Inflection points at $(v, 0.12)$ and $(2\pi - v, 0.12)$. y is rising on $[0, \frac{\pi}{2})$ and $(\frac{\pi}{2}, \pi]$, falling on $(-\frac{\pi}{2}, 0]$ and $[\pi, \frac{3\pi}{2})$. It is concave up on $(-\frac{\pi}{2}, \frac{\pi}{2})$, $(\frac{\pi}{2}, v)$ and $(2\pi - v, \frac{3\pi}{2})$. It is concave down on $(v, 2\pi - v)$.

51. $y = \log_7 \sin x = \ln \sin x / \ln 7$ is periodic of period 2π. We consider only the interval $[0, 2\pi)$. In here we note that the domain is $(0, \pi)$. Graph y in $[0, 2\pi]$ by $[-3, 0]$. $y' = \frac{1}{\ln 7}\frac{\cos x}{\sin x} = \frac{1}{\ln 7}\cot x$. The graph is rising on $(0, \frac{\pi}{2}]$ and falling on $[\frac{\pi}{2}, \pi)$. Rel. max at $(\frac{\pi}{2}, 0)$. $y'' = -\frac{1}{\ln 7}\csc^2 x$. Thus the curve is concave down on $(0, \frac{\pi}{2})$ and $(\frac{\pi}{2}, \pi)$ and there is no point of inflection.

53. We have no analytic method for this integral. NINT$(2^{x^2}, x, 1, 2) = 6.052$.

55. NINT$(x^{\ln x}, x, 1, 3) = 3.591$

57. The equation may be written $7 + 5 = x$ ($a^{\log_a v} = v$ for all $v > 0$), $x = 12$.

59. The graphs of the two functions in $[-2, 5]$ by $[0, 20]$ show that there are 3 solutions (points of intersection). They are $(-0.77, 0.58), (2, 4)$ and $(4, 16)$.

61. a) $\lim_{x\to\infty} \log_2 x = \lim_{x\to\infty} \frac{\ln x}{\ln 2} = \frac{1}{\ln 2}\lim_{x\to\infty} \ln x = \infty$

b) $\lim_{x\to\infty} \log_2(1/x) = \lim_{x\to\infty} \log_2 x^{-1} = -\lim_{x\to\infty} \log_2 x = -\infty$ by part a).

63. a) $\lim_{x\to\infty} 3^x = \infty$ b) $\lim_{x\to\infty} 3^{-x} = \lim_{x\to\infty} \frac{1}{3^x} = 0$

65. a) $\frac{\ln x}{\log x} = \frac{\ln x}{\ln x/\ln 10} = \ln 10.$ $\lim_{x\to\infty} \ln 10 = \ln 10$ b) $\frac{\log_2 x}{\log_3 x} = \frac{\ln x/\ln 2}{\ln x/\ln 3} = \frac{\ln 3}{\ln 2}.$ $\lim_{x\to\infty} \frac{\ln 3}{\ln 2} = \frac{\ln 3}{\ln 2}$

67. Check your result by graphing $y = x^{\sin x}$ in $[0, 40]$ by $[0, 40]$.

69. In each case we have $x^\beta < x^{\sqrt{3}} < x^\alpha$ for $0 < x < 1$ and $x^\alpha < x^{\sqrt{3}} < x^\beta$ for $x > 1$ where $0 < \alpha < \sqrt{3} < \beta$. The closer α and β are, the more we must zoom in to distinguish the curves. For $x > 0$, $(1, 1)$ is the only point of intersection.

71. a) Let $x_1 < x_2$ be in the domain of $f \circ g$. Since g is increasing, $g(x_1) < g(x_2)$. Since f is increasing, $f(g(x_1)) < f(g(x_2))$. But this is the same as $f \circ g(x_1) < f \circ g(x_2)$. Therefore $f \circ g$ is an increasing function. b) Let $f(x) = e^x$ and $g(x) = \sqrt{3}\ln x$. Then $f(x)$ is increasing for all x and $g(x)$ is increasing for $x > 0$. By part a), $f \circ g(x) = f(g(x)) = e^{\sqrt{3}\ln x}$ is an increasing function.

73. Let $u = [H_30^+]$. $-\log_{10} u = 7.37$ leads to $\log_{10} u = -7.37$, $u = 10^{-7.37} = 4.27 \times 10^{-8}$. For the other bound $u = 10^{-7.44} = 3.63 \times 10^{-8}$.

75. k must satisfy $10\log_{10}(kI \times 10^{12}) = 10\log_{10}(I \times 10^{12}) + 10$, $\log_{10} k + \log_{10} I + 12 = \log_{10} I + 12 + 1$, $\log_{10} k = 1$, $k = 10$.

77. $\log_b = \frac{\log_a x}{\log_a b} = (\log_a x)/(\ln b/\ln a) = \frac{\ln a}{\ln b}\log_a x.$

7.4 THE LAW OF EXPONENTIAL CHANGE REVISITED

1. a) $0.99y_0 = y_0e^{1000k}$, $0.99 = e^{1000k}$, $\ln 0.99 = 1000k$, $k = \frac{\ln 0.99}{1000}$.
b) $0.9y_0 = y_0e^{kt}$, $\ln 0.9 = kt$, $t = \frac{\ln 0.9}{k} = 1000\frac{\ln 0.9}{\ln 0.99} \approx 10{,}483$ years. To support graphically, let $y_0 = 1$ and investigate the point of intersection of $y = e^{kt} = e \wedge ((\ln 0.99/1000)x)$ and $y = 0.9$. c) $y = y_0e^{20{,}000k} = y_0e^{20\ln 0.99} = y_0(0.99)^{20} \approx 0.82y_0$, about 82%.

3. From equations (1) and (2), $y = 100e^{-0.6t}$. When $t = 1$ hour, $y = 100e^{-0.6} \approx 54.88$ grams.

5. Let t be the required number of days. Then $0.9y_0 = y_0e^{-0.18t}$, $\ln 0.9 = -0.18t$, $t = \frac{\ln 0.9}{0.18} = 0.59$ day.

7. a) We use $m = 66 + 7 = 73$. Then distance coasted $= \frac{v_0 m}{k} = \frac{9(73)}{3.9} = 168.46\ldots m$. b) $v = v_0 e^{-(k/m)t} = 9e^{-3.9t/73} = 1$ m/sec. $-3.9t/73 = \ln(1/9) = -\ln 9$, $t = 73(\ln 9)/3.9 = 41.1$ sec. .

9. By (8), $T = T_s + (T_0 - T_s)e^{-kt} = 20 + (90 - 20)e^{-kt}$, $T = 20 + 70e^{kt}$. When $t = 10$, $60 = 20 + 70e^{10k}$, $\frac{4}{7} = e^{10k}$, $10k = \ln(4/7)$, and $k = \ln(4/7)/10$ in $T = 20 + 70e^{kt}$. a) $35 = 20 + 70e^{kt}$, $15/70 = e^{kt}$, $kt = \ln(3/14)$, $t = \frac{10\ln(3/14)}{\ln(4/7)} \approx 27.53$ min. so $27.53 - 10 = 17.53$ min. longer. b) Here $T = T_s + (T_0 - T_s)e^{-kt} = -15 + (90 + 15)e^{kt} = -15 + 105e^{kt} = 35$. $e^{kt} = 50/105 = 10/21$, $kt = \ln(10/21)$, $t = (\frac{1}{k})\ln(10/21) = \frac{10\ln(10/21)}{\ln(4/7)} \approx 13.26$ min.

11. $T_s =$ temperature of refrigerator. $T = T_s + (T_0 - T_s)e^{-kt} = T_s + (46 - T_s)e^{-kt}$. $39 = T_s + (46 - T_s)e^{-10k}$ and $33 = T_s + (46 - T_s)e^{-20k} = T_s + (46 - T_s)(e^{-10k})^2$. From the first equation $(e^{-10k})^2 = (\frac{39 - T_s}{46 - T_s})^2 = \frac{33 - T_s}{46 - T_s}$ from the second equation. Multiplying by $(46 - T_s)^2$, we find $T_s = -3°C$.

13. $\frac{dV}{dt} = -\frac{1}{40}V$, $\frac{1}{V}\frac{dV}{dt} = -\frac{1}{40}$ so $\ln|V| = -\frac{1}{40}t + C$, $V = e^{-t/40}e^C$, taking V to be positive and when $t = 0$, $V = e^C = V_0$. Hence $V = V_0 e^{-t/40}$. If t is the desired time, then $0.1V_0 = V_0 e^{-t/40}$, $\ln 0.1 = -\frac{t}{40}$, $t = -40\ln 0.1 \approx 92.10$ sec.

15. From Example 3, $k = \ln 2/5700$. If t is the age $0.445y_0 = y_0 e^{-kt}$, $\ln 0.445 = -kt$, $t = -\frac{1}{k}\ln 0.445 = -\frac{5700\ln 0.445}{\ln 2} \approx 6658.30$ years.

17. $y = y_0 e^{-kt}$ where $k = \ln 2/5700$ as in Example 2. If t is the age of the painting, then $0.995y_0 = y_0 e^{-kt}$, $\ln 0.995 = -kt$, $t = -5700\ln 0.995/\ln 2 \approx 41.22$ years.

19. a) f) $V(t) = 7e^{-2.5t/50} = 1$, $-2.5t/50 = \ln(1/7) = -\ln 7$, $t = 50(\ln 7)/2.5 = 38.918$ sec. g) $s(3) = \frac{350}{2.5}(1 - e^{-(2.5/50)3}) = 19.50088\ldots m$. h) $s(t) = 100$ leads to $\frac{350}{2.5}(1 - e^{-(2.5/50)t}) = 100$, $1 - e^{-(2.5/50)t} = \frac{250}{350} = \frac{5}{7}$, $-(2.5/50)t = \ln\frac{2}{7}$, $t = -\frac{50}{2.5}\ln\frac{2}{7} = 25.055\ldots$sec. b) $s(t) = \frac{350}{2.5}(1 - e^{-(2.5/50)t}) \leqq \frac{350}{2.5} = 140$. So Jenny will never coast 141 m from the finish line.

21. We first find the amount of time, t_1, required for the population to grow from 5,000 to 10,000. $y = y_0 e^{kt}$, $10000 = 5000e^{t_1/4}, 2 = e^{t_1/4}$, $\ln 2 = t_1/4$, $t_1 = 4\ln 2$. Next we find the amount of time, t_2, required for the population to grow from 10,000 to 25,000. $y = 10000e^{t/12}$, $25000 = 10000e^{t_2/12}$, $2.5 = e^{t_2/12}$, $\ln 2.5 = t_2/12$, $t_2 = 12\ln 2.5$. Answer: $t_1 + t_2 = 13.768$ years from now.

7.5 INDETERMINATE FORMS AND L'HÔPITAL'S RULE

1. $\lim_{x\to 2} \frac{x-2}{x^2-4} = \lim_{x\to 2} \frac{1}{2x} = \frac{1}{4}$

3. $\lim_{x\to 1} \frac{x^3-1}{4x^3-x-3}(\text{form } \frac{0}{0}) = \lim_{x\to 1} \frac{3x^2}{12x^2-1} = \frac{3}{11}$

5. $\lim_{t\to 0} \frac{\sin t^2}{t} = \lim_{t\to 0} \frac{2t\cos t}{1} = 0$

7. $\lim_{x\to\infty} \frac{3x^2-1}{2x^2-x+1}(\frac{\infty}{\infty}) = \lim_{x\to\infty} \frac{6x}{4x-1}(\frac{\infty}{\infty}) = \lim_{x\to\infty} \frac{6}{4} = \frac{3}{2}$

9. Graph $y = \frac{2x-\pi}{\cos x}$ in $[\frac{\pi}{2} - 0.1, \frac{\pi}{2} + 0.1]$ by $[-3, 3]$. Use of TRACE suggests $y \to -2$ as $x \to \frac{\pi}{2}$. $\lim_{x\to\pi/2} \frac{2x-\pi}{\cos x}(\frac{0}{0}) = \lim_{x\to\frac{\pi}{2}} \frac{2}{-\sin x} = -2$.

11. $\lim_{x\to\infty} \frac{5x^2-3x}{7x^2+1} = \lim_{x\to\infty} \frac{10x-3}{14x} = \lim_{x\to\infty} \frac{10}{14} = \frac{5}{7}$

13. $\lim_{x\to\pi/2} \frac{1-\sin x}{1+\cos 2x} = \lim_{x\to\pi/2} \frac{-\cos x}{-2\sin 2x} = \lim_{x\to\pi/2} \frac{-\sin x}{4\cos 2x} = \frac{-1}{4(-1)} = \frac{1}{4}$

15. $\lim_{x\to 0^+} \frac{2x}{x+7\sqrt{x}} = \lim_{x\to 0^+} \frac{2}{1+7/(2\sqrt{x})} = 0$ since $\lim_{x\to 0^+} \frac{7}{2\sqrt{x}} = \infty$.

17. $\lim_{t\to 0} \frac{10(\sin t-t)}{t^3} = \lim_{t\to 0} \frac{10(\cos t-1)}{3t^2} = \lim_{t\to 0} \frac{-10\sin t}{6t} = -\frac{5}{3}\lim_{t\to 0} \frac{\cos t}{1} = -\frac{5}{3}$

19. $\lim_{x\to 0}(\frac{1}{\sin x} - \frac{1}{x}) = \lim_{x\to 0} \frac{x-\sin x}{x\sin x}(\frac{0}{0}) = \lim_{x\to 0} \frac{1-\cos x}{x\cos x+\sin x} =$ $\lim_{x\to 0} \frac{\sin x}{-x\sin x+\cos x+\cos x} = \frac{0}{2} = 0$

21. Let $y = x^{(1/\ln x)}$. Then $\lim_{x\to 0^+} \ln y = \lim_{x\to 0^+} \frac{1}{\ln x}\ln x = \lim_{x\to 0^+} 1 = 1$. Therefore, $\lim_{x\to 0^+} y = e^1 = e$.

23. Let $y = (e^x + x)^{1/x}$. Then $\lim_{x\to 0} \ln y = \lim_{x\to 0} \frac{1}{x}\ln(e^x + x)(\frac{0}{0}) = \lim_{x\to 0} \frac{\frac{e^x+1}{e^x+x}}{1} = \frac{e^0+1}{e^0+0} = 2$. Therefore, $\lim_{x\to 0} y = e^2$.

25. Let $y = (\frac{1}{x^2})^x$. Then $\lim_{x\to 0} \ln y = \lim_{x\to 0} x\ln(1/x^2) = -\lim_{x\to 0} x\ln(x^2) = -\lim_{x\to 0} \frac{\ln x^2}{\frac{1}{x}} = -\lim_{x\to 0} \frac{(2x/x^2)}{(-1/x^2)} = \lim_{x\to 0} 2x = 0$. Therefore $\lim_{x\to 0} y = e^0 = 1$.

27. $\lim_{x\to 0^+} x^{\sqrt{2}} = \lim_{x\to 0^+} e^{\sqrt{2}\ln x} = 0$ because $\lim_{x\to 0^+} \ln x = -\infty$. If we define the function to be 0 at 0, then the function is continuous on $[0, \infty)$.

29. $y = x^{\ln x} = e^{\ln x^{\ln x}} = e^{(\ln x)^2} \to \infty$ as $x \to 0^+$. y cannot be defined at 0 in a way that it would be continuous at $x = 0$.

31. $\lim_{x\to 0^+} \ln f(x) = \lim_{x\to 0^+} x\ln x = \lim_{x\to 0^+} \frac{\ln x}{\frac{1}{x}}(-\frac{\infty}{\infty}) = \lim_{x\to 0^+} \frac{\frac{1}{x}}{-\frac{1}{x^2}} = -\lim_{x\to 0^+} x = 0$. Therefore, $\lim_{x\to 0^+} f(x) = e^0 = 1$. Thus the function $F(x)$ defined by $F(x) = f(x)$, $x > 0$ and $F(0) = 1$ is continuous on $[0, \infty)$. b) In part a) we saw that $f(x) = x^x \to 1$ as $x \to 0^+$. Thus $f'(x) = (1+\ln x)x^x$ has the form $(-\infty \cdot 1)$ as $x \to 0^+$ and so $\lim_{x\to 0^+} f'(x) = -\infty$.

33. b) is correct. a) is incorrect because L'Hôpital's rule does not apply to the limit form $\frac{0}{6}$; it is not an indeterminate form.

35. a) The domain of $f(x) = (1+\frac{1}{x})^x$ is the solution set of $1+\frac{1}{x} > 0$ or $\frac{x+1}{x} > 0$. Numerator and denominator have the same sign in $(-\infty, -1) \cup (0, \infty)$. b) $\lim_{x\to -1^-} f(x) = \lim_{x\to -1^-} e^{x\ln(1+\frac{1}{x})} = \infty$ because the exponent has limit form $(-1)(-\infty)$. c) $\lim_{x\to 0^+} \ln f(x) = \lim_{x\to 0^+} x\ln(1+\frac{1}{x}) = \lim_{x\to 0^+} \frac{\ln(1+\frac{1}{x})}{\frac{1}{x}}(\frac{\infty}{\infty}) = \lim_{x\to 0^+} \frac{\frac{1}{1+\frac{1}{x}}(-\frac{1}{x^2})}{-\frac{1}{x^2}} = \lim_{x\to 0^+} \frac{1}{1+\frac{1}{x}} = 0$. Hence $\lim_{x\to 0^+} f(x) = e^0 = 1$. d) $\lim_{x\to \pm\infty} f(x) = e$ was proved in Example 6. $f'(x) = (1+\frac{1}{x})^x[\ln(1+\frac{1}{x}) - \frac{1}{x+1}]$. The second factor $\to 0$ as $x \to \pm\infty$ and it is rising for $x < 0$ and falling for $x > 0$. Thus it, and so $f'(x)$, is positive for all x in the domain. Hence the graph of f is always rising. The graph of NDER$(f'(x), x)$ in $[-4, 4]$ by $[-3, 3]$ shows that $f''(x) > 0$ for $x < 0$ and $f''(x) < 0$ for $x > 0$, confirming the concavity shown in the figure.

37. $f(x) = (1+\frac{2}{x})^x$. By the method of Example 6, $\lim_{x\to\pm\infty} f(x) = e^2$. As in Exercise 35, the domain is $(-\infty, -2) \cup (0, \infty)$ and $\lim_{x\to 0^+} f(x) = 1$. Graph f in $[-10, 10]$ by $[0, 20]$. The graph resembles the one in Fig. 7.30 and may be confirmed as in Exercise 35.

39. $f(x) = x^{(1/\ln x)}$. $\ln f(x) = \frac{1}{\ln x}\ln x = 1$ for $x > 0$, $x \neq 1$, and so $f(x) = e$ for $x > 0$, $x \neq 1$.

41. $\lim_{x\to 0} \frac{3^{\sin x}-1}{x}(\frac{0}{0}) = \lim_{x\to 0} \frac{3^{\sin x}(\cos x)\ln 3}{1} = \ln 3$. Graph $f(x) = \frac{3^{\sin x}-1}{x}$ in $[-2\pi, 2\pi]$ by $[-1, 2]$. f is continuous on the interval except at $x = 0$ where it has a removable discontinuity.

43. We have $A(t) = \int_0^t e^{-x}dx = -e^{-x}|_0^t = 1 - e^{-t}$, $V(t) = \pi\int_0^t e^{-2x}dx = \frac{\pi}{2}(1 - e^{-2t}) = \frac{\pi}{2}(1-e^{-t})(1+e^{-t})$, $V(t)/A(t) = \frac{\pi}{2}(1+e^{-t})$. a) $\lim_{t\to\infty} A(t) = 1$ b) $\lim_{t\to\infty} V(t)/A(t) = \frac{\pi}{2}$ c) $\lim_{t\to 0^+} V(t)/A(t) = \frac{\pi}{2}(1+1) = \pi$

45. a) Let $y = (1+\frac{r}{k})^{kt}$. $\lim_{k\to\infty} \ln y = \lim_{k\to\infty} \frac{\ln(1+\frac{r}{k})}{\frac{1}{kt}}(\frac{0}{0}) = \lim_{k\to\infty} \frac{\frac{1}{1+\frac{r}{k}}(-\frac{r}{k^2})}{\frac{1}{t}(-\frac{1}{k^2})} = \lim_{k\to\infty} \frac{rt}{1+\frac{r}{k}} = rt$ whence $\lim_{k\to\infty} y = e^{rt}$. b) $100e^{0.06} = 106.184$ c) $e^{0.06} \approx$

1.06 yields $1,000,000e^{0.06} = 1,060,000$ d) $1,000,000e^{0.06} = 1061836.55$ e) In computing the product $1000000e^{0.06}$, we find a significant difference if we round off before or after the computation. Investors/bankers should pay careful attention as to when the rounding off takes place.

7.6 THE RATES AT WHICH FUNCTIONS GROW

1. a) $\lim_{x\to\infty} \frac{x+3}{e^x} = \lim_{x\to\infty} \frac{1}{e^x}$ (by L'Hôpital's rule) $= 0$. $x+3$ grows slower than e^x as $x \to \infty$. b) $\lim_{x\to\infty} \frac{x^3-3x+1}{e^x} = \lim_{x\to\infty} \frac{3x^2-3}{e^x} = \lim_{x\to\infty} \frac{6x}{e^x} = \lim_{x\to\infty} \frac{6}{e^x} = 0$. Slower. c) $\lim_{x\to\infty} \frac{\sqrt{x}}{e^x} = \lim_{x\to\infty} \frac{1}{2\sqrt{x}e^x} = 0$. Slower. d) $\lim_{x\to\infty} \frac{e^x}{4^x} = \lim_{x\to\infty} (\frac{e}{4})^x = 0$ since $0 < \frac{e}{4} < 1$. Faster than e^x as $x \to \infty$. e) $\lim_{x\to\infty} \frac{2.5^x}{e^x} = \lim_{x\to\infty} (\frac{2.5}{3})^x = 0$ since $0 < \frac{2.5}{e} < 1$. Slower. f) $\lim_{x\to\infty} \frac{\ln x}{e^x} = \lim_{x\to\infty} \frac{1}{xe^x} = 0$. Slower. g) $\lim_{x\to\infty} \frac{\log_{10} x}{e^x} = \lim_{x\to\infty} \frac{\ln x}{(\ln 10)e^x} = \lim_{x\to\infty} \frac{1}{(\ln 10)xe^x} = 0$. Slower. h) $\lim_{x\to\infty} \frac{e^{-x}}{e^x} = \lim_{x\to\infty} \frac{1}{e^{2x}} = 0$. Slower. i) $\lim_{x\to\infty} \frac{e^{x+1}}{e^x} = \lim_{x\to\infty} e = e$. Same rate. j) $\lim_{x\to\infty} \frac{(1/2)e^x}{e^x} = \lim_{x\to\infty} (1/2) = 1/2$. Same rate.

3. a) $\frac{x^2+4x}{x^2} = 1 + \frac{4}{x} \to 1$ as $x \to \infty$. x^2+4x grows at the same rate as x^2 as $x \to \infty$. b) $\frac{x^2}{x^3+3} = \frac{1}{x+(3/x^2)} \to 0$ as $x \to \infty$. Faster. c) $\frac{x^2}{x^5} = \frac{1}{x^3} \to 0$ as $x \to \infty$. Faster. d) $\frac{15x+3}{x^2} = \frac{15}{x} + \frac{3}{x^2} \to 0$. Slower. e) $\frac{\sqrt{x^4+5x}}{x+1} x)^2 = (1+\frac{1}{x})^2 \to 1$ as $x \to \infty$. Same rate. g) $\lim_{x\to\infty} \frac{\ln x}{x^2} = \lim_{x\to\infty} \frac{(1/x)}{2x} = \lim_{x\to\infty} \frac{1}{2x^2} = 0$. Slower. h) $\lim_{x\to\infty} \frac{\ln(x^2)}{x^2} = \lim_{x\to\infty} = \frac{2\ln x}{x^2} = 0$ (as in g)). Slower. i) $\frac{\ln(10^x)}{x^2} = \frac{x\ln 10}{x^2} = \frac{x\ln 10}{x^2} = \frac{\ln 10}{x} \to 0$. Slower. j) $\lim_{x\to\infty} \frac{2^x}{x^2} = \lim_{x\to\infty} \frac{2^x \ln 2}{2x} = \lim_{x\to\infty} \frac{2^x(\ln 2)^2}{2} = \infty$. Faster.

5. $\frac{e^x}{e^{x/2}} = e^{x/2} \to \infty$ as $x \to \infty$. $\frac{(\ln x)^x}{e^x} = (\frac{\ln x}{e})^x \to \infty$ as $x \to \infty$. $\frac{x^x}{(\ln x)^x} = (\frac{x}{\ln x})^x \to \infty$ as $x \to \infty$ because $\frac{x}{\ln x} \to \infty$ as $x \to \infty$. The order from slowest to fastest is $e^{x/2}, e^x, (\ln x)^x, x^x$.

7. $\lim_{x\to\infty} \frac{\sqrt{10x+1}}{\sqrt{x}} = \lim_{x\to\infty} \sqrt{\frac{10x+1}{x}} = \lim_{x\to\infty} \sqrt{10+(1/x)} = \sqrt{10}$. $\lim_{x\to\infty} \frac{\sqrt{x+1}}{\sqrt{x}} = \lim_{x\to\infty} \sqrt{\frac{x+1}{x}} = \lim_{x\to\infty} \sqrt{1+(1/x)} = \sqrt{1} = 1$.

9. $\frac{\sqrt{x^4+x}}{x^2} = \sqrt{\frac{x^4+x}{x^4}} = \sqrt{1+\frac{1}{x^3}} \to \sqrt{1+0} = 1$ as $x \to \infty$. So $\sqrt{x^4+x}$ and x^2 grow at the same rate. $\frac{\sqrt[3]{x^6+x}}{x^2} = \sqrt[3]{\frac{x^6+x}{x^6}} = \sqrt[3]{1+\frac{1}{x^5}} \to 1$ as $x \to \infty$. So $\sqrt[3]{x^6+x}$ and x^2 grow at the same rate.

11. a) $\frac{x}{x} \to 1 \neq 0$. False b) $\frac{x}{x+5} = \frac{1}{1+(5/x)} \to 1 \neq 0$ as $x \to \infty$. False c) True by the preceding work. d) $\frac{x}{2x} = \frac{1}{2}$. True e) $\frac{e^x}{e^{2x}} = \frac{1}{e^x} \to 0$ as $x \to \infty$. True f) $\frac{x+\ln x}{x} = 1 + \frac{\ln x}{x} \to 1+0$ as $x \to \infty$ by L'Hôpital's rule. True g) $\frac{\ln x}{\ln 2x} = \frac{\ln x}{\ln 2 + \ln x} = \frac{1}{(\ln 2/\ln x)+1} \to \frac{1}{0+1} = 1$ as $x \to \infty$. False h) $\frac{\sqrt{x^2+5}}{x} = \sqrt{\frac{x^2+5}{x^2}} = \sqrt{1+5/x^2} \to 1$ as $x \to \infty$. True

13. By induction, the nth derivative of x^n is $n!$ for any positive integer n. By applying l'Hôpital's rule n times $\lim_{x\to\infty} \frac{x^n}{e^x} = \lim_{x\to\infty} \frac{n!}{e^x} = 0$.

15. a) $\lim_{x\to\infty} \frac{\ln x}{x^{1/n}} = \lim_{x\to\infty} \frac{\frac{1}{x}}{\frac{1}{n}x^{(1/n)-1}} = \lim_{x\to\infty} \frac{n}{x^{1/n}} = 0$ b) $\ln x = x^{10^{-6}}$ implies $\ln(\ln x) = 10^{-6} \ln x$ and $\ln(\ln(\ln x)) = \ln(\ln x) - 6 \ln 10$. Let $u = \ln(\ln x)$. Then the last equation becomes $\ln u = u - 6\ln 10$. Graph $y_1 = \ln x$ and $y_2 = x - 6\ln 10$ in $[0, 20]$ by $[-2, 7]$ and zoom in to the point of intersection. Its x-coordinate is $u = 16.6265089014 = \ln(\ln x)$. Hence $\ln x = e^u$ and $x = e^{e^u}$.

17. $\lim_{x\to\infty} \frac{(2\sqrt{x}-1)^2}{4x} = \lim_{x\to\infty} \frac{4x-4\sqrt{x}+1}{4x} = \lim_{x\to\infty}(1 - \frac{1}{\sqrt{x}} + \frac{1}{4x}) = 1$

19. $\lim_{x\to\infty} \frac{e^x+x^2}{e^x} = \lim_{x\to\infty}(1 + \frac{x^2}{e^x}) = 1$ as in Exercise 1. $\lim_{x\to-\infty} \frac{e^x+x^2}{x^2} = \lim_{x\to-\infty}(\frac{1}{x^2e^{-x}} + 1) = 1$.

21. Since $0 \leq |\frac{\sin x}{x}| \leq \frac{1}{|x|} \to 0$ as $x \to \pm\infty$, $\lim_{x\to\pm\infty} \frac{\sin x}{x} = 0$. Hence $\lim_{x\to\pm\infty} \frac{f(x)}{g(x)} = \lim_{x\to\pm\infty}(1 + \frac{\sin x}{x}) = 1$ and g is an end behavior model for f.

23. $\lim_{x\to\pm\infty} \frac{2x^3-3x^2+x-1}{2x^3} = \lim_{x\to\pm\infty}(1 - \frac{3}{2x} + \frac{1}{2x^2} - \frac{1}{2x^3}) = 1$. g is an end behavior model for f.

25. $\lim_{x\to\infty} \frac{f(x)}{g(x)} = \lim_{x\to\infty} \frac{2^x+x}{x} = \lim_{x\to\infty} \frac{2^x \ln 2+1}{1} = \infty$ (l'Hôpital's rule). $\lim_{x\to-\infty} \frac{2^x+x}{x} = \lim_{x\to-\infty}(\frac{1}{x2^{-x}} + 1) = 1$. g is only a left end behavior model for f.

27. $\lim_{x\to\pm\infty} \frac{\sqrt{x^4+2x-1}}{x^2} = \lim_{x\to\pm\infty} \sqrt{\frac{x^4+2x-1}{x^4}} = \lim_{x\to\pm\infty} \sqrt{1+\frac{2}{x^3}-\frac{1}{x^4}} = \sqrt{1+0-0} = 1$. g is an end behavior model for f.

29. $\lim_{x\to\pm\infty} \frac{\sqrt[3]{x^2-2x-1}}{\sqrt[3]{x^2}} = \lim_{x\to\pm\infty} \sqrt[3]{1 - \frac{2}{x} - \frac{1}{x^2}} = 1$. g is an end behavior model for f.

31. We show only that f_1/g_1 is a right end behavior model for f/g. The other proofs are similar. $\lim_{x\to\infty} \frac{(f/g)}{(f_1/g_1)} = \lim_{x\to\infty} \frac{f}{f_1}\frac{g_1}{g} = 1 \cdot 1 = 1$.

33. Repeat Exercise 31 replacing $x \to \infty$ by $x \to \pm\infty$.

35. By assumption $0 = \lim_{x\to\infty} \frac{x}{a_n x^n + a_{n-1}x^{n-1} + \cdots + a_0}$. Dividing numerator and denominator by x^n, we have $\lim_{x\to\infty} \frac{x^{1-n}}{a_n + \frac{a_{n-1}}{x} + \cdots + \frac{a_0}{x^n}} = 0$. Since the limit of the denominator is a_n, we must have $\lim_{x\to\infty} x^{1-n} = 0$. This is true if and only if $1 - n < 0$ or $n > 1$.

37. The first is the most efficient because the number of steps for each of the others grows faster than n as $n \to \infty$.

39. We are given that $\lim_{x\to\infty} \frac{f}{g} = L$ where L is a non-zero finite number. Hence $\frac{f(x)}{g(x)} \leqq L + 1$ for x sufficiently large. Therefore $f = O(g)$. $\lim_{x\to\infty} \frac{g}{f} = \lim_{x\to\infty} (\frac{f}{g})^{-1} = L^{-1}$. Hence $\frac{g}{f} \leqq \frac{1}{L} + 1$ for x sufficiently large. Therefore $g = O(f)$.

7.7 THE INVERSE TRIGONOMETRIC FUNCTIONS

1. a) $\frac{\pi}{4}$ b) $\frac{\pi}{3}$ c) $\frac{\pi}{6}$

3. a) $-\frac{\pi}{6}$ b) $-\frac{\pi}{4}$ c) $-\frac{\pi}{3}$

5. a) $\frac{\pi}{3}$ b) $\frac{\pi}{4}$ c) $\frac{\pi}{6}$

7. a) $\frac{3\pi}{4}$ b) $\frac{5\pi}{6}$ c) $\frac{2\pi}{3}$

9. a) $\frac{\pi}{4}$ b) $\frac{\pi}{3}$ c) $\frac{\pi}{6}$

11. a) $\frac{3\pi}{4}$ b) $\frac{5\pi}{6}$ c) $\frac{2\pi}{3}$

13. $\alpha = \sin^{-1}(1/2) = \pi/6$. $\cos\alpha = \frac{\sqrt{3}}{2}$, $\tan\alpha = \frac{1}{\sqrt{3}}$, $\sec\alpha = \frac{2}{\sqrt{3}}$, $\csc\alpha = 2$

15. $\alpha = \tan^{-1}(4/3)$ implies $\tan\alpha = 4/3$ and α is in the first quadrant. Cot $\alpha = 3/4$. Sec$^2\alpha = 1 + \tan^2\alpha = 1 + \frac{16}{9} = \frac{25}{9}$ and so $\sec\alpha = \frac{5}{3}$, $\cos\alpha = \frac{3}{5}$. Sin $\alpha = (\cos\alpha)\tan\alpha = \frac{3}{5}\frac{4}{3} = \frac{4}{5}$. Csc $\alpha = \frac{5}{4}$.

17. $\sin(\cos^{-1}\frac{\sqrt{2}}{2}) = \sin\frac{\pi}{4} = \frac{\sqrt{2}}{2}$

19. $\tan(\sin^{-1}(-\frac{1}{2})) = \tan(-\frac{\pi}{6}) = -\frac{1}{\sqrt{3}}$

21. $\csc(\sec^{-1} 2) + \cos(\tan^{-1}(-\sqrt{3})) = \csc\frac{\pi}{3} + \cos(-\frac{\pi}{3}) = \frac{2}{\sqrt{3}} + \frac{1}{2}$

23. $\sin[\sin^{-1}(-\frac{1}{2}) + \cos^{-1}(-\frac{1}{2})] = \sin(-\frac{\pi}{6} + \frac{2\pi}{3}) = \sin\frac{\pi}{2} = 1$

25. $\sec(\tan^{-1} 1 + \csc^{-1} 1) = \sec(\frac{\pi}{4} + \frac{\pi}{2}) = \sec(\frac{3\pi}{4}) = -\sqrt{2}$

27. $\sec^{-1}(\sec(-\frac{\pi}{6})) = \sec^{-1}(\frac{2}{\sqrt{3}}) = \cos^{-1}(\frac{\sqrt{3}}{2}) = \frac{\pi}{6}$

29. $\sec[\tan^{-1}\frac{x}{2}]$. Let $\theta = \tan^{-1}\frac{x}{2}$. Then $-\frac{\pi}{2} < \theta < \frac{\pi}{2}$ and $\tan\theta = \frac{x}{2}$. Since $-\frac{\pi}{2} < \theta < \frac{\pi}{2}$, $\sec\theta > 0$. $\sec^2\theta = 1 + \tan^2\theta = 1 + \frac{x^2}{4} = \frac{x^2+4}{4}$. $\sec\theta = \frac{\sqrt{x^2+4}}{2}$.

31. Let $\theta = \sec^{-1} 3y = \cos^{-1}\frac{1}{3y}$. Then $0 \leqq \theta < \frac{\pi}{2}$ if $3y \geqq 1$, $\frac{\pi}{2} < \theta \leqq \pi$ if $3y \leqq -1$ and $\cos\theta = \frac{1}{3y}$, $\sec\theta = 3y$. $\tan^2\theta = \sec^2\theta - 1 = 9y^2 - 1$. $\tan\theta = \sqrt{9y^2-1}$ if $y \geqq \frac{1}{3}$, $\tan\theta = -\sqrt{9y^2-1}$ if $y \leqq -\frac{1}{3}$.

33. Let $\theta = \sin^{-1} x$. Then $-\frac{\pi}{2} \leqq \theta \leqq \frac{\pi}{2}$ (so $\cos\theta \geqq 0$) and $\sin\theta = x$. $\cos^2\theta = 1 - \sin^2\theta = 1 - x^2$. $\cos\theta = \sqrt{1-x^2}$.

35. We claim $\sin(\tan^{-1} u) = \frac{u}{\sqrt{1+u^2}}$ for any u. Let $\theta = \tan^{-1} u$. Then $-\frac{\pi}{2} < \theta < \frac{\pi}{2}$ (so $\cos\theta > 0$) and $\tan\theta = u$. $\sec^2\theta = \tan^2\theta + 1 = u^2 + 1$, $\sec\theta = \sqrt{u^2+1}$. $\sin\theta = \frac{\sin\theta}{\cos\theta}\cos\theta = u\frac{1}{\sqrt{u^2+1}} = \frac{u}{\sqrt{u^2+1}}$. Hence $\sin[\tan^{-1}\sqrt{x^2-2x}] = \frac{u}{\sqrt{1+u^2}} = \frac{\sqrt{x^2-2x}}{\sqrt{1+(\sqrt{x^2-2x})^2}} = \frac{\sqrt{x^2-2x}}{\sqrt{x^2-2x+1}} = \frac{\sqrt{x^2-2x}}{|x-1|}$.

37. We claim $\cos(\sin^{-1} u) = \sqrt{1-u^2}$ for any u with $|u| \leqq 1$. Let $\theta = \sin^{-1} u$. Then $-\frac{\pi}{2} \leqq \theta \leqq \frac{\pi}{2}$ (so $\cos\theta \geqq 0$) and $\sin\theta = u$. $\cos^2\theta = 1 - \sin^2\theta = 1 - u^2$, $\cos\theta = \sqrt{1-u^2}$. Hence $\cos(\sin^{-1}\frac{2y}{3}) = \sqrt{1 - \frac{4y^2}{9}} = \sqrt{9-4y^2}/3$.

39. We claim $\sin(\sec^{-1} u) = \frac{\sqrt{u^2-1}}{|u|}$ for all u, $|u| \geqq 1$. Let $\theta = \sec^{-1} u = \cos^{-1}(1/u)$. Then $0 \leqq \theta \leqq \pi$ (so $\sin\theta \geqq 0$) and $\cos\theta = \frac{1}{u}$. $\sin^2\theta = 1 - \cos^2\theta = 1 - \frac{1}{u^2} = \frac{u^2-1}{u^2}$, so $\sin\theta = \frac{\sqrt{u^2-1}}{|u|}$. It follows that $\sin(\sec^{-1}\frac{x}{4}) = \frac{\sqrt{\frac{x^2}{4^2}-1}}{(|x|/4)} = \frac{\sqrt{x^2-16}}{|x|}$.

41. $\lim_{x\to 1^-} \sin^{-1} x = \frac{\pi}{2}$

43. $\lim_{x\to\infty} \tan^{-1} x = \frac{\pi}{2}$

45. $\lim_{x\to\infty} \sec^{-1} x = \frac{\pi}{2}$

47. $\lim_{x\to\infty} \csc^{-1} x = 0$

49. $\alpha =$ large angle $-$ small angle $= \cot^{-1}\frac{x}{15} - \cot^{-1}\frac{x}{3}$

51. Let θ be the indicated angle. Then $r = 3\sin\theta$, $h = 3\cos\theta$. $V = \frac{1}{3}\pi r^2 h = \frac{1}{3}\pi 9(\sin^2\theta)3\cos\theta = 9\pi\sin^2\theta\cos\theta$. $\frac{dV}{d\theta} = 9\pi(-\sin^3\theta + 2\sin\theta\cos^2\theta) = 9\pi\sin\theta(2\cos^2\theta - \sin^2\theta) = 9\pi\sin\theta(3\cos^2\theta - 1)$. $\frac{dV}{d\theta} = 0$ leads, under the given circumstances, to $\cos\theta = \frac{1}{\sqrt{3}}$, $\theta = \cos^{-1}\frac{1}{\sqrt{3}} = 0.955$ radians or $54.736°$.

53. $\cot^{-1} 2 = \frac{\pi}{2} - \tan^{-1} 2 = 0.464$, $\sec^{-1}(1.5) = \cos^{-1}(1/1.5) = 0.841$, $\csc^{-1}(1.5) = \sin^{-1}(1/15) = 0.730$

55. Let $\theta = \pi - \cos^{-1} x$. $0 \leqq \cos^{-1} x \leqq \pi$ leads to $0 \geqq -\cos^{-1} x \geqq -\pi$ and $\pi \geqq \pi - \cos^{-1} x \geqq 0$ or 1) $0 \leqq \theta \leqq \pi$. $\cos(\pi - \cos^{-1} x) = (\cos\pi)\cos(\cos^{-1} x) + \sin\pi\sin(\cos^{-1} x) = (-1)x + 0$ and so 2) $\cos\theta = -x$. 1) and 2) together prove $\theta = \cos^{-1}(-x)$ as was to be shown.

57. Let $\theta = -\sin^{-1} x$. Then $-\frac{\pi}{2} \leqq \theta \leqq \frac{\pi}{2}$ and $\sin\theta = \sin(-\sin^{-1} x) = -\sin(\sin^{-1} x) = -x$. These two facts imply $\theta = \sin^{-1}(-x)$.

59. $\lim_{x\to\infty} \frac{\tan^{-1} x}{(\pi/2)} = \frac{(\pi/2)}{(\pi/2)} = 1$. $\lim_{x\to-\infty} \frac{\tan^{-1} x}{(-\pi/2)} = \frac{(-\pi/2)}{(-\pi/2)} = 1$.

61. Graph $y = \sin^{-1}(1/2x) = \sin^{-1}((2x)^{-1})$ in $[-3.5, 3.5]$ by $[-\frac{\pi}{2}, \frac{\pi}{2}]$

63. Graph $y = 2\cos^{-1}(1/3x) = 2\cos^{-1}((3x)^{-1})$ in $[-3, 3]$ by $[0, 2\pi]$.

65. Graph $y = 3 + \cos^{-1}(x - 2)$ in $[1, 3]$ by $[3, 3 + \pi]$.

7.8 DERIVATIVES OF INVERSE TRIGONOMETRIC FUNCTIONS; RELATED INTEGRALS

1. $y = \cos^{-1} x^2$. $y' = -\frac{1}{\sqrt{1-(x^2)^2}}(2x) = -\frac{2x}{\sqrt{1-x^4}}$. To confirm graphically we graph this last result and NDER(y, x) in $[-1, 1]$ by $[-5, 5]$ and see that the two graphs match.

3. $y = 5\tan^{-1} 3x$. $\frac{dy}{dx} = 5\frac{1}{1+(3x)^2}(3) = \frac{15}{1+9x^2}$

5. $y = \sin^{-1}(x/2)$. $y' = \frac{1}{\sqrt{1-(x/2)^2}}\left(\frac{1}{2}\right) = \frac{1}{2\sqrt{\frac{4-x^2}{4}}} = \frac{1}{\sqrt{4-x^2}}$

7. $y = \sec^{-1}(5x)$. $y' = \frac{5}{|5x|\sqrt{25x^2-1}} = \frac{1}{|x|\sqrt{25x^2-1}}$

9. $y = \csc^{-1}(x^2 + 1)$. $y' = \frac{-2x}{|x^2+1|\sqrt{(x^2+1)^2-1}} = \frac{-2x}{(x^2+1)\sqrt{x^4+2x^2}}$

11. $y = \csc^{-1}\sqrt{x} + \sec^{-1}\sqrt{x} = \sin^{-1}\frac{1}{\sqrt{x}} + \cos^{-1}\frac{1}{\sqrt{x}} = \frac{\pi}{2}$, a constant function. $y' = 0$.

13. $y = \cot^{-1}\sqrt{x-1}$. $y' = -\frac{\frac{1}{2}(x-1)^{-1/2}}{1+(\sqrt{x-1})^2} = \frac{-1}{2x\sqrt{x-1}}$

15. $y = \sqrt{x^2-1} - \sec^{-1} x$. $y' = \frac{1}{2}(x^2-1)^{-1/2}(2x) - \frac{1}{|x|\sqrt{x^2-1}} = \frac{x}{\sqrt{x^2-1}} - \frac{1}{|x|\sqrt{x^2-1}} = \frac{x|x|-1}{|x|\sqrt{x^2-1}}$

17. $y = 2x\tan^{-1} x - \ln(x^2+1)$. $y' = 2(x\frac{1}{1+x^2} + \tan^{-1} x) - \frac{2x}{x^2+1} = 2\tan^{-1} x$

19. $\int_0^{1/2} \frac{dx}{\sqrt{1-x^2}} = \sin^{-1} x\big]_0^{1/2} = \sin^{-1}(1/2) - \sin^{-1} 0 = \frac{\pi}{6}$

21. $\int_{\sqrt{2}}^{2} \frac{dx}{x\sqrt{x^2-1}} = \sec^{-1} x\big]_{\sqrt{2}}^{2} = \cos^{-1}(1/x)\big]_{\sqrt{2}}^{2} = \cos^{-1}(1/2) - \cos^{-1}(1/\sqrt{2}) = \frac{\pi}{3} - \frac{\pi}{4} = \frac{\pi}{12}$

23. $\int_{-1}^{0} \frac{4dx}{1+x^2} = 4\tan^{-1} x\big]_{-1}^{0} = 4[\tan^{-1} 0 - \tan^{-1}(-1)] = 4[0 - (-\frac{\pi}{4})] = \pi$

25. Let $I = \int_0^{\sqrt{2}/2} \frac{xdx}{\sqrt{1-x^4}}$. Let $u = x^2$. Then $u^2 = x^4$, $du = 2xdx$, $I = \frac{1}{2}\int_0^{1/2} \frac{du}{\sqrt{1-u^2}} = \frac{1}{2}\sin^{-1} u\big]_0^{1/2} = \frac{1}{2}[\frac{\pi}{6} - 0] = \frac{\pi}{12}$.

27. Let $I = \int_{1/\sqrt{3}}^{1} \frac{dx}{x\sqrt{4x^2-1}}$. Let $u = 2x$, $du = 2dx$. $I = \frac{1}{2}\int_{2/\sqrt{3}}^{2} \frac{2du}{u\sqrt{u^2-1}} = \sec^{-1}|u|\big]_{2/\sqrt{3}}^{2} = \cos^{-1}(1/2) - \cos^{-1}(\sqrt{3}/2) = \frac{\pi}{3} - \frac{\pi}{6} = \frac{\pi}{6}$.

29. Let $I = \int_0^1 \frac{4xdx}{\sqrt{4-x^4}} = \frac{4}{2}\int_0^1 \frac{xdx}{\sqrt{1-(x^4/4)}}$. Let $u = \frac{x^2}{2}$, $du = xdx$. $I = 2\int_0^{1/2} \frac{du}{\sqrt{1-u^2}} = 2\sin^{-1} u\big]_0^{1/2} = \frac{\pi}{3}$.

31. Let $I = \int \frac{dx}{\sqrt{9-x^2}}$, $x = 3u$, $dx = 3du$. Then $I = \int \frac{3du}{\sqrt{9(1-u^2)}} = \int \frac{du}{\sqrt{1-u^2}} = \sin^{-1} u + C = \sin^{-1}\frac{x}{3} + C$.

33. Let $I = \int \frac{dx}{17+x^2}$, $x = \sqrt{17}u$, $dx = \sqrt{17}du$. Then $I = \int \frac{\sqrt{17}du}{17+17u^2} = \frac{1}{\sqrt{17}}\int \frac{du}{1+u^2} = \frac{1}{\sqrt{17}}\tan^{-1} u + C = \frac{1}{\sqrt{17}}\tan^{-1}\frac{x}{\sqrt{17}} + C$.

35. Let $I = \int \frac{dx}{x\sqrt{25x^2-2}}$, (we need $2u^2 = 25x^2$) $u = \frac{5}{\sqrt{2}}x$, $u^2 = \frac{25x^2}{2}$, $x = \frac{\sqrt{2}}{5}u$, $dx = \frac{\sqrt{2}}{5}du$. Then $I = \int \frac{(\sqrt{2}/5)du}{(\sqrt{2}/5)u\sqrt{2u^2-2}} = \frac{1}{\sqrt{2}}\int \frac{du}{u\sqrt{u^2-1}} = \frac{1}{\sqrt{2}}\sec^{-1}|u| + C = \frac{1}{\sqrt{2}}\sec^{-1}(\frac{5}{\sqrt{2}}|x|) + C$.

37. Let $I = \int \frac{ydy}{\sqrt{1-y^4}}$, $u = y^2$, $du = 2ydy$, $\frac{1}{2}du = ydy$. Then $I = \frac{1}{2}\int \frac{du}{\sqrt{1-u^2}} = \frac{1}{2}\sin^{-1} u + C = \frac{1}{2}\sin^{-1}(y^2) + C$.

39. Let $I = \int_{\sqrt[4]{2}}^{\sqrt{2}} \frac{2xdx}{x^2\sqrt{x^4-1}}$, $u = x^2$, $du = 2xdx$. $I = \int_{\sqrt{2}}^{2} \frac{du}{u\sqrt{u^2-1}} = \sec^{-1}|u|\big]_{\sqrt{2}}^{2} = \cos^{-1}\frac{1}{2} - \cos^{-1}\frac{1}{\sqrt{2}} = \frac{\pi}{3} - \frac{\pi}{4} = \frac{\pi}{12}$.

41. Let $I = \int_1^{\sqrt{3}} \frac{2dx}{(1+x^2)\tan^{-1} x}$, $u = \tan^{-1} x$, $du = \frac{dx}{1+x^2}$. $I = 2\int_{\pi/4}^{\pi/3} \frac{du}{u} = 2\ln|u|\big]_{\pi/4}^{\pi/3} = 2[\ln(\pi/3) - \ln(\pi/4)] = 2\ln(4/3)$.

43. Let $I = \int_2^4 \frac{dx}{2x\sqrt{x-1}}$, $u^2 = x$, $u = \sqrt{x}$, $du = \frac{dx}{2\sqrt{x}}$, $\frac{du}{u} = \frac{dx}{2x}$. $I = \int_{\sqrt{2}}^{2} \frac{du}{u\sqrt{u^2-1}} = \sec^{-1} u\big]_{\sqrt{2}}^{2} = \cos^{-1}(\frac{1}{2}) - \cos^{-1}(\frac{1}{\sqrt{2}}) = \frac{\pi}{3} - \frac{\pi}{4} = \frac{\pi}{12}$.

45. $\lim_{x\to 0} \frac{\sin^{-1} x}{x} = \lim_{x\to 0} \frac{\frac{1}{\sqrt{1-x^2}}}{1} = 1$ **47.** $\lim_{x\to 0} \frac{\tan^{-1} x}{x} = \lim_{x\to 0} \frac{\frac{1}{1+x^2}}{1} = 1$

49. $V = \pi \int_{-\frac{\sqrt{3}}{3}}^{\sqrt{3}} \frac{dx}{1+x^2} = \pi \tan^{-1} x]_{-\sqrt{3}/3}^{\sqrt{3}} = \pi[\frac{\pi}{3} - (-\frac{\pi}{6})] = \frac{\pi^2}{2}$

51. We wish to maximize $\alpha = \cot^{-1} \frac{x}{15} - \cot^{-1} \frac{x}{3}$. $\frac{d\alpha}{dx} = -\frac{1}{15}\frac{1}{1+(x/15)^2} + \frac{1}{3}\frac{1}{1+(x/3)^2} = -\frac{15}{225+x^2} + \frac{3}{9+x^2} = 12[\frac{45-x^2}{(9+x^2)(225+x^2)}]$. From this we see that α rises from $x = 0$ to $x = \sqrt{45} = 3\sqrt{5}$ and then falls from larger x. You should sit $3\sqrt{5}$ft ≈ 6.71ft from the wall. You may confirm graphically by zooming in to the maximum of the graph of $\alpha = \tan^{-1} \frac{15}{x} - \tan^{-1} \frac{3}{x}$, $x > 0$.

53. We start with the straight line angle $\pi = \cot^{-1} \frac{x}{1} + \theta + \cot^{-1} \frac{2-x}{1}$, $\theta = \pi - \cot^{-1} x - \cot^{-1}(2-x)$. $\frac{d\theta}{dx} = \frac{1}{1+x^2} - \frac{1}{1+(2-x)^2} = \frac{4(1-x)}{(1+x^2)[1+(2-x)^2]}$. Thus θ rises for $0 < x < 1$ and falls for $x > 1$. Hence θ is a maximum for $x = 1$. When $x = 1$, $\theta = \pi - \cot^{-1} 1 - \cot^{-1} 1 = \pi - \frac{\pi}{4} - \frac{\pi}{4} = \frac{\pi}{2}$.

55. $y = \sec^{-1} |x| + C$. When $x = 2$, $\pi = \sec^{-1} 2 + C$, $\pi = \frac{\pi}{3} + C$, $C = \frac{2\pi}{3}$. $y = \sec^{-1} |x| + \frac{2\pi}{3}$ or $y = \sec^{-1} x + \frac{2\pi}{3}$ since $x > 0$.

57. $y = \cos^{-1} x + C$. When $x = -\sqrt{2}/2$, $\pi/2 = \cos^{-1}(-\sqrt{2}/2) + C = 3\pi/4 + C$, $C = -\pi/4$. $y = \cos^{-1} x - \frac{\pi}{4}$

61. Both answers can be correct because they differ by a constant: $\sin^{-1} x = \frac{\pi}{2} - \cos^{-1} x$.

63. $\frac{d(\cos^{-1} u)}{dx} = \frac{d}{dx}(\frac{\pi}{2} - \sin^{-1} u) = -\frac{d(\sin^{-1} u)}{dx} = -\frac{du/dx}{\sqrt{1-u^2}}$

65. $\frac{d(\cot^{-1} u)}{dx} = \frac{d}{dx}(\frac{\pi}{2} - \tan^{-1} u) = -\frac{d}{dx}(\tan^{-1} u) = -\frac{du/dx}{1+u^2}$ by the preceding exercise.

67. Graph $y_1 = \text{NDER}(\sin^{-1} x, x)$ and $y_2 = \frac{1}{\sqrt{1-x^2}}$ in $[-1, 1]$ by $[0, 5]$ and see that the graphs match. Graph $\text{NDER}(\cos^{-1} x, x)$ and $\frac{-1}{\sqrt{1-x^2}}$ in $[-1, 1]$ by $[-5, 0]$ and see that the graphs coincide.

69. Graph $y_1 = \text{NDER}(\cos^{-1}(1/x), x)$ and $\frac{1}{|x|\sqrt{x^2-1}}$ in $[-3, 3]$ by $[0, 3]$ and see that the graphs coincide. Similarly for $\text{NDER}(\sin^{-1}(1/x), x)$ and $\frac{-1}{|x|\sqrt{x^2-1}}$ in $[-3, 3]$ by $[-3, 0]$.

71. Graph $y = \sec^{-1}(3x) = \cos^{-1}(\frac{1}{3x})$ in $[-5, 5]$ by $[0, \pi]$. Other windows can show that $y = 0$ when $x = \frac{1}{3}$ and $y = \pi$ when $x = -\frac{1}{3}$.

73. Graph $y = \cot^{-1} \sqrt{x^2 - 1} = \frac{\pi}{2} - \tan^{-1} \sqrt{x^2 - 1}$ in $[-10, 10]$ by $[-1, 2]$.

7.9 HYPERBOLIC FUNCTIONS

1. $\sinh x = -\frac{3}{4}$. $x = \sinh^{-1}(-0.75) = -0.693$. Alternatively one may proceed as follows. Let $u = e^x$. Then $\frac{1}{u} = e^{-x}$. $\sinh x = -\frac{3}{4}$ becomes $\frac{e^x - e^{-x}}{2} = -\frac{3}{4}$, $u - \frac{1}{u} = -\frac{3}{2}$, $u^2 - 1 = -\frac{3}{2}u$, $2u^2 - 2 = -3u$, $2u^2 + 3u - 2 = 0$, $(2u-1)(u+2) = 0$, $u = \frac{1}{2}$, $u = -2$ so $e^x = \frac{1}{2}$, $e^x = -2$ (no solution), $x = \ln(\frac{1}{2}) = -\ln 2$. Answer: $x = -\ln 2$. This method may be used in the next 5 exercises.

3. $\cosh x = 2$. $x = \pm\cosh^{-1} 2 = \pm 1.317$.

5. $\operatorname{sech} x = 0.7$. $x = \pm \operatorname{sech}^{-1} 0.7 = \pm\cosh^{-1}(0.7^{-1}) = \pm 0.896$.

7. $2\cosh(\ln x) = 2\frac{e^{\ln x} + e^{-\ln x}}{2} = x + \frac{1}{x}$, $x > 0$. To confirm graph both $2\cosh(\ln x)$ and $x + \frac{1}{x}$ in $[0,4]$ by $[0,5]$ and see that the two graphs coincide.

9. $\cosh 5x + \sinh 5x = \frac{e^{5x} + e^{-5x} + e^{5x} - e^{-5x}}{2} = e^{5x}$.

11. $\cosh 3x - \sinh 3x = \frac{1}{2}(e^{3x} + e^{-3x} - e^{3x} + e^{-3x}) = e^{-3x}$.

13. a) $\sinh 2x = \sinh(x + x) = \sinh x \cosh x + \cosh x \sinh x = 2 \sinh x \cosh x$

b) $\cosh 2x = \cosh(x + x) = \cosh x \cosh x + \sinh x \sinh x = \cosh^2 x + \sinh^2 x$.

15. Graph $y = \sinh 3x$ in $[-3,3]$ by $[-3,3]$. This graph may be obtained from the graph of $y = \sinh x$ by horizontally shrinking it by a factor of $\frac{1}{3}$.

17. Graph $y = 2\tanh\frac{x}{2}$ in $[-4,4]$ by $[-2,2]$. The graph can be obtained from the graph of $y = \tanh x$ by stretching vertically and horizontally by a factor of 2.

19. Graph $y = \ln(\operatorname{sech} x)$ in $[-10,10]$ by $[-10,0]$. $y' = -\frac{\operatorname{sech} x \tanh x}{\operatorname{sech} x} = -\tanh x$ which is positive for $x < 0$ and negative for $x > 0$. $y'' = -\operatorname{sech}^2 x$ which shows that the graph of y is concave down for all x.

21. Graph $y = \sinh^{-1}(2x)$ in $[-5,5]$ by $[-5,5]$. The graph may be obtained from the graph of $y = \sinh^{-1} x$ by shrinking horizontally by a factor of $\frac{1}{2}$.

23. Graph $y = (1-x)\tanh^{-1} x$ in $[-1,1]$ by $[-4,0.5]$. $\lim_{x\to -1^+}(1-x)\tanh^{-1} x = -\infty$ (has the form $2(-\infty)$). $\lim_{x\to 1^-}(1-x)\tanh^{-1} x = \lim_{x\to 1^-} \frac{\tanh^{-1} x}{\frac{1}{1-x}} (\frac{\infty}{\infty}) = \lim_{x\to 1^-} \frac{\frac{1}{1-x^2}}{\frac{1}{(1-x)^2}} = \lim_{x\to 1^-} \frac{1-x}{1+x} = 0$. $y' = \frac{1}{1+x} - \tanh^{-1} x$ and, using technology, we find that the graph of y rises on $(-1, 0.564]$ to a maximum at $x = 0.564$ and then falls on $[0.564, 1)$. $y'' = \frac{-2+x}{1-x^2}$ which is negative in the domain $(-1,1)$ and so the graph of y is concave down on $(-1,1)$.

25. Graph $y = x \operatorname{sech}^{-1} x = x\cosh^{-1}(x^{-1})$ in $[0,1]$ by $[0,1]$. Since $(0,1]$ is the domain of $\operatorname{sech}^{-1} x$, it is the domain of y. $y' = -\frac{1}{\sqrt{1-x^2}} + \operatorname{sech}^{-1} x$. From this, using technology, we find y is rising for $0 < x \leqq 0.552$ and falling for $0.552 \leqq x \leqq 1$. $y'' = -\frac{1}{(1-x^2)^{3/2}} - \frac{1}{x\sqrt{1-x^2}} < 0$ for $0 < x < 1$ so the curve is concave down for $0 < x < 1$.

27. $y = \ln(\operatorname{csch} x + \coth x)$. $y' = \frac{-\operatorname{csch} x \coth x - \operatorname{csch}^2 x}{\operatorname{csch} x + \coth x} = \frac{-\operatorname{csch} x(\coth x + \operatorname{csch} x)}{\operatorname{csch} x + \coth x} = -\operatorname{csch} x$.

29. $y = \frac{1}{2}\ln|\tanh|$. $y' = \frac{1}{2}\frac{\operatorname{sech}^2 x}{\tanh x} = \frac{1}{2\cosh^2 x \frac{\sinh x}{\cosh x}} = \frac{1}{2\cosh x \sinh x} = \frac{1}{\sinh(2x)}$ (by Exercise 13 a)) $= \operatorname{csch}(2x)$.

31. a) $y = \cosh^2 x$. $y' = 2\cosh x \sinh x = \sinh(2x)$. b) $y = \sinh^2 x$. $y' = 2\sinh x \cosh x = \sinh(2x)$. c) $y = \frac{1}{2}\cosh 2x$. $y' = \frac{1}{2}(\sinh 2x)2 = \sinh(2x)$.

33. $y = \sinh^{-1}(\tan x)$. $y' = \frac{\sec^2 x}{\sqrt{1+\tan^2 x}} = \frac{|\sec x|^2}{|\sec x|} = |\sec x|$.

35. $y = \tanh^{-1}(\sin x), -\frac{\pi}{2} < x < \frac{\pi}{2}$. $y' = \frac{\cos x}{1-\sin^2 x} = \sec x$.

37. $y = \operatorname{sech}^{-1}(\sin x),\ 0 < x < \frac{\pi}{2}$. $y' = \frac{-\cos x}{\sin x\sqrt{1-\sin^2 x}} = -\csc x,\ 0 < x < \frac{\pi}{2}$.

39. $\int_{-1}^{1} \cosh 5x\, dx = \frac{\sinh 5x}{5}\Big]_{-1}^{1} = \frac{1}{5}(\sinh(5) - \sinh(-5)) = \frac{2\sinh 5}{5}$.

41. $\int_{-3}^{3} \sinh x\, dx = 0$ since $\sinh x$ is an odd function.

43. $\int_0^{1/2} 4e^x \cosh x\, dx = 2\int_0^{1/2}(e^{2x}+1)dx = 2\left[\frac{e^{2x}}{2} + x\right]_0^{1/2} = 2\left[\frac{e}{2} + \frac{1}{2} - \frac{1}{2}\right] = e$.

45. Let $I = \int_1^2 \frac{\cosh(\ln x)}{x}dx$. Let $u = \ln x$, $du = \frac{1}{x}dx$. $I = \int_0^{\ln 2}\cosh u\, du = \sinh(\ln 2) - \sinh 0 = \frac{e^{\ln 2} - e^{\ln 2^{-1}}}{2} = \frac{2-\frac{1}{2}}{2} = \frac{3}{4}$.

47. $\int_0^{\ln 3} \operatorname{sech}^2 x\, dx = \tanh x\Big]_0^{\ln 3} = \frac{4}{5}$.

49. Let $I = \int_0^4 \frac{\cosh\sqrt{x}}{\sqrt{x}}dx$. Let $u = \sqrt{x}$, $du = \frac{1}{2\sqrt{x}}dx$. $I = 2\int_0^2 \cosh u\, du = 2\sinh u\Big]_0^2 = 2\sinh 2$.

51. $\int \sinh 2x\, dx = \frac{\cosh 2x}{2} + C$.

53. $\int 2e^{2t}\cosh t\, dt = 2\int e^{2t}\frac{e^t + e^{-t}}{2}dt = \int(e^{3t} + e^t)dt = \frac{e^{3t}}{3} + e^t + C$.

55. Let $I = \int \tanh \frac{x}{7} dx$, $u = \frac{x}{7}$, $du = \frac{dx}{7}$. Then $I = 7\int \tanh u\, du = 7\int \frac{\sinh u}{\cosh u} du$. Let $v = \cosh u$, $dv = \sinh u\, du$. $I = 7\int \frac{dv}{v} = 7\ln|v| + C = 7\ln|\cosh u| + C = 7\ln(\cosh \frac{x}{7}) + C$.

57. Since $\frac{d}{dt}(-2 \text{ sech } \sqrt{t}) = -2\left[-\text{sech } \sqrt{t} \tanh \sqrt{t} \frac{1}{2\sqrt{t}}\right] = \frac{\text{sech } \sqrt{t} \tanh \sqrt{t}}{\sqrt{t}}$, $\int \frac{\text{sech } \sqrt{t} \tanh \sqrt{t}}{\sqrt{t}} dt = -2 \text{ sech } \sqrt{t} + C$.

59. One may also verify an identity by showing that both sides have the same derivative (hence they differ by a constant) and then showing that both sides have the same value at a convenient value of x (so the constant is 0).

61. a) $\int_0^1 \frac{dx}{\sqrt{1+x^2}} = \sinh^{-1} x\Big]_0^1 = \sinh^{-1}(1)$ b) $\sinh^{-1}(1) = \ln(1+\sqrt{2})$.

63. a) $\int_{5/4}^{5/3} \frac{dx}{\sqrt{x^2-a}} = \cosh^{-1} x\Big]_{5/4}^{5/3} = \cosh^{-1}\left(\frac{5}{3}\right) - \cosh^{-1}\left(\frac{5}{4}\right)$ b) $\ln\left(\frac{5}{3} + \sqrt{\frac{25}{9} - 1}\right) - \ln\left(\frac{5}{4} + \sqrt{\frac{25}{16} - 1}\right) = \ln \frac{\frac{5}{3}+\frac{4}{3}}{\frac{5}{4}+\frac{3}{4}} = \ln \frac{3}{2}$.

65. a) $\int_{5/4}^{2} \frac{dx}{1-x^2} = \coth^{-1} x\Big]_{5/4}^{2} = \coth^{-1} 2 - \coth^{-1}\left(\frac{5}{4}\right)$ b) $\frac{1}{2}\ln\frac{2+1}{2-1} - \frac{1}{2}\ln\frac{\frac{5}{4}+1}{\frac{5}{4}-1} = \frac{1}{2}\ln 3 - \frac{1}{2}\ln 9 = \frac{1}{2}\ln\frac{1}{3} = -\frac{\ln 3}{2}$.

67. a) Let $I = \int_1^2 \frac{dx}{x\sqrt{4+x^2}}$. Let $x = 2u$, $dx = 2du$. $I = \int_{1/2}^{1} \frac{2du}{2u\sqrt{4(1+u^2)}} = \frac{1}{2}\int_{1/2}^{1} \frac{du}{u\sqrt{1+u^2}} = -\frac{1}{2}\sinh^{-1}\left(\frac{1}{|u|}\right)\Big]_{1/2}^{1} = \frac{1}{2}\left[\sinh^{-1} 2 - \sinh^{-1}(1)\right]$.
b) $\frac{1}{2}\left[\sinh^{-1} 2 - \sinh^{-1}(1)\right] = \frac{1}{2}\left[\ln(2+\sqrt{5}) - \ln(1+\sqrt{2})\right] = \frac{1}{2}\ln\frac{2+\sqrt{5}}{1+\sqrt{2}}$.

69. $\int_1^3 \frac{\sinh x}{x} dx = \text{NINT}(\frac{\sinh x}{x}, x, 1, 3) \approx 3.916$. We used Tol = 1.

71. $\int_1^4 \frac{\cosh x - 1}{x} dx = \text{NINT}(\frac{\cosh x - 1}{x}, x, 1, 4) \approx 7.589$.

73. Graph $y = \text{NINT}(\frac{\sinh t}{t}, t, 1, x)$ in $[-4, 4]$ by $[-10, 10]$.

75. Graph $\text{NINT}((\cosh t - 1)/t, t, 1, x)$ in $[-5, 5]$ by $[-1, 10]$.

77. $V = \pi \int_0^2 (\cosh^2 x - \sinh^2 x) dx = \pi \int_0^2 1 dx = 2\pi$.

79. By symmetry $\bar{x} = 0$. $A = 2\int_0^{\ln\sqrt{3}} \text{sech } x\, dx = 2\text{NINT}((\cosh t)^{-1}, t, 0, \ln\sqrt{3}) \approx 1.0472$. $M_x = \frac{1}{2}\int_{-\ln\sqrt{3}}^{\ln\sqrt{3}} \text{sech}^2 x\, dx = \int_0^{\ln\sqrt{3}} \text{sech}^2 x\, dx = \tanh x\Big]_0^{\ln\sqrt{3}} = \tanh(\ln\sqrt{3}) = \frac{\sqrt{3}-\frac{1}{\sqrt{3}}}{\sqrt{3}+\frac{1}{\sqrt{3}}} = \frac{3-1}{3+1} = \frac{1}{2}$. $\bar{y} = \frac{M_x}{A} \approx 0.477$.

81. Let $f_1(x) = \frac{f(x)+f(-x)}{2}$ and $f_2(x) = \frac{f(x)-f(-x)}{2}$. $f_1(-x) = \frac{f(-x)+f(x)}{2} = f_1(x)$ so $f_1(x)$ is even. $f_2(-x) = \frac{f(-x)-f(x)}{2} = -\frac{f(x)-f(-x)}{2} = -f_2(x)$ so $f_2(x)$ is odd.

83. $v = \sqrt{\frac{mg}{k}}\tanh(\sqrt{\frac{gk}{m}}t)$. a) $\frac{dv}{dt} = \sqrt{\frac{mg}{k}}\sqrt{\frac{gk}{m}}\operatorname{sech}^2(\sqrt{\frac{gk}{m}}t) = g\operatorname{sech}^2(\sqrt{\frac{gk}{m}}t)$. $m\frac{dv}{dt} = mg\operatorname{sech}^2(\sqrt{\frac{gk}{m}}t)$. $mg - kv^2 = mg - mg\tanh^2(\sqrt{\frac{gk}{m}}t) = mg\operatorname{sech}^2(\sqrt{\frac{gk}{m}}t)$. Hence $m\frac{dv}{dt} = mg - kv^2$ and v satisfies the differential equation. When $t = 0$, $v = \sqrt{\frac{mg}{k}}\tanh(0) = 0$. b) Since $\lim_{x\to\infty}\tanh x = 1$, $\lim_{t\to\infty} v = \sqrt{\frac{mg}{k}}$. c) $\sqrt{\frac{160}{0.005}}$ ft/sec. $= 178.885$ ft/sec.

85. $y = 10\cosh(\frac{x}{10})$. $y' = \sinh\frac{x}{10}$. $L = 2\int_0^{10\ln 10}\sqrt{1+\sinh^2 x}\,dx = 2\int_0^{10\ln 10}\cosh x\,dx = 2\sinh x\big]_0^{10\ln 10} = 2\sinh(10\ln 10) = 10^{10} - \frac{1}{10^{10}}$. $L = 2\int_0^{10\ln 10}\sqrt{1+\sinh^2(\frac{x}{10})}\,dx = 2\int_0^{10\ln 10}\cosh(\frac{x}{10})dx = 20\sinh(\frac{x}{10})\Big]_0^{10\ln 10} = 20\sinh(\ln 10) = 20\frac{10-\frac{1}{10}}{2} = 99$.

87. $2\pi\int_0^{\ln 8} 2\cosh(x/2)\sqrt{1+\sinh^2(\frac{x}{2})}\,dx = 4\pi\int_0^{\ln 8}\cosh^2(\frac{x}{2})dx = 2\pi\int_0^{\ln 8}(\cosh x + 1)dx = 2\pi\left[\sinh x + x\right]_0^{\ln 8} = 2\pi\left(\frac{8-\frac{1}{8}}{2} + \ln 8\right) = \pi\left(\frac{63}{8} + 2\ln 8\right)$.

89. a) $A(u)$ = area of triangle $-\int_1^{\cosh u}\sqrt{x^2-1}\,dx = \frac{1}{2}\cosh u\sinh u - \int_1^{\cosh u}\sqrt{x^2-1}\,dx$. b) $A'(u) = \frac{1}{2}(\cosh^2 u + \sinh^2 u) - \sqrt{\cosh^2 u - 1}\,\sinh u = \frac{1}{2}(\cosh^2 u - \sinh^2 u) = \frac{1}{2}$. c) $A(u) = \frac{1}{2}u + C$. Since $A(0) = 0$, $C = 0$ and $A(u) = \frac{1}{2}u$, $u = 2A(u)$.

PRACTICE EXERCISES, CHAPTER 7

1. $y = \ln\sqrt{x} = \frac{1}{2}\ln x$. $\frac{dy}{dx} = \frac{1}{2x}$. Support by graphing NDER($\ln\sqrt{x}, x$) in $[0,2]$ by $[0,10]$.

3. $y = \ln(3x^2+6)$. $\frac{dy}{dx} = \frac{6x}{3x^2+6} = \frac{2x}{x^2+2}$.

5. $y = \frac{1}{e^x} = e^{-x}$. $y' = e^{-x} = -\frac{1}{e^x}$.

7. $y = e^{(1+\ln x)} = e^1 e^{\ln x} = ex$. $\frac{dy}{dx} = e$.

9. $(\ln(\cos x))' = \frac{-\sin x}{\cos x} = -\tan x$.

11. $(\ln(\cos^{-1} x))' = \frac{-\frac{1}{\sqrt{1-x^2}}}{\cos^{-1} x} = -\frac{1}{\sqrt{1-x^2}\cos^{-1} x}$.

13. $(\log_2(x^2))' = (\frac{\ln(x^2)}{\ln 2})' = \frac{1}{\ln 2}\frac{2x}{x^2} = \frac{2}{(\ln 2)x}$.

15. $(8^{-x})' = 8^{-x}(\ln 8)(-1) = -(\ln 8)8^{-x}$.

17. $(\sin^{-1}(\sqrt{1-x})' = \frac{1}{\sqrt{1-(1-x)}}\frac{-1}{2\sqrt{1-x}} = \frac{-1}{2\sqrt{x(1-x)}}$.

19. $y = \cos^{-1}(1/x) - \csc^{-1} x = \cos^{-1}(1/x) - \sin^{-1}(1/x)$, $x > 0$. $y' = \frac{(1/x^2)}{\sqrt{1-(1/x)^2}} + \frac{(1/x^2)}{\sqrt{1-(1/x)^2}} = \frac{2}{|x|\sqrt{x^2-1}} = \frac{2}{x\sqrt{x^2-1}}$, $x > 0$.

21. $(2\sqrt{x-1}\sec^{-1}\sqrt{x})' = 2\sqrt{x-1}\,\frac{\frac{1}{2\sqrt{x}}}{\sqrt{x}\,\sqrt{x-1}} + \frac{1}{\sqrt{x-1}}\sec^{-1}\sqrt{x} = \frac{1}{x} + \frac{\sec^{-1}\sqrt{x}}{\sqrt{x-1}}$.

23. a) $\ln e^{2x} = 2x$ ($\ln e^u = u$ for any u) b) $\ln 2e = \ln 2 + \ln e = 1 + \ln 2$, c) $\ln\frac{1}{e} = \ln e^{-1} = -1$

25. $\ln(y^2 + y) - \ln y = x$, $\ln(\frac{y^2+y}{y}) = x$, $\ln(y+1) = x$, $y + 1 = e^x$, $y = e^x - 1$

27. $e^{2y} = 4x^2$, $x > 0$. $2y = \ln 4x^2$, $y = \frac{1}{2}\ln 4x^2$, $y = \ln(4x^2)^{1/2} = \ln 2x$.

29. $y = \frac{2(x^2+1)}{\sqrt{\cos 2x}}$. $\ln y = \ln 2 + \ln(x^2+1) - \frac{1}{2}\ln\cos 2x$. $\frac{y'}{y} = \frac{2x}{x^2+1} - \frac{1}{2}\frac{(-2\sin 2x)}{\cos 2x}$, $y' = \frac{2(x^2+1)}{\sqrt{\cos 2x}}(\frac{2x}{x^2+1} + \tan 2x)$.

31. $y = \left(\frac{(x+5)(x-1)}{(x-2)(x+3)}\right)^5$. $\ln y = 5[\ln(x+5) + \ln(x-1) - \ln(x-2) - \ln(x+3)]$, $\frac{y'}{y} = 5\left[\frac{1}{x+5} + \frac{1}{x-1} - \frac{1}{x-2} - \frac{1}{x+3}\right]$, $y' = 5\left(\frac{(x+5)(x-1)}{(x-2)(x+3)}\right)^5\left[\frac{1}{x+5} + \frac{1}{x-1} - \frac{1}{x-2} - \frac{1}{x+3}\right]$.

33. $y = (1+x^2)e^{\tan^{-1} x}$. $\ln y = \ln(1+x^2) + \tan^{-1} x$, $\frac{y'}{y} = \frac{2x}{1+x^2} + \frac{1}{1+x^2}$. $y' = e^{\tan^{-1} x}(2x+1)$.

35. $(x - \coth x)' = 1 + \operatorname{csch}^2 x = \coth^2 x$

37. $[\ln(\operatorname{csch} x) + x\coth x]' = \frac{-\operatorname{csch} x\coth x}{\operatorname{csch} x} - x\operatorname{csch}^2 x + \coth x = -x\operatorname{csch}^2 x$.

39. $(\sin^{-1}(\tanh x))' = \frac{\operatorname{sech}^2 x}{\sqrt{1-\tanh^2 x}} = \operatorname{sech} x$.

41. $(\sqrt{1+x^2}\sinh^{-1} x)' = \sqrt{1+x^2}\,\frac{1}{\sqrt{1+x^2}} + \frac{x}{\sqrt{1+x^2}}\sinh^{-1} x = 1 + \frac{x\sinh^{-1} x}{\sqrt{1+x^2}}$.

43. $(1 - \tanh^{-1}(1/x))' = -\frac{(-1/x^2)}{1-(1/x)^2} = \frac{1}{x^2-1}$.

45. $(\operatorname{sech}^{-1}(\cos 2x))' = \frac{-(-2\sin 2x)}{(\cos^2 x)\sqrt{1-\cos^2 2x}} = 2\sec 2x$.

47. Graph $y = e^{\tan^{-1} x}$ in $[-30, 30]$ by $[0, 5]$. Note that $\lim_{x\to\pm\infty} y = e^{\pm\pi/2}$, so that there are two horizontal asymptotes. $y' = \frac{e^{\tan^{-1} x}}{1+x^2} > 0$ so y is always increasing. $y'' = (1-2x)\frac{e^{\tan^{-1} x}}{(1+x^2)^2}$ so that the curve is concave up on $(-\infty, \frac{1}{2})$, concave down on $(\frac{1}{2}, \infty)$ and has an inflection point at $x = \frac{1}{2}$.

49. Graph $y = x\tan^{-1} x - \frac{1}{2}\ln x$ in $[0, 4]$ by $[0, 5]$. $y' = \frac{x}{1+x^2} + \tan^{-1} x - \frac{1}{2x}$. Technology yields that y decreases on $(0, 0.544]$ and increases on $[0.544, \infty)$ with a minimum at $(0.544, 0.578)$. $y'' = \frac{2}{(1+x^2)^2} + \frac{1}{2x^2}$ so the curve is concave up.

51. $\int_{-1}^{1} \frac{dx}{3x-4} = \frac{1}{3}\ln|3x-4|\Big]_{-1}^{1} = \frac{1}{3}(\ln 1 - \ln 7) = -\frac{\ln 7}{3}$.

53. $\int_{\ln 3}^{\ln 4} e^x dx = \left[e^x\right]_{\ln 3}^{\ln 4} = e^{\ln 4} - e^{\ln 3} = 4 - 3 = 1$.

55. $\int_0^{\pi/4} e^{\tan x}\sec^2 x\, dx = \left[e^{\tan x}\right]_0^{\pi/4} = e^1 - e^0 = e - 1$.

57. $\int_0^{\pi} \tan\frac{x}{3}dx = -3\ln|\cos\frac{x}{3}|\Big]_0^{\pi} = -3\left[\ln\frac{1}{2} - \ln 1\right] = 3\ln 2 = \ln 8$.

59. $I = \int_0^4 \frac{2x\,dx}{x^2-25}$. Let $u = x^2 - 25$, $du = 2x\,dx$. $I = \int_{-25}^{-9} \frac{du}{u} = \ln|u|\Big]_{-25}^{-9} = \ln 9 - \ln 25 = \ln(9/25)$.

61. Let $I = \int_0^{\pi/4} \frac{\sec x\tan x + \sec^2 x}{\sec x + \tan x}dx$. Let $u = \sec x + \tan u$. Then $du = (\sec x\tan x + \sec^2 x)dx$, $I = \int_1^{1+\sqrt{2}} \frac{du}{u} = \ln|u|\Big]_1^{1+\sqrt{2}} = \ln(1+\sqrt{2})$.

63. Let $I = \int_1^8 \frac{\log_4 x}{x}dx = \frac{1}{\ln 4}\int_1^8 \frac{\ln x}{x}dx$. Let $u = \ln x$. Then $du = \frac{dx}{x}$, $I = \frac{1}{\ln 4}\int_0^{\ln 8} u\,du = \frac{1}{\ln 4}\frac{u^2}{2}\Big]_0^{\ln 8} = \frac{(\ln 8)^2}{2\ln 4} = \frac{(\frac{3}{2}\ln 4)^2}{2\ln 4} = \frac{9}{8}\ln 4 = \frac{9}{4}\ln 2$.

65. Let $I = \int_0^1 x\,3^{x^2}\,dx$. Let $u = 3^{x^2}$, $du = 2(\ln 3)x\,3^{x^2}\,dx$. $I = \frac{1}{2\ln 3}\int_1^3 du = \frac{1}{2\ln 3}(2) = \frac{1}{\ln 3}$.

67. $\int_{-1/2}^{1/2} \frac{3dx}{\sqrt{1-x^2}} = 6\int_0^{1/2} \frac{dx}{\sqrt{1-x^2}}$ (integral of even function) $= 6\sin^{-1} x\Big]_0^{1/2} = 6\frac{\pi}{6} = \pi$.

69. $\int_{-1}^{1} \frac{dx}{1+x^2} = \tan^{-1} x\Big]_{-1}^{1} = \frac{\pi}{4} - (-\frac{\pi}{4}) = \frac{\pi}{2}$.

71. Let $I = \int_{1/2}^{3/4} \frac{dx}{\sqrt{x}\sqrt{1-x}}$. Let $u = \sqrt{x}$, $du = \frac{dx}{2\sqrt{x}}$. $I = 2\int_{1/\sqrt{2}}^{\sqrt{3}/2} \frac{du}{\sqrt{1-u^2}} = 2\sin^{-1} u\Big]_{1/\sqrt{2}}^{\sqrt{3}/2} = 2\left(\frac{\pi}{3} - \frac{\pi}{4}\right) = \frac{\pi}{6}$.

73. $\int_0^{\ln 2} 4e^x \cosh x\, dx = 4\int_0^{\ln 2} e^x \frac{(e^x+e^{-x})}{2} dx = 2\int_0^{\ln 2}(e^{2x}+1)dx = 2\left(\frac{e^{2x}}{2}+x\right)_0^{\ln 2} = 2\left[\frac{4}{2}+\ln 2-\frac{1}{2}\right] = 3+\ln 4.$

75. $\int_{-\ln 3}^{\ln 3} 3\sqrt{\cosh 2x+1}\ dx = 6\int_0^{\ln 3}\sqrt{2\cosh^2 x} = 6\sqrt{2}\int_0^{\ln 3}\cosh x\ dx = 6\sqrt{2}\sinh x\Big]_0^{\ln 3} = 6\sqrt{2}\ \frac{3-\frac{1}{3}}{2} = 8\sqrt{2}.$

77. $\int_2^4 10\ \mathrm{csch}^2 x \coth x\ dx = -10\ \frac{\mathrm{csch}^2\ x}{2}\Big]_2^4 = 5(\mathrm{csch}^2\ 2-\mathrm{csch}^2\ 4).$

79. Let $I=\int \sec^2(x)e^{\tan x}dx$, $u=\tan x$, $du=\sec^2 x\ dx$. Then $I=\int e^u du = e^u+C=e^{\tan x}+C$.

81. Let $I=\int \frac{\tan(\ln v)}{v}\ dv$, $u=\ln v$, $du=\frac{dv}{v}$. Then $I=\int \tan u\ du = -\ln|\cos u| + C = -\ln|\cos(\ln v)|+C$.

83. Let $I=\int (x)3^{x^2}dx$, $u=x^2$, $du=2x\ dx$, $\frac{1}{2}du=x\ dx$. Then $I=\frac{1}{2}\int 3^u\ du = \frac{1}{2}3^u/\ln 3+C=\frac{3^{x^2}}{2\ln 3}+C$.

85. Let $I=\int \frac{dy}{y\sqrt{4y^2-1}}$, $u=2y$, $y=\frac{u}{2}$, $dy=\frac{1}{2}du$. Then $I=\frac{1}{2}\int\frac{du}{(u/2)\sqrt{u^2-1}} = \int\frac{du}{u\sqrt{u^2-1}} = \sec^{-1}|u|+C=\sec^{-1}|2y|+C$.

87. a) Let $I=\int_0^{\pi/2}\frac{\sin x\ dx}{\sqrt{1+\cos^2 x}}$. Let $u=\cos x$, $du=-\sin x\ dx$. $I=-\int_1^0\frac{du}{\sqrt{1+u^2}} = \sinh^{-1}1$. b) $\ln(1+\sqrt{2})$

89. a) $\int_{1/5}^{1/2}\frac{4\tanh^{-1}x}{1-x^2}dx = 2(\tanh^{-1}x)^2\Big]_{1/5}^{1/2} = 2\left[(\tanh^{-1}(\frac{1}{2}))^2-(\tanh^{-1}(\frac{1}{5}))^2\right]$ b) $2\left[(\frac{1}{2}\ln 3)^2-(\frac{1}{2}\ln\frac{3}{2})^2\right] = \frac{1}{2}\left[(\ln 3)^2-(\ln\frac{3}{2})^2\right] = \frac{1}{2}\left[\ln 3-\ln\frac{3}{2}\right]\left[\ln 3+\ln\frac{3}{2}\right] = \frac{1}{2}(\ln 2)\ln\frac{9}{2}$.

91. a) $\int_{3/5}^{4/5}\frac{2\ \mathrm{sech}^{-1}\ x}{x\sqrt{1-x^2}}dx = -(\mathrm{sech}^{-1}\ x)^2\Big]_{3/5}^{4/5} = (\mathrm{sech}^{-1}(\frac{3}{5}))^2-(\mathrm{sech}^{-1}(\frac{4}{5}))^2$ b) $(\ln 3)^2-(\ln 2)^2=(\ln 3-\ln 2)(\ln 3+\ln 2)=(\ln\frac{3}{2})\ln 6$.

93. Let $I=\int_1^6\log_5 x\ dx = \frac{1}{\ln 5}\int_1^6\ln x\ dx$. Let $u=\ln x$, $dv=dx$, $du=\frac{1}{x}dx$, $v=x$. $(\ln 5)I = x\ln x]_1^6-\int_1^6 x\frac{1}{x}dx = 6\ln 6-5$, $I=\frac{6\ln 6-5}{\ln 5}$.

95. $\int_1^5\frac{e^{\sinh^{-1}x}}{1+x^2}dx \approx \mathrm{NINT}\left(\frac{e^{\sinh^{-1}x}}{1+x^2},x,1,5\right)\approx 2.714.$

97. Graph $\int_1^x\frac{\tanh t}{t}dt = \mathrm{NINT}\left(\frac{\tanh t}{t},t,1,x\right)$ in $[-10,10]$ by $[-5,3]$.

99. $f(x) = e^x + x,\ f'(x) = e^x + 1.\ \ df^{-1}/dx]_{x=f(\ln 2)} = \frac{1}{f'(\ln 2)} = \frac{1}{3}.$

101. $\ln 5x - \ln 3x = \ln \frac{5x}{3x} = \ln \frac{5}{3}.$

103. For any positive $a \neq 1$, $\int_1^e \frac{2\log_a x}{x} = \frac{2}{\ln a}\int_1^e \frac{\ln x}{x}dx = \frac{1}{\ln a}(\ln x)^2\Big]_1^e = \frac{1}{\ln a}$. Thus the two areas are $\frac{1}{\ln 2}$ and $\frac{1}{\ln 4}$ and their ratio is 2.

105. $y = y_0 e^{-kt}$ and $k = \ln 2/5700$ from Example 3 of Section 7.3. Let t be the desired age. Then $0.1y_0 = y_0 e^{-kt}$, $-kt = \ln(0.1)$, $t = -5700\frac{\ln(0.1)}{\ln 2} \approx 18935$ years.

107. We wish to determine k in the model $y = y_0 e^{kt}$. Let $t = 0$ correspond to 1924. Then $y_0 = 250$ and $t = 64$ corresponds to 1988. When $t = 64$, $7500 = 250e^{64k}$, $e^{64k} = 30$, $64k = \ln 30$, $k = (\ln 30)/64 \approx 0.053$. Thus the rate of appreciation is about 5.3%.

109. $p = p_0 e^{0.04t}$. Let t be the required time. Then $\frac{4}{3}p_0 = p_0 e^{0.04t}$, $\ln(4/3) = 0.04t$, $t = \ln(4/3)/0.04 \approx 7.19$ years.

111. a) $p = p_0 e^{kt} = 295.5e^{kt}$ using $t = 0$ for 1980. When $t = 6$, $p = 480.1 = 295.5e^{6k}$. This leads to $k = 0.08\ldots$ or an inflation rate of about 8%.

b) $p(12) = 295.5e^{(0.08)12} \approx 771.8$.

113. $A_0 e^{-kt} = \frac{1}{2}A_0$ leads to $e^{-kt} = \frac{1}{2}$, $-kt = \ln 1 - \ln 2 = -\ln 2$, $t = (\ln 2)/k$.

115. $T - T_s = (T_0 - T_s)e^{-kt}$. After 15 minutes, $180 - 40 = (220-40)e^{-15k}$, $e^{-15k} = \frac{7}{9}$, $k = -\frac{\ln \frac{7}{9}}{15}$. Let t be the total number of minutes to cool to 70°. $70 - 40 = (220-40)e^{-kt}$, $t = \frac{\ln 6}{k} \approx 106.943$ minutes. Thus the answer is $106.94 - 15 = 91.943$ minutes.

117. $\lim_{t\to 0} \frac{t - \ln(1+2t)}{t^2} = \lim_{t\to 0} \frac{1 - \frac{2}{1+2t}}{2t}$. From this we see that the limit is ∞ as $t \to 0^-$ and $-\infty$ as $t \to 0^+$. This can be confirmed by graphing the original function in $[-0.5, 1]$ by $[-50, 50]$.

119. $\lim_{x\to 0} \frac{x \sin x}{1 - \cos x} = \lim_{x\to 0} \frac{x\cos x + \sin x}{\sin x}$ (still $\frac{0}{0}$) $= \lim_{x\to 0} \frac{-x\sin x + 2\cos x}{\cos x} = 2.$

121. $\lim_{x\to 0} \frac{2^{\sin x} - 1}{e^x - 1} = \lim_{x\to 0} \frac{2^{\sin x}(\cos x)\ln 2}{e^x} = \ln 2$. Defining $f(0) = \ln 2$ removes the discontinuity.

123. Let $y = x^{\frac{1}{x}}$. $\lim_{x\to\infty} \ln y = \lim_{x\to\infty} \frac{\ln x}{x} = \lim_{x\to\infty} \frac{1}{x} = 0$. Therefore $\lim_{x\to\infty} y = e^0 = 1$.

125. $\lim_{x\to\infty}(1+\frac{3}{x})^x = e^3$ is found by the method of Example 6 of 7.5. See also Exercise 45 of that section.

127. In all three cases the functions grow at the same rate as $x \to \infty$ because $\lim_{x\to\infty}\frac{f(x)}{g(x)} = L$, L finite and not 0.

129. a) $\frac{\frac{1}{x^2}+\frac{1}{x^4}}{\frac{1}{x^2}} = 1+\frac{1}{x^2} \leqq 2$ for $x \geqq 1$. True. b) $\frac{\frac{1}{x^2}+\frac{1}{x^4}}{\frac{1}{x^4}} = x^2+1 \to \infty$ as $x\to\infty$. False. c) $\frac{\sqrt{x^2+1}}{x} = \sqrt{\frac{x^2+1}{x^2}}$ (for $x>0$) $= \sqrt{1+\frac{1}{x^2}} \leqq \sqrt{2}$ for $x \geqq 1$. True.

131. $g(x)$ is not defined for $x<0$ and $f(x)$ is not defined for $x<-1$. $\lim_{x\to\infty}\frac{f(x)}{g(x)} = \lim_{x\to\infty}\frac{\sqrt[4]{x^3-x}}{\sqrt[4]{x^3}} = \lim_{x\to\infty}\sqrt[4]{1-\frac{1}{x^2}} = 1$. g is a right end behavior model for f.

133. $\frac{f(x)}{g(x)} = \frac{2x^3-x+1}{(4x^2-x+1)(0.5x)} = \frac{2x^3-x+1}{2x^3-0.5x+0.5x} = \frac{2-\frac{1}{x^2}+\frac{1}{x^3}}{2-\frac{0.5}{x}+\frac{0.5}{x^2}} \to \frac{2}{2} = 1$ as $x\to\pm\infty$. g is an end behavior model for f.

135. $\frac{f(x)}{g(x)} = \frac{2^x+2^{-x}}{2^x} = 1+\frac{1}{2^{2x}} \to 1$ as $x\to\infty$. ($\to\infty$ as $x\to-\infty$) g is a right end behavior model for f.

137. Let $f(x) = \tan^{-1}x+\tan^{-1}\frac{1}{x}$. $f'(x) = \frac{1}{1+x^2}+\frac{(-\frac{1}{x^2})}{1+(\frac{1}{x})^2} = \frac{1}{1+x^2}-\frac{1}{x^2+1} = 0$ so $f(x)$ is constant. $f(1) = \frac{\pi}{2}$ but $f(-1) = -\frac{\pi}{2}$. Hence $f(x) = \frac{\pi}{2}$ for $x>0$ and $f(x) = -\frac{\pi}{2}$ for $x<0$.

139. $\theta = \pi-\cot^{-1}\frac{x}{60}-\cot^{-1}\frac{50-x}{30}$, $\frac{d\theta}{dx} = \frac{1}{60}\frac{1}{1+(\frac{x}{60})^2}-\frac{1}{30}\frac{1}{1+(\frac{50-x}{30})^2} = \frac{30(x^2-200x+3200)}{(3600+x^2)[900+(50-x)^2]}$. $\frac{d\theta}{dx} = 0$ when $x = 100\pm20\sqrt{17}$. For the problem situation the solution is $x = 100-20\sqrt{17} \approx 17.538$ m.

141. $y = -\int_1^x\frac{dt}{t\sqrt{1-t^2}}+\int_1^x\frac{t\,dt}{\sqrt{1-t^2}}$. (Note $y=0$ if $x=1$.) $y = \operatorname{sech}^{-1}|t|-\sqrt{1-t^2}\Big|_1^x = \operatorname{sech}^{-1}|x|-\sqrt{1-x^2}-\operatorname{sech}^{-1}1 = \operatorname{sech}^{-1}|x|-\sqrt{1-x^2} = \cosh^{-1}(\frac{1}{|x|})-\sqrt{1-x^2}$. You may graph this to confirm the given figure.

CHAPTER 8

TECHNIQUES OF INTEGRATION

8.1 FORMULAS FOR ELEMENTARY INTEGRALS

1. Let $u = 8x^2 + 1$; $du = 16x^2 dx$. The integral becomes $\int \frac{du}{\sqrt{u}} = \int u^{-1/2} du = \frac{u^{1/2}}{1/2} + C = 2(8x^2 + 1)^{1/2} + C$.

3. Let $u = 8x^2 + 2 : 0 \le x \le 1 \Rightarrow 2 \le u \le 10$, $du = 16x dx$. The integral becomes $\int_2^{10} \frac{du}{u} \approx 1.60944$.

5. Let $u = x^2$, $du = 2x dx \Rightarrow 4x dx = 2du$. The integral becomes $2\int \tan u du = -2\ln|\cos u| + C = \ln(\cos^2 x^2) + C$.

7. Let $u = x/3$; $du = \frac{1}{3}dx$ or $dx = 3du$, $-\pi \le x \le \pi \Rightarrow -\pi/3 \le u \le \pi/3$. The integral becomes $\int_{-\pi/3}^{\pi/3} \sec u(3du) = 3\int_{-\pi/3}^{\pi/3} \sec u du \approx 7.902$.

9. The integral becomes $\int_{\pi/2}^{3\pi/4} \csc u du \approx 0.881$, where $u = x - \pi$.

11. Let $u = e^x + 1$. The integral becomes $\int \csc u du = -\ln|\csc(e^x + 1) + \cot(e^x + 1)| + C$.

13. The integral becomes: $\int_0^{\ln 2} e^u du = 2 - 1 = 1$, where $u = x^2$.

15. The integral becomes: $\int_0^1 3^u du \approx 1.82048$.

17. $u = \sqrt{y} \Rightarrow du = \frac{1}{2\sqrt{y}} dy$ so $\frac{dy}{\sqrt{y}} = 2du$; $1 \le y \le 3 \Rightarrow 1 \le u \le \sqrt{3}$. The integral becomes: $\int_1^{\sqrt{3}} \frac{2 \cdot 6du}{1+u^2} = 12\int_1^{\sqrt{3}} \frac{du}{1+u^2} \approx 3.14159$ (which is obviously π).

19. $u = 3x \Rightarrow du = 3dx \Rightarrow dx = \frac{1}{3}du$. The integral becomes $\int_0^{1/2} \frac{\frac{1}{3}du}{\sqrt{1-u^2}} = \frac{1}{3}\int_0^{1/2} \frac{du}{\sqrt{1-u^2}} = \frac{1}{3}\sin^{-1} u\big|_0^{1/2} \approx 0.1743 = \frac{\pi}{18}$.

21. Let $u = 5x$; $2/(5\sqrt{3}) \le x \le \frac{2}{5} \Rightarrow 2/\sqrt{3} \le u \le 2$; $x = u/5 \Rightarrow dx = \frac{1}{5}du$. The integral becomes $\int_{2/\sqrt{3}}^{2} \frac{6 \cdot \frac{1}{5}du}{\frac{1}{5}u\sqrt{u^2-1}} = 6\int_{2/\sqrt{3}}^{2} \frac{du}{u\sqrt{u^2-1}} \approx 3.14159$, (π).

23. $(-x^2 + 4x - 3) = -(x^2 - 4x) - 3 = -(x^2 - 4x + 4) - 3 + 4 = 1 - (x-2)^2$. The integral becomes $\int \frac{dx}{\sqrt{1-(x-2)^2}}$. Let $u = x - 2$. Then $du = dx$; the integral is $\int \frac{du}{\sqrt{1-u^2}} = \sin^{-1} u + C = \sin^{-1}(x-2) + C$.

25. $x^2 - 2x + 2 = x^2 - 2x + 1 + 1 = (x-1)^2 + 1$. Let $u = x - 1$. Then $\int_1^2 \frac{8dx}{(x-1)^2+1} = \int_0^1 \frac{8du}{u^2+1} \approx 6.28319\ (2\pi)$.

27. $x^2+2x = x^2+2x+1-1 = (x+1)^2-1$. Let $u = x+1$. The integral becomes $\int \frac{du}{u\sqrt{u^2-1}} = \sec^{-1}|u| + C = \sec^{-1}|x+1| + C$.

29. $\int(\csc x - \cot x)^2 dx = \int(\csc^2 x + \cot^2 x - 2\csc x \cot x)dx = \int(2\csc^2 x - 1)dx - 2\int \csc x \cot x\, dx = -2\cot x - x + 2\csc x + C$. $[-2\cot x - x + 2\csc x]_{\pi/4}^{3\pi/4} = [2 - 3\pi/4 + \sqrt{2}] - [-2 - \pi/4 + \sqrt{2}] = [4 - \pi/2]$.

31. $\int(\csc x - \sec x)(\sin x + \cos x)dx = \int(\frac{1}{\sin x} - \frac{1}{\cos x})(\sin x + \cos x)dx = \int(1 + \cot x - \tan x - 1)dx = \ln|\sin x| - \ln|\sec x| + C = \ln|\sin x| + \ln|\cos x| + C$. $[\ln|(\sin x)(\cos x)|]_{\pi/6}^{\pi/3} = [\ln(\frac{1}{2}\frac{\sqrt{3}}{2})] - [\ln(\frac{\sqrt{3}}{2}\cdot\frac{1}{2})] = 0$.

33. $\int_0^{\sqrt{3}/2} \frac{1}{\sqrt{1-x^2}}dx - \int_0^{\sqrt{3}/2} \frac{x}{\sqrt{1-x^2}}dx = \sin^{-1} x|_0^{\sqrt{3}/2} + \frac{1}{2}\int_0^{\sqrt{3}/2}(-2x)(1-x^2)^{-1/2}dx = \pi/3 + \frac{1}{2}[\frac{(1-x^2)^{1/2}}{1/2}]_0^{\sqrt{3}/2} = \frac{\pi}{3} + [\frac{1}{2} - 1] = \frac{\pi}{3} - \frac{1}{2}$.

35. $\int_0^{\pi/4} \frac{1}{\cos^2 x}dx + \int_0^{\pi/4} \frac{\sin x}{\cos^2 x}dx = \int_0^{\pi/4} \sec^2 x dx + \int_{x=0}^{x=\pi/4} \frac{-du}{u^2} = [\tan x]_0^{\pi/4} + [\frac{1}{u}]_{x=0}^{x=\pi/4} = 1 + [\frac{1}{\cos x}]_0^{\pi/4} = 1 + [\sqrt{2} - 1] = \sqrt{2}$; where $u = \cos x$.

37. $y = \ln(\cos x) \Rightarrow y' = \frac{1}{\cos x}(-\sin x)$; $L = \int_0^{\pi/3} \sqrt{1+(y')^2}dx = \int_0^{\pi/3} \sqrt{1+\tan^2 x}dx = \int_0^{\pi/3} \sqrt{\sec^2 x}dx = \int_0^{\pi/3} \sec x\, dx \approx 1.31696$.

39.

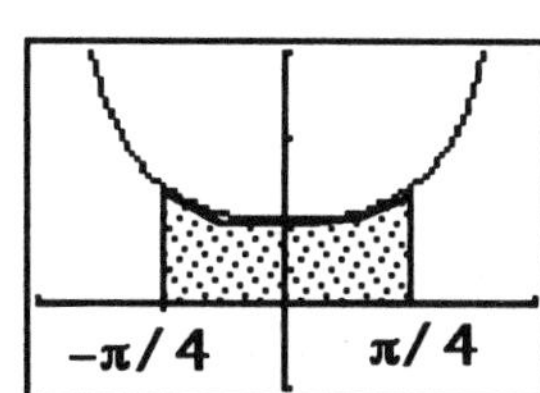

By symmetry $\bar{x} = 0$. $M_x = \int_{-\pi/4}^{\pi/4} \frac{1}{2}(\sec x)(\sec x)dx = \int_0^{\pi/4} \sec^2 x\, dx = [\tan x]_0^{\pi/4} = 1$; $M = 2\int_0^{\pi/4} \sec x\, dx = 2[\ln|\sec x + \tan x|]_0^{\pi/4} = 2\ln(\sqrt{2}+1) - 0$; $\bar{y} = \frac{1}{2\ln(1+\sqrt{2})} = \frac{1}{\ln(1+\sqrt{2})^2} = \frac{1}{\ln(3+2\sqrt{2})}$.

41. $D = 25\int_{30°}^{45°} \sec x\, dx = 25\ln|\sec x + \tan x|]_{30°}^{45°} = 8.30169$ cm.

43. Let $u = \sin x$; $du = \cos x\, dx$; $0 \le x \le \pi/2 \Rightarrow 0 \le u \le 1$. The integral becomes $3\int_0^1 u^{1/2}du = \frac{3u^{3/2}}{\frac{3}{2}}]_0^1 = 2$.

45. Let $u = 2 + \cos x$. $-\pi \le x \le 0 \Rightarrow 1 \le u \le 3$; $du = -\sin x\, dx$. The integral becomes $-\int_1^3 \frac{du}{u} = -\ln|u|]_1^3 = -[\ln 3 - \ln 1] = \ln(\frac{1}{3}) \approx -1.0986$.

47. Let $u = \pi x$; then $du = \pi\, dx$ or $x = (\frac{1}{\pi}du)$. The integral becomes $\frac{1}{\pi}\int_0^{\pi/4} \sec u\, du = \frac{1}{\pi}\ln|\sec u + \tan u|]_0^{\pi/4} = \frac{1}{\pi}[\ln(\sqrt{2}+1) - \ln(1+0)] = \frac{1}{\pi}\ln(\sqrt{2}+1) = 0.28055.$

49. Let $u = \tan x$; then $0 \le u \le \sqrt{3}$ and $du = \sec^2 x\, dx$. The integral becomes $\int_0^{\sqrt{3}} e^u du = e^u]_0^{\sqrt{3}} = e^{\sqrt{3}} - 1 = 4.65223.$

51. Let $u = \sqrt{x}$; then $du = \frac{1}{2\sqrt{x}}dx$. The integral becomes $\int_1^2 2^u du = \frac{1}{\ln 2}[2^u]_1^2 = \frac{1}{\ln 2}[4-2] = \frac{2}{\ln 2} = 2.8854.$

53. Let $u = 3x$; $du = 3dx$ and $0 \le u \le \sqrt{3}$. The integral becomes $3\int_0^{\sqrt{3}} \frac{du}{1+u^2} = 3[\tan^{-1} u]_0^{\sqrt{3}} = 3\cdot[\pi/3 - 0] = \pi.$

55. Let $u = 2x$. The integral becomes $\int_0^{1/2} \frac{du}{\sqrt{1-u^2}} = [\sin^{-1} u]_0^{1/2} = \frac{\pi}{6} - 0 = \frac{\pi}{6}.$

57. Let $u = 2x$; then $du = 2dx$. $\int_{1/\sqrt{2}}^1 \frac{dx}{x\sqrt{4x^2-1}} = \int_{1/\sqrt{2}}^1 \frac{2dx}{2x\sqrt{4x^2-1}} = \int_{2/\sqrt{2}}^2 \frac{du}{u\sqrt{u^2-1}} = [\sec^{-1}|u|]_{\sqrt{2}}^2 = \sec^{-1} 2 - \sec^{-1}\sqrt{2} = \frac{\pi}{3} - \frac{\pi}{4} = \frac{\pi}{12}.$

59. Graph $y1 = \int_0^x \frac{dt}{t^2+2t+2} = \text{NINT}(1/(t^2+2t+2), t, 0, x)$ and $y2 = \tan^{-1}(x+1)$ on $[-2,2]$ by $[-2,2]$. Using trace or trigonometry we see that $y2(0) = \pi/4 \approx .78$. Thus $C = -\pi/4$, and $\tan^{-1}(x+1) - \pi/4 = \int_0^x \frac{dt}{t^2+2t+1}$

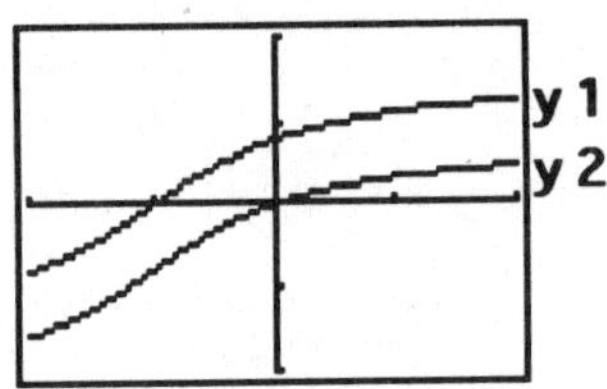

8.2 INTEGRATION BY PARTS

1. Let $u = x$, $dv = \sin x\, dx$. Then $du = dx$, $v = -\cos x$; $\int x \sin x\, dx = x(-\cos x) - \int(-\cos x)dx = -x\cos x + \int \cos x\, dx = -x\cos x + \sin x + C$. To support graphically, compare the graphs of $y = -x\cos x + \sin x + 2$ and $y = \int_0^x t\sin t\, dt = \text{NINT}(t\sin t, t, 0, x)$ on $[-5,5]$ by $[-5,5]$.

3. Let $u = x^2$, $dv = \sin$; $du = 2xdx$, $v = -\cos x$; $\int x^2 \sin x\, dx = -x^2\cos x + 2\int x\cos x\, dx = -x^2\cos x + 2[x\sin x - \int \sin dx] = -x^2\cos x + 2x\sin x + 2\cos x + C$, where $u = x$, $dv = \cos x\, dx$ for the second integration by parts.

Support this answer with an NINT computation, or by comparing the graph of $y = x^2 \sin x$ to the graph of $y =$ NDER $1(-x^2 \cos x + 2x \sin x + 2 \cos x)$. Since the graphs are the same, the answer is correct; this method is often quicker.

5. Let $u = \ln x$, $dv = x\,dx$; $du = \frac{dx}{x}$, $v = \frac{x^2}{2} \Rightarrow \int x \ln x\,dx = \frac{x^2}{2} \ln x - \int \frac{x^2}{2} \frac{dx}{x} = \frac{x^2}{2} \ln x - \frac{1}{4}x^2 + C$.

7. Let $u = \tan^{-1} x$, $dv = x$; $du = \frac{1}{1+x^2}$, $v = x$; $\int \tan^{-1} x\,dx = x \tan^{-1} x - \int \frac{x}{1+x^2} dx = x \tan^{-1} x - \frac{1}{2} \int \frac{2x}{1+x^2} dx = x \tan^{-1} x - \frac{1}{2} \ln(1 + x^2) + C$.

9. Let $u = x$, $dv = \sec^2 x\,dx$; $du = dx$, $v = \tan x$; $\int x \sec^2 x\,dx = x \tan x - \int \tan x\,dx = x \tan x + \ln|\cos x| + C$.

11. Evaluating $\int x^3 e^x dx$ requires using integration by parts three times. This is most easily done by tabular integration (see Example 7)

f and its derivatives		g and its integrals
x^3		e^x
	$(+)$ ↘	
$3x^2$		e^x
	$(-)$ ↘	
$6x$		e^x
	$(+)$ ↘	
6		e^x
	$-$ ↘	
0		e^x

$\int x^3 e^x dx = x^3 e^x - 3x^2 e^x + 6xe^x - 6e^x + C$.

13. Again, use tabular integration:

f and its derivatives		g and its integrals
$x^2 - 5$		e^x
	$(+)$ ↘	
$2x - 5$		e^x
	$(-)$ ↘	
2		e^x
	$(+)$ ↘	
0		e^x

$\int (x^2 - 5x)e^x dx = (x^2 - 5x)e^x - (2x - 5)e^x + 2e^x + C = (x^2 - 7x + 7)e^x + C$.

15. Use tabular integration:

f		g
x^5	(+) ↘	e^x
$5x^4$	·	e^x
$20x^3$	·	e^x
$60x^2$	·	e^x
$120x$	·	e^x
120	· (−) ↘	e^x
		e^x

$\int x^5 e^x dx = x^5e^x - 5x^4e^x + 20x^3e^x - 60x^2e^x + 120xe^x - 120e^x + C.$

17. Use tabular integration $\int_0^{\pi/2} x^2 \sin 2x dx = [x^2(\frac{-\cos 2x}{2}) - 2x(\frac{-\sin 2x}{4}) + 2(\frac{\cos 2x}{8})]_0^{\pi/2} = [\frac{\pi^2}{4}(\frac{-(-1)}{2}) - 2\frac{\pi}{2}(0) + \frac{2(-1)}{8}] - [0 - 0 + 2(\frac{1}{8})] = \frac{\pi^2-4}{8} = 0.7337$

f and its derivatives		g and its integrals
x^2	(+) ↘	$\sin 2x$
$2x$	(−) ↘	$-\frac{1}{2}\cos 2x$
2	(+) ↘	$-\frac{1}{4}\sin 2x$
0		$\frac{1}{8}\cos 2x$

Use NINT to confirm.

19. $\int_1^2 x \sec^{-1} x\, dv = [\frac{x^2}{2}\sec^{-1} x]_1^2 - \frac{1}{2}\int_1^2 \frac{x}{\sqrt{x^2-1}}dx = [\frac{4}{2}\sec^{-1} 2 - \frac{1}{2}\sec^{-1} 1] - \frac{1}{4}\int_1^2 (x^2-1)^{-1/2} 2x dx$; (let $u = x^2 - 1$; then $du = 2xdx$, $x = 1 \Rightarrow u = 0$, $x = 2 \Rightarrow u = 3$) $= [2 \cdot \frac{\pi}{3} - \frac{1}{2} \cdot 0] - \frac{1}{4}\int_0^3 u^{-1/2} du = \frac{2\pi}{3} - \frac{\frac{1}{4}[u^{1/2}]_0^3}{1/2} = \frac{2\pi}{3} - \frac{1}{2}\sqrt{3} = 1.228$, where $u = \sec^{-1} x$, $dv = xdx$, $du = \frac{dx}{|x|\sqrt{x^2-1}} = \frac{dx}{x\sqrt{x^2-1}}$ since $1 \le x \le 2$. To confirm, evaluate NINT$(t\cos^{-1}(1/t), t, 1, 2)$.

21. Use integration by parts twice: $\int_{-1}^3 e^x \sin x\, dx = e^x \sin x|_{-1}^3 - \int_{-1}^3 e^x \cos x\, dx = e^x \sin x|_{-1}^3 - \{e^x \cos x|_{-1}^3 + \int_{-1}^3 e^x \sin x\, dx\}$. Thus $2\int_{-1}^3 e^x \sin x\, dx = (e^x \sin x - e^x \cos x)|_{-1}^3 = 23.227$ or, $\int_{-1}^3 e^x \sin x\, dx = 11.614$.

23. Let $I=\int e^{2x}\cos 3x dx$ and use integration by parts twice. First let $u=\cos 3x$, $dv = e^{2x}dv$, $du=-3\sin 3x$, $v=\frac{1}{2}e^{2x}$; $I=(e^{2x}\cos 3x)/2+\frac{3}{2}\int e^{2x}\sin 3x dx =$ $(e^{2x}\cos 3x)/2+\frac{3}{2}[(e^{2x}\sin 3x)/2-\frac{3}{2}I]$, where $u=\sin 3x$, $dv=e^{2x}dx$, $du=$ $3\cos 3x dx$, $v=\frac{1}{2}e^{2x}$ in the second integration by parts. $(1+\frac{9}{4})I=(\cos 3x+$ $\frac{3}{2}\sin 3x)e^{2x}/2$. $I=\frac{4}{13}(\cos 3x+\frac{3}{2}\sin 3x)\frac{e^{2x}}{2}+C$. Hence, $\int_{-2}^{3}e^{2x}\cos 3x dx =$ $\frac{2e^{2x}}{13}(\cos 3x+\frac{3}{2}\sin 3x)|_{-2}^{3} = 18.186$.

25.

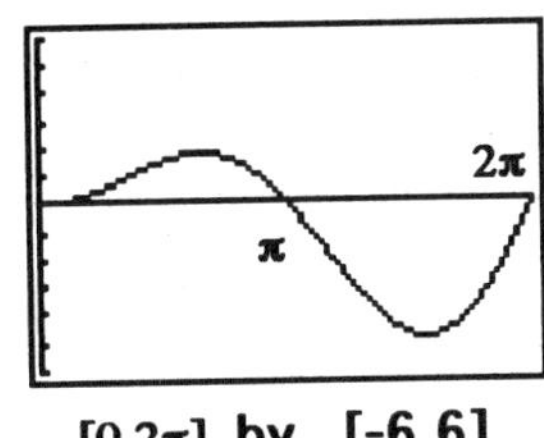

[0,2π] by [-6,6]

From problem 1,

$$\int x\sin x\, dx = \sin x - x\cos x + C$$

a) $\int_0^{\pi} x\sin x\, dx = [\sin x - x\cos x]_0^{\pi} = \pi$

b) $\int_{\pi}^{2\pi} x\sin x\, dx = [\sin x - x\cos x]_{\pi}^{2\pi} =$

$-2\pi-\pi=-3\pi$

Since area, in itself, is taken to be positive, the answer to b) is 3π.

27.

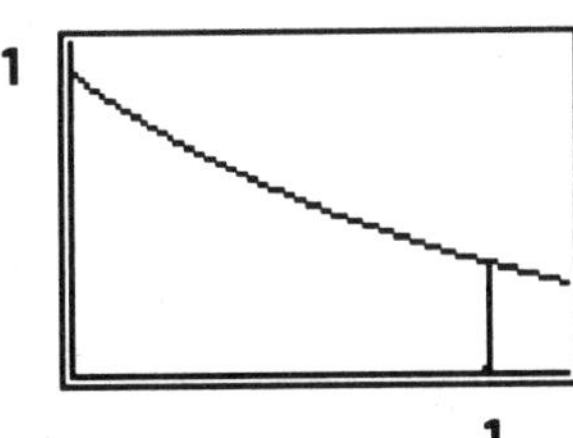

Using cylindrical shells, and tabular integration by parts, $V=2\pi\int$ radius $\cdot$ height dx $=2\pi\int_0^1 xe^{-x}dx = 2\pi[x(-e^{-x})-1\cdot(e^{-x})]_0^1$

f		g
x	(+) ↘	e^{-x}
1	(−) ↘	$-e^{-x}$
0		e^{-x}

$=2\pi[(-e^{-1}-e^{-1})-(-1)]=2\pi[1-2/e]$

29. $M_y=\int_0^{\pi} x(\sin x)(1+x)dx=\int_0^{\pi}(x^2+x)\sin x\, dx$. Use tabular integration to get $=[(x^2+x)(-\cos x)-(2x+1)(-\sin x)+2\cos x]_0^{\pi}=[\pi^2+\pi-2]-[2]=\pi^2+\pi-4$.

31. a) Let $u=(\ln x)^n$ and $dv=dx$. $du=n(\ln x)^{n-1}(\frac{1}{x})dx$; $v=x$. Then $\int(\ln x)^n dx = x(\ln x)^n - n\int(\ln x)^{n-1}(\frac{1}{x})x\, dx = x(\ln x)^n - n\int(\ln x)^{n-1}dx$.

b) $\int_1^3(\ln x)^4 x = [x(\ln x)^4-4[x(\ln x)^3-3[x(\ln x)^2-2[x(\ln x)^1-x(\ln x)^0]]]]^3 =$ $[x(\ln x)^4-4x(\ln x)^3+12x(\ln x)^2-24x(\ln x)+24x]_1^3=[3(\ln 3)^4-12(\ln 3)^3+$ $36(\ln 3)^2-72\ln 3+72-24]$. To evaluate: store $\ln 3$ in x, then evaluate $3x^4-12x^3+36x^2-72x+72-24\approx 0.8086\ldots$.

c) $\text{NINT}((\ln x)^4-4,x,1,3)=0.8086\ldots$.

33. a) $\int_{\ln 2}^{\ln 3} e^x dx = 3 - 2 = 1$; b) $\int_2^3 \ln y\, dy = [y \ln y - y]_2^3 = \ln \frac{27}{4} - 1$

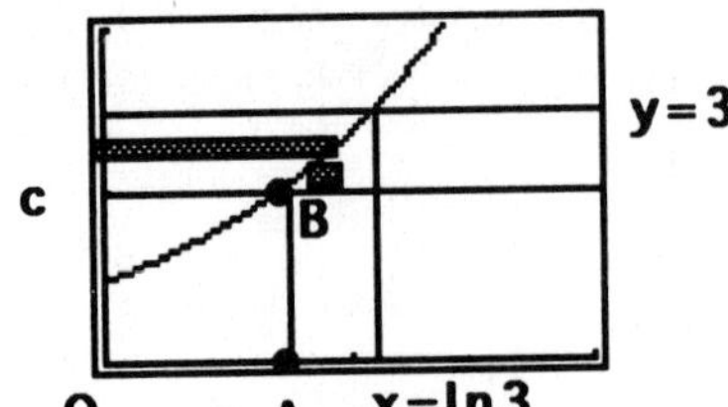

c) $(a)+(b)+\text{area}(OABC) = $ area of rectangle $= 3\ln 3$, or $(a) + (b) = 3\ln 3 - 2\ln 2$

d) Let $y = e^x = u$, $x = v$; $(a) = \int u\, dv$, $(b) = \int x\, dy = \int v\, du$

8.3 INTEGRALS INVOLVING TRIGONOMETRIC FUNCTIONS

1. $\int \sin^5 x\, dx = \int \sin x(\sin^4 x)dx = \int \sin x(\sin^2 x)^2 dx = \int \sin x(1 - \cos^2 x)^2 dx = \int \sin x(1 - 2\cos^2 x + \cos^4 x)dx = -\cos x + 2\frac{\cos^3 x}{3} - \frac{\cos^5 x}{5} + C$

3. $\int \cos^3 x\, dx = \int \cos x(1 - \sin^2 x)dx = -\frac{\sin^3 x}{3} + \sin x + C$

5. $\int \sin^7 y\, dy = \int \sin y[\sin^2 y]^3 dy = \int [1 - \cos^2 y]^3 \sin y\, dy = \int [1 - 3\cos^2 y + 3\cos^4 y - \cos^6 y] \sin y\, dy = -\cos y + \frac{3\cos^3 y}{3} - \frac{3\cos^5 y}{5} + \frac{\cos^7 y}{7} + C$

7. $\int 8\sin^4 x\, dx = 3x - 2\sin 2x + \frac{1}{4}\sin 4x + C$ since $\sin^4 x = (\sin^2 x)^2 = [(1 - \cos 2x)/2]^2 = \frac{1}{4}[1 - 2\cos 2x + \cos^2 2x] = \frac{1}{4}[1 - 2\cos 2x + \frac{1}{2}(1 + \cos 4x)] = \frac{1}{4}[\frac{3}{2} - 2\cos 2x + \frac{1}{2}\cos 4x]$

9. $\int 16\sin^2 x \cos^2 x\, dx = 4\int (2\sin x \cos x)^2 dx = 4\int (\sin 2x)^2 dx = 2\int (1 - \cos 4x)dx = 2[x - \frac{\sin 4x}{4}] + C$

11. $\int 35\sin^4 x \cos^3 x\, dx = \int 35\sin^4 x(1 - \sin^2 x)\cos x\, dx = 35[\frac{\sin^5 x}{5} - \frac{\sin^7 x}{7}] + C$

13. $\int 8\cos^3 2\theta \sin 2\theta d\theta = -4\int \cos^3 2\theta(-\sin 2\theta)2d\theta = -4\frac{\cos^4 2\theta}{4} + C$

15. $\frac{1-\cos x}{2} = \sin^2 \frac{x}{2}$. Since $0 \le x \le 2\pi \Rightarrow 0 \le \frac{x}{2} \le \pi \Rightarrow \sin \frac{x}{2} \ge 0$, we may take a square root. The integral becomes $\int_0^{2\pi} \sin \frac{x}{2} dx = [2(-\cos(\frac{x}{2})]_0^{2\pi} = 2 - (-2) = 4$.

17. $1 - \sin^2 t = \cos^2 t$. Since the integrand is non-negative, $\int_0^\pi \sqrt{1 - \sin^2 t}\, dt = \int_0^\pi |\cos t| dt = \int_0^{\pi/2} \cos t\, dt + \int_{\pi/2}^{\pi}(-\cos t)dt = 2$ (the area under one loop of the cosine curve is 2).

19. $\int_{-\pi/4}^{\pi/4}\sqrt{1+\tan^2 x}\,dx = 2\int_0^{\pi/4}\sqrt{1+\tan^2 x}\,dx = 2\int_0^{\pi/4}\sec x\,dx = [2\ln|\sec x + \tan x|]_0^{\pi/4} = 2\ln|\sqrt{2}+1| = \ln(\sqrt{2}+1)^2 = \ln(3+2\sqrt{2})$

21. On the interval $0 \le \theta \le \pi/2$ the integral becomes $\sqrt{2}\int_0^{\pi/2}\theta\sin\theta\,d\theta = \sqrt{2}[\theta\cos\theta - (-\sin\theta)]_0^{\pi/2} = \sqrt{2}$

θ		$\sin\theta$
	(+) ↘	
1		$-\cos\theta$
	(−) ↘	
0		$-\sin\theta$

23. First, find the indefinite integral $I = \int\sec^3 x\,dx = \int\sec^2 x\sec x\,dx = \sec x\tan x - \int\sec x\tan^2 x\,dx$; $u = \sec x$, $dv = \sec^2 x$, $du = \sec x\tan x$, $v = \tan x$; $I = \sec x\tan x - \int\sec x(\sec^2 x - 1)dx = \sec x\tan x - I + \int\sec x\,dx$; $2I = \sec x\tan + \ln|\sec x + \tan x|$: $I = [\sec x\tan x + \ln|\sec x + \tan x|]/2 + C$. The problem becomes $2I]_{-\pi/3}^{0} = [0 - 2(-\sqrt{3}) + \ln(2-\sqrt{3})] = 2\sqrt{3} + \ln(2-\sqrt{3})$.

25. $\int_0^{\pi/4}\sec^2\theta\sec^2\theta\,d\theta = \int_0^{\pi/4}(\tan^2\theta + 1)\sec^2\theta\,d\theta = \int_0^{\pi/4}\tan^2\theta\sec^2\theta\,d\theta + \int_0^{\pi/4}\sec^2\theta\,d\theta = [\frac{\tan^3\theta}{3} + \tan\theta]_0^{\pi/4} = \frac{4}{3}$

27. $\int_{\pi/4}^{\pi/2}\csc^2\theta\csc^2\theta\,d\theta = \int_{\pi/4}^{\pi/2}(1+\cot^2\theta)\csc^2\theta\,d\theta = [-\cot\theta - \frac{\cot^3\theta}{3}]_{\pi/4}^{\pi/2} = 4/3$

29. $4\int_0^{\pi/4}\tan^3 x\,dx = 4\int_0^{\pi/4}(\sec^2 x - 1)\tan x\,dx = 4\int_0^{\pi/4}(\tan x\sec^2 x - \tan x)dx = 4[\frac{\tan^2 x}{2} + \ln|\cos x|]_0^{\pi/4} = 4[\frac{1}{2} + \ln\frac{1}{\sqrt{2}} - 0] = 2 + \ln(\frac{1}{\sqrt{2}})^4 = 2 + \ln\frac{1}{4} = 2 - \ln 4$

31. $\int_{\pi/6}^{\pi/3}(\csc^2 x - 1)\cot x\,dx = [-\frac{\cot^2 x}{2} - \ln|\sin x|]_{\pi/6}^{\pi/3} = -\{[\frac{1/3}{2} + \ln\frac{\sqrt{3}}{2}] - [\frac{3}{2} + \ln(\frac{1}{2})]\} = -\{-\frac{4}{3} + \ln\frac{\sqrt{3}/2}{1/2}\} = \frac{4}{3} - \ln\sqrt{3}$

33. $\sin 3x\cos 2x = \frac{1}{2}[\sin x + \sin 5x]$. Hence $\int_{-\pi}^{0}\sin 3x\cos 2x dx = \frac{1}{2}[-\cos x - \frac{1}{5}\cos 5x]_{-\pi}^{0} = \frac{1}{2}[-\frac{6}{5} - \frac{6}{5}] = -\frac{6}{5}$

35. $\int_{-\pi}^{\pi}\sin 3x\sin 3x dx = 2\int_0^{\pi}\sin 3x\sin 3x dx = \frac{2}{2}\int_0^{\pi}(\cos\cdot 0x - \cos 6x)dx = [x - \frac{\sin 6x}{x}]_0^{\pi} = \pi$

37. $\int_0^{\pi}\cos 3x\cos 4x dx = \frac{1}{2}\int_0^{\pi}(\cos x + \cos 7x)dx = \frac{1}{2}[\sin x + \frac{\sin 7x}{7}]_0^{\pi} = 0$

39. (a), (b), (c), (e), (g) and (i) are odd functions. The integrals of these functions over an interval centered at 0 are all zero. The others are even, non-negative functions; their integrals are not zero.

41. d) $\int_{-\pi/2}^{\pi/2} x\sin x\,dx = 2\int_0^{\pi/2} x\sin x\,dx = 2[-x\cos x - (-\sin x)]_0^{\pi/2} = 2[1] = 2$

x	$+$ ↘	$\sin x$
1	$-$ ↘	$-\cos x$
0		$-\sin x$

f) $\int_{-\pi/2}^{\pi/2} \cos^3 x\,dx = 2\int_0^{\pi/2}(1-\sin^2 x)\cos x\,dx = 2[\sin x - \frac{\sin^3 x}{3}]_0^{\pi/2} = \frac{4}{3}$

h) $2\int_0^{\pi/2} \sin x \sin 2x dx = 2\int_0^{\pi/2} \cos(1x) - \cos 3x dx = 2[\sin x - \frac{\sin 3x}{3}]_0^{\pi/2} = \frac{4}{3}$

43. $I = \int \csc x\,dx = \int \frac{\csc x(\csc x + \cot x)}{\csc x + \cot x} dx$. If $u = \csc x + \cot x$, then $du = (-\csc x \cot x - \csc^2 x)dx$. Hence $I = -\int \frac{du}{u} = -\ln|u| + C = -\ln|\csc x + \cot x| + C$.

45. a) f even

areas are

doubled

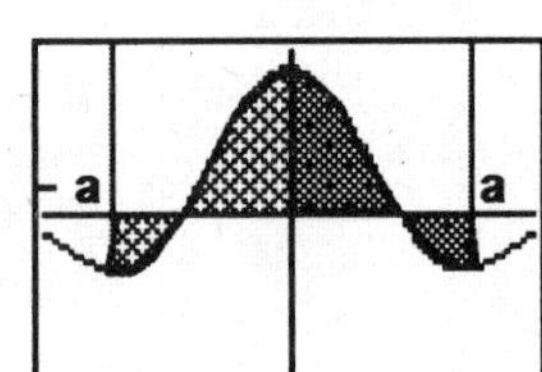

b) f odd

areas are

cancelled

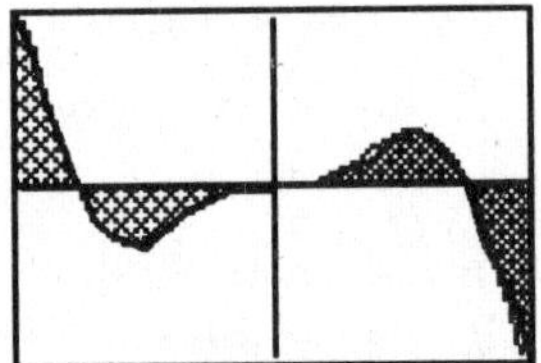

47. $f = x \cdot \sin x = \text{odd} \cdot \text{odd} = \text{even}$; $\int_{-\pi/2}^{\pi/2} x\sin x\,dx = 2\int_0^{\pi/2} x\sin x\,dx = 2(-x\cos x + \sin x)|_0^{\pi/2} = 2 \cdot 1 = 2$

49. $\cos x \sin 3x = \text{even} \cdot \text{odd} = \text{odd}$. $\int_{-\pi/2}^{\pi/2} f(x)dx = 0$, $\int_0^{\pi/2} \cos x \sin 3x dx = 0.5$

51. Let $t = \sin x$; $-t = \sin(-x)$, $-\frac{\pi}{2} \le x \le \pi/2 \Rightarrow -1 \le t \le 1$; $f(-x) = \cos(\sin(-x)) = \cos(-t) = \cos t \Rightarrow f$ is even, $\int_{-\pi/2}^{\pi/2} \cos(\sin x)dx = 2\int_0^{\pi/2} \cos(\sin x)dx = 2 \cdot 1.202 = 2.404$.

8.4 TRIGONOMETRIC SUBSTITUTIONS

1. $\int_{-2}^{2} \frac{dx}{4+x^2} = 2\int_0^2 \frac{dx}{2^2+x^2} = 2[\frac{1}{2}\tan^{-1}\frac{x}{2}]_0^2 = \frac{\pi}{4}$ by formula (8).

3. $\int_0^{3/2} \frac{dx}{\sqrt{9-x^2}} = \int_0^{3/2} \frac{dx}{\sqrt{3^2-x^2}} = [\sin^{-1}\frac{x}{3}]_0^{3/2} = \frac{\pi}{6}$

5. $\int \frac{dx}{\sqrt{x^2-4}} = \int \frac{dx}{\sqrt{x^2-2^2}}$. Set $x = 2\sec\theta$. Then $dx = 2\sec\theta\tan\theta\,d\theta$, $x^2 - 4 = 4\tan^2\theta$. The integral becomes $\int \frac{2\sec\theta\tan\theta\,d\theta}{|2\tan\theta|} = \ln|\sec\theta + \tan\theta| + C' = \ln|\frac{x}{2} + \sqrt{\frac{x^2-4}{4}}| + C' = \ln|x + \sqrt{x^2-4}| + C$.

7. Let $x = 5\sin\theta$. $\int \sqrt{25-x^2}dx = \int 5|\cos\theta| \cdot 5\cos\theta\,d\theta = 25\int \cos^2\theta\,d\theta = 25[\frac{\theta}{2} + \frac{\sin 2\theta}{4}] + C = \frac{25}{2}[\sin^{-1}(\frac{x}{5}) + \frac{2}{4}(\frac{x}{5})\sqrt{1-(\frac{x}{5})^2}] + C = [\frac{25}{2}\sin^{-1}(\frac{x}{5}) + \frac{x}{2}\sqrt{25-x^2}] + C$.

9. Let $x = \sin\theta$. $\int \frac{4x^2}{(1-x^2)^{3/2}}dx = \int \frac{4\sin^2\theta\cos\theta\,d\theta}{|\cos\theta|^3} = 4\int \tan^2\theta\,d\theta = 4[\tan\theta - \theta] + C = 4[\frac{x}{\sqrt{1-x^2}} - \sin^{-1}x] + C$.

11. $\int_0^{\sqrt{3}/2} \frac{2dy}{1+4y^2} = \int_0^{\sqrt{3}/2} \frac{2dy}{1+(2y)^2} = [\frac{1}{1}\tan^{-1}\frac{2y}{1}]_0^{\sqrt{3}/2} = \pi/3$ by (8).

13. Let $u = -x^2 + 2x$; $du = (-2x+2)dx = -2(x-1)dx$; $\int \frac{(x-1)dx}{\sqrt{2x-x^2}} = \frac{1}{-2}\int \frac{du}{\sqrt{u}} = -\frac{1}{2}\int u^{-1/2}du = \frac{-\frac{1}{2}u^{1/2}}{\frac{1}{2}} + C = -\sqrt{2x-x^2} + C$.

15. $\int \frac{dx}{\sqrt{x^2-2x}} = \int \frac{dx}{\sqrt{x^2-2x+1-1}} = \int \frac{dx}{\sqrt{(x-1)^2-1}}$.

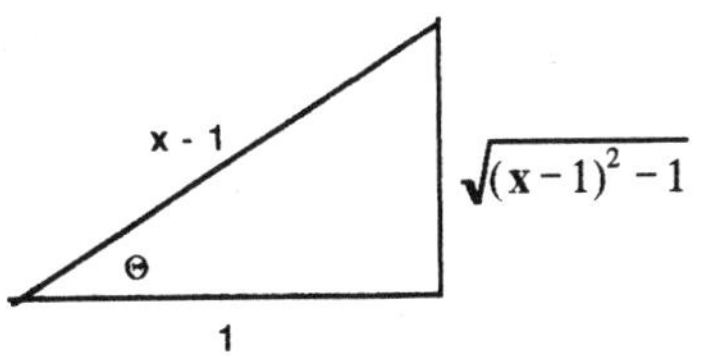

Let $(x-1) = \sec\theta$. The integral becomes

$= \int \frac{\sec\theta\tan\theta\,d\theta}{|\tan\theta|} = \ln|\sec\theta + \tan\theta| + C$

$= \ln|x - 1 + \sqrt{x^2-2x}| + C$.

17. $\int_0^{3\sqrt{2}/4} \frac{dx}{\sqrt{9-4x^2}} = \frac{1}{2}\int_0^{3\sqrt{2}/4} \frac{2dx}{\sqrt{3^2-(2x)^2}} = [\frac{1}{2}\sin^{-1}\frac{2x}{3}]_0^{3\sqrt{2}/4} = \frac{1}{2}\sin^{-1}\frac{\sqrt{2}}{2} = \frac{1}{2}\cdot\frac{\pi}{4} = \pi/8$, by (9).

19. Let $2z = \sec\theta$. Then $z = \frac{1}{2}\sec\theta \Rightarrow dz = \frac{1}{2}\sec\theta\tan\theta\,d\theta$ and $4z^2 - 1 = \sec^2\theta - 1 = \tan^2\theta$. Since $1/\sqrt{3} \le z \le 1$, $2/\sqrt{3} \le 2z \le 2$ and $\pi/6 \le \theta \le \pi/3$. On this range, $\tan\theta > 0$. The integral becomes $\int_{\pi/6}^{\pi/3} \frac{\sec\theta\tan\theta\,d\theta}{\frac{1}{2}\sec\theta\tan\theta} = 2(\pi/3 - \pi/6) = \pi/3$.

21. $\int_0^2 \frac{dx}{\sqrt{2^2+x^2}} = \int_0^{\pi/4} \frac{2\sec^2\theta\,d\theta}{2\sec\theta} = \int_0^{\pi/4}\sec\theta\,d\theta = [\ln(\sec\theta + \tan\theta)]_0^{\pi/4} = \ln(\sqrt{2}+1)$, where $x = 2\tan\theta$.

23. Let $(x-1) = u$. Then $1 \le x \le 2 \Rightarrow 0 \le u \le 1$. The integral becomes, by (9), $\int_0^1 \frac{6du}{\sqrt{2^2-u^2}} = 6[\sin^{-1}\frac{u}{2}]_0^1 = 6\cdot\frac{\pi}{6} = \pi$.

25. $\int_1^3 \frac{dy}{y^2-2y+5} = \int_1^3 \frac{dy}{y^2-2y+1+4} = \int_1^3 \frac{dy}{(y-1)^2+2^2}$; let $u = y - 1$; then $0 \le u \le 2$, $du = dy$, and the integral $= \int_0^2 \frac{du}{u^2+2^2} = [\frac{1}{2}\tan^{-1}\frac{u}{2}]_0^2 = \frac{\pi}{8}$.

27. Let $u = x^2+4x+13$; then $du = (2x+4)dx$. $\int_{-2}^{2} \frac{(x+2)dx}{\sqrt{x^2+4x+13}} = \frac{1}{2}\int_9^{25} u^{-1/2}du = [\frac{\frac{1}{2}u^{1/2}}{\frac{1}{2}}]_9^{25} = 2$. Another method of solution is to write the integrand as $\frac{(x+2)}{\sqrt{(x+2)^2+9}}$ and make the substitution $x+2 = 3\tan\theta$. This gives: $\int_{x=-2}^{x=2} \frac{3\tan\theta 3\sec^2\theta\, d\theta}{3\sec\theta} = [3\sec\theta]_{x=-2}^{x=2} = [\sqrt{(x+2)^2+9}]_{x=-2}^{x=2} = 2$.

29. $A = \int_0^1 \frac{2dx}{x^2-4x+5} = 2\int_0^1 \frac{dx}{(x-2)^2+1} = [2\tan^{-1}\frac{x-2}{1}]_0^1 = 2[\tan^{-1}(-1) - \tan^{-1}(-2)] = 0.64350\ldots$.

31. Let $x = \sin\theta$. Then $dx = \cos\theta\, d\theta$. The integral becomes $\int \frac{4\sin^2\theta\cos\theta\, d\theta}{(\cos^2\theta)^{3/2}} = 4\int \frac{\sin^2\theta}{\cos^2\theta}d\theta$

$= 4\int \frac{1-\cos^2\theta}{\cos^2\theta}d\theta = 4\int(\sec^2\theta - 1)d\theta$

$= 4[\tan\theta - \theta] + C = 4[\frac{x}{\sqrt{1-x^2}} - \sin^{-1}x] + C$.

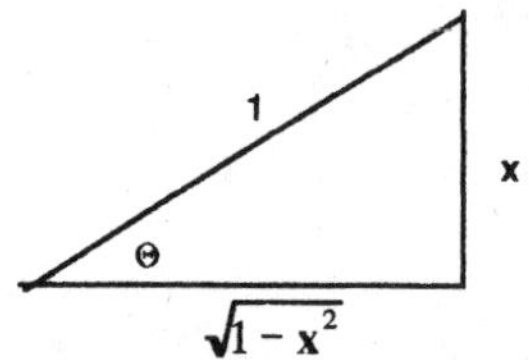

33. Let $x = 3\sin\theta$; $A = \frac{1}{3}\int_0^3 \sqrt{9-x^2}dx = \frac{1}{3}\int_0^{\pi/2} 3\cos\theta\, 3\cos\theta\, d\theta = 3\int_0^{\pi/2}\cos^2\theta\, d\theta = \frac{3}{2}[\theta + \frac{\sin 2\theta}{2}]_0^{\pi/2} = \frac{3\pi}{4}$.

35. The substitution $x = 3\sin\theta$ would mean that $\sin\theta = 4/3$ at the left hand limit, which is impossible. Rewrite the integral as $-\int_4^5 \frac{dx}{x^2-9}$ and let $x = 3\sec\theta$. This leads to $-\int_{x=4}^{x=5} \frac{3\sec\theta\tan\theta\, d\theta}{9\tan^2\theta} = -\frac{1}{3}\int_{x=4}^{x=5}\csc\theta\, d\theta$ which can be done analytically. To confirm, use NINT: $(-1/3)\text{NINT}(1/\sin t, t, \cos^{-1}(3/4), \cos^{-1}(3/5)) = -0.093269\ldots$; $\text{NINT}(1/(9-x^2), x, 4, 5) = -0.093269\ldots$.

8.5 RATIONAL FUNCTIONS AND PARTIAL FRACTIONS

1. $\frac{5x-13}{(x-3)(x-2)} = \frac{A}{x-3} + \frac{B}{x-2}$ or $5x - 13 = A(x-2) + B(x-3)$ for all x. Set $x = 3$ to obtain $15 - 13 = A \cdot 1$ or $A = 2$; set $x = 2$ to obtain $10 - 13 = B(-1)$ or $B = 3$: $\frac{2}{x-3} + \frac{3}{x-2}$.

3. $\frac{x+4}{(x+1)^2} = \frac{A}{x+1} + \frac{B}{(x+1)^2}$, or $x+4 = A(x+1)+B$. Rewrite, equating coefficients of corresponding terms; $x+4 = Ax + (A+B)$, and solve to find $A = 1, A+B = 4 \Rightarrow B = 3$. $\frac{1}{x+1} + \frac{3}{(x+1)^2}$.

5. $\frac{x+1}{x^2(x-1)} = \frac{A}{x} + \frac{B}{x^2} + \frac{C}{x-1} = \frac{-2}{x} - \frac{1}{x^2} + \frac{2}{x-1}$.

7. Use long division, $x^2 - 5x + 6 \,\overline{\big)\, x^2 + 8}$ with quotient $1 \; R\, 5x+2$ to find $\frac{x^2+8}{x^2-5x+6} = 1 + \frac{5x+2}{(x-3)(x-2)} = 1 + \frac{A}{x-3} + \frac{B}{x-2} \Rightarrow 5x + 2 = A(x-2) + B(x-3)$. Set $x = 3$, then set $x = 2$ to get $17 = A$, $12 = -B$: $1 + \frac{17}{x-3} - \frac{12}{x-2}$.

9. $\int_0^{1/2} \frac{dx}{1-x^2} = \frac{1}{2}\int_0^{1/2}(\frac{1}{1-x} + \frac{1}{1+x})dx = \frac{1}{2}[-\ln(1-x) + \ln(1+x)]_0^{1/2} = \frac{1}{2}[-\ln(1/2) + \ln(3/2)] = \frac{1}{2}\ln 3$.

11. $\frac{y}{y^2-2y-3} = \frac{A}{(y-3)} + \frac{B}{(y+1)} = \frac{3/4}{y-3} + \frac{1/4}{y+1}$; $\int_4^8 \frac{y\,dy}{y^2-2y-3} = [\frac{3}{4}\ln|y-3| + \frac{1}{4}\ln y + 1|]_4^8 = \frac{3}{4}\ln 5 + \frac{1}{4}\ln 9 - 0 - \frac{1}{4}\ln 5 = \frac{1}{2}\ln 5 + \frac{1}{2}\ln 3 = \frac{1}{2}\ln 15$.

13. Let $f(t) = \frac{1}{t^3+t^2-2t} = \frac{1}{t(t^2+t-2)} = \frac{A}{t} + \frac{B}{t+2} + \frac{C}{t-1} = \frac{-1/2}{t} + \frac{1/6}{t+2} + \frac{1/3}{t-1} \Rightarrow \int f(t)dt = -\frac{1}{2}\ln|t| + \frac{1}{6}\ln|t+2| + \frac{1}{3}\ln|t-1| + C = \frac{1}{6}\ln|\frac{(t+2)(t-1)^2}{t^3}| + C$.

15. $\frac{x^3}{x^2+1} = \frac{x^3+x-x}{x^2+1} = x - \frac{x}{x^2+1}$; $\int_0^{2\sqrt{2}} \frac{x^3\,dx}{x^2+1} = \int_0^{2\sqrt{2}} x\,dx - \int_0^{2\sqrt{2}} \frac{x\,dx}{x^2+1} = [\frac{x^2}{2}]_0^{2\sqrt{2}} - \frac{1}{2}\int_0^{2\sqrt{2}} \frac{2x\,dx}{x^2+1} = 4 - [\frac{1}{2}\ln(x^2+1)]_0^{2\sqrt{2}} = 4 - \ln 3$.

17. $\frac{5x^2}{x^2+1} = \frac{5x^2+5-5}{x^2+1} = 5 - \frac{5}{x^2+1}$; $\int_0^{\sqrt{3}} \frac{5x^2}{x^2+1}dx = 5\int_0^{\sqrt{3}}[1 - \frac{1}{x^2+1}]dx = 5[x - \tan^{-1}x]_0^{\sqrt{3}} = 5[\sqrt{3} - \frac{\pi}{3}]$.

19. Let $f(x) = \frac{1}{(x^2-1)^2} = \frac{A}{x-1} + \frac{B}{(x-1)^2} + \frac{C}{x+1} + \frac{D}{(x+1)^2} \Rightarrow 1 = A(x-1)(x+1)^2 + B(x+1)^2 + C(x+1)(x-1)^2 + D(x-1)^2$; setting $x = 1, x = -1$ gives $B = 1/4, D = 1/4$. Differentiating gives $0 = A(x+1)^2 + 2A(x-1)(x+1) + 2B(x-1) + C(x-1)^2 + 2C(x-1)(x+1) + 2D(x-1)$; setting $x = 1, x = -1$ gives $A = -1/4, C = 1/4$. Hence $\int f(x)dx = \frac{-1}{4}\ln|x-1| - \frac{1}{4}\frac{1}{(x-1)} + \frac{1}{4}\ln|x+1| - \frac{1}{4}\frac{1}{(x+1)} + C$.

21. $\int \frac{x+4}{x^2+5x-6}dx = \frac{1}{7}\int(\frac{2}{x+6} + \frac{5}{x-1})dx = \frac{1}{7}[2\ln|x+6| + 5\ln|x-1|] + C = \frac{1}{7}\ln|(x+6)^2(x-1)^5| + C$.

23. $\int \frac{2}{x^2-2x+2}dx = 2\int \frac{dx}{(x-1)^2+1} = 2\tan^{-1}(x-1) + C$.

25. $\frac{x^2-2x-2}{x^3-1} = \frac{A}{x-1} + \frac{Bx+C}{x^2+x+1} \Rightarrow x^2 - 2x - 2 = A(x^2+x+1) + (Bx+C)(x-1)$ evaluating at $x = 1 \Rightarrow A = -1$, evaluating at $x = 0 \Rightarrow C = 1$, evaluating at $x = 2 \Rightarrow B = 2$. $\int \frac{x^2-2x-2}{x^3-1}dx = -\int \frac{dx}{x-1} + \int \frac{2x+1}{x^2+x+1}dx = -\ln|x-1| + \ln|x^2+x+1| + C = \ln|\frac{x^2+x+1}{x-1}| + C$.

27. $\frac{2x^4+x^3+16x^2+4x+32}{(x^2+4)^2} = 2 + \frac{x^3+4x}{(x^2+4)^2} = 2 + \frac{x}{x^2+4}$. Hence, the integral is $2x + \frac{1}{2}\ln(x^2+4) + C$.

29. $\int_0^1 \frac{x}{x+1}dx = \int_0^1 (1-\frac{1}{x+1})dx = [x - \ln|x+1|]_0^1 = 1 - \ln 2 = 0.307.$

31. $\int_{\sqrt{2}}^3 \frac{2x^3}{x^2-1}dx = \int_{\sqrt{2}}^3 (2x+\frac{2x}{x^2-1})dx = [x^2+\ln|x^2-1|]_{\sqrt{2}}^3 = 9+\ln 8-(2) = 7+\ln 8.$

33. $V = \pi \int_{1/2}^{5/2} \frac{9}{3x-x^2}dx = 9\pi \int_{1/2}^{5/2} \frac{dx}{(3-x)x} = \frac{9\pi}{3} \int_{1/2}^{5/2} (\frac{1}{3-x}+\frac{1}{x})dx$; $\frac{1}{(3-x)(x)} = \frac{A}{3-x} + \frac{B}{x}$; $B = 1/3,\ A = 1/3$; $V = 3\pi[-\ln|3-x| + \ln|x|]_{1/2}^{5/2} = 3\pi[[-\ln\frac{1}{2} + \ln\frac{5}{2}] - [-\ln\frac{5}{2} + \ln\frac{1}{2}]] = 3\pi[2\ln 5/2 - 2\ln 1/2] = 3\pi \ln 25.$

35. (a) Rewrite the equation as (∗) $\frac{250dx}{x(1000-x)} = dt$. To find the antiderivative of the left-hand-side, use the method of partial fractions to write $\frac{250}{x(1000-x)} = \frac{A}{x} + \frac{B}{1000-x} = \frac{1/4}{x} + \frac{1/4}{1000-x}$. Hence $\int \frac{250}{x(1000-x)}dx = \frac{1}{4}[\ln x - \ln(1000-x)] + C' = \frac{1}{4}\ln[\frac{Cx}{1000-x}]$. Since the integral of $dt = t$; $\frac{1}{4}\ln\frac{Cx}{1000-x} = t$ or $\frac{Cx}{1000-x} = e^{4t}$; when $t = 0,\ x = 2$; hence $\frac{C\cdot 2}{998} = 1$ or $C = 499$. Solve for x: $499x = (1000-x)e^{4t} \Rightarrow x = \frac{1000e^{4t}}{499+e^{4t}}$.

(b) Set $x = 500$ and solve for t: $t = \frac{1}{4}\ln(499) = 1.55$ days.

37. $\int_0^{\pi/2} \frac{dx}{1+\sin x} = \int_0^1 \frac{2dz}{(1+z^2)(1+\frac{2z}{1+z^2})} = 2\int_0^1 \frac{dz}{1+z^2+2z} = 2\int_0^1 \frac{dz}{(z+1)^2} = 2[\frac{(z+1)^{-1}}{-1}]_0^1 = 2(-\frac{1}{2}+1) = 1$

39. $\int \frac{dx}{1-\sin x} = 2\int \frac{dz}{(z-1)^2} = \frac{2}{-1}\frac{1}{(z-1)} + C = \frac{2}{1-\tan(x/2)} + C$

41. $\int \frac{\cos x\, dx}{1-\cos x} = \int \frac{1-z^2}{1+z^2}\frac{1}{[1-\frac{1-z^2}{1+z^2}]}\frac{2dz}{1+z^2} = 2\int \frac{(1-z^2)dx}{(1+z^2-1+z^2)(1+z^2)} = 2\int \frac{(1-z^2)}{2z^2(1+z^2)}dz = 2\int \frac{1+z^2-2z^2}{2z^2(1+z^2)}dz = \int \frac{dz}{z^2} - 2\int \frac{dz}{1+z^2} = -\frac{1}{z} - 2\tan^{-1} z + C = -\cot(x/2) - 2(x/2) + C = -\cot(x/2) - x + C$

43. $\int \frac{dx}{\sin x - \cos x} = \int \frac{2dz}{(1+z^2)}\frac{1}{\frac{2z-(1-z^2)}{(1+z^2)}} = 2\int \frac{dz}{z^2+2z-1} = 2\int \frac{dz}{z^2+2z+1-2} = 2\int \frac{dz}{(z+1)^2-2} = (-2)\frac{1}{2\sqrt{2}}\ln|\frac{z+1+\sqrt{2}}{z+1-\sqrt{2}}| + C = \frac{1}{\sqrt{2}}\ln|\frac{z+1-\sqrt{2}}{z+1+\sqrt{2}}| + C = \frac{1}{\sqrt{2}}\ln|\frac{\tan(x/2)+1-\sqrt{2}}{\tan(x/2)+1+\sqrt{2}}| + C$

8.6 IMPROPER INTEGRALS

1. $\int_0^\infty \frac{dx}{x^2+4} = \lim_{b\to\infty} \frac{1}{2}[\tan^{-1}\frac{x}{2}]_0^b = \frac{1}{2}\lim_{b\to\infty} \tan^{-1}\frac{b}{2} - \frac{1}{2}\tan^{-1} 0 = \frac{1}{2}\cdot\frac{\pi}{2} = \frac{\pi}{4}.$

3. $\int_{-1}^1 \frac{dx}{x^{2/3}} = \int_{-1}^0 \frac{dx}{x^{2/3}} + \int_0^1 \frac{dx}{x^{2/3}} = \lim_{b\to 0^-} \int_{-1}^b \frac{dx}{x^{2/3}} + \lim_{c\to 0^+} \int_c^1 \frac{dx}{x^{2/3}} = \lim_{b\to 0^-}[3x^{1/3}]_{-1}^b + \lim_{c\to 0^+}[3x^{1/3}]_c^1 = 0 - (-3) + 3 - 0 = 6.$

5. $\int_0^4 \frac{dx}{\sqrt{4-x}} = \lim_{b\to 4^-} \int_0^b (4-x)^{-1/2}dx = \lim_{b\to 4^-} -[\frac{2(4-x)^{1/2}}{1}]_0^b = 4.$

7. $\int_0^1 \frac{dx}{x^{0.999}} = \lim_{b\to 0^+} \int_b^1 x^{-0.999}dx = \lim_{b\to 0^+} \frac{x^{0.001}}{0.001}\Big]_b^1 = 1000.$

9. $\int_2^\infty \frac{2}{x^2-x}dx = \lim_{b\to\infty} \int_2^b \frac{2}{x^2-x}dx = \lim_{b\to\infty} \int_2^b [\frac{-2}{x}+\frac{2}{x-1}]dx = \lim_{b\to\infty}[-2\ln|x| + 2\ln|x-1|]_2^b = \lim_{b\to\infty}[2\ln|\frac{x-1}{x}|]_2^b = 0 - 2\ln(\frac{1}{2}) = \ln 4.$

11. $\int_1^\infty \frac{dx}{x^{1/3}} = \lim_{b\to\infty} \frac{x^{2/3}}{2/3}\Big]_1^b$; diverges.

13. $\int_1^\infty \frac{dx}{x^3+1}$ converges by the Domination Test; $\frac{1}{x^3+1} < \frac{1}{x^3}$; see #12.

15. $\int_0^\infty \frac{dx}{x^{3/2}+1} = \int_0^1 \frac{dx}{x^{3/2}+1} + \int_1^\infty \frac{dx}{x^{3/2}+1}$; the first integral is finite, the second is dominated by $\int_1^\infty \frac{dx}{x^{3/2}}$ which converges; converges.

17. $\int_0^{\pi/2} \tan x\, dx = \lim_{b\to\frac{\pi}{2}^-} \int_0^b \frac{\sin x}{\cos x}dx = \lim_{b\to\frac{\pi}{2}^-}[-\ln|\cos x|]_0^b$; diverges.

19. $\int_{-1}^1 \frac{dx}{x^{2/5}} = \int_{-1}^0 \frac{dx}{x^{2/5}} + \int_0^1 \frac{dx}{x^{2/5}}$; both integrals converge; converges.

21. $\int_2^\infty \frac{dx}{\sqrt{x-1}} = \lim_{b\to\infty}[2(x-1)^{1/2}]_2^b$; diverges.

23. $\int_0^2 \frac{dx}{1-x^2} = \int_0^1 \frac{dx}{1-x^2} + \int_1^2 \frac{dx}{1-x^2} = \lim_{b\to 1^-} \frac{1}{2}\int_0^b [\frac{1}{1-x} + \frac{1}{1+x}]dx + \lim_{c\to 1^+} \frac{1}{2}\int_c^2 [\frac{1}{1-x} + \frac{1}{1+x}]dx = \lim_{b\to 1^-}[\frac{1}{2}\ln|\frac{1+x}{1-x}|]_0^b + \lim_{c\to 1^+}[\frac{1}{2}\ln|\frac{1+x}{1-x}|]_c^2$; both integrals diverge; diverges.

25. $\int_0^\infty \frac{dx}{\sqrt{x^6+1}} = \int_0^1 \frac{dx}{\sqrt{x^6+1}} + \int_1^\infty \frac{dx}{\sqrt{x^6+1}}$; the first integral is finite, the second is dominated by $\int_1^\infty \frac{dx}{x^3}$ which converges; converges.

27. $\int_0^\infty x^2 e^{-x}dx = \lim_{b\to\infty}[e^{-x}(-x^2-2x-2)]_0^b = \lim_{b\to\infty}[\frac{-b^2-2b-2}{e^b}] + 2 = 2$; converges.

29. $\frac{1}{x} \le \frac{2+\cos x}{x}$; $\int_\pi^\infty \frac{dx}{x}$ diverges $\Rightarrow \int_\pi^\infty \frac{2+\cos x}{x}dx$ diverges by the Domination Test.

31. Diverges by the Limit Comparison Test since $\int_6^\infty \frac{1}{\sqrt{x}}dx$ diverges and $\lim_{x\to\infty} \frac{\frac{1}{\sqrt{x}}}{\frac{1}{\sqrt{x+5}}} = 1.$

33. Converges by the Limit Comparison test since $\int_2^\infty \frac{dx}{x^2}$ converges and $\lim_{x\to\infty} \frac{\frac{1}{x^2}}{\frac{2}{x^2-1}} = \frac{1}{2}.$

35. $2 \le x \Rightarrow \ln x \le x \Rightarrow \frac{1}{x} \le \frac{1}{\ln x}$; diverges by the Domination Test since $\int_2^\infty \frac{dx}{x}$ diverges.

37. Converges by the Limit Comparison Test since $\int_1^\infty \frac{dx}{e^x}$ converges and $\lim_{x\to\infty} \frac{\frac{1}{e^x}}{\frac{1}{e^x-2^x}} = \lim_{x\to\infty}(1-(\frac{2}{e})^x) = 1.$

39. $\int_0^\infty \frac{dx}{\sqrt{x+x^4}} = \int_0^1 \frac{dx}{\sqrt{x+x^4}} + \int_1^\infty \frac{dx}{\sqrt{x+x^4}}$; $\frac{1}{\sqrt{x+x^4}} < \frac{1}{\sqrt{x^4}} = \frac{1}{x^2}$ and $\int_1^\infty \frac{dx}{x^2}$ converges $\Rightarrow$ the second integral converges. $\frac{1}{\sqrt{x+x^4}} < \frac{1}{\sqrt{x}}$ and $\int_0^1 \frac{1}{\sqrt{x}}$ converges $\Rightarrow$ the first integral converges. Converges

41. $\int_0^\infty e^{-2x}dx = \lim_{b\to\infty}[-\frac{1}{2}e^{-2x}]_0^b$; converges.

43. $\int_{-3}^\infty x^2e^{-2x}dx = \lim_{b\to\infty}[e^{-2x}(-\frac{1}{2}x^2 - \frac{1}{2}x - \frac{1}{4})]_{-3}^b$ converges (use L'Hôpital's Rule).

45. a) 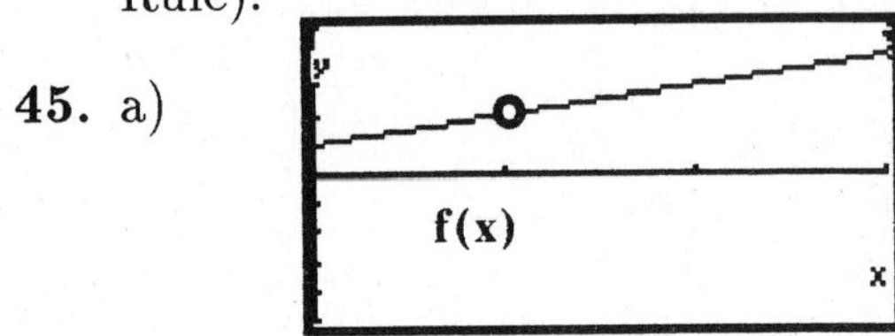

b) $\int_0^3 \frac{x^2-1}{x-1}dx = \int_0^1 \frac{x^2-1}{x-1}dx + \int_1^3 \frac{x^2-1}{x-1}dx$

$= \lim_{b\to1^-}[\frac{x^2}{2} + x]_0^b + \lim_{b\to1^+}[\frac{x^2}{2} + x]_b^3$

$= 7.5.$

47. $x > 3 \Rightarrow e^{-x\cdot x} < e^{-3\cdot x} \Rightarrow \int_3^\infty e^{-x^2}dx < \int_3^\infty e^{-3x}dx = \lim_{b\to\infty}[-\frac{1}{3}e^{-3x}]_3^b = \frac{1}{3}e^{-9} = 0.000042$. Using NINT, $\int_0^3 e^{-x^2}dx = 0.886217\ldots$; thus $\int_0^\infty e^{-x^2}dx = 0.886217$ with an error of at most 0.000042.

49. $\int_1^\infty x^{-p}dx = \lim_{b\to\infty}[\frac{x^{-p+1}}{1-p}]_1^b = -\frac{1}{1-p} = \frac{1}{p-1}$ if $p > 1$; if $p < 1$, $-p+1 > 1$ and the integral is infinite.

51. Let $u = \ln x$; $du = \frac{dx}{x} \Rightarrow \int_1^2 \frac{dx}{x(\ln x)^p} = \int_0^{\ln 2} \frac{du}{u^p}$; the integral converges when $p < 1$ and diverges if $p \geq 1$. (cf #50).

53. $A = \lim_{b\to\infty} \int_0^b e^{-x}dx = \lim_{b\to\infty}[-\frac{1}{e^b} + \frac{1}{e^0}] = 1.$

55. (use shells) $V = 2\pi \int xe^{-x}dx = 2\pi \cdot 1$ (see #54).

57. 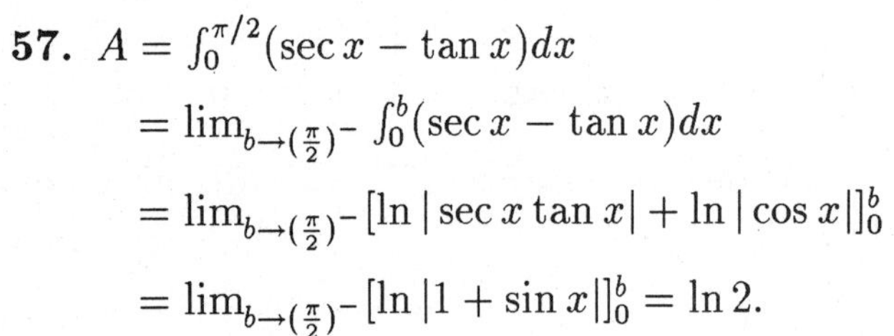

$A = \int_0^{\pi/2}(\sec x - \tan x)dx$

$= \lim_{b\to(\frac{\pi}{2})^-} \int_0^b(\sec x - \tan x)dx$

$= \lim_{b\to(\frac{\pi}{2})^-}[\ln|\sec x \tan x| + \ln|\cos x|]_0^b$

$= \lim_{b\to(\frac{\pi}{2})^-}[\ln|1 + \sin x|]_0^b = \ln 2.$

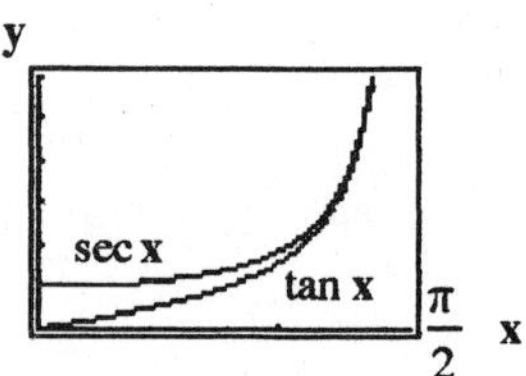

59. $\int_2^\infty \frac{dx}{\sqrt{x^2+1}}$ diverges by the Limit Comparison Theorem since $\lim_{x\to\infty} \frac{\frac{1}{x}}{\frac{1}{\sqrt{x^2+1}}} = 1$ and $\int_2^\infty \frac{dx}{x}$ diverges.

61. Let $u = e^x$; $\int_{-\infty}^\infty \frac{dx}{e^x+e^{-x}} = \int_0^\infty \frac{du}{u^2+1} = \lim_{b\to\infty} \tan^{-1} u|_0^b = \frac{\pi}{2}$; converges.

63. By the Limit Comparison Test, $\int_0^\infty \frac{2x}{x^2+1}dx$ and $\int_0^\infty \frac{2}{x}dx$ both diverge. However $\int_{-b}^b \frac{2x}{x^2+1}dx = 0$ for all finite b since the integrand is odd.

8.7 DIFFERENTIAL EQUATIONS

1. a) $y = x^2$, $y' = 2x$, $y'' = 2 \Rightarrow xy'' - y' = x \cdot 2 - 2x = 0$; b) $y = 1$, $y' = 0$, $y'' = 0 \Rightarrow xy'' - y' = x \cdot 0 - 0 = 0$; c) $y = c_1x^2 + c_2$, $y' = 2c_1x$, $y'' = 2c_1 \Rightarrow xy'' - y' = x \cdot 2c_1 - 2c_1x = 0$.

3. a) $y = e^{-x}$, $y' = e^{-x}$, $y'' = e^{-x} \Rightarrow 2y' + 3y = 2(-e^{-x}) + 3e^{-x} = e^{-x}$; b) $y = e^{-x} + e^{-(3/2)x}$, $y' = -e^{-x} - \frac{3}{2}e^{-(3/2)x}$, $y'' = e^{-x} + \frac{9}{4}e^{-(3/2)x} \Rightarrow 2y' + 3y = 2(-e^{-x} - \frac{3}{2}e^{-(3/2)x}) + 3(e^{-x} + e^{-(3/2)x}) = e^{-x}$; c) $y = e^{-x} + Ce^{-(3/2)x}$, $y' = -e^{-x} - \frac{3}{2}Ce^{-(3/2)x}$, $y'' = e^{-x} + \frac{9}{4}Ce^{-(3/2)x} \Rightarrow 2y' + 3y = 0$.

5. $y = 160x - 16x^2 \Rightarrow y' = 160 - 32x$, $y'' = -32$; $y'' = -32$, thus y is a solution of the differential equation. $y(5) = 160 \cdot 5 - 16 \cdot 25 = 400$; $y'(5) = 160 - 32 \cdot 5 = 0$.

7. $y = 3\cos 2t - \sin 2t \Rightarrow y' = -6\sin 2t - 2\cos 2t$, $y'' = -12\cos 2t + 4\sin 2t$; $y'' + 4y = -12\cos 2t + 4\sin 2t + 12\cos 2t - 4\sin 2t = 0$; $y(0) = 3 \cdot 1 - 0 = 3$, $y'(0) = -6 \cdot 0 - 2 \cdot 1 = -2$.

9. a) $\frac{1}{1000+0.10x}\frac{dx}{dt} \cdot dt = 1 \cdot dt \Rightarrow \int \frac{1}{1000+0.10x}dx = \int 1 \cdot dt \Rightarrow \frac{1}{0.10}\ln(1000+0.10x) = t + C' \Rightarrow \ln(1000 + 0.10x) = 0.10(t + C') = 0.10t + C'' \Rightarrow 1000 + 0.10x = Ce^{0.10t}$. Using $x(0) = 1000$ gives $1000 + 100 = C$, or $x = \frac{1100e^{0.10t} - 1000}{0.10} = 11000e^{0.10t} - 10000$ b) $100000 = 11000e^{0.10t} - 10000 \Rightarrow 10 = e^{0.10t} \Rightarrow t = \frac{\ln 10}{0.10} = 23.026$ yrs.

11. $\frac{dy}{y-95} = 0.017dx \Rightarrow \ln|y - 95| = 0.017x + C' \Rightarrow |y - 95| = Ce^{0.017x}$; $y(0) = 30 \Rightarrow |30 - 95| = 65 = Ce^0$, so that $|y - 95| = 65e^{0.017x}$. Since $y(0) < 95$, for x near 0, $y < 95$ and $|y - 95| = 95 - y \Rightarrow y = 95 - 65e^{0.017x}$. The graph of y agrees with that produced by IMPEULG, with $y1 = 0.017(y-95)$, $y(0) = 30$.

13. Using partial fractions, $\frac{dy}{y(350-y)} = 0.00125dt$ becomes: $\frac{1}{350}[\frac{1}{y} + \frac{1}{350-y}]dy = 0.00125dt \Rightarrow \frac{1}{350}[\ln|y| - \ln|350 - y|] = 0.00125t + C' \Rightarrow \ln|\frac{y}{350-y}| = 0.4375t + C'' \Rightarrow \frac{y}{350-y} = Ce^{0.4375t}$; $\frac{30}{350-30} = C = \frac{3}{32}$. Finally $y = (350 - y)\frac{3}{32}e^{0.4375t} \Rightarrow y = \frac{\frac{350 \cdot 3}{32}e^{0.4375t}}{1+\frac{3}{32}e^{0.4375t}} = \frac{1050e^{0.4375t}}{32+3e^{0.4375t}}$.

15. $y' = e^{-2y} \Rightarrow e^{2y}dy = dx \Rightarrow \frac{1}{2}e^{2y} = x + C$; $y(0) = 0 \Rightarrow \frac{1}{2} \cdot 1 = C \Rightarrow e^{2y} = 2(x + \frac{1}{2})$ or $y = \frac{1}{2}\ln(2x + 1)$.

17. $y' = \frac{1}{x^2+1} \Rightarrow y = \arctan x + C$, $y(0) = 1 \Rightarrow 1 = 0 + C$; $y = \arctan x + 1$.

19. $y' = xe^x \Rightarrow y = xe^x - e^x + C$; $y(0) = 0 \Rightarrow C = 1$; $y = xe^x - e^x + 1$.

21. By example 2, $y = C_1 \cos x + C_2 \sin x$; $y' = -C_1 \sin x + C_2 \cos x$; $y(0) = 0$ and $y'(0) = 1$ give $0 = C_1 + 0$, $1 = 0 + C_2$; hence $y = \sin x$.

23. $y = C_1 e^x \cos x + C_2 e^x \sin x$; $y' = C_1(e^x \cos x - e^x \sin x) + C_2(e^x \sin x + e^x \cos x)$; $y'' = C_1(-2e^x \sin x) + C_2(2e^x \cos x)$; $y'' - 2y' + 2y = C_1(-2e^x \sin x) + C_2(2e^x \cos x) - 2C_1 e^x \cos x + 2C_1 e^x \sin x - 2C_2 e^x \sin x - 2C_2 e^x \cos x + 2C_1 e^x \cos x + 2C_2 e^x \sin x = e^x \sin x(-2C_1 + 2C_1 - 2C_2 + 2C_2) + e^x \cos x(2C_2 - 2C_1 - 2C_2 + 2C_1) = 0$.

25. $y(0) = 1$; $y(0.2) \approx y_1 = 1 + 0.2 \cdot 1 = 1.2$; $y(0.4) \approx y_2 = 1.2 + 0.2(1.2) = 1.44$; $y(0.6) \approx y_3 = 1.44 + 0.2(1.44) = 1.728$; $y(0.8) \approx y_4 = 1.728$; $y(1) \approx y_5 = 2.0736$. If $\frac{dy}{dt} = y$, $y(0) = 1$, then $\frac{dy}{y} = dt \Rightarrow \ln|y| = t + C' \Rightarrow y = Ce^t$; $y(0) = 1 \Rightarrow C = 1 \Rightarrow y(t) = e^t$; $y(1) = e^1 = 2.71828$.

27. Using IMPEULT, gives $y(1) \approx y_5 = 2.7027\ldots$.

29. First solve $y' = y^2$, $y(0) = 1$: $y^{-2}dy = dx \Rightarrow -y^{-1} = x + C$; when $x = 0 - 1^{-1} = 0 + C$, so that $-\frac{1}{y} = x - 1$ or $y = \frac{1}{1-x}$ which becomes infinite at $x = 1$. To compare the functions, set $y1 = x^2 + y^2$, $y^2 = 1/(1-x)$; set the range to $[0, 1]$ by $[-10, 100]$. Run the program RUNKUTG — take $h = 0.02$ to obtain the graph of $y1$. While the dotted graph is showing, use DrawF to graph $y2$.

31. The computed values of $y(2)$ are:

Euler, 1.271428571;

Improved Euler, 1.28571428571;

Runge-Kutta, 1.28571428571

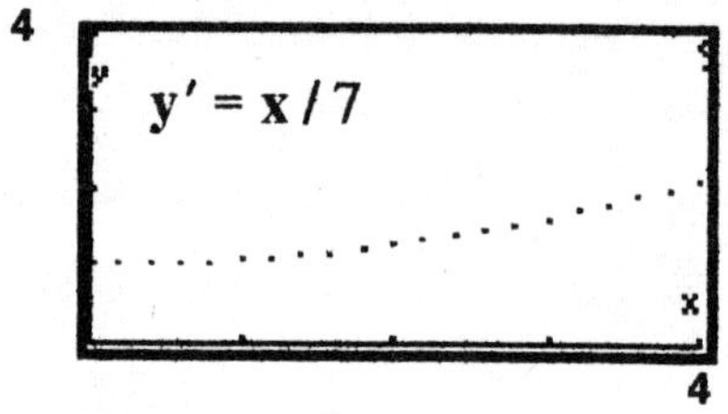

33. The computed values of $y(2)$ are:

Euler, 2.07334340632;

Improved Euler, 2.21157126441;

Runge-Kutta, 2.20283462521

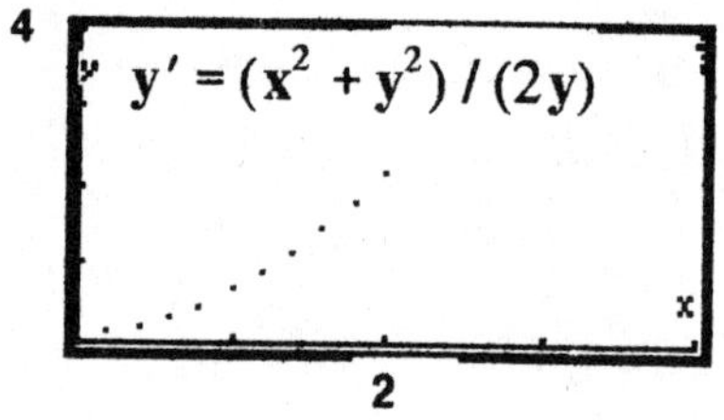

35. a) $y(1) \approx 0.310268270416$ if $h = 0.05$ (20 steps); b) $y(1) \approx 0.310268299767$ if $h = 0.025$ (40 steps).

37. $y = \int_a^x -2t \sec^2(a^2 - t^2)dt$, $y(a) = 0 \Rightarrow y = \tan(a^2 - x^2)$; $y' = -2x \sec^2(a^2 - x^2) = -2x(\tan^2(a^2 - x^2) + 1) = -2x(y^2 + 1)$.

39. $y = \tan(a^2 - x^2)$, $y(1) = 1 \Rightarrow 1 = \tan(a^2 - 1) \Rightarrow a^2 - 1 = \arctan 1 = \frac{\pi}{4} \Rightarrow a = \sqrt{\frac{\pi}{4} + 1} \approx 1.336$.

41. Using step size 0.5, $N = 6$, the corresponding entries for $x = 3$ are:

	x	y(R-K)	y(true)	Difference
$y' - x - y$ $y(0) = 1$	3	2.09981094687	$\cdots$	2.37×10^{-4}
$y' = x - y$ $y(0) = -2$	3	1.95009452656	$\cdots$	-1.18×10^{-4}

The errors are much greater because the step size is five times as large.

43. a) $\frac{dy}{y(250-y)} = 0.0004dt$; $\frac{1}{250}\left[\frac{1}{y} - \frac{1}{250-y}\right] = 0.0004t + C'$; $\ln \frac{y}{250-y} = 0.1t + C'$; $\frac{y}{250-y} = Ce^{0.1t}$; $y(0) = 28 \Rightarrow C = \frac{28}{222} = \frac{14}{111}$; $y = (250 - y)\frac{14}{111}e^{0.1t}$; $111y = (250 - y)14e^{0.1t}$; $y(111 + 14e^{0.1t}) = 250 \cdot 14e^{0.15}$; $y = \frac{3500e^{0.1t}}{111+14e^{0.1t}}$

e) $y = 100$ at $t = 16.65$ years, $y = 200$ at $t = 34.57$ years

f) at 92 years, $y > 249$.

45. $\frac{\pi}{4}a^2 = 1 \Rightarrow a = \pm\frac{2}{\sqrt{\pi}}$.

47. Your company invents a new game. Let $P(t)$ be the number sold at time t. At first the sales are exponential, but eventually the market approaches saturation. If $\frac{dP}{dt} = aP(b - P)$, the inflection point will be when $\frac{dP}{dt}$, the rate at which the game is selling, is the greatest.

8.8 COMPUTER ALGEBRA SYSTEMS

1. (Using *Mathematica*)

```
In[2]:= Integrate[Exp[-x^2],{x,0,Infinity}]

         Sqrt[Pi]
Out[2]= --------
            2
```

3.

```
In[5]:= Integrate[1/(x Sqrt[x-3]),{x,6,9}]

           -Pi        2 ArcTan[Sqrt[2]]
Out[5]= --------- + -----------------
        2 Sqrt[3]       Sqrt[3]
```

5.

```
In[7]:= Integrate[1/(9-x^2)^2,x]

            -x            Log[-3 + x]   Log[3 + x]
Out[7]= ------------  -  -----------  + ----------
                  2          108           108
        18 (-9 + x )
```

7.

```
In[9]:= Integrate[1/(x^2 Sqrt[7+x^2]),{x,3,11}]

         4    8 Sqrt[2]
Out[9]= -- - ---------
        21       77
```

9.

```
In[11]:= Integrate[(Sqrt[x^2-2])/x,{x,-2,-Sqrt[2]}]

            Pi       4 + Pi
Out[11]= ------- - ---------
         Sqrt[2]   2 Sqrt[2]
```

11.

```
In[13]:= Integrate[1/(4+5Sin[2x]),x]

          -Log[2 Cos[x] + Sin[x]]     Log[Cos[x] + 2 Sin[x]]
Out[13]=  -----------------------  +  ----------------------
                     6                          6
```

13.

```
In[15]:= Integrate[x Sqrt[2x-3],x]

                                         2
                          3    x    2 x
Out[15]= Sqrt[-3 + 2 x] (-(-) - - + ----)
                          5    5     5
```

15.

```
In[17]:= Integrate[x^10 Exp[-x],{x,0,Infinity}]

Out[17]= 3628800
```

17.

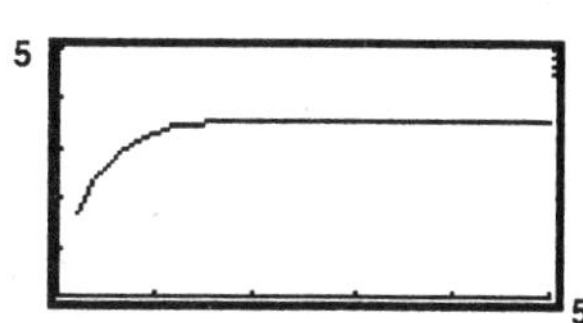

19.

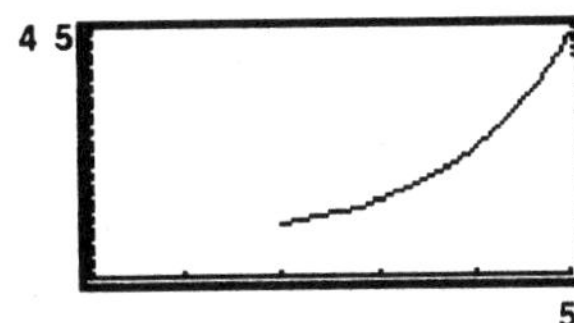

21.

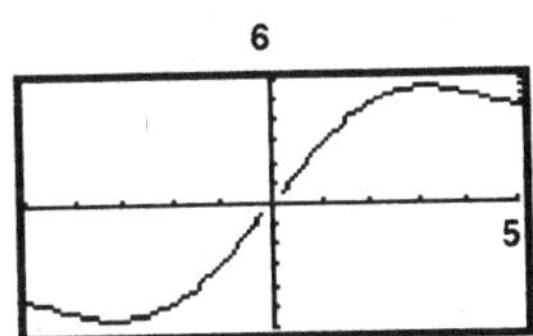

23.

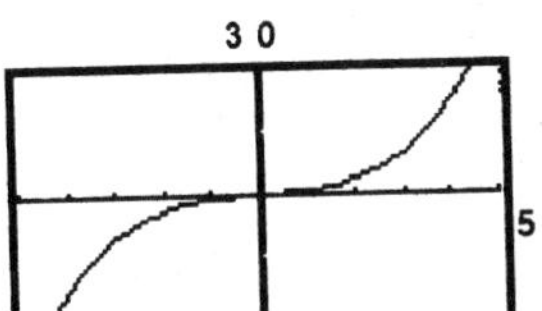

25. Since $y' = \sin^{-1}\sqrt{x}$, $y(0) = 0$, first use IMPEULT on $0 \leq x \leq 3$ to locate the solution. Set $y1 = \sin^{-1}\sqrt{x}$, take $h = 0.1$. The solution appears to be between 0.7 and 0.8. Graph $y1 = 0.5$, $y2 = \text{NINT}(\sin^{-1}\sqrt{t}, t, 0, x)$ on $[0.7, 0.8]$ by $[0.4, 0.6]$. When the graphs are completed, use trace to find $x = 0.7705$.

27. Use IMPEUT on $y' = \sqrt{(1+x^4)}$ with stepsize $= 0.01$ to locate x between 0.49 and 0.50. Graph $y1 = 0.5$, $y2 = \text{NINT}(\sqrt{(1+T\wedge 4)}, T, 0, x)$ on $[0.49, 0.50]$ by $[0.4, 0.6]$ to find $x = 0.498730$.

29. Graph $f' = y = xe^{-x}$ on $[0,5]$ by $[0,0.5]$. Since $f' > 0$, f is always increasing, f' increasing on $[0,1] \Rightarrow f$ is concave up; there is an inflection point at $x = 1$; $f' \to 0$ as $x \to \infty \Rightarrow f$ approaches a constant value.

31. Since $\delta = 1$, $M = \int_0^3 \frac{1}{\sqrt{x+1}} dx$; $M_y = \int_0^3 \frac{x}{\sqrt{x+1}} dx$, $M_y = \int_0^3 \frac{2}{x+1} dx$. Using NINT, $\bar{x} = M_y/M \approx 1.33333333143$, $\bar{y} = 1.38629436004$.

33. Let $y = \sqrt{\frac{2a-x}{x}} = (2a-x)^{1/2}x^{-1/2}$; $y' = \frac{-1}{2}(2a-x)^{-1/2}x^{-1/2} + (2a-x)^{1/2}(-\frac{1}{2})x^{-3/2} = \frac{-1}{2}x^{-1/2}[\frac{1}{\sqrt{2a-x}} + \frac{\sqrt{2a-x}}{x}] = \frac{-1}{2\sqrt{x}}[\frac{x+(2a-x)}{x\sqrt{2a-x}}] = \frac{-a}{x\sqrt{2ax-x^2}}$. Hence $\frac{d}{dx}[-\frac{1}{a}\sqrt{\frac{2a-x}{x}}] = \frac{1}{x\sqrt{2ax-x^2}}$.

35. If $u = ax+b$, $x = (u-b)/a$, $dx = \frac{1}{a}du$. The integral becomes $\frac{1}{a^2}\int \frac{(u-b)du}{u^2} = \frac{1}{a^2}\int [\frac{1}{u} - \frac{b}{u^2}]du = \frac{1}{a^2}[\ln|u| + \frac{b}{u}] + C = \frac{1}{a^2}[\ln|ax+b| + \frac{b}{ax+b}] + C$.

PRACTICE EXERCISES, CHAPTER 8

1. Let $u = 1 + \sin x$, $du = \cos x\, dx$, $0 \le x \le \frac{\pi}{2}$, $1 \le u \le 2$; $\int_0^{\pi/2} \frac{\cos x\, dx}{(1+\sin x)^{1/2}} = \int_1^2 u^{-1/2} du = [\frac{u^{1/2}}{\frac{1}{2}}]_1^2 = 2[\sqrt{2} - 1]$.

3. $\int_{-1}^{1} \frac{2y}{y^4+1} dy = 0$. The integrand is odd.

5. Let $u = \sin^{-1} x$, $du = (1/\sqrt{1-x^2})dx$; $0 \le x \le \sqrt{2}/2 \Rightarrow 0 \le u \le \frac{\pi}{4}$; $\int_0^{\sqrt{2}/2} \frac{\sin^{-1} x}{\sqrt{1-x^2}} dx = \int_0^{\pi/4} u\, du = [\frac{u^2}{2}]_0^{\pi/4} = \frac{\pi^2}{32}$.

7. $\int_{\pi/4}^{\pi/3} \frac{dx}{2\sin x \cos x} = \int_{\pi/4}^{\pi/3} \frac{1}{\sin 2x} dx = \int_{\pi/4}^{\pi/3} \csc 2x dx = \frac{1}{2}[\ln|\csc 2x + \cot 2x|]_{\pi/4}^{\pi/3} = \frac{1}{2}[\ln|\csc \frac{2\pi}{3} + \cot \frac{2\pi}{3}| - \ln|\csc \frac{\pi}{2} + \cot \frac{\pi}{2}|] = \frac{1}{2}[\ln|\frac{2\sqrt{3}}{3} - \frac{\sqrt{3}}{3}| - \ln|1+0|] = \frac{1}{4}\ln 3$.

9. $\int \frac{x+4}{x^2+1} dx = \frac{1}{2}\int \frac{2x}{x^2+1} dx + 4\int \frac{dx}{x^2+1} = \frac{1}{2}\ln(x^2+1) + 4\tan^{-1} x + C$.

11. Let $u = \ln x$; $dv = x^2 dx$; $du = \frac{dx}{x}$; $v = \frac{x^3}{3}$. $\int x^2 \ln x\, dx = \frac{x^3}{3}\ln x - \int \frac{x^2}{3} dx = \frac{x^3}{3}\ln x - \frac{x^3}{9} + C$.

13. $\int x^5 \sin x\, dx = -x^5 \cos x + 5x^4 \sin x + 20x^3 \cos x - 60x^2 \sin x + 120x(-\cos x) + 120 \sin x + C$, using tabular integration.

15. Let $u = \cos 2x$; $dv = e^x dv$; $du = -2\sin 2x$; $v = e^x \Rightarrow I = \int e^x \cos 2x\,dx = e^x \cos 2x + 2\int e^x \sin 2x dx = e^x \cos 2x + 2[e^x \sin 2x - 2I]$, after another integration by parts, $I = \frac{1}{5}[e^x \cos 2x + 2e^x \sin 2x] + C$.

17. $\int \sin^3 y\,dy = \int (1-\cos^2 y)\sin y\,dy = -\cos y + \frac{\cos^3 y}{3} + C$.

19. $\int \sin^4 x(1-\sin^2 x)dx = \int \sin^4 x\,dx - \int \sin^6 x\,dx$; use formula (60) $= \int \sin^4 x\,dx - [-\frac{\sin^5 x\cos x}{6} + \frac{5}{6}\int \sin^4 x\,dx] = \frac{\sin^5 x\cos x}{6} + \frac{1}{6}\int \sin^4 x\,dx = \frac{\sin^5 x\cos x}{6} + \frac{1}{6}[-\frac{\sin^3 x\cos x}{4} + \frac{3}{4}\int \sin^2 x\,dx] = \frac{\sin^5 x\cos x}{6} - \frac{1}{24}\sin^3 x\cos x + \frac{1}{8}\cdot\frac{1}{2}[x - \frac{\sin 2x}{2}] + C$.

21. $\int_0^\pi \sqrt{\frac{1+\cos 2x}{2}} = \int_0^\pi \sqrt{\cos^2 x}dx = \int_0^\pi |\cos x|dx = 2$.

23. $\int_0^{\pi/3} \tan^3 t\,dt = \int_0^{\pi/3}(\sec^2 t - 1)\tan t\,dt = \int_0^{\pi/3} \sec t(\sec t\tan t)dt - \int_0^{\pi/3} \tan t\,dt = [\frac{\sec^2 t}{2} + \ln|\cos t|]_0^{\pi/3} = 2 + \ln\frac{1}{2} - [\frac{1}{2} + 0] = \frac{3}{2} - \ln 2$.

25. Let $z = 4\tan\theta$; $0 \le z \le 3 \Rightarrow 0 \le \theta \le \arctan(3/4)$; $\int_0^3 \frac{dz}{(16+z^2)^{3/2}} = \frac{1}{16}\int_0^{\arctan(3/4)} \frac{\sec^2\theta\,d\theta}{\sec^3\theta} = \frac{1}{16}\int_0^{\arctan(3/4)} \cos\theta\,d\theta = [\frac{\sin\theta}{16}]_0^{\arctan(3/4)} = \frac{1}{16}\cdot\frac{3}{5} = \frac{3}{80}$.

27. Let $x = \sin\theta$; $\int \frac{dx}{x^2\sqrt{1-x^2}} = \int \frac{\cos\theta\,d\theta}{\sin^2\theta\cos\theta}$

$= \int \csc^2\theta\,d\theta = -\cot\theta + C = -\frac{\sqrt{1-x^2}}{x} + C$

29. Let $x = \sec\theta$; $\int_{5/4}^{5/3} \frac{12dx}{(x^2-1)^{3/2}} = \int_{x=5/4}^{x=5/3} \frac{12\sec\theta\tan\theta\,d\theta}{(\tan^2\theta)^{3/2}} = 12\int_{x=5/4}^{x=5/3} \frac{\sec\theta}{\tan^2\theta}d\theta = 12\int_{x=5/4}^{x=5/3} \frac{\cos\theta}{\sin^2\theta}d\theta = \frac{12(\sin\theta)^{-1}}{-1}\Big]_{x=5/4}^{x=5/3} = -12[\frac{1}{4/5} - \frac{1}{3/5}] = 5$.

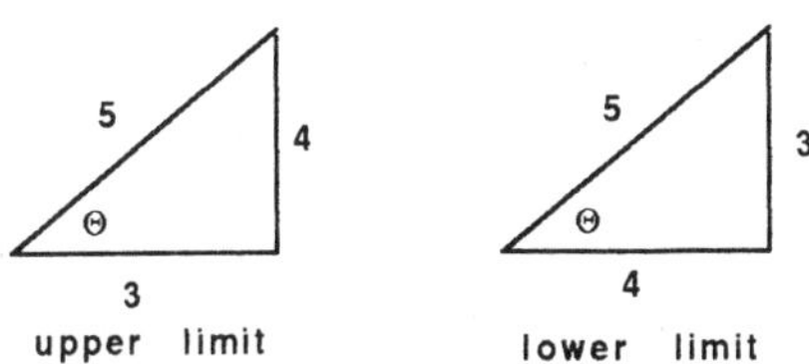

31. $\int_{1/3}^1 \frac{3dx}{9x^2-6x+1+4} = \int_{1/3}^1 \frac{3dx}{(3x-1)^2+2^2} = \int_0^2 \frac{dt}{t^2+2^2} = \frac{1}{2}[\tan^{-1}\frac{t}{2}]_0^2$; let $t = 3x - 1$; $= \frac{1}{2}[\frac{\pi}{4} - 0] = \frac{\pi}{8}$.

33. $\int_0^1 \frac{dx}{(x+1)\sqrt{x^2+2x}} = \int_0^1 \frac{dx}{(x+1)\sqrt{(x+1)^2-1}}$; let $t = x + 1$; $= \int_1^2 \frac{dt}{t\sqrt{t^2-1}} = [\sec^{-1}|t|]_1^2 = \pi/3$.

35. Reduce the integrand: $\frac{x^3+x^2}{x^2+x-2} = x + \frac{2x}{x^2+x-2} = x + \frac{2x}{(x+2)(x-1)} = x + \frac{A}{x+2} + \frac{B}{x-1}$. Since $2x = A(x-1) + B(x+2)$, $-3A = -4$, $3B = 2$. The problem becomes

$\int_2^6 (x + \frac{4/3}{x+2} + \frac{2/3}{x-1})dx = [\frac{x^2}{2} + \frac{4}{3}\ln|x+2| + \frac{2}{3}\ln|x-1|]_2^6 = [18 + \frac{4}{3}\ln 8 + \frac{2}{3}\ln 5] - [2 + \frac{4}{3}\ln 4 + \frac{2}{3}\ln 1] = 16 + \frac{4}{3}\ln\frac{8}{4} + \frac{2}{3}\ln 5 = 16 + \frac{1}{3}\ln(16 \cdot 25) = 16 + (\ln 400)/3$.

37. $\int \frac{x\,dx}{(x-1)^2} = \int \frac{x-1+1}{(x-1)^2}dx = \int(\frac{1}{x-1} + \frac{1}{(x-1)^2})dx = \ln|x-1| - \frac{1}{x-1} + C$.

39. $\frac{4}{x^3+4x} = \frac{4}{x(x^2+4)} = 4[\frac{A}{x} + \frac{Bx+C}{x^2+4}]$. Since $1 = A(x^2+4) + (Bx+C)x$, $x = 0 \Rightarrow A = \frac{1}{4}$; $1 = x^2(A+B) + x \cdot C + 4A \Rightarrow B = -A = -\frac{1}{4}$, $C = 0$. The problem becomes: $4\int[\frac{1/4}{x} + \frac{-1/4x}{x^2+4}]dx = \int[\frac{1}{x} - \frac{x}{x^2+4}]dx = \ln|x| - \frac{1}{2}\ln|x^2+4| + C = \ln\frac{|x|}{\sqrt{x^2+4}} + C$.

41. First, rewrite the integrand $\frac{2}{x(x-1)} = \frac{-1}{x} + \frac{1}{x-2}$. Then write a definite integral with finite upper limit: $\lim_{b\to\infty}\int_3^b[\frac{-1}{x} + \frac{1}{x-2}]dx = \lim_{b\to\infty}[-\ln|x| + \ln|x-2|]_3^b = \lim_{b\to\infty}[\ln|\frac{x-2}{x}|]_3^b = \lim_{b\to\infty}[\ln|1-\frac{2}{x}|]_3^b = \lim_{b\to\infty}[\ln|1-\frac{2}{b}| - \ln|1-\frac{2}{3}|] = -\ln\frac{1}{3} = \ln 3$.

43. $M_x = \int_1^e[\frac{1+\ln x}{2}]2[1-\ln x]dx = \int_1^e[1-(\ln x)^2]dx = \int_1^e dx - \int_1^e(\ln x)^2dx = [(e-1) - [x(\ln x)^2 - 2\int_1^e \ln x\,dx]_1^e$; let $u = (\ln x)^2$, $du = \frac{2\ln x}{x}dx$; $dv = dx$, $v = x$; $= (e-1) - [x(\ln x)^2 - 2[x\ln x - x]]_1^e = (e-1) - [e \cdot 1 - 2(e-e) - (2)] = 1$.

45. Use shells. $V = 2\pi\int_0^1 x \cdot 3x\sqrt{1-x}\,dx = 6\pi\int_0^1 x^2(1-x)^{1/2}dx$; let $u = 1-x$; $= 6\pi\int_1^0(1-u)^2u^{1/2}(-du) = 6\pi\int_0^1(u^{1/2} - 2u^{3/2} + u^{5/2})du = 6\pi[\frac{2}{3} - 2\cdot\frac{2}{5} + \frac{2}{7}] = 32\pi/35$.

47. $y' = \frac{(3x^2+5)}{2\sqrt{x^3+5x}}$; $s = \int_1^8\sqrt{1 + \frac{(3x^2+5)^2}{4(x^3+5x)}}dx$. Use NINT to find $S = 22.25369$. If y is graphed in $[0, 8]$ by $[-10, 30]$, the graph is very close to being the hypoteneuse of a right triangle with sides $8 - 2 = 6$ and 21.7, which is about 22.

49. $V = \pi\int_0^1(-\ln x)^2dx = \pi\lim_{b\to 0^+}\int_b^1(\ln x)^2dx$ (see problem 43) $= \pi\lim_{b\to 0^+}[x(\ln x)^2 - 2(x\ln x - x)]_b^1 = \pi\lim_{b\to 0^+}[2 - b(\ln b)^2 + 2b\ln b - 2b] = \pi[2 - 0 + 0 - 0] = 2\pi$.

51. $\int_{-\infty}^0 \frac{4x}{x^2+1}dx + \int_0^\infty \frac{4x}{x^2+1}dx$ is the sum of divergent integrals. Therefore, $\int_{-\infty}^\infty \frac{4x}{x^2+1}dx$ diverges.

53. Rewrite the integral as $\lim_{c\to-\infty}\int_c^0 \frac{e^{-x}dx}{e^{-x}+e^x} + \lim_{b\to\infty}\int_0^b \frac{e^{-x}dx}{e^{-x}+e^x}$. We will show that the first is divergent: let $u = -e^{-x}$, then $\frac{1}{u} = -e^x$, $du = e^{-x}dx$, and, since $\frac{e^{-x}dx}{e^{-x}+e^x} = \frac{-e^{-x}dx}{-e^{-x}-e^x}$ the integral becomes $\lim_{c\to-\infty}\int_{-e^{-c}}^{-1}\frac{-du}{u+1/u} = \lim_{c\to-\infty}\int_{-1}^{-e^{-c}}\frac{u\,du}{u^2+1} = \lim_{c\to-\infty}[\frac{1}{2}\ln(u^2+1)]_{-1}^{-e^{-c}}$ as $c \to -\infty$, the logarithm becomes infinite.

55. Rewrite as $\frac{dy}{dx} = \frac{e^{y-2}}{e^{x+2y}} = e^{-(2+x+y)}$, $y(0) = -2$. Since $y(2)$ is asked for, set the range to $[0, 2]$ by $[-5, 5]$. Run IMPEULG with $h = 0.1$. To estimate $y(2)$, sue IMPEULT. $y(2) \approx -1.377$. Problems 56-62 are done in a similar fashion. you may need to adjust the Y-range values.

57. Taking $h = 0.1$, $y(6) = -7.349$.

59. $\frac{dy}{dx} = \frac{-(x^2+y)}{e^y+x}$; taking $h = 0.1$ gives $y(3) = -2.691$.

61. $\frac{dy}{dx} = \frac{x-2y}{x+1}$; $y(3) = 0.907$.

63. $\frac{dy}{y-22} = -0.15dx \Rightarrow \ln|y-22| = -0.15x + C' \Rightarrow |y-22| = Ce^{-0.15x}$. $y(0) = 50 \Rightarrow |y-22| = y-22 \Rightarrow y-22 = Ce^{-0.15x}$. Applying the initial condition gives $y = 28e^{-0.15x} + 22$.

65. $\frac{dP}{(500-P)P} = 0.002dt \Rightarrow \frac{1}{500}[\frac{1}{500-P} + \frac{1}{P}]dP = 0.002dt$; $[\frac{dP}{500-P} + \frac{dP}{P}] = 1 \cdot dt \Rightarrow -\ln|500-P| + \ln|P| = t + C' \Rightarrow |\frac{P}{500-P}| = Ce^t$. $P(0) = 20 \Rightarrow \frac{P}{500-P} = Ce^t$ where $C = \frac{20}{480} = \frac{1}{24}$. Solving for P gives $P = \frac{500e^t}{24+e^t}$.

67. $\frac{dx}{x(a-x)} = kdt \Rightarrow \frac{1}{a}[\frac{1}{x} + \frac{1}{a-x}]dx = kdt$ or $\frac{dx}{x} + \frac{dx}{a-x} = kadt$. Assume $x(t_0) = x_0$. Then $\ln|\frac{x}{a-x}| = kat + C'$, or, $|\frac{x}{a-x}| = Ce^{akt}$. Assuming $x_0 < a$, this leads to $x = (a-x)[\frac{x_0}{a-x_0}]e^{akt}$. Multiplying by $(a-x_0)$ gives $x[(a-x_0) + x_0e^{akt}] = ax_0e^{akt}$. Before solving for x, multiply by e^{-akt} to get $x = \frac{ax_0}{x_0+(a-x_0)e^{-akt}}$.

CHAPTER 9

INFINITE SERIES

9.1 LIMITS OF SEQUENCES OF NUMBERS

1. $a_1 = 0,\ a_2 = -0.25,\ a_3 = -0.22222\ldots = -2/9,\ a_4 = -0.1875 = -3/16$

3. $a_1 = (-1)^2/(2\cdot 1 - 1) = 1/1 = 1,\ a_2 = -1/3,\ a_3 = 1/5,\ a_4 = -1/7$

5. $x_1 = 1,\ x_2 = 1 + \frac{1}{2} = \frac{3}{2},\ x_3 = \frac{3}{2} + \frac{1}{4} = \frac{7}{4},\ x_4 = \frac{7}{4} + \frac{1}{8} = \frac{15}{8},\ x_5 = \frac{31}{16},\ x_6 = \frac{63}{32},\ x_7 = \frac{127}{64},\ x_8 = \frac{255}{128},\ x_9 = \frac{511}{256},\ x_{10} = \frac{1023}{512}$

7. $x_1 = 2,\ x_2 = \frac{2}{2} = 1, x_3 = \frac{1}{2},\ x_4 = \frac{1}{2^2}, \ldots, x_{10} = \frac{1}{2^8}$

9. $x_1 = 1,\ x_2 = 1,\ x_3 = 2,\ x_4 = 3,\ x_5 = 5,\ x_6 = 8,\ x_7 = 13,\ x_8 = 21,\ x_9 = 34,\ x_{10} = 55$

11. $\lim_{n\to\infty}(0.1)^n = 0,\ \lim_{n\to\infty} 2 = 2 \Rightarrow \lim_{n\to\infty} a_n = 2$; the sequence converges to 2.

13. $\lim_{n\to\infty} a_n = \lim_{n\to\infty} 5 = 5$; converges

15. $\lim_{n\to\infty} a_n = \infty$; the sequence diverges

17. $a_n = \frac{1+2n-4n}{1+2n} = 1 - 4\left(\frac{n}{1+2n}\right) = 1 - 4\left(\frac{1}{\frac{1}{n}+2}\right)$. Since $\lim_{n\to\infty} 1/n = 0$, $\lim_{n\to\infty} a_n = 1 - 4/2 = -1$. The sequence converges.

19. $a_n = n\left[\frac{n-2+\frac{1}{n}}{n-1}\right]$; the bracketed factor converges to 1. Hence $\lim_{n\to\infty} a_n = \infty$. The sequence diverges.

21. $a_n = \frac{-5+\frac{1}{n^4}}{1+\frac{8}{n}}$; $\lim_{n\to\infty} a_n = -5$; the sequence converges

23. $a_n = 1 + \frac{(-1)^n}{n}$; $\lim_{n\to\infty} a_n = 1$; converges

25. $\lim_{n\to\infty}\left(\frac{n+1}{2n}\right) = \lim_{n\to\infty}\left(\frac{1+\frac{1}{n}}{2}\right) = \frac{1}{2}$; $\lim_{n\to\infty}\left(1 - \frac{1}{n}\right) = 1$. The limits of both factors exist. Hence $\lim_{n\to\infty} a_n = \frac{1}{2}\cdot 1 = \frac{1}{2}$.

27. $|a_n| = \frac{1}{2n-1}$; hence $\lim_{n\to\infty} |a_n| = 0$. This implies $\lim_{n\to\infty} a_n = 0$.

29. $\lim_{n\to\infty} \frac{\sin n}{n} = 0.$

31. $\lim_{n\to\infty} \frac{2n}{n+1} = 2$. Since $\sqrt{\frac{2x}{x+1}}$ is a continuous function of x, $\lim_{n\to\infty} \sqrt{\frac{2n}{n+1}} = \sqrt{\lim_{n\to\infty} \frac{2n}{n+1}} = \sqrt{2}.$

33. Since $\tan^{-1} x$ is a continuous function, $\lim_{n\to\infty} \tan^{-1} n = \tan^{-1}(\lim_{n\to\infty} n) = \pi/2.$

35. Graphing $y = x/2^x$ on $[0, 100]$ by $[-2, 2]$ suggests the sequence converges to 0. Using L'Hôpital's rule gives $\lim_{n\to\infty} \frac{n}{2^n} = \lim_{n\to\infty} \frac{1}{2^n \ln 2} = 0.$

37. Graphing $y = \frac{\ln(x+1)}{x}$ suggests that $\{a_n\}$ converges to 0. Analytically, $\lim_{n\to\infty} a_n = \lim_{n\to\infty} \frac{\frac{1}{x+1}}{1} = 0.$

39. Graphing $y = 8^{1/x}$ on $[0, 100]$ by $[-2, 2]$ suggests $\{a_n\}$ converges to 1. Analytically, $\lim_{n\to\infty} 8^{1/n} = 8^{\lim(1/n)} = 8^0 = 1.$

41. $\lim_{n\to\infty} \left(1 + \frac{7}{n}\right)^n = e^7$, using Table 9.1, #5.

43. $\lim_{n\to\infty} (0.9)^n = 0$; the sequence diverges

45. Graphing $y = (10x)^{(1/x)}$ suggests that $\{a_n\}$ converges to 1. Analytically, let $w = (10x)^{1/x}$; $\ln w = \frac{1}{x}\ln(10x)$; $\lim_{x\to\infty} \ln w = \lim_{x\to\infty} \frac{\frac{10}{10x}}{1} = 0$. $\ln w \to 0 \Rightarrow w \to 1.$

47. Let $w = (\frac{3}{x})^{1/x}$; $\ln w = \frac{\ln(3/x)}{x} = \frac{\ln 3 - \ln x}{x}$; $\lim_{x\to\infty} \ln w = \lim_{x\to\infty} \frac{-\frac{1}{x}}{1} = 0.$ Thus, $\{a_n\}$ converges to 1.

49. Recall that $\lim_{n\to\infty} n^{1/n} = 1$. Since the numerator grows large without bound, $\lim_{n\to\infty} a_n = \lim_{n\to\infty} \frac{\ln n}{1} = \infty$; $\{a_n\}$ diverges.

51. $\lim_{n\to\infty} x^n = 0$ if $|x| < 1$. $\{(\frac{1}{3})^n\}$ converges to 0.

53. By the squeeze theorem, $0 < \frac{1}{n!} < \frac{1}{n} \Rightarrow \{\frac{1}{n!}\}$ converges to 0.

55. Graphing the function $y = (\frac{1}{x})^{1/\ln x}$ and using trace indicates that the sequence converges to 0.3678794. Analytically, let $w = (\frac{1}{x})^{1/\ln x}$; $\ln w = \frac{1}{\ln x}\ln(\frac{1}{x}) = \frac{-\ln x}{\ln x} = -1$. $\lim_{x\to\infty} w = e^{-1} = 0.3678\ldots.$

57. $a_{10} = 3.6\ E - 54$; $a_{100} = 9.3\ E - 443$ certainly suggests that $\lim_{n\to\infty} a_n = 0.$

59. Graph $y_1 = \text{abs}(0.5\hat{}\,(1/x) - 1)$ and $y_2 = 0.001$ on $0 \le x \le 1000$; find their intersection (at $x = 692.8$). For $n \ge 693$, the inequality will hold.

61. Graph $y = 0.9\hat{}\ x$ and $y = 0.001$ on $[0, 100]$; the curves intersect at $x = 65.563$. $n \geq N = 66 \Rightarrow 0.9^n < 10^{-3}$.

63. The sequence can be written $2^{\frac{1}{2}}, 2^{\frac{1}{4}}, 2^{\frac{1}{8}}, \ldots$. Graph $y = 2\hat{}(1/2\hat{}x)$ and $y = 1$ on $[0, 6]$ by $[0, 2]$.

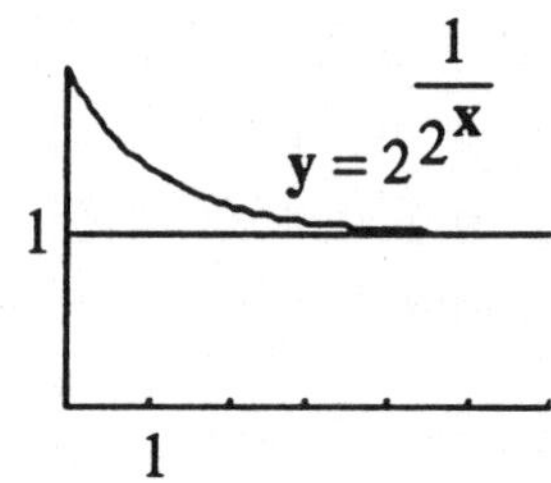

65. Equation 5. Let $a_n(x) = (1 + \frac{x}{n})^n$; $a_{10}(0.5) = 1.629$, $a_{100}(0.5) = 1.647$, $e^{0.5} = 1.649$; $a_{10}(2) = 6.192$, $a_{100}(x) = 7.245$, $a_{1000}(2) = 7.374$, $e^2 = 7.389$.

67. $f(x) = \sin x - x^2$, $x_{n+1} = x_n - \frac{f(x_n)}{f'(x_n)} \Rightarrow a_{n+1} = a_n - \frac{\sin a_n - a_n^2}{\cos a_n - 2a_n}$,

$a_1 = 1 \rightarrow 0.891, 0.877, 0.877, \ldots$; $a_1 = 2 \rightarrow 1.300, 0.989, 0.889, \ldots$;

$a_1 = -1 \rightarrow -0.275, -0.045, -0.002, \ldots$.

69. The sequence stabilizes at 1.57079632679 at x_4.

71. Graph $y = 0.5x$ and $y = \sin x$ on $[-5, 5]$ by $[-2, 2]$. There are two solutions, $0.5x = \sin x \Rightarrow x = 2\sin x$ or $x_{n+1} = 2\sin x_n$. The iteration will converge to the solutions. $x_1 = 1 \Rightarrow \{x_n\}$ converges to $1.895494\ldots$. $x_1 = -1 \Rightarrow \{x_n\}$ converges to $-1.895494\ldots$.

9.2 INFINITE SERIES

1. $s_n = 2(1 + \frac{1}{3} + \cdots + \frac{1}{3^{n-1}}) = \frac{2(1-(\frac{1}{3})^n)}{1-\frac{1}{3}} \rightarrow 3$, $s_{20} = 2.99999999914$

3. $s_n = \frac{1-(1/2)^n}{1-(-1/2)} \rightarrow 2/3$. $s_{20} = 0.6666660308$

5. Recall that $\frac{1}{n(n+1)} = \frac{1}{n} - \frac{1}{n+1}$. $s_n = \left[\frac{1}{2} - \frac{1}{3}\right] + \left[\frac{1}{3} - \frac{1}{4}\right] + \cdots + \left[\frac{1}{n+1} - \frac{1}{n+2}\right] = \frac{1}{2} - \frac{1}{n+2} \rightarrow \frac{1}{2}$

7. $s_1 = \frac{1}{4^0}$, $s_2 = \frac{1}{4^0} + \frac{1}{4^1}$, etc. $s_5 = 1.3320\ldots$, $s_n = 1 + \frac{1}{4} + \frac{1}{4^2} + \cdots + \frac{1}{4^{n-1}} = \frac{1-(1/4)^n}{1-1/4} = 4/3$

9. $s_1 = \frac{7}{4}, \ldots, s_5 = 2.33105\ldots, s_n = 7(\frac{1}{4}+\frac{1}{16}+\cdots+\frac{1}{4^n}) = 7\cdot\frac{1}{4}\left[\frac{1-(1/4)^{n+1}}{1-1/4}\right] = 7/3$

11. $s_1 = 5+1$; $s_5 = 11.181327\ldots$. Break up s_n into two parts: $s_n = 5(1+\frac{1}{2}+\cdots+\frac{1}{2^{n-1}})+(1+\frac{1}{3}+\cdots+\frac{1}{3^{n-1}}) = 5(1-(\frac{1}{2})^n)/(1-\frac{1}{2})+(1-(\frac{1}{3})^n)/(1-\frac{1}{3}) \to 10+3/2 = 11.5$

13. $s_n = (1+\frac{1}{2}+\cdots+\frac{1}{2^{n-1}})+(1-\frac{1}{5}+\frac{1}{25}+\cdots+\frac{(-1)^{n-1}}{5^{n-1}}) = \frac{1-(1/2)^n}{1-1/2}+\frac{1-(-1)^n/5^n}{1-(-1/5)} = 2+\frac{5}{6} = \frac{17}{6}$

15. $\frac{4}{(4n-3)(4n+1)} = \frac{1}{4n-3}-\frac{1}{4n+1}$. Hence $s_n = (\frac{1}{1}-\frac{1}{5})+(\frac{1}{5}-\frac{1}{9})+(\frac{1}{9}-\frac{1}{13})+\cdots \to 1$.

17. See #15. $s_n = (\frac{1}{9}-\frac{1}{13})+(\frac{1}{13}-\frac{1}{17})+\cdots \to \frac{1}{9}$. $s_{50} = 0.10613\ldots$; $s_{100} = 0.10861\ldots$.

19. This is a geometric series with $r = 1/\sqrt{2}$, $0<r<1$. $\lim_{n\to\infty} s_n = \frac{1}{1-1/\sqrt{2}} = \frac{\sqrt{2}}{\sqrt{2}-1} = \frac{\sqrt{2}(\sqrt{2}+1)}{1} = 2+\sqrt{2}$. ($s_{20} = 3.410879\ldots$)

21. $s_n = -3\sum_{k=1}^{n}(-\frac{1}{2})^k \to -3(-\frac{1}{2})(1)/(1+\frac{1}{2}) = 1$. This is a multiple of a convergent geometric series.

23. The series is $1-1+1-1+1\ldots$; the nth term fails to approach 0; the series diverges

25. Convergent geometric series: $\sum_{n=0}^{\infty}(\frac{1}{e^2})^n \to \frac{1}{1-1/e^2} = \frac{e^2}{e^2-1}$, $s_{30} = 1.15651764275$

27. Diverges; $\lim_{n\to\infty} a_n \neq 0$.

29. This is the sum of two convergent, geometric series; $\sum_{n=0}^{\infty}(\frac{2}{3})^n - \sum_{n=0}^{\infty}(\frac{1}{3})^n \to \frac{1}{1-2/3}-\frac{1}{1-1/3} = 3-\frac{3}{2} = \frac{3}{2}$ ($s_{10} = 1.4480008\ldots$)

31. By #6, Table 9.1, $\lim_{n\to\infty}\frac{x^n}{n!} = 0$. Hence $\lim_{n\to\infty} a_n = \infty$. The series diverges.

33. $\sum_{n=0}^{\infty}(-1)^n x^n = 1-x+x^2-\cdots = 1(1-x+x^2-\cdots)$. $a=1$, $r=-x$

35. a) The sum of the series is $\frac{1}{6} = 0.1666666666\ldots$. If the series is written as $a_0+a_1+\cdots$, then $a_n = \frac{1}{3^{n+2}}$, $s_{17} = 0.166666666235$, i.e., 18 terms needed.

b) The sum of the series is $\frac{8}{3}$. If $a_n = 4(-\frac{1}{2})^n$, then $s_n = a_0+a_1+\cdots+a_n$; $s_{30} = 2.66666666791$, i.e., 31 terms needed.

37. The ball travels $4+2(0.75)4+2(0.75)^2 4+\cdots$ m $= 8[1+0.75+0.75^2+\cdots]-4 = 8\left[\frac{1}{1-0.75}\right]-4 = 28$ m.

39. Let $x = 0.234234\ldots$, then $1000x = 234.234234\ldots$ and $999x = 234 \Rightarrow x = \frac{234}{999} = \frac{26}{111}$

41. The denominator of the first term must be $2 \cdot 3$. If $n = -2$, the $n+4 = 2$, $n+5 = 3$. Hence a) $\sum_{n=-2}^{\infty} \frac{1}{(n+4)(n+5)}$, b) $\sum_{n=0}^{\infty} \frac{1}{(n+2)(n+3)}$, c) $\sum_{n=5}^{\infty} \frac{1}{(n-3)(n-2)}$

43. Corollary 1. Let $\sum a_n$ be divergent, let $b_n = ca_n$. Then $s_k = \sum_{n=1}^{k} b_n = \sum_{n=1}^{k} ca_n = c\sum_{n=1}^{k} a_n$, which diverges.

Corollary 2. Let $s_k = \sum_{n=1}^{k}(a_n \pm b_n) = \sum_{n=1}^{k} a_n \pm \sum_{n=1}^{k} b_n$. The first series converges, the second diverges. Hence $\{s_k\}$ diverges.

45. The partial sums of the series can be written $s_k = 1 + \frac{1}{2} + \frac{1}{4} + \frac{1}{8} + \cdots + \frac{1}{2^k} = \sum_{n=0}^{k} \frac{1}{2^n}$. $s_{10} = 1.9990\ldots$, $s_{15} = 1.99999694\ldots$, $s_{20} = 2.000000$; the sum appears to be 2.

47. $s_k = 0 + 1 + 3 + 5 + \cdots (2k-1)$, which diverges.

49. Present value $= \sum_{n=1}^{\infty} a_n$, where $a_n = (1.07)^{-1} a_{n-1}$, $a_0 = 5000$. $\sum_{n=1}^{100} a_n = 71,346.25$; $\sum_{n=1}^{1000} a_n = 71,428.571$; $\sum_{n=1}^{1500} a_n = 71,428.571$.

51. a) The money needed is $a_1 + a_2 + \cdots + a_{50}$; $a_0 = 500$, $a_{n+1} = 1.07^{-1} a_n$, $N = 50$. $6,900.38 is needed, b) $7,134.63

53. If a square has side s (area s^2),

the next square has side

$\sqrt{2(\frac{s}{2})^2} = \frac{s}{\sqrt{2}}$; (area $\frac{s^2}{2}$).

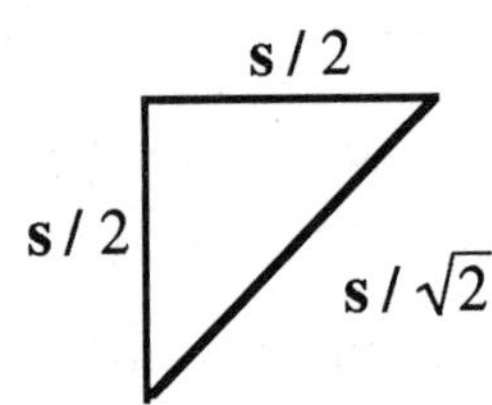

The next smaller square has side

$\frac{s/\sqrt{2}}{\sqrt{2}} = \frac{s}{2}$; (area $\frac{s^2}{4}$).

$s = 2 \Rightarrow$ the first square has area 4.

The sum of the areas is

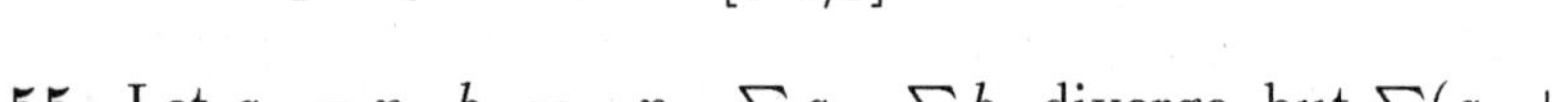

$4(1 + \frac{1}{2} + \frac{1}{4} + \cdots) = 4\left[\frac{1}{1-1/2}\right] = 8 \text{ m}^2$.

55. Let $a_n = n$, $b_n = -n$. $\sum a_n$, $\sum b_n$ diverge, but $\sum(a_n + b_n) = \sum 0$ converges.

57. Let $a_n = \frac{1}{3^n}, \sum_{n=0}^{\infty} a_n = \frac{3}{2}$; let $b_n = \frac{1}{2^n}$, $\sum_{n=0}^{\infty} b_n = 2$, $\frac{a_n}{b_n} = (\frac{2}{3})^n$, $\sum(\frac{2}{3})^n = \frac{1}{1-2/3} = 3$, but $A/B = \frac{3}{4}$.

9.3 SERIES WITHOUT NEGATIVE TERMS: COMPARISON AND INTEGRAL TESTS

1. Converges; geometric series with $r = 1/10$; sum is $10/9$

3. $\lim_{n\to\infty} a_n = \lim_{n\to\infty}(1 - \frac{2}{n+2}) \neq 0$; diverges.

5. $\frac{\sin^2 n}{2^n} > 0$; $\frac{\sin^2 n}{2^n} \leq \frac{1}{2^n}$. Converges by the comparison test; sum $\approx s_{100} = s_{666} = 0.637$

7. For $n \geq 3$, $\ln n > 1$; $\frac{\ln n}{n} > \frac{1}{n}$. Diverges by comparison test.

9. $\sum_1^\infty \frac{1}{n^{3/2}}$ converges by the integral test, since $\int_1^\infty \frac{dx}{x^{3/2}}$ converges.

11. Diverges by the Limit Comparison Test with $\sum \frac{1}{n}$; $\lim_{n\to\infty} \frac{\frac{2}{n+1}}{\frac{1}{n}} = 2 \neq 0$.

13. Diverges by the Limit Comparison Test with $\sum \frac{1}{n}$; $\lim_{n\to\infty} \frac{\frac{1}{n}}{\frac{1}{2n-1}} = 2 \neq 0$.

15. Diverges; $\lim_{n\to\infty} a_n = e \neq 0$

17. Diverges by limit comparison with $\sum \frac{1}{n}$: $\lim_{n\to\infty} \frac{\frac{n}{n^2+1}}{\frac{1}{n}} = 1$.

19. Compare with $\sum \frac{1}{n^{3/2}}$ which is a convergent p-series. $\lim_{n\to\infty} \frac{\frac{1}{\sqrt{n^3+2}}}{\frac{1}{\sqrt{n^3}}} = \lim_{n\to\infty} \sqrt{\frac{n^3}{n^3+2}} = 1$. Converges

21. $\sum_{n=1}^\infty \frac{1+n}{n\cdot 2^n} = \sum_{n=1}^\infty \frac{1}{n\cdot 2^n} + \sum_{n=1}^\infty \frac{1}{2^n}$; the second series is a convergent geometric series; the first converges by the comparison test: $\frac{1}{n2^n} < \frac{1}{2^n}$.

23. Converges by the Comparison Test; $\frac{1}{3^{n+1}+1} < \frac{1}{3^{n-1}}$; $\sum_{n=1}^\infty \frac{1}{3^{n-1}} = 3\sum_1^\infty \frac{1}{3^n}$ which is a geometric series with $r = (\frac{1}{3})$.

25. $n = 14$.

27. $n \approx 365 * 24 * 60^2 * 13 \times 10^9 \approx 4.09968 \times 10^{17}$; $\ln(n+1) = 40.5548\cdots \leq s_n \leq 41.5548$.

29. $0 \leq \frac{a_n}{n} \leq a_n$; by the comparison test, $\sum(a_n/n)$ converges.

31. If $\{s_n\}$ is nonincreasing, with $M \leq s_n$ for every n, then $\{-s_n\}$ is nondecreasing, $-s_n \leq -M \Rightarrow \{-s_n\}$ is bounded above. A bounded, nondecreasing sequence converges. If $\{-s_n\} \to s$ then $\{s_n\} \to -s$. Similarly, if $\{s_n\}$ is nonincreasing and not bounded below, the sequence $\{-s_n\}$ diverges.

9.4 SERIES WITH NONNEGATIVE TERMS: RATIO AND ROOT TESTS

1. Converges, by the ratio test: $\lim_{n\to\infty}\left|\frac{(n+1)^2}{2^{n+1}}\cdot\frac{2^n}{n^2}\right| = \frac{1}{2} < 1$. The partial sums stabilize at 6.

3. Converges, by the ratio test: $\lim_{n\to\infty}\frac{(n+1)^{10}}{10^{n+1}}\cdot\frac{10^n}{n^{10}} = \frac{1}{10} < 1$. $s_{20} = 376.1794\ldots$; s_{40} is essentially the same.

5. Diverges, by the ratio test: $\lim_{n\to\infty}\frac{(n+1)!e^{-(n+1)}}{n!e^{-n}} = \lim_{n\to\infty}\frac{n+1}{e} = \infty$.

7. $a_n = (1-\frac{2}{n})^n$; $a_n \to e^{-2} \neq 0$. The series diverges.

9. Diverges: $\lim_{n\to\infty} a_n = \lim_{n\to\infty}(1-\frac{3}{n})^n = e^{-3} \neq 0$.

11. Diverges: $\sum\frac{1}{n} - \sum\frac{1}{n^2}$ is the difference between a divergent series and a convergent series.

13. For $n \geq 3$, $\ln n > 1$. Series diverges by the comparison test: $n \geq 3 \Rightarrow a_n > \frac{1}{n}$.

15. Converges, by the ratio test: $\lim_{n\to\infty}\frac{(n+2)(n+3)}{(n+1)!}\frac{n!}{(n+1)(n+2)} = \lim_{n\to\infty}\frac{n+2}{(n+1)^2} = 0$. Estimated value is $17.0280\ldots$.

17. Converges, by the ratio test: $\lim_{n\to\infty}\frac{(n+4)!}{3!(n+1)!3^{n+1}}\cdot\frac{3!n!3^n}{(n+3)!} = \lim_{n\to\infty}\frac{n+4}{3(n+1)} = \frac{1}{3} < 1$. Estimated sum is 4.0625.

19. The ratio test can be applied to $\sum_{n=1}^{\infty}\frac{n^2}{2^n}$, which is a series with nonnegative terms: $\lim_{n\to\infty}\frac{(n+1)^2}{2^{n+1}}\cdot\frac{2^n}{n^2} = \frac{1}{2} < 1$. The series converges; the estimated value is -6.

21. When there are n's in the exponents, and no factorials, the root test is likely to be successful. $\lim_{n\to\infty}\sqrt[n]{a_n} = \lim_{n\to\infty}\frac{\sqrt[n]{n}}{\ln n} = 0$. The series converges; estimated sum is $8.25271035\ldots$.

23. Converges, by the comparison test: $a_n = \frac{1}{(n+1)(n+2)} < \frac{1}{n^2}$; $\sum\frac{1}{n^2}$ converges. Since $s_{500} = 0.498007$, and $s_{2000} = 0.499500$, a reasonable estimate is that $s = 0.5$.

25. Diverges, by the ratio test: $\lim_{n\to\infty}\frac{3^{n+1}}{(n+1)^3 2^{n+1}}\cdot\frac{n^3 2^n}{3^n} = \frac{3}{2} > 1$.

27. Converges, by the ratio test: $\lim_{n\to\infty}\frac{a_{n+1}}{a_n} = \lim_{n\to\infty}\frac{1+\sin n}{n} = 0$. The series stabilizes at 2.680118.

29. The ratio test is inconclusive: $\lim_{n\to\infty} \frac{a_{n+1}}{a_n} = 1$. Generating the first few terms gives $a_1 = 3$, $a_2 = \frac{1}{2}\cdot 3$, $a_3 = \frac{2}{3}\cdot\frac{3}{2} = \frac{3}{3}$, $a_4 = \frac{3}{4}\cdot\frac{3}{3} = \frac{3}{4}, \ldots, a_n = \frac{3}{n}$, the series diverges.

31. Converges, by the ratio test: $\lim_{n\to\infty} \frac{1+\ln n}{n} = 0$. Estimated sum is -2.119527.

33. Converges, by the ratio test: $\lim_{n\to\infty} \frac{2^{n+1}(n+1)!(n+1)!}{(2n+2)!} \frac{(2n)!}{2^n n! n!} = \lim_{n\to\infty} \frac{2(n+1)^2}{(2n+1)(2n+2)} = \frac{1}{2}$. The sum stabilizes at $2.5707963\ldots$.

35. Following the hint $a_1 = 1$, $a_2 = \frac{2\cdot 1}{4\cdot 3}$, $a_3 = \frac{3\cdot 2\cdot 2\cdot 1}{5\cdot 4\cdot 4\cdot 3}$, $a_4 = \frac{3\cdot 2\cdot 2\cdot 1}{6\cdot 5\cdot 5\cdot 4}$, $a_5 = \frac{3\cdot 2\cdot 2\cdot 1}{7\cdot 6\cdot 6\cdot 5}$. In general, $a_n = \frac{3\cdot 2\cdot 2\cdot 1}{(n+2)(n+1)^2 n} = \frac{12}{(n+2)(n+1)^2 n} < \frac{12}{n^4}$. The series converges, by comparison test. The estimated sum is $1.26079119\ldots$.

9.5 ALTERNATING SERIES AND ABSOLUTE CONVERGENCE

1. Converges by absolute convergence theorem since $\sum \frac{1}{n^2}$ converges. The absolute value of the error, $|s_n - L| < \frac{1}{(n+1)^2}$. Solving $\frac{1}{(n+1)^2} < 0.001$ gives $\sqrt{\frac{1}{0.001}} < n+1$ or $n \geq 31$. $s_{32} = 0.82199\ldots < L < 0.82297\ldots = s_{31}$.

3. Diverges; $\lim_{n\to\infty} a_n \neq 0$.

5. Converges by Alternating Series Theorem. Let $f(x) = \frac{\sqrt{x}+1}{x+1}$; $\lim_{n\to\infty} f(x) = \lim_{n\to\infty} a_n = 0$; to show f is decreasing, compute $f'(x) = \frac{(1-x-2\sqrt{x})}{2\sqrt{x}(x+1)^2} < 0$. To find sums within 0.001, try this approach: for large values of n, $\frac{\sqrt{n}+1}{n+1} \approx \frac{\sqrt{n}}{n} = \frac{1}{\sqrt{n}}$. Solving $\frac{1}{\sqrt{n+1}} < 0.001$ gives $n \approx 1{,}000{,}000$.

7. Converges: $a_n = f(n)$, where $f(x) = (\ln x)/x$. $f'(x) = \frac{1-\ln x}{x^2} < 0$ for $x > e$. Hence $a_n > a_{n+1}$. $\text{Lim}_{x\to\infty} \frac{\ln x}{x} = \lim_{x\to\infty} \frac{\frac{1}{x}}{1} = 0$. The series converges very slowly; look at the graph of $(\ln x)/x$ on $[0, 10{,}000]$ by $[0, 0.01]$. The series needs at least 10,000 terms to be within 0.001 of its sum.

9. $|a_n| = \frac{3\sqrt{n+1}}{\sqrt{n}+1} = \frac{3\sqrt{n}\sqrt{1+1/n}}{\sqrt{n}(1+1/\sqrt{n})} \to 3$. Diverges.

11. Converges absolutely since $\sum |a_n|$ is a convergent p-series.

13. Converges absolutely; let $c_n = \frac{1}{n^2}$; then $\lim_{n\to\infty} \frac{|a_n|}{c_n} = \lim_{n\to\infty} \frac{n^3}{n^3+1} = 1$.

15. Converges conditionally: $|a_n| = \frac{1}{n+3} > |a_{n+1}| = \frac{1}{n+4}$; $a_n \to 0$. However $\sum |a_n|$ diverges by the Limit Comparison Test when compared to $\sum \frac{1}{n}$.

17. Converges absolutely: $0 \le |(-1)^n \frac{\sin n}{n^2}| \le \frac{1}{n^2}$; $\sum \frac{1}{n^2}$ is a convergent $p-$series.

19. Converges conditionally: $|a_n| = \frac{1+n}{n^2} > \frac{2+n}{(n+1)^2} = |a_{n+1}|$. This can be established directly by comparing $(n+1)^3 = n^3 + 3n^2 + 3n + 1 > n^3 + 2n^2$, or by setting $f(x) = \frac{1+x}{x^2}$ and showing that $f'(x) = \frac{x^2 - 2x(1+x)}{x^4} < 0$. $|a_n| \to 0$, $\sum |a_n| = \sum(\frac{1}{n^2} + \frac{1}{n}) = \sum \frac{1}{n^2} + \sum \frac{1}{n}$, the sum of a convergent and a divergent series, is divergent.

21. Converges absolutely by the Ratio Test: $\left|\frac{a_{n+1}}{a_n}\right| = \frac{(n+1)^2}{n^2}\frac{2}{3} \to \frac{2}{3}$.

23. $\int_0^\infty \frac{\tan^{-1} x}{x^2+1} dx = \int_0^{\pi/2} u du$ where $u = \tan^{-1} x$. The integral converges; $\sum a_n$ is absolutely convergent.

25. Diverges: $a_n = \frac{1}{2n}$, $\frac{1}{2}\sum \frac{1}{n}$ diverges.

27. Diverges: $\lim_{n\to\infty} |a_n| = 1 \ne 0$.

29. $a_n = \frac{-1}{(n+1)^2}$; $0 < -a_n < \frac{1}{n^2}$. The series $\sum -a_n$ converges, hence $\sum a_n$ converges.

31. Converges absolutely since $|5^{-n}| = (\frac{1}{5})^n$; $\sum(\frac{1}{5})^n$ is a geometric series with $|r| < 1$.

33. $0 \le |a_n| < \frac{1}{n^{3/2}}$ since $|\cos x| \le 1$. The series is absolutely convergent.

35. The series converges conditionally because $|a_n| = \frac{1}{\sqrt{n}+\sqrt{n+1}} > \frac{1}{\sqrt{n+1}+\sqrt{n+2}} = |a_{n+1}|$ and $a_n \to 0$. To show that the series does not converge absolutely, write $|a_n| = \frac{1}{\sqrt{n}(1+\sqrt{1+1/n})}$ and compare it with the divergent series $\sum \frac{1}{\sqrt{n}} = \sum d_n$. By the Limit Comparison Test $\frac{|a_n|}{d_n} = \frac{1}{1+\sqrt{1+1/n}} \to \frac{1}{2}$.

37. By Theorem 8, $|s_4 - s| < \frac{1}{5} < 0.2$

39. $|\text{error}| < |a_5| = \frac{(0.01)^5}{5} = 2E - 11 = 2 \times 10^{-11}$

41. The first omitted term must satisfy $\frac{1}{(2n)!} < 5 \cdot 10^{-6}$, i.e. $10^6 < 5 \cdot (2n)!$. This is true for $n = 5$. Thus s_4 will suffice. Using the program PARTSUM, $s_4 = 0.5403023\ldots$.

43. a) The absolute values of the terms are not not strictly decreasing, b) If the terms are rearranged the series becomes $\left[\frac{1}{3} + \frac{1}{9} + \frac{1}{27} + \cdots\right] - \left[\frac{1}{2} + \frac{1}{4} + \frac{1}{8} + \cdots\right]$. It is, in fact, permitted to rearrange the terms of an absolutely convergent series without affecting its sum. Hence the given series converges to $\frac{1}{2} - 1 = -\frac{1}{2}$.

45. Assume the series is written $a_1 - a_2 + a_3 - a_4 \ldots$ where all the a_i are positive. If $a_n > 0$ the series can be written $(a_1 - a_2 + \cdots + a_n) - a_{n+1} + a_{n+2} - \cdots$ and the remainder is $-(a_{n+1} - a_{n+2}) - (a_{n+3} - a_{n+4}) - \ldots$. All the terms enclosed in parentheses are positive. Hence the remainder is negative, as is the $n + 1^{st}$ term. If $a_n < 0$ we have $(a_1 - a_2 + \cdots - a_n) + [(a_{n+1} - a_{n+2}) + (a_{n+3} - a_{n+4}) + \cdots]$ and both the remainder and the $(n+1)^{st}$ term are positive.

47.

	x	Approximate Value of n	Underestimate	Overestimate
a)	0.2	$\frac{0.2^n}{n} < 0.001 \Rightarrow n \geq 4$	$s_4 = 0.18226$	$s_5 = 0.18233$
b)	0.5	$\frac{0.5^n}{n} < 0.001 \Rightarrow n \geq 8$	$s_8 = 0.4053$	$s_9 = 0.4055$
c)	0.8	$\frac{0.8^n}{n} < 0.001 \Rightarrow n \geq 19$	$s_{20} = 0.5875$	$s_{19} = 0.58811$
d)	0.9	$\frac{0.9^n}{n} < 0.001 \Rightarrow n \geq 34$	$s_{34} = 0.64147$	$s_{35} = 0.64218$

9.6 POWER SERIES

1. $\lim_{n\to\infty} \left|\frac{(x+1)^{n+1}}{(x+1)^n}\right| < 1 \Rightarrow |x+1| < 1 \Rightarrow -2 < x < 0$; when $x = -2$ we have $\sum(-1)^n(-1)^n = \sum 1$, divergent. At $x = 0$ we have $\sum(-1)^n$, divergent.

3. $\lim_{n\to\infty} \left|\frac{x^{n+1}}{(n+1)\sqrt{n+1}} \frac{n\sqrt{n}}{x^n}\right| < 1 \Rightarrow |x| < 1 \Rightarrow -1 < x < 1$; at $x = -1$, we have $\sum \frac{(-1)^n}{n^{3/2}}$ which is convergent by the alternating series test; at $x = 1$, we have $\sum \frac{1}{n^{3/2}}$, a convergent p-series. Hence the series is absolutely convergent on $[-1, 1]$.

5. $\lim_{n\to\infty} \left|\frac{x^{2n+3}}{(n+1)!} \frac{n!}{x^{2n+1}}\right| < 1 \Rightarrow 0 < 1$, true for all x.

7. $\lim_{n\to\infty} \left|\frac{x^{n+1}}{\sqrt{(n+1)^2+3}} \cdot \frac{\sqrt{n^2+3}}{x^n}\right| < 1 \Rightarrow |x| \lim_{n\to\infty} \sqrt{\frac{n^2+3}{(n+1)^2+3}} < 1 \Rightarrow |x| \cdot 1 < 1 \Rightarrow -1 < x < 1$; at $x = -1$ the (alternating) series converges; at $x = 1$ we have $\sum \frac{1}{\sqrt{n^2+3}}$ which diverges.

9. $\lim_{n\to\infty} \left|\frac{(n+1)x^{n+1}}{(n+1)^2+1} \cdot \frac{n^2+1}{(n+1)x^n}\right| < 1 \Rightarrow |x| \lim_{n\to\infty} \frac{n^2+1}{(n+1)^2+1} < 1 \Rightarrow |x| < 1 \Rightarrow -1 < x < 1$; when $x = -1$, the (alternating) series converges; when $x = 1$ we have $\sum \frac{n}{n^2+1}$ which diverges.

11. $\lim_{n\to\infty}\left|\frac{\sqrt{n+1}x^{n+1}}{3^{n+1}}\cdot\frac{3^n}{\sqrt{n}x^n}\right| < 1 \Rightarrow |x|\cdot\frac{1}{3} < 1 \Rightarrow -3 < x < 3$; at $x = -3$ we have $\sum(-1)^n\sqrt{n}$ which diverges ($\lim a_n \neq 0$); at $x = 3$ we have $\sum\sqrt{n}$ which diverges.

13. $\lim_{n\to\infty}\left|\frac{(1+\frac{1}{n+1})^{n+1}x^{n+1}}{(1+\frac{1}{n})^n x^n}\right| < 1 \Rightarrow |x|\cdot\frac{e}{e} < 1 \Rightarrow -1 < x < 1$, at $x = -1$ and $x = 1$, the series diverge ($\lim|a_n| = e$)

15. $\lim_{n\to\infty}\left|\frac{x^{n+1}}{x^n}\right| < 1 \Rightarrow |x| < 1 \Rightarrow -1 < x < 1$; diverges for $x = \pm 1$. For $-1 \le x < 1$, the series converges to $\frac{1}{1-x}$. Use a calculator to compare P_{20} and $\frac{1}{1-x}$: set $y_1 = \text{sum seq}(x\hat{\ }N, N, 0, 20, 1)$. $y_2 = 1/(1-x)$; $y_3 = \text{abs}(y_1 - y_2)$ and graph y_3 on $[-1, 1]$ by $[0, 0.01]$. Investigation shows $-0.8 \le x \le 0.7 \Rightarrow |\text{sum} - P_{20}| < 0.01$.

17. Absolute convergence when $\left|\frac{x-2}{10}\right| < 1 \Rightarrow -8 < x < 12$. The series converges to $\frac{1}{1-(\frac{x-2}{10})} = \frac{10}{12-x}$. Error is less than 0.0.1 when $-0.8 < \frac{x-2}{10} < 0.7 \Rightarrow -6 < x < 9$.

19. a) Use sumseq, and the identity $(-1)^{2n-1} = -1$ to calculate $P_{20} = \text{sumseq}(-1/N, N, 1, 20, 1) = -3.597\ldots$, $P_{30} = -3.994\ldots$, $P_{50} = -4.4992$, $P_{100} = -5.187\ldots$. These are the partial sums of the harmonic series, which diverges. b) $P_{10}(-0.9) = -2.1187$; $\ln(1 - 0.9) = -2.302$. The maximum error is 0.19. c)

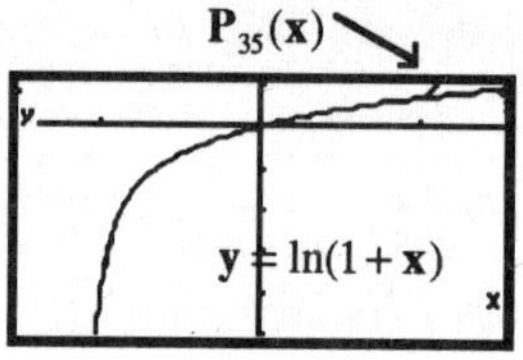

	a) Convergent	b) Absolutely Convergent	c) Alternating on	d) Error < 0.01
21.	$[1,3)$	$(1,3)$	$(1,2)$	P_{30} on $[1.1, 2]$ Compare with $\ln(3-1.1)$
23.	$(2,4)$	$(2,4)$	$(2,3)$	P_{110} on $[2.1, 3]$ Compare with $(2.1-3)/(4-2.1)^2$; series is $(x-3)D_x\sum(x-3)^n$

25. a) Using sum seq $(((-1)\hat{}N(\pi/4)\hat{}(2n+1))/(2N+1)!, N, 0, 5, 1)$ etc. we get

x	$P_5(x)$	$P_{11}(x)$	$P_{17}(x)$
$\pi/4$	$0.707106\ldots$	$0.707106\ldots$	$0.707106\ldots$
$-\pi/4$	$-0.707106\ldots$		
$\pi/2$	-0.99999994	1	
$-\pi/2$	-0.99999994	-1	
2π	-3.19507	$-5.494\,E-06$	$-4E-13$
-2π	3.19507		

b) For positive and negative values of x, the sums are the same. Using the Alternating series estimate and graphing we find:

$|\text{error } P_{11}(x)| < \left|\frac{x^{2\cdot12+1}}{(2\cdot12+1)!}\right| = \frac{x^{25}}{25!} < 0.01 \Rightarrow |x| < 8.4;$

$|\text{error } P_{21}(x)| < 0.01 \Rightarrow \frac{x^{2\cdot22+1}}{45!} < 0.01 \Rightarrow |x| < 15.9;$

$|\text{error } P_{31}(x)| < 0.01 \Rightarrow |x| < 23.3.$

27. $\left|\frac{x^{n+1}}{(n+1)!}\cdot\frac{n!}{x^n}\right| = \frac{|x|}{n+1} \to 0$; the series converges for all x. Experimenting with $-10 \to x : e\hat{}x -$ sum seq$(x\hat{}N/N!, N, 0, 20, 1) \Rightarrow -7 < x < 7$.

29. a) The series is $\sum \frac{(3x)^n}{n!}$ which sums to e^{3x}, it converges for all x;

b) $-7 < 3x < 7 \Rightarrow -7/3 < x < 7/3$

31. a) $s_n = x \cdot \frac{1-x^{n+1}}{1-x} \to \frac{x}{1-x}$, b) $1 + \sum_1^\infty x^n = \frac{1}{1-x}$; sum $= \frac{1}{1-x} - 1 = \frac{x}{1-x}$,

c) $|x| < 1$, d) for $x < 0$ the series alternates; $|P_{20}(x) - \frac{1}{1-x}| < |x^{21}|$;

solving $|x|^{21} < 0.01 \Rightarrow |x| < 0.80$. Evaluating $-0.80 \to x : x/(1-x)-$ sumseq$(x\hat{}N, N, 1, 20, 1)$ for positive values of x shows the error < 0.01 when $-0.80 < x < 0.74$.

33. a) $-\ln(1-x) = -\sum_{n=1}^\infty \frac{(-1)^{n-1}(-x)^n}{n} = -\sum_{n=1}^\infty -1\,\frac{(-1)^n(-x)^n}{n} = \sum_{n=1}^\infty \frac{x^n}{n}$;
b) $-\ln(3-x) = -\ln(1-(x-2)) = \sum \frac{(x-2)^n}{n}$ by a); c) $-\ln(3+x) = -\ln(1+(2+x)) = -\sum_{n=1}^\infty \frac{(-1)^{n-1}(2+x)^n}{n}$ by a) $= \sum_{n=1}^\infty \frac{(-1)^n(x+2)^n}{n}$

35. On a grapher, set $y1 = 1 + x + x^2/2$, $y2 = y1 + x^3/6$, $y3 = y2 + x^4/24$, $y4 = y1 - y2$. Graph the polynomials on $[-2, 2]$ by $[-0.01, 0.01]$. Use Trace to find the results: a) $(-0.38, 0.38)$ b) $(-0.698, 0.698)$ For c), observe that $|P_9(x) - P_{10}(x)| = x^{10}/10!$; graphing that on $[-3, 3]$ by $[-0.01, 0.01]$ gives $(-2.85, 2.85)$ d) the series diverges for all $x \neq 0$.

37. $\lim_{n\to\infty}\frac{(n+1)^{n+1}x^{n+1}}{n^n x^n} = |x|\lim_{n\to\infty}\left(\frac{n+1}{n}\right)^n(n+1) = |x|\cdot e\lim_{n\to\infty}(n+1)$; this series converges if and only if $x=0$.

39. Let $f(x) = 1-\frac{1}{2}(x-3)+\frac{1}{4}(x-3)^2+\cdots = \sum_{n=0}^{\infty}\left(\frac{-(x-3)}{2}\right)^n$; this is a geometric series with $|r| = \frac{|x-3|}{2}$ which converges, for $1<x<5$, to $\frac{1}{1+\frac{(x-3)}{2}} = \frac{2}{x-1}$; when $x=1$ the series is $\sum 1^n$ which diverges, at $x=5$, the series on $\sum(-1)^n$ which diverges. $f' = \sum_{n=1}^{\infty}\frac{n}{2}\left[\frac{-(x-3)}{2}\right]^{n-1} = \sum_{k=0}^{\infty}\frac{(k+1)}{2}\left[\frac{-(x-3)}{2}\right]^k$; by calculus the sum of this series is $\frac{-2}{(x-1)^2}$ which also converges for $1<x<5$.

41. a) Graph $P_7(x) - \tan x$: error is < 0.01 on $-0.873 < x < 0.873$; b) $\ln|\sec x| = \int\tan x\,dx = \frac{x^2}{2}+\frac{x^4}{12}+\frac{x^6}{45}+\frac{17x^8}{8\cdot 315}+C$; evaluating at $x=0$ gives $C=0$; the series converges absolutely for the same values as the series for $\tan x$; $-\frac{\pi}{2}<x<\frac{\pi}{2}$; c) differentiating the series for $\tan x$ gives $\sec^2 x = 1+x^2+\frac{2x^4}{3}+\cdots$; d) $(1+\frac{x^2}{2}+\frac{5}{24}x^4+\cdots)(1+\frac{x^2}{2}+\frac{5}{24}x^4+\cdots) = 1+\frac{2x^2}{2}+x^4(\frac{2\cdot 5}{24}+\frac{1}{4})+\cdots = 1+x^2+\frac{2x^4}{3}+\cdots$.

43. a) The graphs are nearly vertical near $x=\pm\pi/2$; b) Graph s_{30} and the lines $y=\pm 0.99$ on $[-\pi,\pi]$ by $[-1.01, 1.01]$; use Trace to estimate $|s_{30}-f(x)| < 0.01$ for $0\le|x|<1.172$ or $1.97<|x|<\pi$. (This is called a "square wave".)

45. $|a_n| = \frac{|\sin(n!x)|}{n^2} \le \frac{1}{n^2}$; $\sum\frac{1}{n^2}$ converges so the series converges absolutely for all x.

47. The values appear to differ in the hundredths or thousandths place.

49. Zooming in near a peak reveals a host of subpeaks and valleys.

51. The graph of the exact derivative of $s_{13}(x)$ is a set of vertical lines. This suggests that $s'_{13}(x)$ alternates between very large positive and negative numbers, and that $s(x)$ will not have a derivative.

53. If $|x| = 1+C$, by the Ratio Test, $\lim_{n\to\infty}\left|\frac{a_{n+1}}{a_n}\right| = 1+C$, hence the series diverges. Plotting the points $(k, \sum_{n=1}^{k}a_n)$ will show the divergence.

9.7 TAYLOR SERIES AND MACLAURIN SERIES

1. $f(x)=\ln x$, $f'(x)=\frac{1}{x}$, $f''(x)=-\frac{1}{x^2}$, $f'''(x)=\frac{2}{x^3}$; $f(1)=0$, $f'(1)=1$, $f''(1)=-1$, $f'''(1)=2$; $P_1(x)=(x-1)$; $P_2(x)=(x-1)-(x-1)^2/2$; $P_3(x)=(x-1)-(x-1)^2/2+2(x-1)^3/6$. Graph P_3-f on $[0,3]$ by $[-0.01, 0.01]$; $|P_3-f|<0.01$ when $0.60<x<1.47$.

3. $f(x) = \frac{1}{x}$, $f'(x) = -\frac{1}{x^2}$, $f''(x) = \frac{2}{x^3}$, $f'''(x) = -\frac{6}{x^4}$; $f(2) = \frac{1}{2}$, $f'(2) = -\frac{1}{4}$, $f''(2) = \frac{1}{4}$, $f'''(2) = \frac{-3}{8}$; $P_1 = \frac{1}{2} - \frac{1}{4}(x-2)$; $P_2 = \frac{1}{2} - \frac{1}{4}(x-2) + \frac{1}{4}(x-2)^2/2$; $P_3 = \frac{1}{2} - \frac{1}{4}(x-2) + \frac{1}{8}(x-2)^2 - \frac{3}{8}(x-2)^3/6 = \frac{1}{2} - \frac{1}{4}(x-2) + \frac{1}{8}(x-2)^2 - \frac{1}{16}(x-2)^3$. $|P_3 - f| < 0.01$ when $1.34 < x < 2.78$.

5. $f(x) = \sin x$, $f'(x) = \cos x$, $f''(x) = -\sin x$, $f'''(x) = -\cos x$; $f(\frac{\pi}{4}) = \frac{1}{\sqrt{2}}$, $f'(\frac{\pi}{4}) = \frac{1}{\sqrt{2}}$, $f''(\frac{\pi}{4}) = -\frac{1}{\sqrt{2}}$, $f'''(\frac{\pi}{4}) = -\frac{1}{\sqrt{2}}$. $P_1 = \frac{1}{\sqrt{2}} + \frac{1}{\sqrt{2}}(x - \frac{\pi}{4})$; $P_2 = \frac{1}{\sqrt{2}}\left[1 + (x - \frac{\pi}{4})\right] - \frac{1}{\sqrt{2}}(x - \frac{\pi}{4})^2/2$;
$P_3 = \frac{1}{\sqrt{2}}\left[1 + (x - \frac{\pi}{4}) - (x - \frac{\pi}{4})^2/2 - (x - \frac{\pi}{4})^3/6\right]$. $|P_3 - f| < 0.01$ when $-0.008 < x < 1.515$.

7. $f(x) = x^{\frac{1}{2}}$, $f'(x) = \frac{1}{2}x^{-\frac{1}{2}}$, $f''(x) = -\frac{1}{4}x^{-\frac{3}{2}}$, $f'''(x) = \frac{3}{8}x^{-\frac{5}{2}}$; $f(4) = 2$, $f'(4) = \frac{1}{4}$, $f''(4) = \frac{-1}{32}$, $f'''(4) = \frac{3}{256} \Rightarrow P_3(x) = 2 + \frac{1}{4}(x-4) - \frac{1}{32}(x-4)^2/2 + \frac{3}{256}(x-4)^3/6$; $|P_3 - f| < 0.01$ when $1.91 < x < 6.60$.

9. $e^x = \sum_{n=0}^{\infty} x^n/n! \Rightarrow e^{-x} = \sum_{n=0}^{\infty}(-x)^n/n!$. $|R_{10}(x)| \le \frac{|e^x||x^{11}|}{11!} \le 0.01$ for $-5 \le x \le 2.5$.

11. $\sin 3x = \sum_{n=0}^{\infty} \frac{(-1)^n(3x)^{2n+1}}{(2n+1)!}$; $P_{10} = \sum_{n=0}^{4} \frac{(-1)^n(3x)^{2n+1}}{(2n+1)!}$; graphically, $|\sin 3x - P_{10}(x)| < 0.01$ when $|x| < 1.083$; analytically, $|R_{10}| < 0.01$, when $\frac{3^{11}|x|^{11}}{11!} \cdot 1 < 0.01$, i.e., $|x| < 0.95$.

13. $\cos(-x) = \sum_{n=0}^{\infty} (-1)^n(-x)^{2n}/(2n)! = \sum_{n=0}^{\infty} (-1)^n x^{2n}/(2n)!$; $|R_{10}| \le 1 \cdot |x|^{11}/(11)! \le 0.01$ for $|x| < 3.22$; graphically, $|P_{10}(x) - \cos(-x)| < 0.01$ for $|x| < 3.624$.

15. $\cosh x = \frac{1}{2}\left[1 + x + \frac{x^2}{2!} + \frac{x^3}{3!} + \cdots + 1 - x + \frac{x^2}{2!} - \frac{x^3}{3!} + \cdots\right] = \left[1 + \frac{x^2}{2!} + \frac{x^4}{4!} + \cdots\right] = \sum_{n=0}^{\infty} \frac{x^{2n}}{(2n)!}$; $P_{10}(x) = \sum_{n=0}^{5} \frac{x^{2n}}{(2n)!}$; graphically, $|P_{10}(x) - \cosh x| < 0.01$ when $|x| < 3.581$; $|R_{10}(x)| < 0.01$ when $\left|\frac{x^{12}}{12!}\cosh c\right| < 0.01$; if we assume $|x| < 4$, then $|\cosh c| < \cosh 4 < 27.4$; $\left[\frac{(0.01)(12!)}{27.4}\right]^{1/12} = 2.734$.

17. $\cos x - \left[1 - \frac{x^2}{2}\right] = \left[1 - \frac{x^2}{2!} + \frac{x^4}{4!} - \frac{x^6}{6!} + \cdots\right] - \left[1 - \frac{x^2}{2}\right] = \sum_{n=2}^{\infty} (-1)^n x^{2n}/(2n)!$; $P_{10} = \sum_{n=2}^{5} \frac{(-1)^n x^{2n}}{(2n)!}$; graphically, $|P_{10} - \text{function}| < 0.01$ when $|x| < 3.624$; $|R_{10}| = \frac{|x^{12}|}{12!}|\sin c| \le \frac{|x^{12}|}{12!} < 0.01$ when $|x| < 3.603$.

19. $f(x) = \frac{1}{1+x}$; $f'(x) = \frac{-1}{(1+x)^2}$; $f''(x) = \frac{2}{(1+x)^3}$; $f'''(x) = \frac{-6}{(1+x)^4} \Rightarrow f(0) = 1$, $f'(0) = -1$, $f''(0) = 2$, $f'''(c) = \frac{-6}{(1+c)^4}$; $\frac{1}{1+x} = P_2(x) + R_2(x) = 1 - x + \frac{2x^2}{2} - \frac{6x^3}{(1+c)^4}\frac{1}{3!} = [1 - x + x^2] - x^3/(1+c)^4$, where c is between 0 and x.

21. $f(x) = \ln(1+x)$; $f'(x) = \frac{1}{1+x}$; $f''(x) = \frac{-1}{(1+x)^2}$; $f'''(x) = \frac{2}{(1+x)^3}$; $\Rightarrow f(0) = 0$, $f'(0) = 1$, $f''(0) = -1$, $f'''(c) = 2/(1+c)^3$. $\ln(1+x) = P_2(x) + R_2(x) = [+x - x^2/2] + [2x^3/(1+c)^3]/3!$

23. From Example 5, $\sin x = x - \frac{x^3 \cos c}{3!}$.

25. All derivatives of f, evaluated at a, are $f^{(k)}(a) = e^a$. $e^x = \sum_{n=0}^{\infty} (x-a)^n e^a/n! = e^a[\sum_{n=0}^{\infty} (x-a)^n/n!]$.

27. $x - x^3/6$ is actually P_4. Using the Taylor remainder, $|R_4| = \frac{|x|^5}{5!}|\cos c| \le \frac{|x|^5}{120} \le 5 \times 10^{-4}$ for $|x| \le (5 \cdot 120 \times 10^{-4})^{1/5} = 0.56\ldots$.

29. $|\sin x - x| = |R_3(x)| \le \frac{|x|^3}{6} \cdot 1 \le 10^{-9}/6 < 1.67 \times 10^{-10}$. For $x > 0$, $x > \sin x$.

31. $|R_2| = \frac{|x|^3}{3!} e^c \le \frac{(0.1)^3 e^{0.1}}{6} = 1.84 \times 10^{-4}$.

33. $|R_4| = \frac{|x|^5 |\cos c|}{5!} \le \frac{(0.5)^5(1)}{5!} = 2.6 \times 10^{-4}$.

35. This is the series for $\sin(0.1) = 0.0998334\ldots$.

37. $\sin x = x - \frac{x^3}{3!} + \frac{x^5}{5!} - \frac{x^7}{7!} + \cdots$; differentiating each term gives $1 - \frac{x^2}{2!} + \frac{x^4}{4!} - \frac{x^6}{6!} + \cdots = \cos x$. Formally, $D_x e^x = D_x \sum_{n=0}^{\infty} x^n/n! = \sum_{n=0}^{\infty} D_x x^n/n! = \sum_{n=1}^{\infty} x^{n-1}/(n-1)! = \sum_{k=0}^{\infty} x^k/k!$.

39. $e^x = 1 + x + \frac{x^2}{2} + \frac{x^3}{6} + \frac{x^4}{24} + \frac{x^5}{5!} + \cdots$, $\sin x = x - \frac{x^3}{3!} + \frac{x^5}{5!} + \cdots$; $e^x \sin x = x + x^2 + x^3(\frac{1}{2} - \frac{1}{3!}) + x^4(\frac{1}{6} - \frac{1}{3!}) + x^5(\frac{1}{24} - \frac{1}{2 \cdot 3!} + \frac{1}{5!}) + x^6(\frac{1}{5!} - \frac{1}{6 \cdot 3!} + \frac{1}{5!}) + \cdots = x + x^2 + \frac{1}{3}x^3 + 0 \cdot x^4 + x^5(-\frac{1}{30}) + x^6(-\frac{1}{90}) + \cdots$

41. $P_3 = x - \frac{x^3}{6} = \sin x - R_3 = \sin x - \frac{x^4}{4!}\sin c < \sin x$ if $0 < |x| < 1$; similarly $\sin x < x$ for $0 < |x| < 1$. Hence $x - \frac{x^3}{6} < \sin x < x \Rightarrow 1 - \frac{x^2}{6} < \frac{\sin x}{x} < 1$ if $x > 0$. If $x < 0$, $x = -t$, say, then $\frac{\sin x}{x} = \frac{-\sin t}{-t} = \frac{\sin t}{t}$ and $1 - \frac{t^2}{6} < \frac{\sin t}{t} < 1$, or $1 - \frac{x^2}{6} < \frac{\sin x}{x} < 1$.

43. a) $e^{i\pi} = \cos \pi + i \sin \pi = -1$, b) $e^{i\pi/4} = \cos\frac{\pi}{4} + i \sin\frac{\pi}{4} = \frac{1}{\sqrt{2}}(1+i)$; c) $e^{-i\pi/2} = \cos(-\frac{\pi}{2}) + i\sin(-\frac{\pi}{2}) = -i$

45. $e^{i\theta} = 1 + i\theta + \frac{(i)^2\theta^2}{2} + \frac{(i)^3\theta^3}{3!} + \frac{(i)^4\theta^4}{4!} + \cdots$, $e^{-i\theta} = 1 - i\theta + \frac{(i)^2\theta^2}{2} + \frac{(-i)^3\theta^3}{3!} + \frac{(i)^4\theta^4}{4!} + \cdots$, $\frac{1}{2}[e^{i\theta} + e^{-i\theta}] = \frac{1}{2}[2 + \frac{2(i)^2\theta^2}{2!} + \frac{2(i)^4\theta^4}{4!} + \cdots] = \frac{2}{2}[1 - \frac{\theta^2}{2!} + \frac{\theta^4}{4!} + \cdots] = \cos\theta$; $\frac{1}{2i}[e^{i\theta} - e^{-i\theta}] = \frac{1}{2i}[2i\theta - \frac{i\theta^3}{3!} + \cdots] = \sin\theta$.

47. $\int e^{(a+ib)x}dx = \int e^{ax}\cos bx\ dx + i\int e^{ax}\sin bx\ dx = \frac{1}{a^2+b^2}[a-ib][e^{ax}][\cos bx + i\sin bx] + c = \frac{e^{ax}}{a^2+b^2}[a\cos bx + b\sin bx] + \frac{ie^{ax}}{a^2+b^2}[a\sin bx - b\cos bx] + c_1 + ic_2$. Equating real and imaginary parts gives $\int e^{ax}\cos bx\ dx = \frac{e^{ax}}{a^2+b^2}[a\cos bx + b\sin bx] + c_1$ and $\int e^{ax}\sin bx\ dx = \frac{e^{ax}}{a^2+b^2}[a\sin bx - b\cos bx] + c_2$.

9.8 FURTHER CALCULATIONS WITH TAYLOR SERIES

1. $\cos \approx \cos 1 + (x-1)(-\sin 1) + \frac{(x-1)^2}{2}(-\cos 1) + \frac{(x-1)^3}{6}\sin 1$. $|\text{error}| \le \frac{|x-1|^4}{4!}\cdot 1$.

3. $e^x \approx e^{0.4} + (x-0.4)e^{0.4} + \frac{(x-0.4)^2}{2}e^{0.4} + \frac{(x-0.4)^3}{6}e^{0.4}$; $|\text{error}| \le \frac{|x-0.4|^4}{4!}e^{0.4}$.

5. $\cos x \approx \cos 69 + (x-69)(-\sin 69) + \frac{(x-69)^2}{2}(-\cos 69) + \frac{(x-69)^3}{6}\sin 69$; $|\text{error}| \le \frac{|x-69|^4}{4!}\cdot 1$.

7. $f = (1+x)^3$; $f' = 3(1+x)^2$, $f'' = 6(1+x)$; $f''' = 6$, $f^{\text{iv}} \equiv 0$; $f(0) = 1$, $f'(0) = 3$, $f''(0) = 6$, $f'''(0) = 6 \Rightarrow (1+x)^3 = 1 + x\cdot 3 + \frac{x^2}{2}6 + \frac{x^3}{6}\cdot 6 + \frac{x^4}{4!}\cdot 0$.

9. – 13. Enter the functions: $y1 = 1 + Mx + (M(M-1)/2)x^2 + (M(M-1)(M-2)/6)x^3 + (M(M-1)(M-2)(M-3)/24)x^4 + (M(M-1)(M-2)(M-3)(M-4)/120)x^5$ and $y2 = (1+x)^\wedge M$. Store the desired value of M, then graph $y1 - y2$ on $[a,b]$ by $[-0.01, 0.01]$; for part b) graph $y1$ and $y2$

	x-Range	Viewing Window
9.	$(-0.88, 1.14)$	$(-2,6)$ by $(-2,15)$
11.	$(-0.35, 0.4)$	$(-1,6)$ by $(0,3)$
13.	$(-0.27, 0.26)$	$(-1,4)$ by $(0,1.5)$

15. $\frac{\sin t}{t} - P_{21} = R$, where $|R| = \frac{t^{22}}{(23)!}|\cos c| \le \frac{t^{22}}{23!}$; $\left|\int_0^x \frac{\sin t}{t}dt - \int_0^x P_{21}(t)dt\right| \le \int_0^x \left|\frac{t^{22}}{23!}\right|dt = \frac{x^{23}}{(23)(23)!}$; $\frac{5^{23}}{(23)(23)!} < 2.005\times 10^{-8}$, $\frac{8^{23}}{(23)(23)!} < 9.93\times 10^{-4}$, $\frac{10^{23}}{(23)(23)!} < 0.169$.

17. $\frac{1-\cos t}{t^2} = \frac{\frac{t^2}{2!} - \frac{t^4}{4!} + \frac{t^6}{6!} + \cdots}{t^2} = -\sum_{k=1}^{\infty} \frac{(-1)^k t^{2k-2}}{(2k)!} = \sum_{k=1}^{\infty} \frac{(-1)^{k+1} t^{2k-2}}{(2k)!}$;

$\int_0^x \frac{1-\cos t}{t^2}dt = \sum_{k=1}^{\infty} \frac{(-1)^{k+1}x^{2k-1}}{(2k-1)(2k)!}$.

19. $\int_0^{0.1} P_{10}(x)dx = \int_0^{0.1} \sum_{k=1}^{6} \frac{(-1)^{k+1}x^{2k-1}}{(2k-1)!x}dx = \sum_1^6 \frac{(-1)^{k+1}}{(2k-1)!}\int_0^{0.1} x^{2k-2}dx = \sum_1^6 \frac{(-1)^{k+1}}{(2k-1)(2k-1)!}(0.1)^{2k-1} = 0.999444612$; using NINT gives 0.0999444601.

21. $\int_0^{0.1} \frac{1-\cos x}{x^2} dx = -\int_0^{0.1} \sum_{n=1}^{\infty} \frac{(-1)^n x^{2n-2}}{(2n)!} dx$; $\int_0^{0.1} P_{10}(x) dx =$ $\sum_{n=1}^{5} \int_0^{0.1} \frac{(-1)^{n+1} x^{2n-2}}{(2n)!} dx = \sum_{n=1}^{5} \left. \frac{(-1)^{n+1} x^{2n-1}}{(2n-1)(2n)!} \right]_0^{0.1} = 0.49986114\ldots$.

23. $\ln(1+x) = 1 + x + \frac{x^2}{2} + \frac{x^3}{3} + \frac{x^4}{4} + \cdots$, $\ln(1-x) = 1 - x + \frac{x^2}{2} - \frac{x^3}{3} + \frac{x^2}{4} + \cdots$, $\ln(1+x) - \ln(1-x) = \ln\frac{1+x}{1-x} = 2[x + \frac{x^3}{3} + \frac{x^5}{5} + \cdots]$.

PRACTICE EXERCISES, CHAPTER 9

1. $\lim_{n\to\infty} a_n = \lim_{n\to\infty} 1 + \lim_{n\to\infty} \frac{(-1)^n}{n} = 1$; converges to 1.

3. This sequence is $1, 0, -1, 0, 1, 0, -1, 0, \ldots$; diverges.

5. $\lim_{n\to\infty} a_n = 2\lim_{n\to\infty} \frac{\ln n}{n} = 2\lim_{n\to\infty} \frac{\frac{1}{n}}{1} = 0$; converges to 0.

7. $\lim_{n\to\infty} a_n = \lim_{n\to\infty} \frac{3}{\sqrt[n]{n}} = \frac{3}{1} = 3$; converges to 3.

9. $\lim_{n\to\infty} a_n = \lim_{n\to\infty} \frac{(-4)^n}{n!} = 0$, $\left(\frac{x^n}{n!} \to 0 \text{ for all } x\right)$; converges to 0.

11. $\lim_{n\to\infty} a_n = \lim_{n\to\infty}(n+1) = +\infty$; diverges.

13. $\sum^k \ln\left(\frac{n}{n+1}\right) = \ln\left(\prod^k \frac{n}{n+1}\right) = \ln\left(\frac{1}{2} \cdot \frac{2}{3} \cdot \frac{3}{4} \cdot \cdots \frac{k}{k+1}\right) = \ln 1 - \ln k$; $\lim_{k\to\infty} \ln 1 - \ln k = -\infty$; diverges.

15. $1 + \left(\frac{1}{e}\right) + \left(\frac{1}{e}\right)^2 + \cdots$ converges to $\frac{1}{1-1/e} = \frac{e}{e-1}$.

17. $a_n = 1.05^{-1} a_{n-1} = 1.05^{-2} a_{n-2} = \cdots = (1.05)^{-n} a_0$; $\sum_{n=0}^{\infty} a_n = a_0 \sum_{n=0}^{\infty} \left(\frac{1}{1.05}\right)^n = 125\left(\frac{1}{1-1/1.05}\right) = 2625$.

19. $\sum_{n=1}^{\infty} \frac{1}{n^p}$ diverges if $p \le 1$; here $p = \frac{1}{2}$; diverges.

21. $\lim_{n\to\infty} a_n = 0$; by the Alternating Series Test, the series converges. By #17, it does not converge absolutely; conditionally convergent. $|\text{error of } S_{9999}| < |a_{10000}| = \frac{1}{100} = 0.01$; on the other hand $S_{1001} = -0.621 < S < S_{1000} = -0.589$. Taking the average, $S = 0.605$ with an error less than 0.02.

23. The series converges by the Alternating Series Test ($\ln x$ is increasing $\Rightarrow$ $1/\ln x$ is decreasing). $\ln(n+1) < n+1 \Rightarrow \frac{1}{\ln(n+1)} > \frac{1}{n+1}$; since $\sum \frac{1}{n+1}$ diverges so does $\sum 1/\ln(n+1)$; conditionally convergent. $S_{1000} = -0.85\ldots$, $S_{1001} = -0.997$; since $|a_{10001}| = 0.087$; the series converges very slowly.

25. By the Limit Comparison Test, $\lim_{n\to\infty} = \left(\frac{1}{n\sqrt{n^2+11}}\right)/(1/n^2) = \lim_{n\to\infty}\frac{n}{\sqrt{n^2+1}} = 1$; since $\sum\frac{1}{n^2}$ converges, the series converges absolutely. $S_{100} = -0.55155$; $|\text{error}| \le \frac{1}{11\sqrt{122}} = 0.008$.

27. $\sum_{n=1}^{\infty}\frac{n+1}{n!} = \sum_{n=1}^{\infty}\frac{1}{(n-1)!} + \sum_{n=1}^{\infty}\frac{1}{n!} = \sum_{k=0}^{\infty}\frac{1}{k!} + \sum_{n=1}^{\infty}\frac{1}{n!} = e^1 + (e^1 - 1) = 2e^1 - 1 = 2e - 1 = 4.4365\ldots$.

29. Since $e^{-3} = \sum_{n=0}^{\infty}\frac{(-3)^n}{n!} = 1 + \sum_{n=1}^{\infty}\frac{(-3)^n}{n!}$, the series is convergent; sum is $e^{-3} - 1$.

31. $\lim_{n\to\infty}\left|\frac{a_{n+1}}{a_n}\right| = \lim_{n\to\infty}\left|\frac{(x+2)^{n+1}}{3^{n+1}(n+1)}\cdot\frac{3^n n}{(x+2)^n}\right| = \frac{|x+2|}{3}\lim_{n\to\infty}\frac{n}{n+1} = \frac{|x+2|}{3} < 1$ for $-3 < x+2 < 3$ or $-5 < x < 1$. When $x = -5$ we have $\sum\frac{(-1)^n 3^n}{3^n n} = \sum(-1)^n/n$ which converges; when $x = 1$ we have $\sum\frac{3^n}{3^n n} = \sum\frac{1}{n}$ which diverges. S_{20} will have its greatest error at -3; solving $\frac{(x+2)^{21}}{3^{21}\cdot 21} < 0.01$ gives $|x+2| \le 2.78$; if $x+2 = -2+0.78$, $x = -4.78$.

33. $\lim_{n\to\infty}\left|\frac{x^{n+1}}{(n+1)^{n+1}}\cdot\frac{n^n}{x^n}\right| = |x|\lim_{n\to\infty}\left|\frac{n^n}{(n+1)^{n+1}}\right| = |x|\lim_{n\to\infty}\left|\left(\frac{n}{n+1}\right)^n\cdot\frac{1}{n+1}\right| = |x|\left(\frac{1}{e}\right)\cdot 0 = 0 \Rightarrow$ the series converges absolutely for all x.

35. $\lim_{n\to\infty}\left|\frac{a_{n+1}}{a_n}\right| = \lim_{n\to\infty}\left|\frac{(x-1)^{n+1}}{(n+1)^2}\cdot\frac{n^2}{(x-1)^n}\right| = |x-1|\lim_{n\to\infty}\frac{n^2}{(n+1)^2} = |x-1|\cdot 1 \Rightarrow$ the series converges for $|x-1| < 1$, or $0 < x < 2$; at $x = 0$ we have $\sum\frac{(-1)^{n-1}}{n^2}(-1)^n = \sum\frac{(-1)}{n^2}$ which converges; at $x = 2$, $\sum\frac{(-1)^{n-1}}{n^2}$ converges by the alternating series test.

37. This is the series for $\frac{1}{1+x}$, where $x = \frac{1}{4}$; sum is $\frac{1}{1+1/4} = 0.8$.

39. This is the series for $\sin x$, at $x = \pi$; $\sin\pi = 0$.

41. This is the series for e^x at $x = \ln 2$, sum is $e^{\ln 2} = 2$.

43. $f = (3+x^2)^{1/2}$; $f' = \frac{1}{2}(3+x^2)^{-1/2}(2x)$; $f'' = (3+x^2)^{-1/2} + x(-\frac{1}{2})(3+x^2)^{-3/2}\ (2x) = (3+x^2)^{-1/2} - x^2(3+x^2)^{-3/2}$; $f''' = -\frac{1}{2}(3+x^2)^{-3/2}2x - 2x(3+x^2)^{-3/2} + \frac{3x^2}{2}(3+x^2)^{-5/2}2x = -3x(3+x^2)^{-3/2} + 3x^3(3+x^2)^{-5/2}$; $f(-1) = 2$, $f'(-1) = -\frac{1}{2}$, $f''(-1) = \frac{3}{8}$, $f'''(-1) = \frac{3}{8} - \frac{3}{32} = \frac{9}{32}$. $f(x) = 2 + (-\frac{1}{2})(x+1) + (\frac{3}{8})\frac{1}{2}(x+1)^2 + (\frac{9}{32})\frac{1}{6}(x+1)^3 + \cdots$. Graphing abs ($\sqrt{3+x^2}$ - first four non-zero terms) shows the error is less than 0.01 for $-2.143 < x < 0.238$.

45. $(\sin x)2(\cos x) = 2(x - \frac{x^3}{3!} + \frac{x^5}{5!} - \frac{x^7}{7!})(1 - \frac{x^2}{2!} + \frac{x^4}{4!} - \frac{x^6}{6!} + \cdots) =$ $2\left[x + x^3(-\frac{1}{3!} - \frac{1}{2!}) + x^5(\frac{1}{5!} + \frac{1}{2!3!} + \frac{1}{4!}) + x^7(-\frac{1}{7!} - \frac{1}{2!5!} - \frac{1}{3!4!} - \frac{1}{6!})\right] =$ $2x - 2 \cdot \frac{4}{6}x^3 + 2\left[\frac{1}{120} + \frac{1}{12} + \frac{1}{24}\right]x^5 - \frac{2}{7!}\left[1 + \frac{7\cdot6}{2} + \frac{7\cdot6\cdot5}{6} + 7\right]x^7 + \cdots =$ $2x - \frac{2^3x^3}{3!} + \frac{2^5x^5}{5!} - \frac{2^7x^7}{7!} + \ldots.$

47. $\int_0^{1/2} e^{-x^3}dx = \int_0^{1/2}(1 - x^3 + \frac{x^6}{2!} - \frac{x^9}{3!} + \frac{x^{12}}{4!})dx + \int_0^{1/2} R_4\, dx$ where $|R_4| \leq \int_0^{1/2} \frac{x^{14}}{5!}\, dx = \left.\frac{x^{15}}{15\cdot5!}\right|_0^{1/2} = 1.69\,E{-}08$ which is a little too large for the problem. For error $< 10^{-8}$ take $\int_0^{1/2}(1-x^3+\frac{x^6}{2!}-\frac{x^9}{3!}+\frac{x^{12}}{4!}-\frac{x^{15}}{5!})dx = 0.48491714$ (NINT).

49. a) The MacLaurin series is $\sum_{n=0}^{\infty} \frac{x^n}{n!}\, 0 = 0$ for all x. It converges for all x; it converges to $f(x)$ only at $x = 0$. b) $e^{-1/x^2} = 0 + R_n(x)$; hence $R_n = e^{-1/x^2}$

CHAPTER 10

PLANE CURVES, PARAMETRIZATIONS, AND POLAR COORDINATES

10.1 CONIC SECTIONS AND QUADRATIC EQUATIONS

1. $y = \frac{x^2}{4p} = \frac{x^2}{16}$ **3.** $y = -\frac{x^2}{4p} = -\frac{x^2}{12}$ **5.** $x = -\frac{y^2}{4p} = -\frac{y^2}{12}$. $x = -\frac{y^2}{12}$

7. $y = 4x^2 = \frac{x^2}{4(\frac{1}{16})}$, $p = \frac{1}{16}$. Focus: $(0, \frac{1}{16})$, directrix: $y = -\frac{1}{16}$.

9. $y = -3x^2 = \frac{x^2}{4(-\frac{1}{12})}$. Focus: $(0, -\frac{1}{12})$. Directrix: $y = \frac{1}{12}$.

11. Graph $y = \frac{x^2}{2}$ in $[-10.6, 10.6]$ by $[0, 12.5]$ to check you result. (We have used the "screen-squaring" feature of our calculator to help determine the viewing rectangle.)

13. Graph $y = \sqrt{8x}$ and $y = -\sqrt{8x}$ together in $[-9.4, 14.4]$ by $[-7, 7]$ to check your result.

15. $\frac{x^2}{4} + \frac{y^2}{9} = 1$. $\frac{y^2}{9} = 1 - \frac{x^2}{4} = \frac{4-x^2}{4}$, $y^2 = \frac{9}{4}(4 - x^2)$, $y = \pm\frac{3}{2}\sqrt{4 - x^2}$. Graph $y = \frac{3}{2}\sqrt{4 - x^2}$ and $y = -\frac{3}{2}\sqrt{x - x^2}$ in $[-5.1, 5.1]$ by $[-3, 3]$.

17. $\frac{y^2}{4} - x^2 = 1$. Graph $y_1 = 2\sqrt{1 + x^2}$ and $y_2 = -y_1$ in $[-13.6, 13.6]$ by $[-8, 8]$.

19. $64x^2 - 36y^2 = 2304$, $\frac{64x^2}{2304} - \frac{36y^2}{2304} = 1$, $\frac{x^2}{6^2} - \frac{y^2}{8^2} = 1$. $c = \sqrt{6^2 + 8^2} = 10$, $e = \frac{c}{a} = \frac{10}{a} = \frac{5}{3}$. Graph $y_1 = (4/3)\sqrt{x^2 - 36}$, $y_2 = -y_1$, $y_3 = (4/3)x$ and $y_4 = -y_3$ in $[-34, 34]$ by $[-20, 20]$. Foci: $(\pm 10, 0)$.

21. $8y^2 - 2x^2 = 16$, $\frac{y^2}{2} - \frac{x^2}{8} = 1$. $c = \sqrt{2+8} = \sqrt{10}$. $e = \frac{\sqrt{10}}{\sqrt{2}} = \sqrt{5}$. Foci: $(0, \pm\sqrt{10})$, asymptotes: $y = \pm\frac{\sqrt{2}x}{2\sqrt{2}} = \pm\frac{x}{2}$. Graph $y_1 = \frac{\sqrt{x^2+8}}{2}$, $y_2 = -y_1$, $y_3 = \frac{x}{2}$ and $y_4 = -\frac{x}{2}$ in $[-8.5, 8.5]$ by $[-5, 5]$.

23. $169x^2 + 25y^2 = 4225$, $\frac{x^2}{25} + \frac{y^2}{169} = 1$. $c = \sqrt{169 - 25} = 12$, $e = \frac{c}{a} = \frac{12}{13}$, foci: $(0, \pm 12)$. Graph $y_1 = \frac{13}{5}\sqrt{25 - x^2}$ and $y_2 = -y_1$ in $[-22, 22]$ by $[-13, 13]$.

25. $8x^2 - 2y^2 = 16$, $\frac{x^2}{2} - \frac{y^2}{8} = 1$. $c = \sqrt{2+8} = \sqrt{10}$, $e = \frac{c}{a} = \frac{\sqrt{10}}{\sqrt{2}} = \sqrt{5}$. Foci: $(\pm\sqrt{10}, 0)$, asymptotes: $y = \pm\frac{\sqrt{8}}{\sqrt{2}}x = \pm 2x$. Graph $y = 2\sqrt{x^2 - 2}$, $y = -2\sqrt{x^2 - 2}$, $y = 2x$ and $y = -2x$ in $[-17, 17]$ by $[-10, 10]$.

27. $9x^2+10y^2=90$, $\frac{x^2}{10}+\frac{y^2}{9}=1$. $c=\sqrt{10-9}=1$, $e=\frac{c}{a}=\frac{1}{\sqrt{10}}$, foci: $(\pm 1,0)$. Graph $y_1=\frac{3}{\sqrt{10}}\sqrt{10-x^2}$ and $y_2=-y_1$ in $[-5.1,5.1]$ by $[-3,3]$.

29. $x^2-y^2=1$, $c=\sqrt{1+1}=\sqrt{2}$, $e=\frac{c}{a}=\sqrt{2}$, foci: $(\pm\sqrt{2},0)$. Graph $y=\sqrt{x^2-1}$, $y=-\sqrt{x^2-1}$, $y=x$ and $y=-x$ in $[-8.2,8.2]$ by $[-4.9,4.9]$.

31. $y^2-x^2=4$, $\frac{y^2}{4}-\frac{x^2}{4}=1$, $c=\sqrt{4+4}=2\sqrt{2}$, $e=\frac{2\sqrt{2}}{2}=\sqrt{2}$, foci: $(0,\pm 2\sqrt{2})$. Graph $y=\sqrt{x^2+4}$, $y=-\sqrt{x^2+4}$, $y=x$ and $y=-x$ in $[-8.2,8.2]$ by $[-4.9,4.9]$.

33. $3x^2+2y^2=6$, $\frac{x^2}{2}+\frac{y^2}{3}=1$, $c=\sqrt{3-2}=1$, $e=\frac{1}{\sqrt{3}}$. Graph $y=\sqrt{1.5}\sqrt{2-x^2}$ and $y=-\sqrt{1.5}\sqrt{2-x^2}$ in $[-2.9,2.9]$ by $[-\sqrt{3},\sqrt{3}]$. Foci: $(0,\pm 1)$.

35.

	$y=-x^2/4$	$x=-y^2/4$
Focal axis:	The y-axis	The x-axis
Focus:	$(0,-1)$	$(-1,0)$
Vertex:	$(0,0)$	$(0,0)$
Directrix:	$y=1$	$x=1$

37. Volume of parabolic solid $=2\pi\int_0^{b/2}x(h-\frac{4h}{b^2}x^2)dx=2\pi h\int_0^{b/2}(x-\frac{4}{b^2}x^3)dx=2\pi h[\frac{x^2}{2}-\frac{x^4}{3b^2}]_0^{b/2}=\frac{\pi h b^2}{8}$. $\frac{3}{2}$ (volume of cone) $=\frac{3}{2}(\frac{1}{3}\pi r^2 h)=\frac{1}{2}\pi(\frac{b}{2})^2 h=\frac{\pi h b^2}{8}$.

39. We require $e=\frac{c}{a}=\frac{4}{5}$. Let us try to find a solution with $c=4$, $a=5$. $4=c=\sqrt{a^2-b^2}=\sqrt{25-b^2}$ yielding $b=3$. Thus the graph of $\frac{x^2}{25}+\frac{y^2}{9}=1$ is an example of an ellipse with $e=\frac{4}{5}$.

41. Because of the symmetry with respect to both axes and the origin, the coordinate axes break the rectangle into four smaller rectangles, one lying in each of the four quadrants. The rectangle in the first quadrant has vertices $(0,0),(0,1),(2,0)$ and $(x,y)=(x,\frac{\sqrt{4-x^2}}{2})$ because it lies on the upper half of the ellipse. The rectangle in the first quadrant has area $A(x)=xy=\frac{1}{2}x\sqrt{4-x^2}$. $A'(x)=\frac{1}{2}[-\frac{x^2}{\sqrt{4-x^2}}+\sqrt{4-x^2}]=\frac{2-x^2}{\sqrt{4-x^2}}$. $A(x)$ is maximized when $x=\sqrt{2}$ and $A(\sqrt{2})=1$. The dimensions of the original rectangle of maximal area are $2x=2\sqrt{2}$ by $2y=\sqrt{4-x^2}=\sqrt{2}$ and it has area $4A(\sqrt{2})=4$.

43. Volume $=2$(Volume of top half) $=2\pi\int_0^3(1+y^2)dy=2\pi[y+\frac{y^3}{3}]_0^3=24\pi$.

45. a) $y^2 = 4px$. $2yy' = 4p$, $y' = \frac{2p}{y}$. Tan β is the slope of L. The slope of L is the slope of the parabola at (x_0, y_0), so $\tan\beta = \frac{2p}{y_0}$. b) It is clear that $\tan\phi = \frac{opp}{adj} = \frac{y_0}{x_0-p}$. c) We see that $\beta + \alpha + (\frac{\pi}{2} - \phi) = \frac{\pi}{2}$ and so $\alpha = \phi - \beta$. Tan $\alpha = \tan(\phi - \beta) = \frac{\tan\phi - \tan\beta}{1+\tan\phi\tan\beta} = \frac{\frac{y_0}{x_0-p} - \frac{2p}{y_0}}{1+\frac{y_0}{x_0-p}\frac{2p}{y_0}} = \frac{2p}{y_0}$ using $y_0^2 = 4px_0$.

10.2 THE GRAPHS OF QUADRATIC EQUATIONS IN x AND y

1. $x^2 - y^2 - 1 = 0$. $B^2 - 4AC = 0^2 - 4(1)(-1) = 4 > 0$: the equation represents a hyperbola. Graph $y_1 = \sqrt{x^2-1}$ and $y_2 = -y_1$ in $[-17, 17]$ by $[-10, 10]$.

3. $y^2 - 4x - 4 = 0$. $B^2 - 4AC = 0^2 - 4(0)(1) = 0$. Parabola. Graph $y_1 = 2\sqrt{x+1}$ and $y_2 = -y_1$ in $[-5.8, 7.8]$ by $[-4, 4]$. Shift the graph of $x = \frac{y^2}{4}$ horizontally left one unit.

5. $x^2 + 4y^2 - 4x - 8y + 4 = 0$. $B^2 - 4AC = 0^2 - 4(1)(4) = -16 < 0$, ellipse. $4y^2 - 8y + (x^2 - 4x + 4) = 0$. Let $y_1 = \sqrt{64 - 16(x^2 - 4x + 4)}$. We graph $y_2 = (8 + y_1)/8$ and $y_3 = (8 - y_1)/8$ in $[0, 4]$ by $[-3.8, 2.3]$.

7. $x^2 + 4xy + 4y^2 - 3x = 6$. $B^2 - 4AC = 16 - 4(1)4 = 0$, parabola. $4y^2 + 4xy + (x^2 - 3x - 6) = 0$. Let $y_1 = \sqrt{(4x)^2 - 16(x^2 - 3x - 6)}$. Graph $y_2 = (-4x + y_1)/8$ and $y_3 = (-4x - y_1)/8$ in $[-8.4, 13.6]$ by $[-9.1, 3.9]$. $\theta = \frac{1}{2}\tan^{-1}(\frac{B}{A-C}) = \frac{1}{2}\tan^{-1}(\frac{4}{1-4}) = \frac{1}{2}\tan^{-1}(-\frac{4}{3}) = -0.464$.

9. $xy + y^2 - 3x = 5$. $B^2 - 4AC = 1 - 0 = 1 > 0$, hyperbola. $y^2 + xy + (-3x - 5) = 0$. Let $y_1 = \sqrt{x^2 + 12x + 20}$. Graph $y_2 = (-x + y_1)/2$ and $y_3 = (-x - y_1)/2$ in $[-49, 43]$ by $[-27.4, 26.8]$, $\theta = \frac{1}{2}\tan^{-1}(\frac{B}{A-C}) = \frac{1}{2}\tan^{-1}(-1) = -\frac{\pi}{8}$.

11. $x^2 - y^2 = 1$. $B^2 - 4AC = 0 - 4(1)(-1) = 4 > 0$, hyperbola. Graph $y = \sqrt{x^2-1}$ and $y = -\sqrt{x^2-1}$ in $[-8.5, 8.5]$ by $[-5, 5]$.

13. $xy = 1$. $B^2 - 4AC = 1^2 - 4(0)(0) = 1 > 0$, hyperbola. Graph $y = \frac{1}{x}$ in $[-5.1, 5.1]$ by $[-3, 3]$. Since $A = C$, $\theta = \frac{\pi}{4}$.

15. $2x^2 + xy + x - y + 1 = 0$. $B^2 - 4AC = 1^2 - 4(2)(0) = 1 > 0$, hyperbola. Solving for y, we obtain $y = -(2x^2 + x + 1)/(x - 1)$. Graph y in dot format in $[-8, 8.7]$ by $[-19, 9.1]$. $\theta = \frac{1}{2}\tan^{-1}\frac{B}{A-C} = \frac{1}{2}\tan^{-1}\frac{1}{2} = 0.232$.

17. $x^2 - 3xy + 3y^2 + 6y = 7$. $B^2 - 4AC = 9 - 4(1)3 = -3 < 0$, ellipse. $3y^2 + (6-3x)y + (x^2-7) = 0$. Let $y_1 = \sqrt{(6-3x)^2 - 4(3)(x^2-7)}$. Graph $y_2 = \frac{-(6-3x)+y_1}{6}$ and $y_3 = \frac{-(6-3x)-y_1}{6}$ in $[-20.3, 8.2]$ by $[-12.9, 3.9]$. $\theta = \frac{1}{2}\tan^{-1}\frac{-3}{1-3} = \frac{1}{2}\tan^{-1}(1.5) = 0.491$.

19. $6x^2+3xy+2y^2+17y+2 = 0$. $B^2-4AC = 9-48 < 0$, ellipse. $2y^2+(3x+17)y+(6x^2+2) = 0$. Let $y_1 = \sqrt{(3x+17)^2 - 4(2)(6x^2+2)}$. Graph $y_2 = \frac{-(3x+17)+y_1}{4}$ and $y_3 = \frac{-(3x+17)-y_1}{4}$ in $[-11.9, 14.4]$ by $[-12.6, 2.9]$. $\theta = \frac{1}{2}\tan^{-1}\frac{3}{6-2} = 0.322$.

21. $xy = 2$. $B^2 - 4AC = 1^2 - 0 > 0$, hyperbola. $y = \frac{2}{x}$ may be graphed in $[-8.5, 8.5]$ by $[-5, 5]$. $\theta = \frac{1}{2}\cot^{-1}(\frac{A-C}{B}) = \frac{1}{2}\cot^{-1} 0 = \frac{\pi}{4}$. $x' = \frac{\sqrt{2}}{2}(x-y)$, $y' = \frac{\sqrt{2}}{2}(x+y)$. (x', y') satisfies $xy = 2$ so we obtain $x'y' = 2$, $\frac{\sqrt{2}}{2}(x-y)\frac{\sqrt{2}}{2}(x+y) = 2$, $x^2 - y^2 = 4$.

23. $x^2 - \sqrt{3}xy + 2y^2 = 1$. $B^2 - 4AC = 3 - 4(1)2 = -5 < 0$, ellipse. $2y^2 - \sqrt{3}xy + (x^2-1) = 0$. Let $y_1 = \sqrt{3x^2 - 4(2)(x^2-1)}$. Graph $y_2 = (\sqrt{3}x + y_1)/4$ and $y_3 = (\sqrt{3}x - y_1)/4$ in $[-2.6, 2.1]$ by $[-1.5, 1.3]$. $\theta = \frac{1}{2}\tan^{-1}\frac{B}{A-c} = \frac{1}{2}\tan^{-1}(\sqrt{3}) = \frac{\pi}{6}$. $x' = x\cos\frac{\pi}{6} - y\sin\frac{\pi}{6} = \frac{\sqrt{3x}-6}{2}$, $y' = x\sin\frac{\pi}{6} + y\cos\frac{\pi}{6} = \frac{x+\sqrt{3}y}{2}$. Substituting this into the original equation, $x'^2 - \sqrt{3}x'y' + 2y'^2 = 1$, and simplifying, we obtain $x^2 + 5y^2 = 2$.

25. $x^2 - 3xy + y^2 = 5$. $B^2 - 4AC = 9 - 4 = 5 > 0$, hyperbola. Let $y_1 = \sqrt{9x^2 - 4(x^2-5)}$. Graph $y_2 = (3x + y_1)/2$ and $y_3 = (3x - y_1)/2$ in $[-17, 17]$ by $[-10, 10]$. $\theta = \frac{\pi}{4}$ since $A = C$. Substituting $x' = \frac{\sqrt{2}}{2}(x-y)$, $y' = \frac{\sqrt{2}}{2}(x+y)$, we obtain $5y^2 - x^2 = 10$.

27. $3x^2 + 2xy + 3y^2 = 19$. $B^2 - 4AC = 4 - 4(3)3 < 0$, ellipse. Let $y_1 = \sqrt{4x^2 - 12(3x^2 - 19)}$. Graph $y_2 = (-2x + y_1)/6$ and $y_3 = (-2x - y_1)/6$ in $[-5.1, 5.1]$ by $[-3, 3]$. Since $A = C = 3$, $\theta = \frac{\pi}{4}$, $x' = \frac{\sqrt{2}}{2}(x-y)$, $y' = \frac{\sqrt{2}}{2}(x+y)$. After substituting and simplifying, we obtain $4x^2 + 2y^2 = 19$.

29. The equations give the coordinates of a point after it is rotated through an angle θ. To get the original coordinates back, we can use the same equations to rotate (x', y') through an angle $-\theta$: $x = x'\cos(-\theta) - y'\sin(-\theta) = x'\cos\theta + y'\sin\theta$, $y = x'\sin(-\theta) + y'\cos(-\theta) = -x'\sin\theta + y'\cos\theta$.

Exercises 31-33. Some results of this section may be combined to arrive at the following statement. If x is replaced by $x\cos\theta + y\sin\theta$ and y is replaced by $-x\sin\theta + y\cos\theta$ throughout an equation, the graph of the new equation is the graph of the original equation rotated through an angle θ.

31. $2x^2 - y^2 - 8 = 0$, $\theta = \frac{\pi}{4}$. x is replaced by $x\cos(\pi/4) + y\sin(\pi/4) = (x+y)/\sqrt{2}$ and y is replaced $-x\sin(\pi/4) + y\cos(\pi/4) = (y-x)/\sqrt{2}$: $2(x+y)^2/2 - (y-x)^2/2 - 8 = 0$, $x^2 + 2xy + y^2 + (-y^2 + 2xy - x^2)/2 - 8 = 0$, $\frac{x^2}{2} + 3xy + \frac{y^2}{2} - 8 = 0$ or $x^2 + 6xy + y^2 - 16 = 0$.

33. $4x^2 - y^2 - 10 = 0$, $\theta = \pi/3$. x is replaced by $x\cos(\pi/3) + y\sin(\pi/3) = (x + \sqrt{3}y)/2$ and y is replaced by $-x\sin(\pi/3) + y\cos(\pi/3) = (y - \sqrt{3}x)/2$: $4(x + \sqrt{3}y)^2/4 - (y - \sqrt{3}x)^2/4 - 10 = 0$. This reduces to $x^2 + 10\sqrt{3}xy + 11y^2 - 40 = 0$.

35. a) and b) are combined. The distance between $(-3,-3)$ and $(3,3)$ is $6\sqrt{2}$. Since the sum of the lengths of two sides of a triangle exceeds the length of the 3rd side, for any point (x,y), $\sqrt{(x+3)^2 + (y+3)^2} + \sqrt{(x-3)^2 + (y-3)^2} \geqq 6\sqrt{2}$. Thus $\sqrt{(x+3)^2 + (y+3)^2} + \sqrt{(x-3)^2 + (y-3)^2} = 6$ is impossible. (Call this fact (A)). The equation $\sqrt{(x+3)^2 + (y+3)^2} - \sqrt{(x-3)^2 + (y-3)^2} = \pm 6$ is equivalent to $\sqrt{(x+3)^2 + (y+3)^2} = \sqrt{(x-3)^2 + (y-3)^2} \pm 6$. The last equation is equivalent to $(x+3)^2 + (y+3)^2 = (\sqrt{(x-3)^2 + (y-3)^2} \pm 6)^2$ using fact (A). The last equation is equivalent to $(x-3) + y = \pm\sqrt{(x-3)^2 + (y-3)^2}$. This is equivalent to $[(x-3) + y]^2 = (x-3)^2 + (y-3)^2$ which in turn is equivalent to $2xy = 9$.

10.3 PARAMETRIC EQUATIONS FOR PLANE CURVES

1. $x = \cos t$, $y = \sin t$, $0 \leqq t \leqq \pi$. Graph $x_1 = \cos t$, $y_1 = \sin t$, $0 \leqq t \leqq \pi$ in $[-1.7, 1.7]$ by $[-1, 1]$. $x^2 + y^2 = \cos^2 t + \sin^2 t = 1$. The upper half of the unit circle is traced out in the counterclockwise direction.

3. Graph $x_1 = \sin 2\pi t$, $y_1 = \cos 2\pi t$, $0 \leqq t \leqq 1$ in $[-1.7, 1.7]$ by $[-1, 1]$. The unit circle is traced out once in the clockwise direction starting at $(0, 1)$.

5. $x = 4\cos t$, $y = 2\sin t$, $0 \leqq t \leqq 2\pi$. Graph this with t-step $= 0.1$ in $[-4, 4]$ by $[-2.4, 2.4]$. $(\frac{x}{4})^2 + (\frac{y}{2})^2 = \cos^2 t + \sin^2 t = 1$. The ellipse $\frac{x^2}{16} + \frac{y^2}{4} = 1$ is traced out once in the counterclockwise direction.

7. $x = 4\cos t$, $y = 5\sin t$, $0 \leqq t \leqq \pi$. Graph in $[-8.5, 8.5]$ by $[-5, 5]$. The upper half of the ellipse $\frac{x^2}{16} + \frac{y^2}{25} = 1$ is traced out in the counterclockwise direction.

9. $x = 3t$, $y = 9t^2$, $-\infty < t < \infty$. Graph for $-1 \leqq t \leqq 1$ in $[-7.6, 7.6]$ by $[0, 9]$. $y = 9t^2 = x^2$. The parabola $y = x^2$ is traced out from left to right.

11. $x = t$, $y = \sqrt{t}$, $t \geqq 0$. Graph for $0 \leqq t \leqq 10$ in $[0, 10]$ by $[-1.4, 4.5]$. $x = t = y^2$. The upper half of the parabola $y^2 = x$ is traced out from left to right.

13. $x = -\sec t$, $y = \tan t$, $-\pi/2 < t < \pi/2$. Graph in $[-17, 17]$ by $[-10, 10]$. $x^2 - 1 = \sec^2 t - 1 = \tan^2 t = y^2$. The left branch of the hyperbola $x^2 - y^2 = 1$ is traced out from bottom to top.

15. $x = 2t - 5$, $y = 4t - 7$, $-\infty < t < \infty$. Graph for $-5 \leqq t \leqq 5$ in $[-16.6, 17.4]$ by $[-7, 13]$. $t = \frac{x+5}{2} = \frac{y+7}{4}$. The graph of the line $y = 2x + 3$ is traced out from left to right.

17. $x = t$, $y = 1 - t$, $0 \leqq t \leqq 1$. Graph in $[-2, 3]$ by $[-1, 2]$. $y = 1 - x$. The line segment from $(0, 1)$ to $(1, 0)$ is traced out from left to right.

19. $x = t$, $y = \sqrt{1 - t^2}$, $-1 \leqq t \leqq 1$. Graph in $[-1, 1]$ by $[-0.1, 1.1]$. The upper half of the unit circle, $y = \sqrt{1 - x^2}$, is traced out from left to right.

21. $x = t^2$, $y = \sqrt{t^4 + 1}$, $t \geqq 0$. Graph for $0 \leqq t \leqq 4$ in $[-6.9, 21.9]$ by $[0, 17]$. $y = \sqrt{x^2 + 1}$. The top half of the hyperbola $y^2 - x^2 = 1$ for $x \geqq 0$ is traced out from left to right.

23. $x = \cosh t$, $y = \sinh t$, $-\infty < t < \infty$. Graph for $-3 \leqq t \leqq 3$ in $[-17, 17]$ by $[-10, 10]$. $x^2 - y^2 = \cosh^2 t - \sinh^2 t = 1$. The right branch of the hyperbola $x^2 - y^2 = 1$ is traced out from bottom to top.

25. a) $x = a\cos t$, $y = -a\sin t$, $0 \leqq t \leqq 2\pi$ b) $x = a\cos t$, $y = a\sin t$, $0 \leqq t \leqq 2\pi$ c) $x = a\cos(2t)$, $y = -a\sin(2t)$, $0 \leqq t \leqq 2\pi$ d) $x = a\cos t$, $y = a\sin t$, $0 \leqq t \leqq 4\pi$

27. $x = 5\sin 2t$, $y = 5\sin 3t$. Take $a = 2\pi$. We obtain more and more of the graph until t reaches 2π. Note that $\sin[2(t+2\pi)] = \sin 2t$ and $\sin[3(t+2\pi)] = \sin 3t$. We also get a closed curve for $a = \pi$.

29. $x = (5\sin 2t)\cos t$, $y = (5\sin 2t)\sin t$, $0 \leqq t \leqq a = 2\pi$ for a complete graph. If $a = \pi/2$, a closed curve is obtained.

31. Use $a = 4\pi$ for a complete graph. Note that $\sin[1.5(t+4\pi)] = \sin(1.5t+6\pi) = \sin 1.5t$. A closed curve is obtained if $a = \pi/1.5$.

33. No such a exists. Note that $x^2 + y^2 = t^2(\sin^2 t + \cos^2 t) = t^2$. Thus the distance from (x, y) to $(0, 0)$ is $|t|$. So for $t > 0$, (x, y) can't get back to $(0, 0)$.

35.

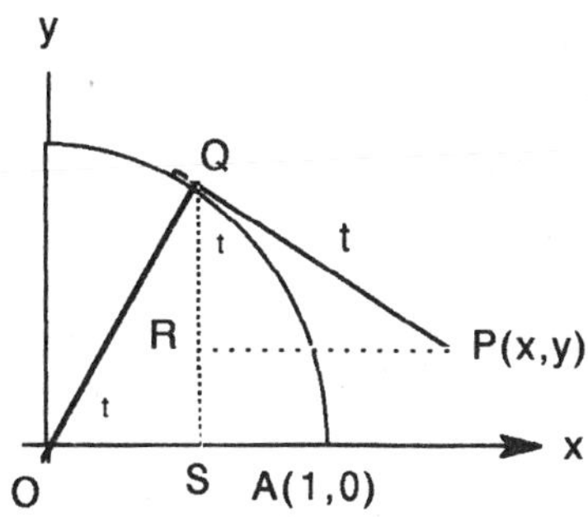

$\angle OQP$ is known to be a right angle. Segment QP = length arc $AQ = r\theta = 1 \cdot t = t$. $x = OS + RP = \cos t + t\sin t$, $y = SQ - RQ = \sin t - t\cos t$.

37. We wish to minimize the square of the distance $D(t) = (x-2)^2 + (y-\frac{1}{2})^2 = (t-2)^2 + (t^2 - \frac{1}{2})^2$. $D'(t) = 2(t-2) + 2(t^2 - \frac{1}{2})2t = 4(t^3 - 1)$. $D'(t) < 0$ for $t < 1$ and $D'(t) > 0$ for $t > 1$. $D(t)$ therefore has its minimal value when $t = 1$ and so $(1, 1)$ is the closest point.

39. $x = x_0 + (x_1 - x_0)t$, $y = y_0 + (y_1 - y_0)t$. a) By taking $t = 0$ and $t = 1$, we see that the curve passes through the two points. $t = \frac{x-x_0}{x_1-x_0} = \frac{y-y_0}{y_1-y_0}$, a linear equation in x and y and so the curve is a line. b) Take $(x_0, y_0) = (0, 0)$: $x = x_1 t$, $y = y_1 t$. c) Take $(x_0, y_0) = (-1, 0)$ and $(x_1, y_1) = (0, 1)$: $x = -1 + t$, $y = t$.

41. a) Use $[-18, 32]$ by $[-15, 15]$. b) Use $[-0.13, 2]$ by $[-0.55, 0.55]$. c) Use $[-50, 1]$ by $[-55, 15]$.

43. The three graphs can be compared in the viewing window $[0, 4\pi]$ by $[-2.7, 4.7]$.

45. Graph in $[-5.1, 5.9]$ by $[-3.2, 3.2]$. The new equations amount to $x = -2\cos t + \cos(2t)$, $y = -2\sin t + \sin(2t)$. Graph in the same window. The original curve had three cusps. The new curve appears to be a cardioid.

47. Graph a), b), c) in $[0, 128]$ by $[-21, 54]$. In d) the curve is part of the y-axis traced from $(0,0)$ to $(0,64)$ and back down to $(0,0)$.

49. Suppose a curve is given parametrically as $x = x(t)$, $y = y(t)$. Let the curve be rotated counterclockwise through an angle θ. According to Section 10.2 the new curve is given by $x_1 = x(t)\cos\theta - y(t)\sin\theta$, $y_1 = x(t)\sin\theta + y(t)\cos\theta$. In this problem, $y = x^2+1$, $\theta = 30°$. Thus, $x = t$, $y = t^2+1$. The rotated curve is given by $x_1 = (\sqrt{3}t - (t^2+1))/2$, $y_1 = (t + \sqrt{3}(t^2+1))/2$. Both curves may be graphed for $-4 \leqq t \leqq 4$ in $[-16.1, 16.1]$ by $[-2, 17]$. If we rotate a curve through an angle θ, we can obtain an equation of the new curve by replacing x by $x\cos\theta + y\sin\theta$ and replacing y by $-x\sin\theta + y\cos\theta$. Here, with $\theta = 30°$, x is replaced by $(\sqrt{3}x + y)/2$ and y is replaced by $(-x + \sqrt{3}y)/2$. After simplifying, we obtain $3x^2 + 2\sqrt{3}xy + y^2 + 2x - 2\sqrt{3}y + 4 = 0$.

51. $y^2 = x$, $\theta = \frac{\pi}{2}$. The rotated curve is clearly $y = x^2$. Graph $x_1 = t^2$, $y_1 = t$, $x_2 = t$, $y_2 = t^2$, $-3 \leqq t \leqq 3$ in $[-15.3, 15.3]$ by $[-9, 9]$.

53. $\frac{x^2}{4} + \frac{y^2}{9} = 1$, $\theta = \frac{\pi}{3}$. Graph $x_1 = 2\cos t$, $y_1 = 3\sin t$, $x_2 = 2\cos t\cos\frac{\pi}{3} - 3\sin t\sin\frac{\pi}{3} = \cos t - (3\sqrt{3}/2)\sin t$, $y_2 = 2\cos t\sin\frac{\pi}{3} + 3\sin t\cos\frac{\pi}{3} = \sqrt{3}\cos t + (3/2)\sin t$, $0 \leqq t \leqq 2\pi$ in $[-5.1, 5.1]$ by $[-3, 3]$. Using the formulas in the solution of Exercise 49, we find an equation of the rotated curve to be $21x^2 + 10\sqrt{3}xy + 31y^2 = 144$.

55. $\frac{x^2}{16} - \frac{y^2}{25} = 1$, $\theta = 60°$. Graph $x_1 = 4\sec t$, $y_1 = 5\tan t$, $x_2 = 4\sec t\cos 60° - 5\tan t\sin 60° = (4\sec t - 5\sqrt{3}\tan t)/2$, $y_2 = 4\sec t\sin 60° + 5\tan t\cos 60° = (4\sqrt{3}\sec t + 5\tan t)/2$, $0 \leqq t \leqq 2\pi$ in $[-34, 34]$ by $[-20, 20]$ (dot format is suggested). Using the formulas in the solution of Exercise 49, we find an equation of the rotated curve to be $-23x^2 + 82\sqrt{3}xy + 59y^2 = 1600$.

10.4 THE CALCULUS OF PARAMETRIC EQUATIONS

1. $x = 2\cos t$, $y = 2\sin t$, $t = \pi/4$. $\frac{dy}{dx} = \frac{dy/dt}{dx/dt} = \frac{2\cos t}{-2\sin t} = -\cot t$. $m = \frac{dy}{dx}\big|_{t=\pi/4} = -\cot(\pi/4) = -1$. $y - y_0 = m(x - x_0)$ becomes $y - \sqrt{2} = (-1)(x - \sqrt{2})$ or $y = -x + 2\sqrt{2}$. $\frac{d^2y}{dx^2} = \frac{dy'/dt}{dx/dt} = \frac{\csc^2 t}{-2\sin t} = -\frac{1}{2\sin^3 t}$. $\frac{d^2y}{dx^2}\big|_{t=\pi/4} = -\sqrt{2}$. Graph $x_1 = 2\cos t$, $y_1 = 2\sin t$, $x_2 = t$, $y_2 = -t + 2\sqrt{2}$, $-3.5 \leqq t \leqq 9$, Tstep $= 0.05$ in $[-10.5, 10, 5]$ by $[-6.1, 6.1]$.

3. $\frac{dy}{dx} = \frac{-2\sin t}{4\cos t} = -\frac{1}{2}\tan t$, $m = \frac{dy}{dx}\big|_{x=\pi/4} = -\frac{1}{2}$. Tangent line: $y = -\sqrt{2} = -\frac{1}{2}(x - 2\sqrt{x})$ or $y = -\frac{1}{2}x + 2\sqrt{2}$. $\frac{d^2y}{dx^2} = \frac{dy'/dt}{dx/dt} = \frac{-\frac{1}{2}\sec^2 t}{4\cos t} = -\frac{1}{8}\sec^3 t$.

$\frac{d^2y}{dx^2}|_{t=\pi/4} = -\frac{\sqrt{2}}{4}$. Graph $x_1 = 4\sin t,\ y_1 = 2\cos t,\ x_2 = t,\ y_2 = -\frac{1}{2}x + 2\sqrt{2},\ -10.5 \leqq t \leqq 10.5$ in $[-10.5, 10.5]$ by $[-6.1, 6.1]$.

5. $x = \sec^2 t - 1,\ y = \tan t,\ t = -\frac{\pi}{4} (x = y^2)$. $\frac{dy}{dx} = \frac{\sec^2 t}{2\sec^2 t \tan t} = \frac{1}{2\tan t} (= -\frac{1}{2}$ at $t = -\frac{\pi}{4})$. Tangent line: $y+1 = -\frac{1}{2}(x-1)$ or $y = -\frac{1}{2}x - \frac{1}{2}$. $\frac{d^2y}{dx^2} = \frac{-\frac{1}{2}\csc^2 t}{2\sec^2 t \tan t} = -\frac{1}{4}\cot^3 t (= \frac{1}{4}$ at $t = -\frac{\pi}{4})$. Graph $x_1 = \sec^2 t - 1,\ y_1 = \tan t,\ x_2 = t,\ y_2 = -\frac{1}{2}t - \frac{1}{2},\ -7 \leqq t \leqq 15$ in $[-6.2, 14.2]$ by $[-8, 4]$.

7. $\frac{dy}{dx} = \frac{1/(2\sqrt{t})}{1} = \frac{1}{2\sqrt{t}} = \frac{1}{2}t^{-1/2}$. $m = \frac{dy}{dx}|_{t=1/4} = 1$. Tangent line: $y - \frac{1}{2} = (1)(x - \frac{1}{4})$ or $y = x + \frac{1}{4}$. $\frac{d^2y}{dx^2} = -\frac{1}{4}t^{-3/2}$, $\frac{d^2y}{dx^2}|_{t=1/4} = -2$. Graph $x_1 = t,\ y_1 = \sqrt{t},\ x_2 = t,\ y_2 = t + 0.25,\ -10.5 \leqq t \leqq 10.5$ in $[-10.5, 10.5]$ by $[-6.1, 6.1]$.

9. $\frac{dy}{dx} = \frac{4t^3}{4t} = t^2$. $m = \frac{dy}{dx}|_{t=-1} = 1$. Tangent line: $y - 1 = (1)(x-5)$ or $y = x - 4$. $\frac{d^2y}{dx^2} = \frac{dy'/dt}{dx/dt} = \frac{2t}{4t} = \frac{1}{2}$ for all $t \neq 0$. Graph $x_1 = 2t^2 + 3,\ y_1 = t^4,\ x_2 = t,\ y_2 = t - 4,\ -2 \leqq t \leqq 20$ in $[-13.2, 24.2]$ by $[-6, 16]$.

11. $\frac{dy}{dx} = \frac{\sin t}{1-\cos t}$. $m = \frac{dy}{dx}|_{t=\pi/3} = \sqrt{3}$. Tangent line: $y - \frac{1}{2} = \sqrt{3}[x - (\frac{\pi}{3} - \frac{\sqrt{3}}{2})]$ or $y = \sqrt{3}x + 2 - \frac{\sqrt{3}}{3}\pi$. $\frac{d^2y}{dx^2} = -\frac{1}{(1-\cos t)^2}$, $\frac{d^2y}{dx^2}|_{t=\pi/3} = -4$. Graph $x_1 = t - \sin t,\ y_1 = 1 - \cos t,\ x_2 = t,\ y_2 = \sqrt{3}\,t + 2 - \frac{\sqrt{3}}{3}\,\pi,\ -17 \leqq t \leqq 17$ in $[-17, 17]$ by $[-10, 10]$.

13. Graph in $[-1, 1]$ by $[0, \pi]$. $(\frac{dx}{dt})^2 + (\frac{dy}{dt})^2 = \sin^2 t + (1 + \cos t)^2 = 2 + 2\cos t = 2(1 + \cos t)\frac{(1-\cos t)}{1-\cos t} = \frac{2\sin^2 t}{1-\cos t}$. On $0 \leqq t \leqq \pi$, $\sqrt{(\frac{dx}{dt})^2 + (\frac{dy}{dt})^2} = \frac{\sqrt{2}\sin t}{\sqrt{1-\cos t}}$. Let $u = 1 - \cos t,\ du = \sin t\, dt$. $\sqrt{2}\int_0^{\pi} \frac{\sin t\, dt}{\sqrt{1-\cos t}} = \sqrt{2}\int_0^2 u^{-1/2}du = 2\sqrt{2}u^{1/2}|_0^2 = 4$.

15. Graph in $[0, 8]$ by $[1/3, 9]$. $(\frac{dx}{dt})^2 + (\frac{dy}{dt})^2 = t^2 + [\frac{1}{2}(2t+1)^{1/2}2]^2 = t^2 + 2t + 1 = (t+1)^2$. $\int_0^4 (t+1)dt = \frac{t^2}{2} + t|_0^4 = 12$.

17. Graph in $[8, 4\pi]$ by $[0, 8]$. $\frac{dx}{dt} = 8(-\sin t + \sin t + t\cos t) = 8t\cos t$. Similarly, $\frac{dy}{dt} = 8t\sin t$. $\int_0^{\pi/2} \sqrt{8^2t^2(\cos^2 t + \sin^2 t)}dt = 8\int_0^{\pi/2} t\, dt = 4t^2|_0^{\pi/2} = \pi^2$.

19. $\frac{dx}{dt} = -\sin t$, $\frac{dy}{dt} = \cos t$. $2\pi\int_0^{2\pi} y\sqrt{(\frac{dx}{dt})^2 + (\frac{dy}{dt})^2}dt = 2\pi\int_0^{2\pi}(2 + \sin t)dt = 2\pi(2t - \cos t)|_0^{2\pi} = 8\pi^2$.

21. $\frac{dx}{dt} = 1$, $\frac{dy}{dt} = t + \sqrt{2}$. $2\pi\int_{-\sqrt{2}}^{\sqrt{2}} x\sqrt{(\frac{dx}{dt})^2 + (\frac{dy}{dt})^2}dt = 2\pi\int_{-\sqrt{2}}^{\sqrt{2}}(t + \sqrt{2})\sqrt{t^2 + 2\sqrt{2}t + 3dt} = \pi\int_1^9 u^{1/2}du$ (where $u = t^2 + 2\sqrt{2}t + 3$, $du = 2(t + \sqrt{2})dt) = \frac{2}{3}\pi u^{3/2}]_1^9 = \frac{52\pi}{3}$.

23. $\frac{dx}{dt} = 2$, $\frac{dy}{dt} = 1$. $2\pi \int_0^1 y\sqrt{(\frac{dx}{dt})^2 + (\frac{dy}{dt})^2}dt = 2\pi \int_0^1 (t+1)\sqrt{5}dt = 2\pi\sqrt{5}[\frac{t^2}{2}+t]_0^1 = 3\sqrt{5}\pi$. Check: $r_1 = 1$, $r_2 = 2$, slant height $=$ distance between $(0,1)$ and $(2,2) = \sqrt{5}$, area $= \pi(1+2)\sqrt{5} = 3\sqrt{5}\pi$.

25. a) $\frac{dx}{dt} = -2\sin 2t$, $\frac{dy}{dt} = 2\cos 2t$. $\int_0^{\pi/2} \sqrt{(\frac{dx}{dt})^2 + (\frac{dy}{dt})^2}dt = \int_0^{\pi/2} \sqrt{4(\sin^2 2t + \cos^2 2t)}dt = \pi$ b) $\frac{dx}{dt} = \pi\cos\pi t$, $\frac{dy}{dt} = -\pi\sin\pi t$. $\int_{-1/2}^{1/2} \sqrt{\pi^2(\cos^2\pi t + \sin^2\pi t)}dt = \pi\int_{-1/2}^{1/2} dt = \pi$.

27. Let $x_1 = (8\sin 2t)\cos t$, $y_1 = (8\sin 2t)\sin t$, $x_2 = \text{NDER}(x_1, t)$, $y_2 = \text{NDER}(y_1, t)$. $\text{NINT}(\sqrt{x_2^2 + y_2^2}, t, 0, \pi/2) = 19.377$.

29. Let $x_1 = 2t\cos t$, $y_1 = 2t\sin t$, $x_2 = \text{NDER}(x_1, t)$, $y_2 = \text{NDER}(y_1, t)$. $\text{NINT}(\sqrt{x_2^2 + y_2^2}, t, 0, 50) = 2505.105$.

31. Let $x_1 = 3\sin t$, $y_1 = 5 + 3\sin 2t$, $0 \leqq t \leqq 2\pi$. The graph in $[-4, 4]$ by $[0, 9]$ suggests that the curve is symmetric with respect to the y-axis. Indeed if we replace t by $t + \pi$, (x, y) is replaced by $(-x, y)$. The right half of the curve which is all we need is traced out as t goes from 0 to π. Let $x_2 = \text{NDER}(x_1, t)$, $y_2 = \text{NDER}(y_1, t)$. S. area $= 2\pi\text{NINT}(x_1\sqrt{x_2^2 + y_2^2}, t, 0, \pi) = 159.485$.

33. Let $x_1 = \sin t$, $y_1 = \sin 2t$. Graph x_1, y_1, $0 \leqq t \leqq 2\pi$ in $[-1.86, 1.86]$ by $[-1.1, 1.1]$. $\frac{dy}{dx} = \frac{dy/dt}{dx/dt} = \frac{2\cos 2t}{\cos t}$. $\frac{dy}{dx} = 0$ when $t = \frac{\pi}{4}$ (first quadrant) and at that point $(x, y) = (\frac{\sqrt{2}}{2}, 1)$. For suitable values at the origin we take $t = 0$ and $t = \pi$. $\frac{dy}{dx}|_{t=0} = 2$, $\frac{dy}{dx}|_{t=\pi} = -2$. The tangent lines at the origin are $y = \pm 2x$. To confirm, we graph x_1, y_1, $x_2 = t$, $y_2 = 2t$, $x_3 = t$, $y_3 = -2t$, $-2\pi \leqq t \leqq 2\pi$ in the viewing rectangle given above.

35 through 41. For each of these we may use $0 \leqq t \leqq 2\pi$ in $[-1.86, 1.86]$ by $[-1.1, 1.1]$. For 38 and 39, $\frac{\pi}{2} \leqq t \leqq \frac{3\pi}{2}$ suffices.

43. Graph $x = 12\cos t + 6\cos(ct)$, $y = 12\sin t + 6\sin(ct)$, $0 \leqq t \leqq 2\pi$ in $[-34, 34]$ by $[-20, 20]$ for each $c = 2, 4, 6, 8$.

45. Let $x_1 = 12\cos t + 6\cos(8t)$, $y_1 = 12\sin t + 6\sin(8t)$, $x_2 = \text{NDER}(x_1, t)$, $y_2 = \text{NDER}(y_1, t)$. Length $= \int_0^{2\pi} \sqrt{(\frac{dx_1}{dt})^2 + (\frac{dy_1}{dt})^2}dt = \text{NINT}(\sqrt{x_2^2 + y_2^2}, t, 0, 2\pi) = 306.324$.

10.5 POLAR COORDINATES

1. {a,c}, {b,d}, {e,k}, {f,h}, {g,j}, {i,l}, {m,o}, {n,p}

3.

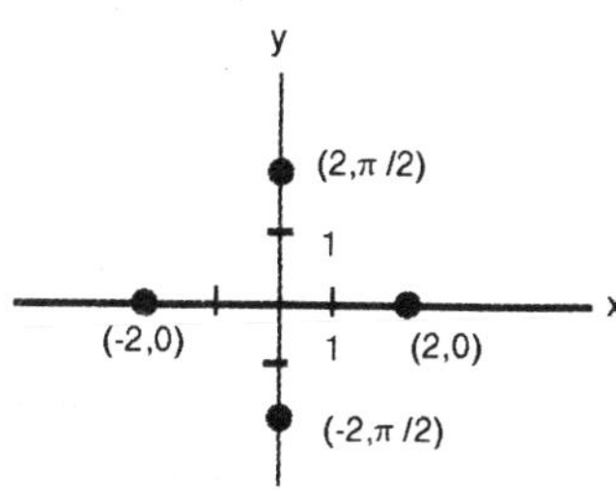

a) $(2, \frac{\pi}{2} + 2n\pi)$, $(-2, -\frac{\pi}{2} + 2n\pi)$ b) $(2, 2n\pi)$, $(-2, (2n+1)\pi)$ c) $(-2, \frac{\pi}{2} + 2n\pi)$, $(2, -\frac{\pi}{2} + 2n\pi)$ d) $(-2, 2n\pi)$, $(2, (2n+1)\pi)$. $n = 0, \pm 1, \pm 2, \ldots$

5. In each case we use $x = r\cos\theta$, $y = r\sin\theta$. a) $x = \sqrt{2}\cos\frac{\pi}{4} = 1$, $y = \sqrt{2}\sin\frac{\pi}{4} = 1$. $(1,1)$ b) $x = 1 \cdot \cos 0 = 1$, $y = 1 \cdot \sin 0 = 0$. $(1,0)$ c) $(0,0)$ d) $x = -\sqrt{2}\cos\frac{\pi}{4} = -1$, $y = -\sqrt{2}\sin\frac{\pi}{4} = -1$. $(-1,-1)$ e) $x = -3\cos\frac{5\pi}{6} = \frac{3\sqrt{3}}{2}$, $y = -3\sin\frac{5\pi}{6} = -\frac{3}{2} \cdot (\frac{3\sqrt{3}}{2}, -\frac{3}{2})$ f) $x = 5\cos\theta = 5(\frac{3}{5}) = 3$, $y = 5\sin\theta = 5(\frac{4}{5}) = 4$. $(3,4)$ g) $x = -\cos 7\pi = 1$, $y = 0$. $(1,0)$ h) $x = 2\sqrt{3}\cos\frac{2\pi}{3} = -\sqrt{3}$, $y = 2\sqrt{3}\sin\frac{2\pi}{3} = 2\sqrt{3}(\frac{\sqrt{3}}{2}) = 3$. $(-\sqrt{3}, 3)$.

7. Graph $r = 2$, $0 \leqq \theta \leqq 2\pi$ is a square window containing $[-2,2]$ by $[-2,2]$ in polar mode.

9.

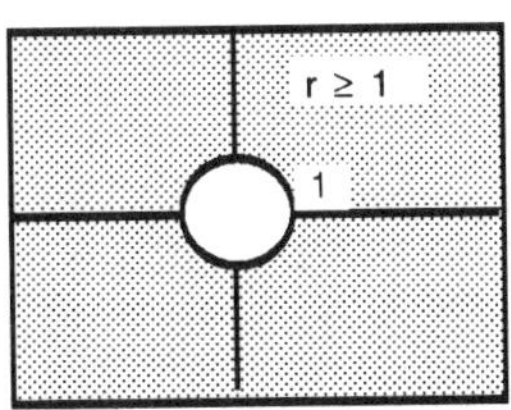

11.

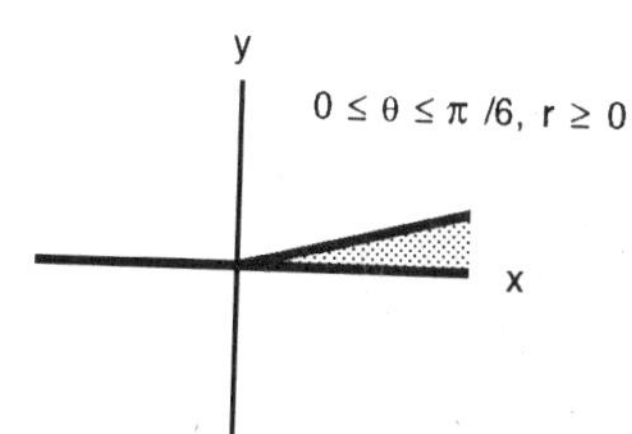

13.

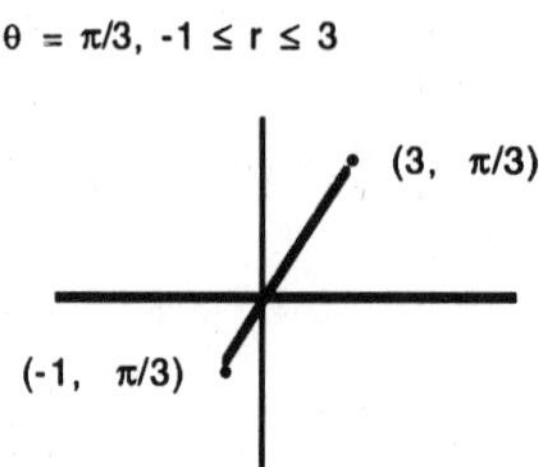

15. The graph consists of the origin and the positive y-axis.

17. The graph consists of the upper half of the unit circle including $(-1, 0)$ and $(1, 0)$.

19.

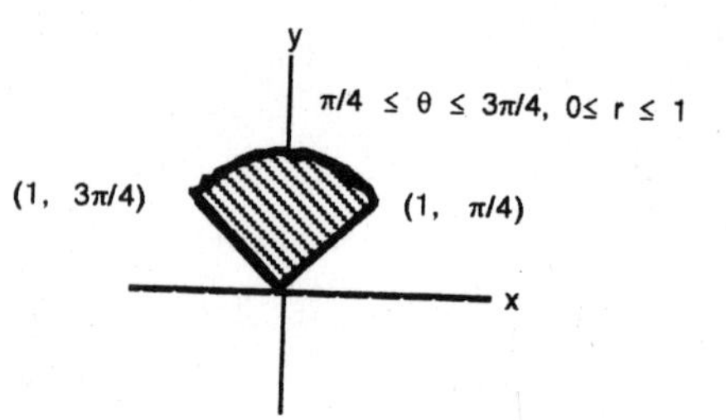

21.

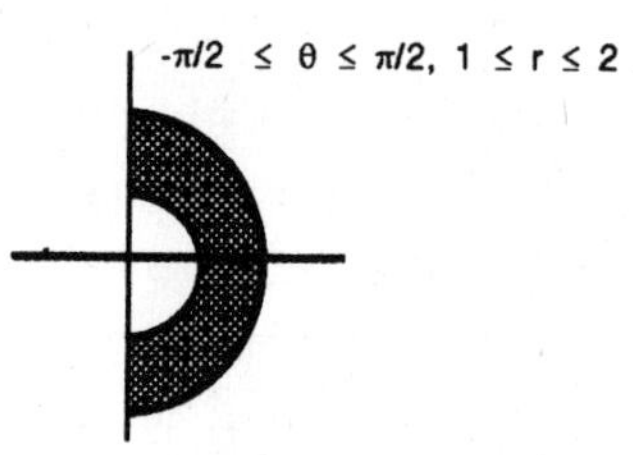

23. $r\cos\theta = 2 \Rightarrow x = 2$. Vertical line consisting of all points with x-coordinate 2.

25. $r\sin\theta = 4$ is equivalent to $y = 4$, the horizontal line through $(1, 4)$.

27. $r\sin\theta = 0$. $y = 0$, the x-axis.

29. $r\cos\theta + r\sin\theta = 1$. $x + y = 1$, the line through $(1, 0)$ and $(0, 1)$.

31. $r^2 = 1$. $x^2 + y^2 = 1$, the unit circle.

33. $r = \frac{5}{\sin\theta - 2\cos\theta}$. $r\sin\theta - 2r\cos\theta = 5$, $y - 2x = 5$, $y = 2x + 5$. Line with slope 2 through $(0, 5)$.

35. $x = 7$. $r\cos\theta = 7$, $r = 7\sec\theta$. Graph $r = 7(\cos\theta)^{-1}$, $-\frac{\pi}{2} \leqq \theta \leqq \frac{\pi}{2}$ in $[0, 10]$ by $[-50, 50]$ obtaining the vertical line through $(7, 0)$.

37. $x = y$. $r\cos\theta = r\sin\theta$, $\cos\theta = \sin\theta$, $\tan\theta = 1$, $\theta = \frac{\pi}{4}$.

39. $x^2 + y^2 = 4$, $r = 2$, circle.

41. $\frac{x^2}{9} + \frac{y^2}{4} = 1$. $\frac{r^2\cos^2\theta}{9} + \frac{r^2\sin^2\theta}{4} = 1$, $(4\cos^2\theta + 9\sin^2\theta)r^2 = 36$, $r = \frac{\pm 6}{\sqrt{4\cos^2\theta + 9\sin^2\theta}}$, ellipse.

43. $y^2 = 4x$. $r^2\sin^2\theta = 4r\cos\theta$, $r = \frac{4\cos\theta}{\sin^2\theta}$. To confirm, graph r for $0 \leqq \theta \leqq 2\pi$ in $[-8.5, 25.5]$ by $[-10, 10]$. Parabola.

45. $x = r\cos\theta = f(\theta)\cos\theta$, $y = r\sin\theta = f(\theta)\sin\theta$.

10.6 GRAPHING IN POLAR COORDINATES

1. In Exercises 1 through 12 the student should use the method of Example 1 including a table of values, use of symmetries and "slope at $(0, \theta_0) = \tan\theta_0$" as a guide as to how the curve goes into and out of the origin. The student's results can then be checked on a grapher. One way of carrying out this check is given in the answers. It is assumed that a graphing utility with a polar graphing mode and with a screen "squaring" function is being used.

Graph $r = 1 + \cos\theta$, $0 \leqq \theta \leqq 2\pi$, θStep $= 0.1$ in a "squared" rectangle containing $[-0.25, 2]$ by $[-1.3, 1.3]$, for example $[-1.33, 3.08]$ by $[-1.3, 1.3]$.

3. Graph $r = 1 - \sin\theta$, $0 \leqq \theta \leqq 2\pi$ in $[-2.4, 2.4]$ by $[-2.2, 0.7]$.

5. Graph $r = 2 + \sin\theta$, $0 \leqq \theta \leqq 2\pi$ in $[-3.5, 3.6]$ by $[-1.2, 3]$.

7. Graph $r = 2\sqrt{\cos 2\theta}$ and $r = -2\sqrt{\cos 2\theta}$ in $[-2, 2]$ by $[-1.2, 1.2]$. Use θStep $= 0.01$, $-\frac{\pi}{4} \leqq \theta \leqq \frac{\pi}{4}$.

9. Graph $r = \theta$ in $[-33, 33]$ by $[-18.4, 20]$ first using $0 \leqq \theta \leqq 20$ then $-20 \leqq \theta \leqq 0$ and then $-20 \leqq \theta \leqq 20$.

11. Graph $r = 8\cos 2\theta$, $0 \leqq \theta \leqq 2\pi$ in $[-13.6, 13.6]$ by $[-8, 8]$. A complete graph requires a minimum of 2π for the range of θ. The factor 8 stretches the graph of $r = \cos 2\theta$ away from the origin by a factor of 8. The range, $\frac{\pi}{4} \leqq \theta \leqq \frac{3\pi}{4}$, for example, produces a closed curve. Replacing θ by 2θ produces 3 more leaves.

13. $r = -1 + \cos\theta$. Slope $= \frac{r'\sin\theta + r\cos\theta}{r'\cos\theta - r\sin\theta} = \frac{-\sin^2\theta - \cos\theta + \cos^2\theta}{-\sin\theta\cos\theta + \sin\theta - \cos\theta\sin\theta}$. At $\theta = \frac{\pi}{2}$, $r = -1$ and slope $= \frac{-1-0+0}{1} = -1$ and the tangent at the rectangular point $(0, -1)$ is $y = -x - 1$ or $r(\sin\theta + \cos\theta) = -1$, $r = \frac{-1}{\sin\theta + \cos\theta}$. At $\theta = -\frac{\pi}{2}$, $r = -1$ and slope $= \frac{-1}{-1} = 1$ and the tangent at the rectangular point $(0, 1)$ is $y = x + 1$ or $r(\sin\theta - \cos\theta) = 1$, $r = \frac{1}{\sin\theta - \cos\theta}$. Graph $r = -1 + \cos\theta$ and the two tangents, $-\pi \leqq \theta \leqq \pi$ in $[-2.6, 4.3]$ by $[-2, 2]$.

15. At the origin $r = \sin 2\theta = 0$ which implies $2\theta = n\pi$, $\theta = n\pi/2$. But $\tan(n\pi/2) = \pm\infty$ or 0 so the tangent line at the origin is either horizontal or vertical, i.e., it is either the x-axis or the y-axis. Slope $= \frac{r'\sin\theta + r\cos\theta}{r'\cos\theta - r\sin\theta} = \frac{2\cos 2\theta\sin\theta + \sin 2\theta\cos\theta}{2\cos 2\theta\cos\theta - \sin 2\theta\sin\theta}$. We give the details for the case $\theta = -3\pi/4$. $r = \sin[2(-\frac{3\pi}{4})] = \sin(-\frac{3\pi}{2}) = 1$: $(1, -3\pi/4)$ is the point on the curve ($x = r\cos\theta = -\frac{\sqrt{2}}{2}$, $y = r\sin\theta = \frac{-\sqrt{2}}{2}$). When $\theta = -\frac{3\pi}{4}$, slope $= \frac{2(0)(-\frac{\sqrt{2}}{2}) + (1)(-\frac{\sqrt{2}}{2})}{2(0)(-\frac{\sqrt{2}}{2}) - (1)(-\frac{\sqrt{2}}{2})} = -1$. Tangent line: $y + \frac{\sqrt{2}}{2} = -1(x + \frac{\sqrt{2}}{2})$, $x + y = -\sqrt{2}$, $r(\cos\theta + \sin\theta) = -\sqrt{2}$, $r = -\sqrt{2}/(\sin\theta + \cos\theta)$. The remaining cases are similar: $\theta = \frac{\pi}{4}$, $m = -1$; $\theta = -\frac{\pi}{4}$, $m = 1$; $\theta = \frac{3\pi}{4}$, $m = 1$. Graph $r = \sin 2\theta$, $r = \pm\sqrt{2}/(\sin\theta + \cos\theta)$, $r = \pm\sqrt{2}/(\sin\theta - \cos\theta)$, $0 \leqq \theta \leqq 2\pi$ in $[-1.9, 1.9]$ by $[-1.2, 1.2]$ and regard the x- and y-axes as tangent lines also.

17. Graph $r = \pm 2\sqrt{\cos 2\theta}$, $-\frac{\pi}{4} \leqq \theta \leqq \frac{\pi}{4}$ in $[-2, 2]$ by $[-1.2, 1.2]$. Use θ Step $=$ 0.01. Because of the $\pm$ sign, $\frac{\pi}{2}$ is a minimum range of θ, but it must be over an interval in which $\cos 2\theta$ is non-negative.

19. a) Graph $r = \frac{1}{2} + \cos\theta$, $0 \leqq \theta \leqq 2\pi$, θStep $= 0.1$ in $[-0.9, 2.3]$ by $[-0.94, 0.94]$. b) Graph $r = 0.5 + \sin\theta$, $0 \leqq \theta \leqq 2\pi$ in $[-1.5, 1.5]$ by $[-0.2, 1.6]$.

21. a) Graph $r = 1.5 + \cos\theta$, $0 \leqq \theta \leqq 2\pi$ in $[-2, 4]$ by $[-1.8, 1.9]$. b) Graph $1.5 - \sin\theta$, $0 \leqq \theta \leqq 2\pi$ in $[-3.1, 3.1]$ by $[-2.7, 1]$.

23. Graph $r = 2 - 2\cos\theta$, $0 \leqq \theta \leqq 2\pi$ in $[-6.5, 3.2]$ by $[-2.8, 2.8]$. The region consists of this closed curve (a cardioid) and every point inside it.

25. $(2, 3\pi/4)$ also has coordinates $(-2, -\pi/4)$ and these coordinates satisfy the equation.

27. We first solve the equations simultaneously: $r = 1 + \cos\theta = 1 - \cos\theta \Rightarrow \cos\theta = 0 \Rightarrow \theta = (2k+1)\frac{\pi}{2}$. We obtain the points $(1, \frac{\pi}{2})$, $(1, \frac{3\pi}{2})$. We graph both curves, $0 \leqq \theta \leqq 2\pi$ in $[-2.5, 2.5]$ by $[-1.5, 1.5]$ and find that the origin is the only other point of intersection.

29. $r^2 = (1-\sin\theta)^2 = 4\sin\theta \Rightarrow 1 - 2\sin\theta + \sin^2\theta = 4\sin\theta$, $\sin^2\theta - 6\sin\theta + 1 = 0$, $\sin\theta = \frac{6\pm\sqrt{36-4}}{2} = 3 \pm 2\sqrt{2}$. Since $\sin\theta \leqq 1$, $\sin\theta = 3 - 2\sqrt{2}$, $\theta = \sin^{-1}(3-2\sqrt{2})$ or $\pi - \sin^{-1}(3-2\sqrt{2})$; $r = 1-\sin\theta = 1-(3-2\sqrt{2}) = 2(\sqrt{2}-1)$. Thus $(2(\sqrt{2}-1), \sin^{-1}(3-2\sqrt{2}))$ and $(2(\sqrt{2}-1), \pi - \sin^{-1}(3-2\sqrt{2}))$ are points of intersection. A careful study of the graphs shows that the origin and $(2, \frac{3\pi}{2})$ are the only other points of intersection.

31. $r^2 = \sin 2\theta = \cos 2\theta$, $\tan 2\theta = 1$ and $\sin 2\theta > 0$. This yields the possibilities $2\theta = \frac{\pi}{4} + 2n\pi$, $\theta = \frac{\pi}{8} + n\pi$. This yields only the distinct points $(\pm 2^{-1/4}, \frac{\pi}{8})$. From the graph of the two curves we see that $(0,0)$ is the only other point of intersection.

33. Graph $r = 1$, $r = 2\sin 2\theta$, $0 \leqq \theta \leqq 2\pi$ in $[-3.4, 3.4]$ by $[-2, 2]$. We see that there are two points of intersection in each quadrant. Four can be found using the above system and four can be found using $r = -1$, $r = 2\sin 2\theta$. $1 = 2\sin 2\theta$ leads to $\sin 2\theta = \frac{1}{2}$, $2\theta = \frac{\pi}{6} + 2n\pi$ or $\frac{5\pi}{6} + 2n\pi$, $\theta = \frac{\pi}{12} + n\pi$ and $\theta = \frac{5\pi}{12} + n\pi$. Similarly, $-1 = 2\sin 2\theta$ leads to $\theta = \frac{7\pi}{12} + n\pi$ and $\frac{11\pi}{12} + n\pi$. The points are: $(1, \frac{\pi}{12}), (1, \frac{5\pi}{12}), (-1, \frac{19\pi}{12}), (-1, \frac{23\pi}{12}), (1, \frac{13\pi}{12}), (1, \frac{17\pi}{12}), (-1, \frac{7\pi}{12}), (-1, \frac{11\pi}{12})$.

35. Graph $y = 5\sin\theta$, $0 \leqq \theta \leqq \pi$ in $[-5, 5]$ by $[-0.44, 5.44]$. Period π.

37. Graph $r = 5\sin 2\theta$, $0 \leqq \theta \leqq 2\pi$ in $[-8.5, 8.5]$ by $[-5, 5]$. Period 2π.

39. Graph $r = 5\sin 5\theta$, $0 \leqq \theta \leqq \pi$ in $[-8.5, 8.5]$ by $[-5, 5]$. Period π.

41. Graph $r = 5\sin(2.5\theta)$, $0 \leqq \theta \leqq 4\pi$ in $[-8.5, 8.5]$ by $[-5, 5]$. Period 4π.

43. Graph $r = 5\sin 1.5\theta$, $0 \leqq \theta \leqq 4\pi$ in $[-8.5, 8.5]$ by $[-5, 5]$. Period 4π.

45. Graph $r = 1 - 2\sin 3\theta$, $0 \leqq \theta \leqq 2\pi$ in $[-4.7, 4.7]$ by $[-2.1, 3.4]$. Period 2π.

47. a) Graph $r = e^{\theta/10}$, $-20 \leqq \theta \leqq 10$ in $[-4.3, 3.6]$ by $[-2, 2.7]$. b) Graph $r = 8/\theta$, $-20 \leqq \theta \leqq 20$ in the same window. c) Graph $r = 10/\sqrt{\theta}$ and $r = -10/\sqrt{\theta}$, $0 \leqq \theta \leqq 200$, θStep $= 0.5$ in the same window.

49. Infinite period

51. Graph $r = 1.75 + (0.06/2\pi)\theta$, $0 \leqq \theta \leqq 10\pi$ in $[-3, 3]$ by $[-3, 3]$.

53. $r = r_0 + (\frac{\theta}{2\pi})b = 1.75 + (0.06/2\pi)\theta$. $\frac{dr}{d\theta} = 0.06/2\pi$. $S = \int_0^{80\pi} \sqrt{(1.75 + (0.06/2\pi)\theta)^2 + (0.06/2\pi)^2}\, d\theta =$ NINT$(\sqrt{(1.75 + (0.06/2\pi)x)^2 + (0.06/2\pi)^2}, x, 0, 80\pi) = 741.420$cm.

55. a) We use the approximation S_a of Exercise 54. $S_a = \int_0^a r\,d\theta = \int_0^\alpha [r_0 + (\theta/2\pi)b]d\theta = r_0\alpha + (b/4\pi)\alpha^2$. Letting $\alpha = 2\pi n$ in this equation, we can arrive at the equation $bn^2 + 4\pi n - 2S_a = 0$. Using the appropriate solution from the quadratic formula, we obtain $n = (-2\pi + \sqrt{4\pi^2 + 2bS_a})/b$. b) The speed of the take-up reel steadily decreases.

10.7 POLAR EQUATIONS OF CONIC SECTIONS

1. $r\cos(\theta - \theta_0) = r_0$ becomes $r\cos(\theta - \frac{\pi}{6}) = 5$. $5 = r[\cos\theta\cos\frac{\pi}{6} + \sin\theta\sin\frac{\pi}{6}] = \frac{\sqrt{3}}{2}(r\cos\theta) + \frac{1}{2}(r\sin\theta)$, $5 = \frac{\sqrt{3}}{2}x + \frac{1}{2}y$, $\sqrt{3}x + y = 10$.

3. Graph $r = \frac{\sqrt{2}}{\cos(\theta - \frac{\pi}{4})}$, $0 \leqq \theta \leqq \pi$ in $[-5.8, 7.8]$ by $[-3, 5]$. $r[\cos\theta\cos\frac{\pi}{4} + \sin\theta\sin\frac{\pi}{4}] = \sqrt{2}$ leads to $x + y = 2$.

5. $r = 2a\cos\theta = 8\cos\theta$

7. $r = 4\cos\theta = 2(2)\cos\theta$. Center: $(2, 0)$, radius $= 2$. Check your sketch by graphing for $0 \leqq \theta \leqq \pi$ in $[-1.4, 5.4]$ by $[-2, 2]$.

9. $r = \frac{ke}{1+e\cos\theta} = \frac{2}{1+\cos\theta}$

11. $r = \frac{ke}{1+e\cos\theta} = \frac{8}{1+2\cos\theta}$

13. $r = \frac{ke}{1+e\cos\theta} = \frac{\frac{1}{2}}{1+\frac{1}{2}\cos\theta} = \frac{1}{2+\cos\theta}$

15. $r = \frac{ke}{1-e\sin\theta} = \frac{10(\frac{1}{5})}{1-(\frac{1}{5})\sin\theta} = \frac{10}{5-\sin\theta}$

17. $r = \frac{1}{1+\cos\theta}$. Directrix: $x = 1$, vertex: $(\frac{1}{2}, 0)$. Graph for $-\pi \leqq \theta \leqq \pi$ in $[-7.4, 2.8]$ by $[-3, 3]$. Include the directrix $r = 1/\cos\theta$.

19. $r = \frac{25}{10-5\cos\theta} = \frac{5}{2-\cos\theta} = \frac{5(\frac{1}{2})}{1-(\frac{1}{2})\cos\theta}$. $e = \frac{1}{2}$, directrix: $x = -5$. Let $P(u, \pi)$, $u > 0$, be the vertex concerned. $PF = ePD \Rightarrow u = \frac{1}{2}(5 - u) \Rightarrow u = \frac{5}{3}$, vertex: $(\frac{5}{3}, \pi)$. $1 - e^2 = \frac{3}{4}$. $5(\frac{1}{2}) = ke = a(1 - e^2) = \frac{3a}{4}$, $a = \frac{10}{3}$, center at $x = ea = \frac{5}{3}$, vertices at $x = \frac{5}{3} \pm \frac{10}{3}$, i.e., at $x = -\frac{5}{3}$ and 5. Vertices: $(\frac{5}{3}, \pi)$ and $(5, 0)$. Center at $(\frac{5}{3}, 0)$. Graph $r = \frac{5}{2-\cos\theta}$, $-\pi \leqq \theta \leqq \pi$ in $[-8.5, 8.5]$ by $[-5, 5]$. Include the directrix $r = -5\sec\theta$.

21. $r = \frac{400}{16+8\sin\theta} = \frac{25}{1+\frac{1}{2}\sin\theta} = \frac{50(\frac{1}{2})}{1+\frac{1}{2}\sin\theta}$. $e = \frac{1}{2}$, directrix $y = 50$. Let $P(v, \frac{\pi}{2})$ be the corresponding vertex. $PF = ePD$, $v = (\frac{1}{2})(50 - v)$, $2v = 50 - v$, $v = \frac{50}{3}$. (See Table 10.4-3). Let $Q(w, \frac{\pi}{2})$ be the other vertex. $QF = eQD$ becomes $-w = \frac{1}{2}(-w + 50)$, $-2w = -w + 50$, $w = -50$. If $(r, \frac{\pi}{2})$ is the center, it is midway between the vertices: $r = \frac{1}{2}(\frac{50}{3} - 50) = -\frac{50}{3}$. Graph $r = 25/(1 + 0.5\sin\theta)$, $0 \leqq \theta \leqq 2\pi$ in $[-93, 93]$ by $[-50, 60]$. Also include $r = 50/\sin\theta$.

23. $r = \frac{8}{2-2\sin\theta} = \frac{4\cdot 1}{1-(1)\sin\theta}$. $e = 1$, directrix $y = -4$. Graph $r = \frac{4}{1-\sin\theta}$ and $r = \frac{-4}{\sin\theta}$, $0 \leqq \theta \leqq 2\pi$ in $[-13, 13]$ by $[-5, 12]$. Vertex: $(2, \frac{3\pi}{2})$.

25. Graph the ellipses sequentially in $[-2, 1.4]$ by $[-1, 1]$, $0 \leqq \theta \leqq 2\pi$, θStep $=$ 0.1. The last two require a larger rectangle. As e increases, the center moves to the left, the ellipse increases in size. The ellipse also flattens out horizontally as can be seen by graphing in $[-11.3, 2.3]$ by $[-4, 4]$.

27. Graph these sequentially in $[-27, 27]$ by $[-16, 16]$. As k becomes more negative, the parabola opens up wider and wider to the right. As k becomes more and more positive, the parabola opens to the left wider and wider.

29. Graph $r = 3\sec(\theta - \pi/3) = 3/\cos(\theta - \pi/3)$, $0 \leqq \theta \leqq 2\pi$ in $[-11, 19]$ by $[-6, 11.7]$. $x + \sqrt{3}y = 6$ in rectangular coordinates.

31. Graph $r = 8/(4 + \cos\theta)$, $0 \leqq \theta \leqq 2\pi$ in $[-4.9, 3.9]$ by $[-2.6, 2.6]$.

33. Graph $r = 1/(1 + 2\sin\theta)$, $0 \leqq \theta \leqq 2\pi$ in $[-2.7, 2.7]$ by $[-0.93, 2.27]$ in dot format, θStep $= 0.01$.

35. Graph $r = -2\cos\theta$, $0 \leqq \theta \leqq \pi$ in $[-2.7, 0.7]$ by $[-1, 1]$.

37. Graph $r = 1/(1 + \cos\theta)$, $0 \leqq \theta \leqq 2\pi$, θStep $= 0.1$ in $[-7.7, 5.9]$ by $[-4, 4]$.

39. Graph $r = 2\cos\theta$, $0 \leqq \theta \leqq \pi$ in $[-0.7, 2.7]$ by $[-1, 1]$. The region consists of the circle and all points within it.

41. a) Without loss of generality we may assume that the major axis coincides with part of the x-axis and that a focus (the sun) is located at the origin. By Equation (15) an equation is $r = \frac{a(1-e^2)}{1+e\cos\theta}$. r is minimized when $\cos\theta = 1$, $\theta = 0$, $r = \frac{a(1-e^2)}{1+e} = a(1-e)$. r is maximized when $\cos\theta = -1$, $\theta = \pi$, $r = \frac{a(1-e^2)}{1-e} = a(1+e)$.

b)

Planet	$a(1-e)AU$	$a(1+e)AU$
Mercury	0.3075	0.4667
Venus	0.7184	0.7282
Earth	0.9833	1.017
Mars	1.382	1.666
Jupiter	4.951	5.455
Saturn	9.021	10.057
Uranus	18.30	20.06
Neptune	29.81	30.31
Pluto	29.65	49.23

43. a) $r = 2\sin\theta$ leads to $r^2 = 2r\sin\theta$, $x^2+y^2 = 2y$, $x^2+y^2-2y+1 = 1$, $x^2+(y-1)^2 = 1$. $r = \csc\theta$ yields $r\sin\theta = 1$, $y = 1$. b) Graph $r = 2\sin\theta$, $r = 1/\sin\theta$, $0 \leqq \theta \leqq \pi$ in $[-1.7, 1.7]$ by $[0, 2]$. Label the points of intersection $(1,1)$, $(\sqrt{2}, \frac{\pi}{4})$ and $(-1,1)$, $(\sqrt{2}, \frac{3\pi}{4})$.

45. Use Fig. 10.69 with $r\cos\theta = 4$, i.e., $x = k = 4$. In the parabola $FP = PD$ or $r = 4 - FB = 4 - r\cos\theta$, $r + r\cos\theta = 4$, $r = \frac{4}{1+\cos\theta}$.

47. We restrict ourselves here to the case where a and b are both nonzero. Let L be the line with equation $ax + by = c$ or $y = -\frac{a}{b}x + \frac{c}{b}$. L has slope $-\frac{a}{b}$ so the line through the origin perpendicular to L has equation $y = \frac{b}{a}x$. Solving simultaneously, we find that the point of intersection of L and $y = \frac{b}{a}x$ is $P_0(\frac{ca}{a^2+b^2}, \frac{cb}{a^2+b^2})$. $r^2 = x^2 + y^2$ and $\theta = \tan^{-1}\frac{y}{x}$ but since we must have $r_0 \geqq 0$, we find $r_0 = \frac{|c|}{\sqrt{a^2+b^2}}$ and $\theta_0 = \tan^{-1}\frac{b}{a}$ or $\theta_0 = \pi + \tan^{-1}\frac{b}{a}$. (Since $-\frac{\pi}{2} \leqq \tan^{-1}\frac{b}{a} \leqq \frac{\pi}{2}$, we must take $\theta_0 = \pi + \tan^{-1}\frac{b}{a}$ if P_0 is in the 2nd or 3rd quadrant.) $r = \frac{r_0}{\cos(\theta-\theta_0)}$ becomes $r = \frac{|c|}{\sqrt{a^2+b^2}\cos(\theta-\theta_0)}$ where θ_0 is as stated above.

49. $3x + 2y = 6$. Here P_0 must be in the first quadrant. From Exercise 47, $r = \frac{6}{\sqrt{13}\cos(\theta - \tan^{-1}\frac{2}{3})}$.

51. $4x + 3y = -12$. From the intercepts $(-3, 0)$ and $(0, -4)$ we see that P_0 lies in the 3rd quadrant. From Exercise 47, $r = \frac{12}{5\cos(\theta - (\pi + \tan^{-1}(3/4)))}$.

53. The first equation "1." has been derived in the text. For "2.", the equation $PF = e \cdot PD$ becomes $r = e[x - (-k)]$ which leads to $r = e(r\cos\theta + k)$, $(1 - e\cos\theta)r = ke$, $r = ke/(1 - e\cos\theta)$ as required. For "3.", $PF = e \cdot PD$ becomes $r = e(k - y) = e(k - r\sin\theta)$. This leads to $r(1 + e\sin\theta) = ke$ and $r = ke/(1 + e\sin\theta)$. For "4.", $PF = e \cdot PD$ becomes $r = e[y - (-k)] = e(r\sin\theta + k)$. This leads to $r(1 - e\sin\theta) = ke$, $r = ke/(1 - e\sin\theta)$.

10.8 INTEGRATION IN POLAR COORDINATES

1. Graph $r = \cos\theta$, $0 \leqq \theta \leqq \frac{\pi}{4}$ in $[0, 1]$ by $[0, 0.5]$. Then draw a line segment connecting $(0, 0)$ and the rectangular point $(\frac{1}{2}, \frac{1}{2})$. $A = \frac{1}{2}\int_0^{\pi/4}\cos^2\theta\, d\theta = \frac{1}{2}\int_0^{\pi/4}(\frac{1+\cos 2\theta}{2})d\theta = \frac{1}{4}[\theta + \frac{\sin 2\theta}{2}]_0^{\pi/4} = \frac{1}{4}(\frac{\pi}{4} + \frac{1}{2}) = \frac{\pi+2}{16}$.

3. Graph $r = 4 + 2\cos\theta$, $0 \leqq \theta \leqq 2\pi$ in $[-8.2, 12.2]$ by $[-6, 6]$. $A = \frac{1}{2}\int_0^{2\pi}[2(2 + \cos\theta)]^2 d\theta = 2\int_0^{2\pi}(4 + 4\cos\theta + \frac{1}{2} + \frac{\cos 2\theta}{2})d\theta = 2[\frac{9}{2}\theta + 4\sin\theta + \frac{\sin 2\theta}{4}]_0^{2\pi} = 18\pi$.

5. Graph $r = \cos 2\theta$, $-\frac{\pi}{4} \leqq \theta \leqq \frac{\pi}{4}$ in $[-0.09, 1.12]$ by $[-0.35, 0.35]$ for one leaf. $A = \frac{1}{2}\int_{-\pi/4}^{\pi/4}\cos^2(2\theta)d\theta = \frac{1}{4}\int_{-\pi/4}^{\pi/4}(1 + \cos 4\theta)d\theta = \frac{1}{4}[\theta + \frac{\sin 4\theta}{4}]_{-\pi/4}^{\pi/4} = \frac{\pi}{8}$.

7. For the purpose of graphing let $a = 2$. For the entire graph use $r = \sqrt{8\cos 2\theta}$, $0 \leqq \theta \leqq 2\pi$ in $[-3, 3]$ by $[-1.7, 1.7]$. $A = 4$(area in 1st quadrant) $= 4(\frac{1}{2})\int_0^{\pi/4} 2a^2\cos 2\theta\, d\theta = 4a^2\frac{\sin 2\theta}{2}]_0^{\pi/4} = 2a^2$.

9. Graph $r_1 = \sqrt{2\sin 3\theta}$ and $r_2 = -r_1$, $0 \leqq \theta \leqq 2\pi$ in $[-2.5, 2.5]$ by $[-1.5, 1.5]$. Area $= 6$(one leaf) $= 3\int_0^{\pi/3}(2\sin 3\theta)d\theta = -6\frac{\cos 3\theta}{3}]_0^{\pi/3} = 4$.

11. Graph $r = 1$ and $r = 2\sin\theta$, $0 \leqq \theta \leqq 2\pi$ in $[-2.5, 2.5]$ by $[-1, 2]$. $2\sin\theta = 1$ yields $\sin\theta = \frac{1}{2}$, $\theta = \frac{\pi}{6}, \frac{5\pi}{6}$. $A =$ area of top circle−area of top part of top circle $= \pi(1)^2 - \frac{1}{2}\int_{\pi/6}^{5\pi/6}(4\sin^2\theta - 1)d\theta = \pi - \frac{1}{2}\int_{\pi/6}^{5\pi/6}(2 - 2\cos 2\theta - 1)d\theta = \pi - \frac{1}{2}[\theta - \sin 2\theta]_{\pi/6}^{5\pi/6} = \pi - \frac{1}{2}[\frac{5\pi}{6} - (-\frac{\sqrt{3}}{2}) - (\frac{\pi}{6} - \frac{\sqrt{3}}{2})] = \pi - \frac{\pi}{3} - \frac{\sqrt{3}}{2} = \frac{2\pi}{3} - \frac{\sqrt{3}}{2}$.

13. Graph $r = 2(1 + \cos\theta)$ and $r = 2(1 - \cos\theta)$, $0 \leqq \theta \leqq 2\pi$ in $[-6.8, 6.8]$ by $[-4, 4]$. Since both curves are symmetric with respect to the x-axis and one is the reflection of the other through the y-axis, we need only take 4 times the area in the first quadrant. The latter area is determined by the second curve. $A = 4[\frac{1}{2}\int_0^{\pi/2}[2(1 - \cos\theta)]^2 d\theta] = 8\int_0^{\pi/2}(1 - 2\cos\theta + \frac{1}{2} + \frac{\cos 2\theta}{2})d\theta = 8[\frac{3}{2}\theta - 2\sin\theta + \frac{\sin 2\theta}{4}]_0^{\pi/2} = 8[\frac{3\pi}{4} - 2] = 6\pi - 16$.

15. Graph $r = 3a\cos\theta$ and $r = a(1+\cos\theta)$, with $a = 2$, $0 \leqq \theta \leqq 2\pi$ in $[-3, 8.3]$ by $[-3.3, 3.3]$. Points of intersection: $r = 3a\cos\theta = a(1+\cos\theta)$, $3\cos\theta = 1+\cos\theta$, $\cos\theta = \frac{1}{2}$, $\theta = \pm\frac{\pi}{3}$. $A = \frac{1}{2}\int_{-\pi/3}^{\pi/3}[(3a\cos\theta)^2 - (a(1+\cos\theta))^2]d\theta = a^2\int_0^{\pi/3}[9\cos^2\theta - 1 - 2\cos\theta - \cos^2\theta]d\theta = a^2\int_0^{\pi/3}(4+4\cos 2\theta - 1 - 2\cos\theta)d\theta = a^2[3\theta + 2\sin 2\theta - 2\sin\theta]_0^{\pi/3} = a^2[\pi + \sqrt{3} - \sqrt{3}] = \pi a^2$.

17. a) $A_1 = 2$ (area of top half) $= 2[\frac{1}{2}\int_0^{2\pi/3}(2\cos\theta + 1)^2 d\theta] = \int_0^{2\pi/3}(4\cos^2\theta + 4\cos\theta + 1)d\theta = \int_0^{2\pi/3}(2 + 2\cos 2\theta + 4\cos\theta + 1)d\theta = 3\theta + \sin 2\theta + 4\sin\theta]_0^{2\pi/3} = 2\pi - \frac{\sqrt{3}}{2} + 4(\frac{\sqrt{3}}{2}) = 2\pi + \frac{3\sqrt{3}}{2}$. b) $A_2 = A_1 - (\pi - \frac{3\sqrt{3}}{2}) = \pi + 3\sqrt{3}$.

19. Graph $r = \theta^2$, $0 \leqq \theta \leqq \sqrt{5}$ in $[-4.9, 2.5]$ by $[0, 4.4]$. $L = \int_0^{\sqrt{5}}\sqrt{\theta^4 + 4\theta^2}\,d\theta = \int_0^{\sqrt{5}}\theta\sqrt{\theta^2 + 4}\,d\theta$. Let $u = \theta^2 + 4$, $du = 2\theta\, d\theta$. $L = \frac{1}{2}\int_4^9 u^{1/2}du = \frac{1}{2}(\frac{2}{3})u^{3/2}]_4^9 = \frac{1}{3}(27 - 8) = \frac{19}{3}$.

21. $r = \sec\theta = 1/\cos\theta \Rightarrow r\cos\theta = 1$, $x = 1$, $0 \leqq \theta \leqq \frac{\pi}{4}$. The initial point is $(1,0)$ and the terminal point is $(\sqrt{2}, \pi/4)$. In rectangular coordinates $(1,0)$ to $(1,1)$ on $x = 1$ which has length 1.

23. Graph $r = 1 + \cos\theta$, $0 \leqq \theta \leqq 2\pi$ in $[-1.7, 3.4]$ by $[-1.5, 1.5]$. $r' = -\sin\theta$. $L = 2\int_0^{\pi}\sqrt{1 + 2\cos\theta + \cos^2\theta + \sin^2\theta}\,d\theta = 2\int_0^{\pi}\sqrt{2}\sqrt{1+\cos\theta}\,d\theta = 2\sqrt{2}\int_0^{\pi}\sqrt{(1+\cos\theta)\frac{(1-\cos\theta)}{1-\cos\theta}}\,d\theta = 2\sqrt{2}\int_0^{\pi}\frac{\sin\theta\, d\theta}{\sqrt{1-\cos\theta}}$. Let $u = 1 - \cos\theta$, $du = \sin\theta\, d\theta$. $L = 2\sqrt{2}\int_0^2 u^{-1/2}du = 2\sqrt{2}(2u^{1/2})]_0^2 = 8$.

25. $A = \int_{\alpha}^{\beta} 2\pi x\, ds = 2\pi\int_0^{\pi/4} r\cos\theta\sqrt{\cos 2\theta + (\frac{-2\sin 2\theta}{2\sqrt{\cos 2\theta}})^2}\,d\theta = 2\pi\int_0^{\pi/4}\sqrt{\cos 2\theta}\cos\theta\sqrt{\cos 2\theta + \frac{\sin^2 2\theta}{\cos 2\theta}}\,\theta = 2\pi\int_0^{\pi/4}\cos\theta\sqrt{\cos^2 2\theta + \sin^2 2\theta}\,d\theta = 2\pi\int_0^{\pi/4}\cos\theta\, d\theta = 2\pi\sin\theta]_0^{\pi/4} = \sqrt{2}\pi$.

27. $r^2 = \cos 2\theta$, $2r\frac{dr}{d\theta} = -2\sin 2\theta$, $\frac{dr}{d\theta} = -\frac{\sin 2\theta}{r}$, $(\frac{dr}{d\theta})^2 = \frac{\sin^2 2\theta}{r^2} = \frac{\sin^2 2\theta}{\cos 2\theta}$. The desired area is twice that generated by the arc $r = \sqrt{\cos 2\theta}$, $0 \leqq \theta \leqq \frac{\pi}{4}$. $A = 4\pi\int_0^{\pi/4} r\sin\theta\sqrt{\cos 2\theta + \frac{\sin^2 2\theta}{\cos 2\theta}}\,d\theta = 4\pi\int_0^{\pi/4}\sin\theta\sqrt{\cos^2 2\theta + \sin^2 2\theta}\,d\theta = -4\pi\cos\theta]_0^{\pi/4} = -4\pi[\frac{\sqrt{2}}{2} - 1] = 2(2 - \sqrt{2})\pi$.

29. b) $y_1 \to \infty$ as $x \to 0^+$ c) The integral is improper because y_1 is not defined at the endpoint $x = 0$.

31. $A = a^2\int_0^{2\pi}(1 + 2\cos\theta + \cos^2\theta)d\theta = a^2\int_0^{2\pi}(1 + 2\cos\theta + \frac{1}{2} + \frac{\cos 2\theta}{2})d\theta = a^2[\frac{3\theta}{2} + 2\sin\theta + \frac{\sin 2\theta}{4}]_0^{2\pi} = 3\pi a^2$. $\bar{x} = \frac{1}{A}(\frac{2}{3})\int_0^{2\pi} r^3\cos\theta\, d\theta = \frac{2a^3}{3A}\int_0^{2\pi}(1+\cos\theta)^3\cos\theta\, d\theta$. We use NINT and obtain $\bar{x} = (0.8333\ldots)a$ ($5a/6$ converting the decimal answer to a fraction). Since the graph is symmetric with respect to the x-axis, $\bar{y} = 0$.

PRACTICE EXERCISES, CHAPTER 10

1. $x = \frac{y^2}{8}$, $y^2 = 4(2)x$. Focus: $(2,0)$, directrix: $x = -2$. Graph $y = \pm 2\sqrt{2x}$ in $[-11.8, 16.8]$ by $[-8.4, 8.4]$. Use the line-drawing feature to include $x = -2$.

3. $16x^2 + 7y^2 = 112$, $\frac{x^2}{7} + \frac{y^2}{16} = 1$, vertices: $(0, \pm 4)$. $c = \sqrt{16-7} = 3$. Foci: $(0, \pm 3)$. $e = c/a = 3/4$. $a/e = 4(\frac{4}{3}) = \frac{16}{3}$. Directrices: $y = \pm\frac{16}{3}$. Graph $y = \pm 4\sqrt{1 - \frac{x^2}{7}}$, $y = \pm\frac{16}{3}$ in $[-10.2, 10.2]$ by $[-6, 6]$.

5. $3x^2 - y^2 = 3$, $x^2 - \frac{y^2}{3} = 1$. Vertices: $(\pm 1, 0)$. $c = \sqrt{1+3} = 2$. Foci: $(\pm 2, 0)$. $e = \frac{c}{a} = 2$, directrices: $x = \pm\frac{a}{e} = \pm\frac{1}{2}$. Graph $y = \pm\sqrt{3(x^2-1)}$ in $[-15, 15]$ by $[-10, 10]$.

7. $B^2 - 4AC = -3 < 0$ indicates ellipse but there is no solution and no graph. When we solve the quadratic for y, the discriminant is negative for all x.

9. $B^2 - 4AC = 0$, parabola. $4y^2 + (4x+1)y + (x^2 + x + 1) = 0$. Graph $y = \frac{-(4x+1) \pm \sqrt{-8x-15}}{8}$ in $[-13.5, 3.5]$ by $[0, 10]$.

11. $2x^2 + xy + 2y^2 - 15 = 0$. $B^2 - 4AC = 1 - 16 = -15 < 0$, ellipse. Since $A = C$, $\theta = \frac{\pi}{4}$, $x' = \frac{\sqrt{2}}{2}(x-y)$, $y' = \frac{\sqrt{2}}{2}(x+y)$. Simplifying $2x'^2 + x'y' + 2y'^2 = 15$, we obtain $\frac{x^2}{6} + \frac{y^2}{10} = 1$. The original equation is $2y^2 + xy + (2x^2 - 15) = 0$. We graph $y = \pm\sqrt{10(1 - \frac{x^2}{6})}$ and $y = \frac{-x \pm \sqrt{x^2 - 8(2x^2-15)}}{4}$ in $[-7.5, 7.5]$ by $[-4, 5]$.

13. Since $A = C(= 0)$, $\theta = \frac{\pi}{4}$. $x'y' = 2$ becomes $\frac{1}{\sqrt{2}}(x-y)\frac{1}{\sqrt{2}}(x+y) = \frac{x^2}{2} - \frac{y^2}{2} = 2$, $\frac{x^2}{4} - \frac{y^2}{4} = 1$. $e = \frac{c}{a} = \frac{\sqrt{4+4}}{2} = \sqrt{2}$.

15. a) $V = \pi\int_{-2}^{2} y^2 dx = 2\pi\int_0^2 \frac{36-9x^2}{4} dx = \frac{\pi}{2}[36x - 3x^3]_0^2 = \frac{\pi}{2}(72 - 24) = 24\pi$.
b) $V = \pi\int_{-3}^{3} x^2 dy = 2\pi\int_0^3 \frac{36-4y^2}{9} dy = \frac{2\pi}{9}[36y - \frac{4}{3}y^3]_0^3 = 16\pi$.

17. We may minimize the distance squared: $d^2 = (x-0)^2 + (y-3)^2 = 4t^2 + (t^2-3)^2 = t^4 - 2t^2 + 9 = (t^2-1)^2 + 8$. We see that this is minimal when $t = \pm 1$ which corresponds to the points $(\pm 2, 1)$.

19. $x = t/2$, $y = t + 1$, $-\infty < t < \infty$. $t = 2x = y - 1$ or $y = 2x + 1$. The entire line is traced out from left to right.

21. $x = (1/2)\tan t$, $y = (1/2)\sec t$; $-\pi/2 < t < \pi/2$. $y^2 = \frac{\sec^2 t}{4} = \frac{1+\tan^2 t}{4} = \frac{1}{4} + (\frac{\tan t}{2})^2 = \frac{1}{4} + x^2$, $y^2 - x^2 = \frac{1}{4}$, $\frac{y^2}{(0.5)^2} - \frac{x^2}{(0.5)^2} = 1$, hyperbola. The graph is the upper branch traced out from left to right. Confirm in parametric mode in $[-25.5, 25.5]$ by $[-10, 20]$.

23. $x = -\cos t$, $y = \cos^2 t$, $0 \leqq t \leqq \pi$. $y = x^2$, $-1 \leqq x \leqq 1$. The portion of the parabola $y = x^2$ determined by $-1 \leqq x \leqq 1$ is traced out from left to right.

25. $16x^2 + 9y^2 = 144$, $\frac{x^2}{9} + \frac{y^2}{16} = 1$. Let $x = 3\cos t$, $y = 4\sin t$, $0 \leqq t \leqq 2\pi$.

27. $x = (1/2)\tan t$, $y = (1/2)\sec t$; $t = \pi/3$. $\frac{dy}{dx} = \frac{dy/dt}{dx/dt} = \frac{\sec t \tan t}{\sec^2 t} = \frac{\tan t}{\sec t} = \sin t$. $\frac{dy}{dx}\big]_{t=\pi/3} = \sqrt{3}/2$. Tangent line: $y - 1 = (\sqrt{3}/2)(x - \sqrt{3}/2)$. $d^2y/dx^2 = (dy'/dt)/(dx/dt) = \cos t/((1/2)\sec^2 t) = 2\cos^3 t$. $[d^2y/dx^2]_{t=\pi/3} = 1/4$.

29. Graph $x = e^{2t} - t/8$, $y = e^t$; $0 \leqq t \leqq \ln 2$ in $[-0.16, 4.4]$ by $[-0.26, 2.5]$. $L = \int_0^{\ln 2} \sqrt{(dx/dt)^2 + (dy/dt)^2}dt = \int_0^{\ln 2} \sqrt{(2e^{2t} - 1/8)^2 + e^{2t}}dt = \int_0^{\ln 2} \sqrt{(2e^{2t} + 1/8)^2}dt = \int_0^{\ln 2} (2e^{2t} + 1/8)dt = e^{2t} + t/8]_0^{\ln 2} = 4 + \ln 2/8 - 1 = 3 + \ln 2/8$.

31. $A = 2\pi \int y\, ds = 2\pi \int_0^{\sqrt{5}} 2t\sqrt{t^2 + 2^2}dt$. Let $u = t^2 + 4$, $du = 2tdt$. $A = 2\pi \int_4^9 u^{1/2}du = 2\pi(\frac{2}{3})u^{3/2}]_4^9 = (4\pi/3)(27 - 8) = 76\pi/3$.

33. Graph $r = \cos 2\theta$, $0 \leqq \theta \leqq 2\pi$ in $[-1.7, 1.7]$ by $[-1, 1]$. Period $= 2\pi$. Four-leaved rose.

35. Graph $r = 6/(1 - 2\cos\theta)$, $0 \leqq \theta \leqq 2\pi$ in $[-30, 22]$ by $[-13, 18]$ in dot format. Period $= 2\pi$. Hyperbola.

37. Graph $r = \theta$, $-4\pi \leqq \theta \leqq 4\pi$ in $[-18.6, 18.6]$ by $[-12.3, 9.7]$. Infinite period. Spiral.

39. Graph $r = 1 + \cos\theta$, $0 \leqq \theta \leqq 2\pi$ in $[-1.6, 3.2]$ by $[-1.4, 1.4]$. Period 2π. Cardioid.

41. Graph $r = 2/(1 - \cos\theta)$, $0 \leqq \theta \leqq 2\pi$ in $[-10, 17]$ by $[-8, 8]$. Period 2π. Parabola.

43. Graph $r = -\sin\theta$, $0 \leqq \theta \leqq 2\pi$ in $[-1, 1]$ by $[-1.1, 0.1]$. Period π. Circle.

45. $2\sqrt{3} = r\cos(\theta - \frac{\pi}{3}) = r(\cos\theta\cos\frac{\pi}{3} + \sin\theta\sin\frac{\pi}{3}) = \frac{1}{2}x + \frac{\sqrt{3}}{2}y$, $x + \sqrt{3}y = 4\sqrt{3}$, $\sqrt{3}y = -x + 4\sqrt{3}$, $y = -\frac{1}{3}x + 4$. Graph $r = 2\sqrt{3}/\cos(\theta - \frac{\pi}{3})$, $0 \leqq \theta \leqq 2\pi$ in $[-17, 17]$ by $[-5, 15]$ and $y = -\frac{1}{\sqrt{3}}x + 4$ in the same window.

47. $r = 2\sin\theta$, $r^2 = 2r\sin\theta$, $x^2 + y^2 = 2y$, $x^2 + y^2 - 2y + 1 = 1$, $x^2 + (y - 1)^2 = 1$. Center $(0, 1)$, radius $= 1$.

49. $r = 6\cos\theta$, $r^2 = 6r\cos\theta$, $x^2+y^2 = 6x$, $x^2-6x+9+y^2 = 9$, $(x-3)^2+y^2 = 9$. The graph of $r = 6\cos\theta$ is the graph of the circle with center $(3,0)$ and radius 3. The region defined consists of all points on and within this circle. Graph the circle $r = 6\cos\theta$, $0 \leqq \theta \leqq \pi$ in $[-2.1, 8.1]$ by $[-3, 3]$.

51. $r = \sin\theta$, $r = 1+\sin\theta$. We first solve the equations simultaneously and then check the graphs to see if any points have been missed. $r = \sin\theta = 1+\sin\theta$, $0 = 1$, no solution. Graph $r = \sin\theta$, $r = 1+\sin\theta$, $0 \leqq \theta \leqq 2\pi$ in $[-2, 2]$ by $[-0.25, 2]$. We see that $(0,0)$ is the only point of intersection.

53. $r = 1+\sin\theta = -1+\sin\theta$, $1 = -1$, no solution. Graph for $0 \leqq \theta \leqq 2\pi$ in $[-1.9, 1.9]$ by $[-0.25, 2]$. We see that the two graphs are identical. Answer: all points on the graph. Analytically, $r = f(\theta)$ is equivalent to $-r = f(\theta-\pi)$. So $r = -1+\sin\theta$ is equivalent to $-r = -1+\sin(\theta-\pi) = -1-\sin\theta$ or $r = 1+\sin\theta$.

55. Graph the parabola $r = 2/(1+\cos\theta)$, $0 \leqq \theta \leqq 2\pi$ in $[-26.5, 7.5]$ by $[-10, 10]$. $(1,0)$ is the vertex.

57. $r = \frac{6}{1-2\cos\theta} = \frac{3(2)}{1-2\cos\theta}$, $e = 2$, hyperbola, $(0,0)$ is a focus and $x = -3$ is the corresponding directrix. Vertices: $(-6,0)$, $(2,\pi)$. Graph r, $0 \leqq \theta \leqq 2\pi$ in $[-13, 4]$ by $[-5, 5]$ in dot format.

59. $e = 2$, $x = r\cos\theta = k = 2$. From Equation (11) or Example 4 of 10.7, $r = \frac{ke}{1+e\cos\theta} = \frac{4}{1+2\cos\theta}$.

61. $e = \frac{1}{2}$, $r\sin\theta = y = k = 2$. From Table 10.4, $r = \frac{ke}{1+e\sin\theta} = \frac{1}{1+\frac{1}{2}\sin\theta} = \frac{2}{2+\sin\theta}$. This may be confirmed by graphing $r = 2/(2+\sin\theta)$, $0 \leqq \theta \leqq 2\pi$ in $[-2.3, 2.3]$ by $[-2, 0.7]$.

63. We use (4) of Section 10.6. $r^2 = \cos 2\theta$. $r = 0$ if and only if $\cos 2\theta = 0$, $2\theta = \frac{\pi}{2}+n\pi$, $\theta = \frac{\pi}{4}+\frac{n\pi}{2}$. For this curve $\theta = \frac{\pi}{4}$ and $\theta = \frac{3\pi}{4}$ cover all cases. In rectangular coordinates, $y = \pm x$.

65. Graph $r = \sin 2\theta$, $0 \leq \theta \leq 2\pi$ in $[-1.3, 1.3]$ by $[-0.77, 0.77]$. $\frac{dr}{d\theta} = 2\cos 2\theta$, $\frac{d^2r}{d\theta^2} = -4\sin 2\theta$. From this we see that r is extreme when $\theta = \frac{\pi}{4}, \frac{3\pi}{4}, \frac{5\pi}{4}$ and $\frac{7\pi}{4}$. Slope at $= \frac{r'\sin\theta+r\cos\theta}{r'\cos\theta-r\sin\theta} = \frac{2\cos 2\theta\sin\theta+\sin 2\theta\cos\theta}{2\cos 2\theta\cos\theta-\sin 2\theta\sin\theta}$. Slope at $(1, \frac{\pi}{4}) = \frac{\sqrt{2}/2}{-\sqrt{2}/2} = -1$. The tangent line at $(1, \frac{\pi}{4})$ must be perpendicular to $y = x$ or $\theta = \frac{\pi}{4}$. By Equation (3) of 10.7, $r\cos(\theta-\frac{\pi}{4}) = 1$ is an equation for this tangent line. By symmetry the other tangent lines are:. at $(1, \frac{3\pi}{4})$, $r\cos(\theta-\frac{3\pi}{4}) = 1$; at $(1, \frac{5\pi}{4})$, $r\cos(\theta-\frac{5\pi}{4}) = 1$; at $(1, \frac{7\pi}{4})$, $r\cos(\theta-\frac{7\pi}{4}) = 1$.

67. Graph $r = 2 - \cos\theta$, $0 \leqq \theta \leqq 2\pi$ in $[-4.7, 2.7]$ by $[-2.2, 2.2]$. $A = \frac{1}{2}\int_0^{2\pi}(2 - \cos\theta)^2 d\theta = \frac{1}{2}\int_0^{2\pi}(4 - 4\cos\theta + \frac{1}{2} + \frac{\cos 2\theta}{2})d\theta = \frac{1}{2}[\frac{9\theta}{2} - 4\sin\theta + \frac{\sin 2\theta}{4}]_0^{2\pi} = \frac{9\pi}{2}$.

69. Graph $r = 1 + \cos 2\theta$ and $r = 1$, $0 \leqq \theta \leqq 2\pi$ in $[-2, 2]$ by $[-1.2, 1.2]$. $r = 1 + \cos 2\theta = 1$, $\cos 2\theta = 0$, $2\theta = \frac{\pi}{2} + n\pi$, $\theta = \frac{\pi}{4} + \frac{n\pi}{2}$. $A = 4$ (area in first quadrant) $= 4(\frac{1}{2})\int_0^{\pi/4}[(1 + \cos 2\theta)^2 - 1]d\theta = 2\int_0^{\pi/4}(2\cos 2\theta + \frac{1}{2} + \frac{1}{2}\cos 4\theta)d\theta = 2[\sin 2\theta + \frac{\theta}{2} + \frac{\sin 4\theta}{8}]_0^{\pi/4} = 2(1 + \pi/8) = 2 + \pi/4$.

71. $r = \sqrt{\cos 2\theta}$, $r' = \frac{-2\sin 2\theta}{2\sqrt{\cos 2\theta}} = -\frac{\sin 2\theta}{\sqrt{\cos 2\theta}}$. $A = 2\pi\int_0^{\pi/4} r\sin\theta\sqrt{\cos 2\theta + \frac{\sin^2 2\theta}{\cos 2\theta}}d\theta = 2\pi\int_0^{\pi/4}\sin\theta\, d\theta = -2\pi\cos\theta]_0^{\pi/4} = -2\pi(\sqrt{2}/2 - 1) = (2 - \sqrt{2})\pi$.

73. Graph $r = -1 + \cos\theta$, $0 \leqq \theta \leqq 2\pi$ in $[-1.3, 3]$ by $[-1.3, 1.3]$. $L = \int_0^{2\pi}\sqrt{r^2 + r'^2}d\theta = 2\int_0^{\pi}\sqrt{(-1 + \cos\theta)^2 + \sin^2\theta d\theta} = 2\int_0^{\pi}\sqrt{2 - 2\cos\theta}\sqrt{\frac{1+\cos\theta}{1+\cos\theta}}d\theta = 2\sqrt{2}\int_0^{\pi}\frac{\sqrt{1-\cos^2\theta}}{\sqrt{1+\cos\theta}}d\theta = 2\sqrt{2}\int_0^{\pi}\frac{\sin\theta\, d\theta}{\sqrt{1+\cos\theta}} = -4\sqrt{2}\sqrt{1 + \cos\theta}]_0^{\pi} = 4\sqrt{2}\sqrt{2} = 8$.

75. Graph $r = \cos^3(\theta/3)$, $0 \leqq \theta \leqq \pi/4$ in $[0.34, 1.4]$ by $[0, 0.64]$. $L = \int_0^{\pi/4}\sqrt{\cos^6(\theta/3) + (\cos^2(\theta/3)\sin(\theta/3))^2 d\theta} = \int_0^{\pi/4}\sqrt{\cos^4(\theta/3)}\sqrt{\cos^2(\theta/3) + \sin^2(\theta/3)}d\theta = \int_0^{\pi/4}\cos^2(\theta/3)d\theta = \frac{1}{2}\int_0^{\pi/4}(1 + \cos(2\theta/3))d\theta = \frac{1}{2}[\theta + \frac{3}{2}\sin(2\theta/3)]_0^{\pi/4} = \frac{1}{2}[\frac{\pi}{4} + \frac{3}{2}\sin(\pi/6)] = \frac{1}{8}(\pi + 3)$.

77. Graph $x_1 = t$, $y_1 = t^2 - 1$, $x_2 = x_1\cos(\pi/6) - y_1\sin(\pi/6)$, $y_2 = x_1\sin(\pi/6) + y_1\cos(\pi/6)$, $-3 \leqq t \leqq 3$ in $[-6.6, 3]$ by $[-1, 8.4]$. Let $P(x, y)$ be a point on the rotated curve. If we rotate P back $30°(\theta = -\pi/6)$, the coordinates of the point obtained, $(\frac{\sqrt{3}}{2}x + \frac{1}{2}y, -\frac{1}{2}x + \frac{\sqrt{3}}{2}y)$ (by (4) of 10.2), will satisfy the original equation, $y = x^2 - 1$. Substituting these in and simplifying, we obtain $3x^2 + 2\sqrt{3}xy + y^2 + 2x - 2\sqrt{3}y - 4 = 0$.

79. a) $\frac{1}{2\pi}\int_0^{2\pi}a(1 - \cos\theta)d\theta = \frac{a}{2\pi}[\theta - \sin\theta]_0^{2\pi} = a$. b) $\frac{1}{2\pi}\int_0^{2\pi}a\, d\theta = a$. c) $\frac{1}{(\pi/2)-(-\pi/2)}\int_{-\pi/2}^{\pi/2}a\cos\theta\, d\theta = \frac{1}{\pi}a\sin\theta]_{-\pi/2}^{\pi/2} = \frac{2a}{\pi}$.

81. We use Fig. 10.13. $\frac{r_{\max} - r_{\min}}{r_{\max} + r_{\min}} = \frac{(c+a)-(a-c)}{(c+a)+(a-c)} = \frac{2c}{2a} = \frac{c}{a} = e$.

83. $\int_\alpha^\beta\sqrt{(2f(\theta))^2 + (2f'(\theta))^2 d\theta} = \int_\alpha^\beta\sqrt{4(f(\theta))^2 + 4(f'(\theta))^2 d\theta} = \sqrt{4}\int_\alpha^\beta\sqrt{(f(\theta))^2 + (f'(\theta))^2 d\theta} = 2L$.

CHAPTER 11

VECTORS AND ANALYTIC GEOMETRY IN SPACE

11.1 VECTORS IN THE PLANE

1.

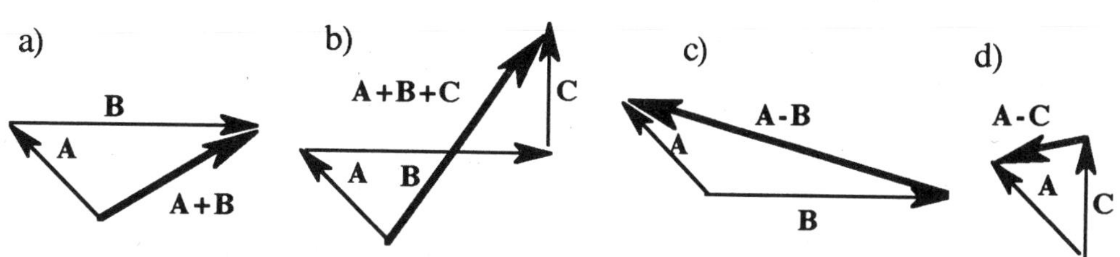

3. $3\boldsymbol{u} = 9\boldsymbol{i} - 6\boldsymbol{j}$; the scalar components are 9 and 6.

5. $\boldsymbol{u} + \boldsymbol{v} = \boldsymbol{i} + 3\boldsymbol{j}$; the scalar components are 1 and 3.

7. $2\boldsymbol{u} - 3\boldsymbol{v} = (6\boldsymbol{i} - 4\boldsymbol{j}) - (-6\boldsymbol{i} + 15\boldsymbol{j}) = 12\boldsymbol{i} - 19\boldsymbol{j}$; the scalar components are 12 and -19.

9. $\overrightarrow{P_1P_2} = (2-1)\boldsymbol{i} + (-1-3)\boldsymbol{j} = \boldsymbol{i} - 4\boldsymbol{j}$

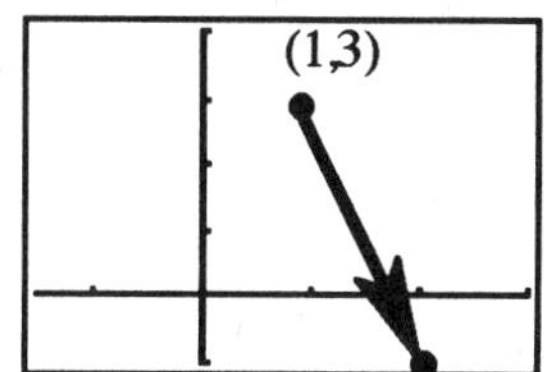

11. $-2\boldsymbol{i} - 3\boldsymbol{j}$

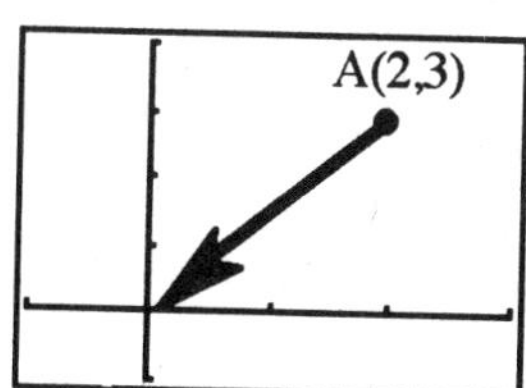

13. $\theta = \frac{\pi}{6}$: $\boldsymbol{u} = \frac{\sqrt{3}}{2}\boldsymbol{i} + \frac{1}{2}\boldsymbol{j}$;

$\theta = \frac{2\pi}{3}$: $\boldsymbol{u} = -\frac{1}{2}\boldsymbol{i} + \frac{\sqrt{3}}{2}\boldsymbol{j}$

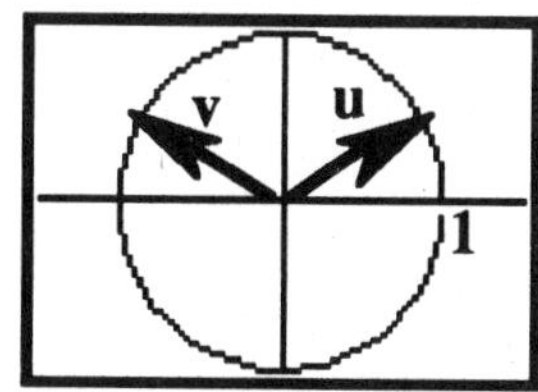

15. $\boldsymbol{j} = \cos 90°\boldsymbol{i} + \sin 90°\boldsymbol{j}$; $\boldsymbol{u} = \cos(90° - 120°)\boldsymbol{i} + \sin(90° - 120°)\boldsymbol{j} = \frac{\sqrt{3}}{2}\boldsymbol{i} - \frac{1}{2}\boldsymbol{j}$.

17. $|\boldsymbol{v}|^2 = 4 + 9 \Rightarrow |\boldsymbol{v}| = \sqrt{13}$.

19. $|\boldsymbol{v}|^2 = \frac{9}{25} + \frac{16}{25} \Rightarrow |\boldsymbol{v}| = \sqrt{\frac{25}{25}} = 1$.

21. $|\boldsymbol{v}|^2 = \frac{25}{13^2} + \frac{144}{13^2} \Rightarrow |\boldsymbol{v}| = 1$.

23. $\boldsymbol{u} = \frac{1}{\sqrt{3^2+4^2}}[3\boldsymbol{i} + 4\boldsymbol{j}] = \frac{3}{5}\boldsymbol{i} + \frac{4}{5}\boldsymbol{j}$.

25. $\boldsymbol{u} = \frac{1}{\sqrt{144+25}}[12\boldsymbol{i} - 5\boldsymbol{j}] = \frac{12}{13}\boldsymbol{i} - \frac{5}{13}\boldsymbol{j}$.

27. $\boldsymbol{u} = \frac{1}{\sqrt{4+9}}[2\boldsymbol{i} + 3\boldsymbol{j}] = \frac{2}{\sqrt{13}}\boldsymbol{i} + \frac{3}{\sqrt{13}}\boldsymbol{j}$.

29. length $= \sqrt{1^2 + 1^2} = \sqrt{2}$; direction $= \frac{1}{\sqrt{2}}(\boldsymbol{i} + \boldsymbol{j})$; $\boldsymbol{i} + \boldsymbol{j} = \sqrt{2}\left(\frac{1}{\sqrt{2}}\boldsymbol{i} + \frac{1}{\sqrt{2}}\boldsymbol{j}\right)$.

31. length $= \sqrt{3 + 1^2} = 2$; direction $= \frac{1}{2}(\sqrt{3}\boldsymbol{i} + \boldsymbol{j})$; $\sqrt{3}\boldsymbol{i} + \boldsymbol{j} = 2\left(\frac{\sqrt{3}}{2}\boldsymbol{i} + \frac{1}{2}\boldsymbol{j}\right)$.

33. length $= \sqrt{5^2 + 12^2} = 13$; direction $= \frac{5}{13}(2\boldsymbol{i} + \frac{12}{13}\boldsymbol{j})$; answer $= 13\left[\frac{5}{13}\boldsymbol{i} + \frac{12}{13}\boldsymbol{j}\right]$.

35. Direction of $\boldsymbol{A} = \frac{1}{|\boldsymbol{A}|}\boldsymbol{A} = \frac{1}{\sqrt{9+36}}(3\boldsymbol{i} + 6\boldsymbol{j}) =$
$\frac{3}{\sqrt{45}}\boldsymbol{i} + \frac{6}{\sqrt{45}}\boldsymbol{j} = \frac{1}{\sqrt{5}}\boldsymbol{i} + \frac{2}{\sqrt{5}}\boldsymbol{j}$;
Direction of $\boldsymbol{B} = \frac{1}{|\boldsymbol{B}|}\boldsymbol{B} = \frac{1}{\sqrt{1+4}}(-\boldsymbol{i} - 2\boldsymbol{j}) =$
$-\frac{1}{\sqrt{5}}\boldsymbol{i} - \frac{2}{\sqrt{5}}\boldsymbol{j}$.

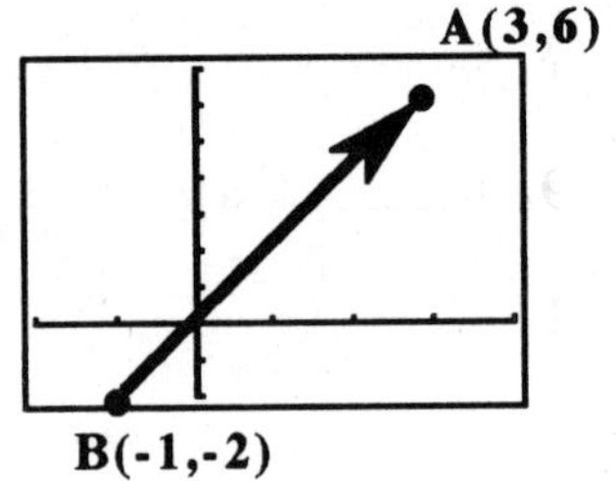

37. $y' = 2x$; $y'(2) = 4$. $\boldsymbol{i} + 4\boldsymbol{j}$ is tangent $\Rightarrow$
$\pm\frac{1}{\sqrt{5}}(\boldsymbol{i} + 4\boldsymbol{j})$ are unit tangent vectors;
$4\boldsymbol{i} - \boldsymbol{j}$ is normal $\Rightarrow \pm\frac{1}{\sqrt{5}}(4\boldsymbol{i} - \boldsymbol{j})$
are unit normal vectors.

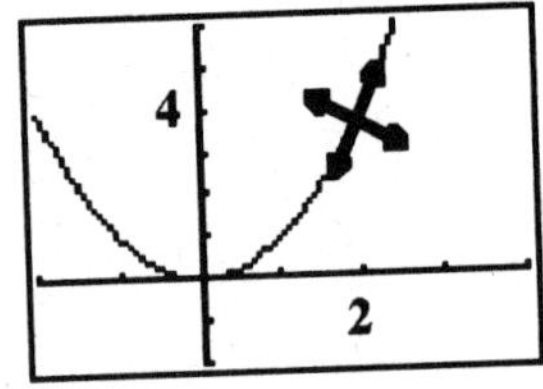

39. $2x + 4yy' = 0 \Rightarrow y'(2) = \frac{-2}{2} = -1$; $\boldsymbol{i} - \boldsymbol{j}$
is tangent and $\boldsymbol{i} + \boldsymbol{j}$ is normal to the curve.
Unit tangent vectors are $\pm\frac{1}{\sqrt{2}}(\boldsymbol{i} - \boldsymbol{j})$;
unit normal vectors are $\pm\frac{1}{\sqrt{2}}(\boldsymbol{i} + \boldsymbol{j})$.

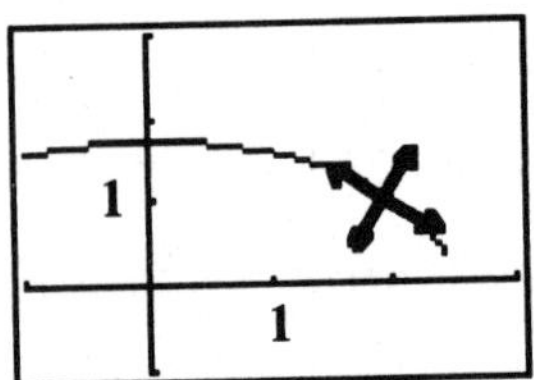

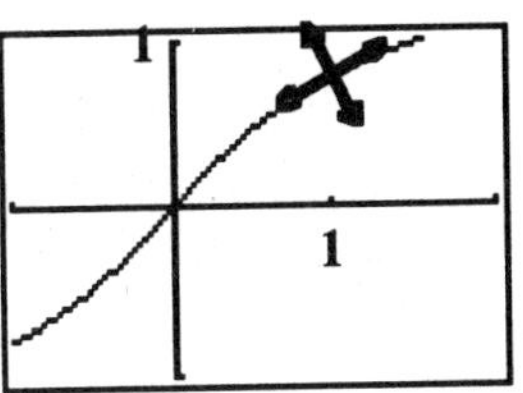

41. $y' = \frac{1}{1+x^2} \Rightarrow y'(1) = \frac{1}{2}$; $2\boldsymbol{i} + \boldsymbol{j}$ is tangent and $\boldsymbol{i} - 2\boldsymbol{j}$ is normal to the curve.
Unit tangent vectors are $\pm\frac{1}{\sqrt{5}}(2\boldsymbol{i} + \boldsymbol{j})$;
unit normal vectors are $\pm\frac{1}{\sqrt{5}}(\boldsymbol{i} - 2\boldsymbol{j})$.

43. The same line segment is used in representing both $\boldsymbol{v}$ and $-\boldsymbol{v}$; the slopes of the vectors are the same.

11.2 CARTESIAN (RECTANGULAR) COORDINATES AND VECTORS IN SPACE

1. A line through $(2,3,0)$ parallel to the z-axis.

3. The x-axis.

5. Circle in xy-plane, center at $(0,0,0)$, radius 2.

7. Circle in xz-plane, center at $(0,0,0)$, radius 2.

9. Circle in yz-plane, center at $(0,0,0)$, radius 1.

11. This is a sphere - center at $(0,0,-3)$, radius 5 - sliced by a plane. The set forms the circle $x^2 + y^2 + 3^2 = 25^2$ or $x^2 + y^2 = 4^2$ in the xy-plane.

13. a) The first quadrant in the xy-plane b) The fourth quadrant in the xy-plane

15. a) The interior and surface of the unit sphere (center at $(0,0,0)$),
b) All of 3-space <u>but</u> the interior and surface of the unit sphere

17. a) The surface of the top half ($z \geq 0$) of the unit sphere, b) The interior and surface of the top half of the unit sphere

19. a) $x = 3$, b) $y = -1$, c) $z = -2$

21. a) $z = 1$, b) $x = 3$, c) $y = -1$

23. a) $(x-0)^2 + (y-2)^2 = 2^2$, $z = 0$; b) $(y-2)^2 + (z-0)^2 = 2^2$, $x = 0$; c) $(x-0)^2 + (z-0)^2 = 2^2$, $y = 2$

25. a) x can be anything; $y = 3$, $z = -1$; b) $x = 1$, $z = -1$, c) $x = 1$, $y = 3$

27. The plane is $z = 3$; the sphere is $x^2 + y^2 + z^2 = 25$. Hence: $x^2 + y^2 + 9 = 25$, $z = 3$.

29. $0 \le z \le 1$

31. $z \le 0$

33. a) $(x-1)^2 + (y-1)^2 + (z-1)^2 < 1$; b) $(x-1)^2 + (y-1)^2 + (z-1)^2 > 1$

35. $|\boldsymbol{A}| = \sqrt{2^2 + 1^2 + 2^2} = 3$, direction $\frac{\boldsymbol{A}}{|\boldsymbol{A}|} = \frac{2}{3}\boldsymbol{i} + \frac{1}{3}\boldsymbol{j} - \frac{2}{3}\boldsymbol{k}$.

37. $|\boldsymbol{A}| = \sqrt{1^2 + 4^2 + 8^2} = 9$, $\frac{\boldsymbol{A}}{|\boldsymbol{A}|} = \frac{1}{9}\boldsymbol{i} + \frac{4}{9}\boldsymbol{j} - \frac{8}{9}\boldsymbol{k}$

39. $|\boldsymbol{A}| = 5$, $\frac{\boldsymbol{A}}{|\boldsymbol{A}|} = \boldsymbol{k}$

41. $|\boldsymbol{A}| = 4$, $\frac{\boldsymbol{A}}{|\boldsymbol{A}|} = -\boldsymbol{j}$

43. $|\boldsymbol{A}| = \sqrt{(\frac{1}{3})^2 + (\frac{1}{4})^2} = \frac{5}{12}$; $\frac{\boldsymbol{A}}{|\boldsymbol{A}|} = \frac{12}{5}(-\frac{1}{3}\boldsymbol{j} + \frac{1}{4}\boldsymbol{k}) = -\frac{4}{5}\boldsymbol{j} + \frac{3}{5}\boldsymbol{k}$

45. $|\boldsymbol{A}| = \sqrt{3(\frac{1}{6})} = \sqrt{\frac{1}{2}} = \frac{1}{\sqrt{2}}$; $\frac{\boldsymbol{A}}{|\boldsymbol{A}|} = \sqrt{2}\left(\frac{1}{\sqrt{6}}\boldsymbol{i} - \frac{1}{\sqrt{6}}\boldsymbol{j} - \frac{1}{\sqrt{6}}\boldsymbol{k}\right) = \frac{1}{\sqrt{3}}\boldsymbol{i} - \frac{1}{\sqrt{3}}\boldsymbol{j} - \frac{1}{\sqrt{3}}\boldsymbol{k}$

47. a) Distance $= |\overrightarrow{P_1P_2}| = |2\boldsymbol{i} + 2\boldsymbol{j} - \boldsymbol{k}| = \sqrt{4+4+1} = 3$; b) Direction $= \frac{2}{3}\boldsymbol{i} + \frac{2}{3}\boldsymbol{j} - \frac{1}{3}\boldsymbol{k}$; c) Midpoint is $\left(\frac{1+3}{2}, \frac{1+3}{2}, \frac{1}{2}\right) = \left(2, 2, \frac{1}{2}\right)$

49. a) Distance $= |\overrightarrow{P_1P_2}| = |3\boldsymbol{i} - 6\boldsymbol{j} + 2\boldsymbol{k}| = 7$; b) Direction $= \frac{3}{7}\boldsymbol{i} - \frac{6}{7}\boldsymbol{j} + \frac{2}{7}\boldsymbol{k}$, c) Midpoint $= \left(\frac{5}{2}, 1, 6\right)$

51. a) Distance $= |\overrightarrow{P_1P_2}| = |2\boldsymbol{i} - 2\boldsymbol{j} - 2\boldsymbol{k}| = \sqrt{12} = 2\sqrt{3}$; b) Direction $= \frac{1}{\sqrt{3}}\boldsymbol{i} - \frac{1}{\sqrt{3}}\boldsymbol{j} - \frac{1}{\sqrt{3}}\boldsymbol{k}$, c) Midpoint $= (1, -1, -1)$

53. a) $2\boldsymbol{i}$, b) $-\sqrt{3}\boldsymbol{k}$, c) $\frac{3}{10}\boldsymbol{j} + \frac{4}{10}\boldsymbol{k}$, d) $6\boldsymbol{i} - 2\boldsymbol{j} + 3\boldsymbol{k}$

55. $|\boldsymbol{A}| = \sqrt{12^2 + 5^2} = 13 \Rightarrow \boldsymbol{v} = \frac{7}{13}\boldsymbol{A} = \frac{84}{13}\boldsymbol{i} - \frac{35}{13}\boldsymbol{j}$

57. $|\boldsymbol{A}| = \sqrt{4 + 9 + 36} = 7 \Rightarrow \boldsymbol{v} = -\frac{5}{7}\boldsymbol{A} = -\frac{10}{7}\boldsymbol{i} + \frac{15}{7}\boldsymbol{j} - \frac{30}{7}\boldsymbol{k}$

59. a) $C(-2, 0, 2)$, radius $= \sqrt{8} = 2\sqrt{2}$, b) $C(-\frac{1}{2}, -\frac{1}{2}, -\frac{1}{2})$, radius $= \frac{\sqrt{21}}{2}$, c) $C(\sqrt{2}, \sqrt{2}, -\sqrt{2})$, radius $= \sqrt{2}$, d) $C(0, -\frac{1}{3}, \frac{1}{3})$, radius $= \frac{\sqrt{29}}{3}$

61. $x^2 + 4x + y^2 + z^2 - 4z = 0 \Rightarrow x^2 + 4x + 4 + y^2 + z^2 - 4z + 4 = 8 \Rightarrow (x+2)^2 + y^2 + (z-2)^2 = 8$; Center $(-2, 0, 2)$, radius $\sqrt{8}$

63. $x^2 + y^2 + z^2 - 2z + 1 = 0 + 1$; Center $(0, 0, 1)$, radius 1

65. The curves of intersection are all circles:
a) $x^2 + z^2 = 5$ in the plane $y = -2$,
$x^2 + z^2 = 9$ in the plane $y = 0$,
$x^2 + z^2 = 8$ in the plane $y = 1$
b) the areas are 5π, 9π, 8π

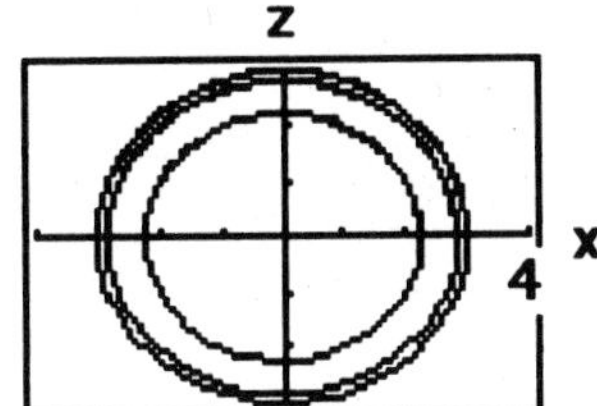

67. a) In $z = 2$, $x^2+y^2+4x = 8-4$ or $x^2+4x+4+y^2 = 8$, i.e., $(x+2)^2+y^2 = 8$; in $z = 4$, $x^2+4x+y^2 = 16-16$, or $(x+2)^2+y^2 = 4$

b)

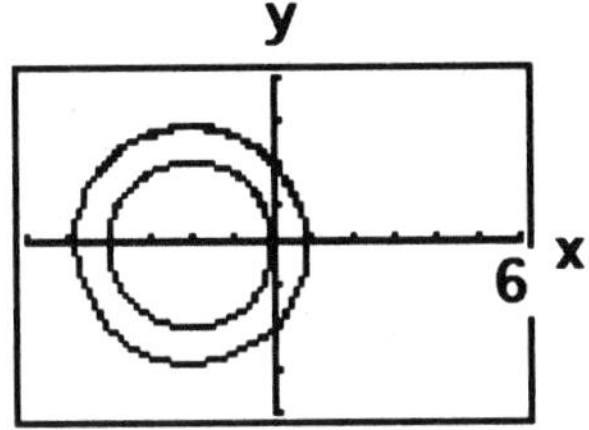

69. a) $\sqrt{y^2+z^2}$,

b) $\sqrt{x^2+z^2}$,

c) $\sqrt{x^2+y^2}$

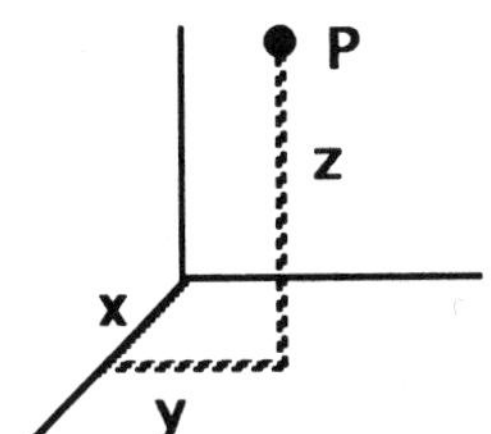

11.3 DOT PRODUCTS

1. $\boldsymbol{A}\cdot\boldsymbol{B} = 3\cdot 0 + 2\cdot 5 + 0\cdot 1 = 10$, $|\boldsymbol{A}| = \sqrt{3^2+2^2+0^2} = \sqrt{13}$, $|\boldsymbol{B}| = \sqrt{0^2+5^2+1^2} = \sqrt{26}$, $\cos\theta = \frac{\boldsymbol{A}\cdot\boldsymbol{B}}{|\boldsymbol{A}||\boldsymbol{B}|} = \frac{10}{\sqrt{13}\sqrt{26}} = \frac{10}{13\sqrt{2}}$; $|\boldsymbol{B}|\cos\theta = \frac{\sqrt{26}\cdot 10}{13\sqrt{2}} = \frac{10}{\sqrt{13}}$. $\text{Proj}_{\boldsymbol{A}}\boldsymbol{B} = \frac{\boldsymbol{A}\cdot\boldsymbol{B}}{\boldsymbol{A}\cdot\boldsymbol{A}}\boldsymbol{A} = \frac{10}{13}[3\boldsymbol{i}+2\boldsymbol{j}]$.

Problems 3-11 are done in exactly the same way.

	$A \cdot B$	$\|A\|$	$\|B\|$	$\cos\theta$	$\|B\|\cos\theta$	$\text{Proj}_{A} B = \frac{A \cdot B}{A \cdot A} A$
1.	10	$\sqrt{13}$	$\sqrt{26}$	$\frac{10}{13\sqrt{2}}$	$\frac{10}{\sqrt{13}}$	$\frac{10}{13}[3i + 2j]$
3.	4	$\sqrt{14}$	2	$\frac{2}{\sqrt{14}}$	$\frac{4}{\sqrt{14}}$	$\frac{2}{7}[3i - 2j - k]$
5.	2	$\sqrt{34}$	$\sqrt{3}$	$\frac{2}{\sqrt{3}\sqrt{34}}$	$\frac{2}{\sqrt{34}}$	$\frac{1}{17}[5j - 3k]$
7.	$\sqrt{3} - \sqrt{2}$	$\sqrt{2}$	3	$\frac{\sqrt{3}-\sqrt{2}}{3\sqrt{2}}$	$\frac{\sqrt{3}-\sqrt{2}}{2}$	$\frac{\sqrt{3}-\sqrt{2}}{2}[-i + j]$
9.	-25	5	5	-1	-5	$-2i + 4j - \sqrt{5}k$
11.	25	15	5	$\frac{1}{3}$	$\frac{5}{3}$	$\frac{1}{9}[10i + 11j - 2k]$

13. $B = \frac{A \cdot B}{A \cdot A} A + (B - \frac{A \cdot B}{A \cdot A} A) = \frac{3}{2} A + (B - \frac{3}{2} A) = \frac{3}{2}(i + j) + [3j + 4k - \frac{3}{2} i + j] = [\frac{3}{2}(i + j)] + [-\frac{3}{2} i + \frac{3}{2} j + 4k]$

15. $B = \frac{A \cdot B}{A \cdot A} A + [B - \frac{A \cdot B}{A \cdot A} A] = \frac{28}{6} A + [B - \frac{28}{6} A] = \frac{14}{3}[i + 2j - k] + [8i + 4j - 12k - \frac{14}{3}(i + 2j - k)] = \frac{14}{3}[i + 2j - k] + [\frac{10}{3} i - \frac{16}{3} j - \frac{22}{3} k]$

17. $N \cdot \overrightarrow{P_0P} = 1(x - 2) + 2 \cdot (y - 1) = 0$

$\Rightarrow x + 2y = 4$

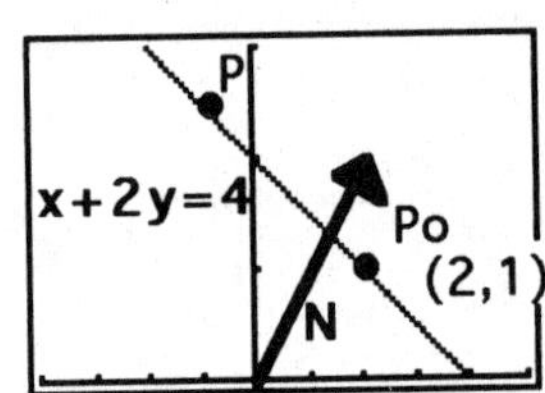

19. $N \cdot \overrightarrow{P_0P} = -2(x + 1) - 1(y - 2) = 0$

$\Rightarrow -2x - y = 0$

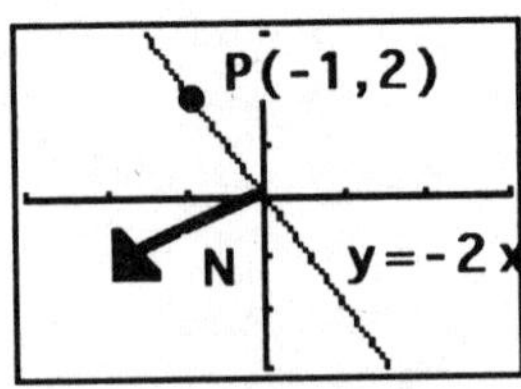

21. The point $P(0, 2)$ is on the line. Then $\overrightarrow{PS} = (2-0)i + (8-2)j = 2i + 2j$; $N = i + 3j$. Distance $= |\text{Proj}_{N} \overrightarrow{PS}| = \left|\overrightarrow{PS} \cdot \frac{N}{|N|}\right| = \frac{2 \cdot 1 + 6 \cdot 3}{\sqrt{1^2 + 3^2}} = \frac{20}{\sqrt{10}} = 2\sqrt{10}$.

23. Let P be $(1,0)$; $\vec{PS} = \boldsymbol{i}+\boldsymbol{j}$; $\boldsymbol{N} = \boldsymbol{i}+\boldsymbol{j}$. Distance $= \left|\frac{1\cdot 1+1\cdot 1}{\sqrt{2}}\right| = \sqrt{2}$.

25. $\boldsymbol{A}\cdot\boldsymbol{B} = \frac{1}{\sqrt{3}}\frac{1}{\sqrt{6}}(1(-2)-1(-1)+1(1)) = 0$; $\boldsymbol{A}\cdot\boldsymbol{B} = \frac{1}{\sqrt{3}}\frac{1}{\sqrt{2}}(0-1(1)+1(1)) = 0$; $\boldsymbol{B}\cdot\boldsymbol{C} = \frac{1}{\sqrt{2}}\frac{1}{\sqrt{6}}(0-1+1) = 0$

27. Since $\boldsymbol{i}\cdot\boldsymbol{j} = 0$ and $\boldsymbol{i}\cdot\boldsymbol{k} = 0$, we have $\boldsymbol{i}\cdot\boldsymbol{j} = \boldsymbol{i}\cdot\boldsymbol{k}$. However, $\boldsymbol{j} \neq \boldsymbol{k}$.

29. $\vec{AB} = 3\boldsymbol{i}+\boldsymbol{j}-3\boldsymbol{k}$, $\vec{AC} = 2\boldsymbol{i}-2\boldsymbol{j}$, $\vec{BC} = -\boldsymbol{i}-3\boldsymbol{j}+3\boldsymbol{k}$. Recalling that $\vec{BA} = -\vec{AB}$, $\vec{CA} = -\vec{AC}$, we have $< A = \cos^{-1}\left[\frac{\vec{AB}\cdot\vec{AC}}{|\vec{AB}|\cdot|\vec{AC}|}\right] = \cos^{-1}\frac{4}{\sqrt{19}\sqrt{8}} = 71.068°$, $< B = \cos^{-1}\left[\frac{\vec{BA}\cdot\vec{BC}}{|\vec{BA}|\cdot|\vec{BC}|}\right] = \cos^{-1}\frac{15}{\sqrt{19}\sqrt{19}} = 37.864°$, $< C = 180 - (< A + < B) = 71.068°$

31. For the unit cube, the corner diagonally opposite the origin is at $A(1,1,1)$. The diagonal corner of the face in the xz-plane is at $B(1,0,1)$. The angle between $\vec{OA}$ and $\vec{OB}$ is $\cos^{-1}\left[\frac{1+1}{\sqrt{3}\sqrt{2}}\right] = 35.264°$.

33. Work $= \boldsymbol{F}\cdot\vec{PQ} = (-5\boldsymbol{k})\cdot(\boldsymbol{i}+\boldsymbol{j}+\boldsymbol{k}) = -5\ N\cdot m$

35. Work $= |\boldsymbol{F}||\vec{PQ}|\cos\theta = (200)(20)\cos 41° = 3018.838\ N\cdot m$.

37. These curves are straight lines; their tangents (and, thus, their normals) always have the same directions. Find the angle between the normals: $\boldsymbol{N}_1 = 3\boldsymbol{i}+\boldsymbol{j}$, $\boldsymbol{N}_2 = 2\boldsymbol{i}-\boldsymbol{j}$. $\theta = \cos^{-1}\frac{6-1}{\sqrt{10}\sqrt{5}} = \cos^{-1}\frac{1}{\sqrt{2}} = 45°$ or $135°$.

39. The curves intersect at $x = -1.92630321991$ (use technology), at that point, by evaluating the derivative, the slope of the tangent to $y = x^2-2$ is $B = -3.85260643982$; the other tangent has slope $C = -0.292287566682$. A set of tangent vectors is $\boldsymbol{i}+B\boldsymbol{j}$ and $\boldsymbol{i}+C\boldsymbol{j}$; $\theta = \cos^{-1}\left[\frac{1+B\cdot C}{\sqrt{B^2+1}\sqrt{C^2+1}}\right] = \cos^{-1} 0.512699775941 = 59.156°$.

41. The curves intersect at $x = 0$ and $x = -1.406$; at $x = 0$, the tangents have slopes 0 and 2; $\theta = \cos^{-1}\left[\frac{1}{\sqrt{5}}\right] = 63.4°$. At $x = -1.406$, the tangents have slope -0.811 and 0.986; $\theta = \cos^{-1}\left[\frac{1+(-0.811)(0.986)}{\sqrt{(0.811)^2+1}\sqrt{(0.986)^2+1}}\right] = 1.460$ radians $\approx 83.7°$. See #39 for more details.

43. $\boldsymbol{A}\cdot\boldsymbol{B} = |\boldsymbol{A}||\boldsymbol{B}|\cos\theta = \frac{|\boldsymbol{A}|^2+|\boldsymbol{B}|^2-|\boldsymbol{C}|^2}{2} = \frac{1}{2}\left[a_1^2+a_2^2+a_3^2+b_1^2+b_2^2+b_3^2 - \{(b_1-a_1)^2+(b_2-a_2)^2+(b_3-a_3)^2\}\right] = \frac{1}{2}\left[2a_1b_1+2a_2b_2+2a_3b_3\right] = a_1b_1+a_2b_2+a_3b_3$.

11.4 CROSS PRODUCTS

1. $\boldsymbol{A} \times \boldsymbol{B} = \begin{vmatrix} \boldsymbol{i} & \boldsymbol{j} & \boldsymbol{k} \\ 2 & -2 & -1 \\ 1 & 0 & -1 \end{vmatrix} = \boldsymbol{i}(2-0) + \boldsymbol{j}(-1+2) + \boldsymbol{k}(0+2) = 2\boldsymbol{i} + \boldsymbol{j} + 2\boldsymbol{k}$;
$|\boldsymbol{A} \times \boldsymbol{B}| = \sqrt{4+1+4} = 3$, direction $= [\frac{2}{3}\boldsymbol{i} + \frac{1}{3}\boldsymbol{j} + \frac{2}{3}\boldsymbol{k}]$. $\boldsymbol{B} \times \boldsymbol{A} = -(\boldsymbol{A} \times \boldsymbol{B})$; length $= 3$, direction $= -[\frac{2}{3}\boldsymbol{i} + \frac{1}{3}\boldsymbol{j} + \frac{2}{3}\boldsymbol{k}]$. In Problems 2-8, $\boldsymbol{A} \times \boldsymbol{B}$ and $\boldsymbol{B} \times \boldsymbol{A}$ have the same length and (when it exists) opposite direction.

3. $\boldsymbol{A} \times \boldsymbol{B} = \begin{vmatrix} \boldsymbol{i} & \boldsymbol{j} & \boldsymbol{k} \\ 2 & -2 & 4 \\ -1 & 1 & -2 \end{vmatrix} = (4-4)\boldsymbol{i} + (-4+4)\boldsymbol{j} + (2-2)\boldsymbol{k} = \boldsymbol{0}.$
Length $= 0$; $\boldsymbol{A} \times \boldsymbol{B}$ has no direction. $\boldsymbol{B} \times \boldsymbol{A} = 0$, has length 0 and no direction.

5. $\boldsymbol{A} \times \boldsymbol{B} = \begin{vmatrix} \boldsymbol{i} & \boldsymbol{j} & \boldsymbol{k} \\ 2 & 0 & 0 \\ 0 & -3 & 0 \end{vmatrix} = -6\boldsymbol{k}$; length $= 6$, direction $= -\boldsymbol{k}$.

7. $\boldsymbol{A} \times \boldsymbol{B} = \begin{vmatrix} \boldsymbol{i} & \boldsymbol{j} & \boldsymbol{k} \\ -8 & -2 & -4 \\ 2 & 2 & 1 \end{vmatrix} = (-2+8)\boldsymbol{i} + (-8+8)\boldsymbol{j} + (-16+4)\boldsymbol{k} = 6\boldsymbol{i} - 12\boldsymbol{k}$;
length $= \sqrt{36+144} = \sqrt{180} = 6\sqrt{5}$; direction is $\frac{1}{\sqrt{5}}\boldsymbol{i} - \frac{2}{\sqrt{5}}\boldsymbol{k}$; $\boldsymbol{B} \times \boldsymbol{A}$ has length $6\sqrt{5}$, opposite direction.

9. $\boldsymbol{A} \times \boldsymbol{B} = \boldsymbol{k}$

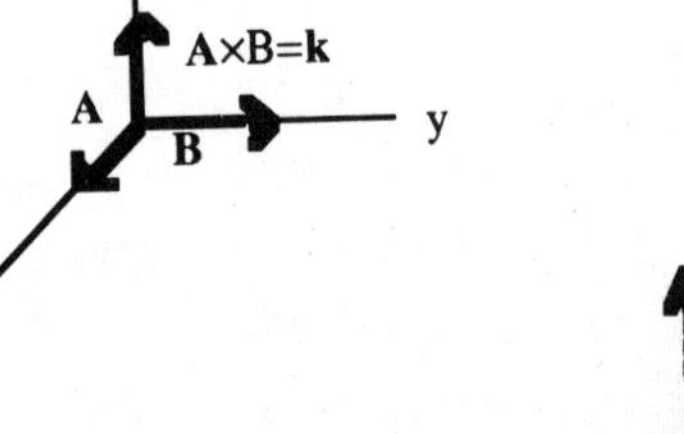

11. $\boldsymbol{A} \times \boldsymbol{B} = \begin{vmatrix} \boldsymbol{i} & \boldsymbol{j} & \boldsymbol{k} \\ 1 & 0 & -1 \\ 1 & 1 & 1 \end{vmatrix} = \boldsymbol{i} - \boldsymbol{j} + \boldsymbol{k}$

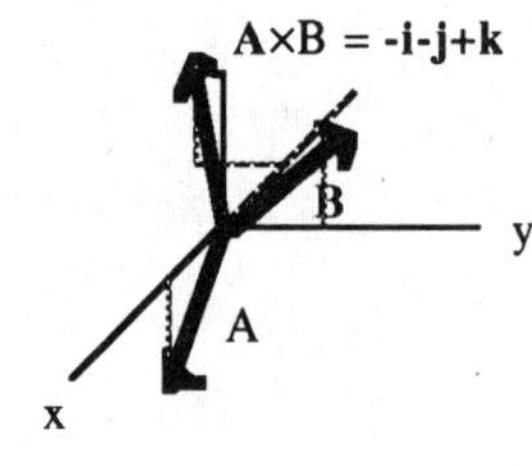

13. $\boldsymbol{A} \times \boldsymbol{B} = \begin{vmatrix} \boldsymbol{i} & \boldsymbol{j} & \boldsymbol{k} \\ 1 & 3 & 2 \\ 0 & 0 & 1 \end{vmatrix} = 3\boldsymbol{i} - \boldsymbol{j}$

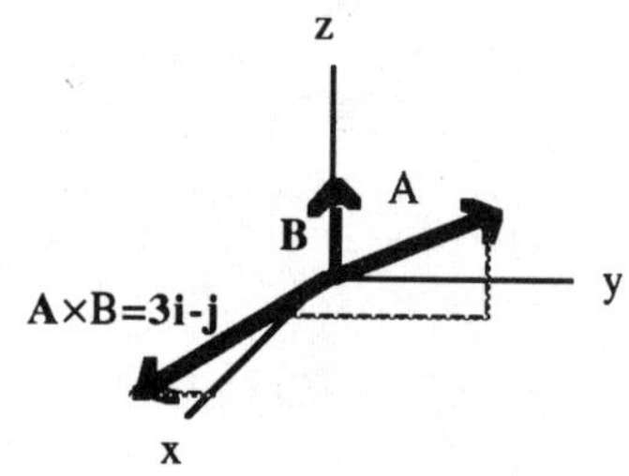

15. a) $N = \pm \overrightarrow{PQ} \times \overrightarrow{PR} = \pm \begin{vmatrix} \boldsymbol{i} & \boldsymbol{j} & \boldsymbol{k} \\ 1 & 1 & -3 \\ -1 & 3 & -1 \end{vmatrix} = \pm(8\boldsymbol{i} + 4\boldsymbol{j} + 4\boldsymbol{k})$;

b) $\frac{1}{2} \mid \overrightarrow{PQ} \times \overrightarrow{PR} \mid = \frac{1}{2}\sqrt{64+16+16} = \frac{1}{2}(4\sqrt{6}) = 2\sqrt{6}$;

c) $\frac{\boldsymbol{N}}{|\boldsymbol{N}|} = \frac{1}{\sqrt{6}}(2\boldsymbol{i} + \boldsymbol{j} + \boldsymbol{k})$

17. a) $\boldsymbol{N} = \pm \overrightarrow{PQ} \times \overrightarrow{PR} = \pm \begin{vmatrix} \boldsymbol{i} & \boldsymbol{j} & \boldsymbol{k} \\ 1 & 1 & 1 \\ 1 & 1 & 0 \end{vmatrix} = \pm(-\boldsymbol{i} + \boldsymbol{j})$; b) $\frac{1}{2}|\boldsymbol{N}| = \frac{1}{2}\sqrt{2} = 1/\sqrt{2}$; c) $\frac{\boldsymbol{N}}{|\boldsymbol{N}|} = \pm\frac{1}{\sqrt{2}}(-\boldsymbol{i} + \boldsymbol{j})$

19. $\boldsymbol{A} \cdot \boldsymbol{B} = 5 \cdot 0 + (-1)(1) + (1)(-5) = -6$; $\boldsymbol{A} \cdot \boldsymbol{C} = 5 \cdot 15 + (-1)3 + 1(-3) \neq 0$; $\boldsymbol{B} \cdot \boldsymbol{C} = 0(-15) + 1(3) - 5 \cdot 3 \neq 0$. No two are perpendicular. $\boldsymbol{A} \times \boldsymbol{B} = \begin{vmatrix} \boldsymbol{i} & \boldsymbol{j} & \boldsymbol{k} \\ 5 & -1 & 1 \\ 0 & 1 & -5 \end{vmatrix} = 4\boldsymbol{i} + 25\boldsymbol{j} + 5\boldsymbol{k} \neq \boldsymbol{0}$; $\boldsymbol{A} \times \boldsymbol{C} = \begin{vmatrix} \boldsymbol{i} & \boldsymbol{j} & \boldsymbol{k} \\ 5 & -1 & 1 \\ -15 & 3 & -3 \end{vmatrix} = \boldsymbol{0}$;

$\boldsymbol{B} \times \boldsymbol{C} = \begin{vmatrix} \boldsymbol{i} & \boldsymbol{j} & \boldsymbol{k} \\ 0 & 1 & -5 \\ -15 & 3 & -3 \end{vmatrix} = 12\boldsymbol{i} + \cdots$(not $\boldsymbol{0}$). $\boldsymbol{A}$ and $\boldsymbol{C}$ are parallel.

21. $\boldsymbol{A} \times \boldsymbol{B} = \begin{vmatrix} \boldsymbol{i} & \boldsymbol{j} & \boldsymbol{k} \\ 2 & -1 & 0 \\ 1 & 3 & -2 \end{vmatrix} = 2\boldsymbol{i} + 4\boldsymbol{j} + 7\boldsymbol{k}$. $(\boldsymbol{A} \times \boldsymbol{B}) \cdot \boldsymbol{A} = 2 \cdot 2 + 4(-1) + 7(0) = 0$; $(\boldsymbol{A} \times \boldsymbol{B}) \cdot \boldsymbol{B} = 2(1) + 4(3) + 7(-2) = 0$.

23. a) $\text{Proj}_{\boldsymbol{B}}\boldsymbol{A} = \frac{\boldsymbol{A} \cdot \boldsymbol{B}}{\boldsymbol{B} \cdot \boldsymbol{B}}\boldsymbol{B}$, b) $\boldsymbol{A} \times \boldsymbol{B}$, c) $\frac{\sqrt{\boldsymbol{A} \cdot \boldsymbol{A}}}{\sqrt{\boldsymbol{B} \cdot \boldsymbol{B}}}\boldsymbol{B}$, d) $(\boldsymbol{A} \times \boldsymbol{B}) \times \boldsymbol{C}$, e) $(\boldsymbol{B} \times \boldsymbol{C}) \times \boldsymbol{A}$

27. – 29.

	$\boldsymbol{A} \times \boldsymbol{B}$	$(\boldsymbol{A} \times \boldsymbol{B}) \cdot \boldsymbol{C}$	$\boldsymbol{B} \times \boldsymbol{C}$	$(\boldsymbol{B} \times \boldsymbol{C}) \cdot \boldsymbol{A}$	$\boldsymbol{C} \times \boldsymbol{A}$	$(\boldsymbol{C} \times \boldsymbol{A}) \cdot \boldsymbol{B}$,	Vol.
27.	$4\boldsymbol{k}$	8	$4\boldsymbol{i}$	8	$4\boldsymbol{j}$	8	8
29.	$\boldsymbol{i} - 2\boldsymbol{j} - 4\boldsymbol{k}$	-7	$-2\boldsymbol{i} - 3\boldsymbol{j} + \boldsymbol{k}$	-7	$-2\boldsymbol{i} + 4\boldsymbol{j} + \boldsymbol{k}$	-7	7

11.5 LINES AND PLANES IN SPACE

1. $x-3=t,\ y+4=t,\ z+1=t,$ or $x=3+1\cdot t,\ y=-4+1\cdot t,\ z=-3+1\cdot t$

3. Direction of $\vec{PQ}$ is $-2\boldsymbol{i}-2\boldsymbol{j}+2\boldsymbol{k}$; using $P: x-1=-2t,\ y-2=-2t,\ z+1=2t,$ or $x=1-2t,\ y=2-2t,\ z=-1+2t.$

5. $\vec{PQ}=-\boldsymbol{j}-\boldsymbol{k}$; using $P: x-1=0,\ y-2=-t,\ z-0=-t,$ or $x=1,\ y=2-t,\ z=-t.$

7. Write the vector as $0\boldsymbol{i}+2\boldsymbol{j}+\boldsymbol{k}$. The line is: $x=0+0t,\ y=0+2t,\ z=0+1\cdot t,$ or $x=0,\ y=2t,\ z=t.$

9. The line is parallel to $0\boldsymbol{i}+0\boldsymbol{j}+\boldsymbol{j}$; $x=1,\ y=1,\ z=1+t.$

11. The line is parallel to the normal of the plane: $\boldsymbol{i}+2\boldsymbol{j}+2\boldsymbol{k}$. $x=0+1\cdot t,\ y=-7+2t,\ z=0+2t,$ or $x=t,\ y=-7+2t,\ z=2t.$

13. The x-axis is parallel to $\boldsymbol{i}$; the origin lies on the axis. $x=0+1\cdot t,\ y=0+0\cdot t,\ x=0+0\cdot t,$ or $x=t,\ y=0,\ z=0.$

15. Let P be $(0,0,0)$, Q be $(1,1,1)$. Then $\vec{PQ}=\boldsymbol{i}+\boldsymbol{j}+\boldsymbol{k}$ and the line through P and Q has equations $x=t,\ y=t,\ z=t$. When $t=0,\ (x,y,z)=(0,0,0)$ or P; when $t=1,\ (x,y,z)=(1,1,1)$ or Q the segment is: $x=t,\ y=t,\ z=t,\ 0\le t\le 1.$ Problems 17-21 are similar to this one.

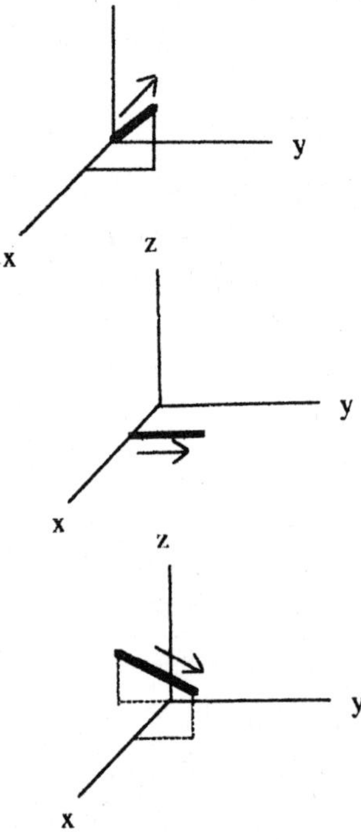

17. See #15. $\vec{PQ}=0\boldsymbol{i}+\boldsymbol{j}+0\boldsymbol{k}=\boldsymbol{j}$. The segment is $x=1,\ y=t,\ z=0,\ 0\le t\le 1.$

19. See #15. $\vec{PQ}=2\boldsymbol{j}$. The segment is $x=0,\ y=-1+2t,\ z=1,\ 0\le t\le 1.$

21. See #15. $\overrightarrow{PQ} = -\boldsymbol{i} - 2\boldsymbol{k}$. The segment is $x = 2 - t,\ y = 2,\ z = -2t,\ 0 \le t \le 1$.

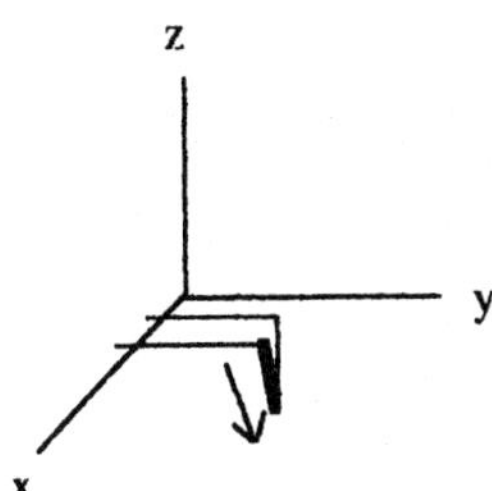

23. $3(x-0) - 2(y-2) - 1(z+1) = 0 \Rightarrow$ $3x - 2y - z = -3$

25. Label the points $A(1,1-1)$, $B(2,0,2)$ and $C(0,-2,1)$. $\overrightarrow{AB} = \boldsymbol{i} - \boldsymbol{j} + 3\boldsymbol{k}$, $\overrightarrow{BC} = -2\boldsymbol{i} - 2\boldsymbol{j} - \boldsymbol{k}$. The vector $\overrightarrow{AB} \times \overrightarrow{BC} = \begin{vmatrix} \boldsymbol{i} & \boldsymbol{j} & \boldsymbol{k} \\ 1 & -1 & 3 \\ -2 & -2 & -1 \end{vmatrix} = 7\boldsymbol{i} - 5\boldsymbol{j} - 4\boldsymbol{k}$ is normal to the plane. Since the plane passes through A, it has equation $7(x-1) - 5(y-1) - 4(z+1) = 0$, or $7x - 5y - 4z = 6$.

27. The vector $1\boldsymbol{i} + 3\boldsymbol{j} + 4\boldsymbol{k}$ is normal to the plane. $(x-2) + 3(y-4) + 4(z-5) = 0$, or $x + 3y + 4z = 34$.

29. Let $Q(4t, -2t, 2t)$ be on the line; let $f(t) = |\overrightarrow{QP}|^2 = (4t)^2 + (-2t)^2 + (2t-12)^2$. Then $f'(t) = 2 \cdot 4t \cdot 4 + 2(-2t)(-2) + 2(2t-12) \cdot 2 = 32t + 8t + 8t - 48 = 48(t-1)$. $f'(t) = 0$ at $t = 1$. $\sqrt{f(1)} = \sqrt{16 + 4 + 100} = \sqrt{120}$.

31. The point is <u>on</u> the line $\Rightarrow$ distance $= 0$.

33. Let S be $(2,-3,4)$; then the point $P(13,0,0)$ is on the plane. Distance $= \left|\overrightarrow{PS} \cdot \frac{\boldsymbol{N}}{|\boldsymbol{N}|}\right| = \left|(-11\boldsymbol{i} - 3\boldsymbol{j} + 4\boldsymbol{k}) \cdot \frac{(\boldsymbol{i} + 2\boldsymbol{j} + 2\boldsymbol{k})}{\sqrt{9}}\right| = \frac{|-11-6+8|}{3} = 3$.

35. See # 33: $S(0,1,1)$. Let P be $(0,0,-4)$. Distance $= \left|\overrightarrow{PS} \cdot \frac{\boldsymbol{N}}{|\boldsymbol{N}|}\right| = \left|(\boldsymbol{j} + 5\boldsymbol{k}) \cdot \frac{(4\boldsymbol{j} + 3\boldsymbol{k})}{\sqrt{25}}\right| = \frac{|4+15|}{5} = \frac{19}{5}$.

37. See #33: $S(0,-1,0)$. Let $P = (0,4,0)$. Distance $= \left|\overrightarrow{PS} \cdot \frac{\boldsymbol{N}}{|\boldsymbol{N}|}\right| = \left|(-5\boldsymbol{j}) \cdot \frac{(2\boldsymbol{i} + \boldsymbol{j} + 2\boldsymbol{k})}{\sqrt{9}}\right| = \frac{|-5|}{3} = \frac{5}{3}$.

39. $2(1-t)-(3t)+3(1+t)=6 \Rightarrow -2t=1 \Rightarrow t=-\frac{1}{2} \Rightarrow x=\frac{3}{2},\ y=-\frac{3}{2},\ z=\frac{1}{2}.$

41. $(1+2t)+(1+5t)+(3t)=2 \Rightarrow 10t=0 \Rightarrow t=0 \Rightarrow x=1,\ y=1,\ z=0.$

43. Find the angle between the normals $\boldsymbol{N}_1=\boldsymbol{i}+\boldsymbol{j}$ and $\boldsymbol{N}_2=2\boldsymbol{i}+\boldsymbol{j}-2\boldsymbol{k}$. $\theta=\cos^{-1}\left[\frac{\boldsymbol{N}_1\cdot\boldsymbol{N}_2}{|\boldsymbol{N}_1||\boldsymbol{N}_2|}\right]=\cos^{-1}\left[\frac{3}{\sqrt{2}\sqrt{9}}\right]=\cos^{-1}\frac{1}{\sqrt{2}}=45°.$

45. See #43: $\theta=\cos^{-1}\left[\frac{\boldsymbol{N}_1\cdot\boldsymbol{N}_2}{|\boldsymbol{N}_1||\boldsymbol{N}_2|}\right]=\cos^{-1}\left[\frac{4-4-2}{\sqrt{12}\sqrt{9}}\right]=\cos^{-1}\left[\frac{-2}{\sqrt{108}}\right]=101.096°.$

47. See #43: $\theta=\cos^{-1}\left[\frac{\boldsymbol{N}_1\cdot\boldsymbol{N}_2}{|\boldsymbol{N}_1||\boldsymbol{N}_2|}\right]=\cos^{-1}\left[\frac{2+4-1}{\sqrt{9}\sqrt{6}}\right]=\cos^{-1}\frac{5}{\sqrt{54}}=47.124°.$

49. The line is parallel to the cross product $\boldsymbol{v}$ of the normals: $\boldsymbol{v}=\begin{vmatrix}\boldsymbol{i}&\boldsymbol{j}&\boldsymbol{k}\\1&1&1\\1&1&0\end{vmatrix}=$ $-\boldsymbol{i}+\boldsymbol{j}$. Find a point on the line by setting $x=0$ and solving $y+z=1,\ y=2$ to get $(0,2,-1)$ is on the line: $x=-t,\ y=2+t,\ z=-1.$

51. $\boldsymbol{v}=\boldsymbol{N}_1\times\boldsymbol{N}_2=\begin{vmatrix}\boldsymbol{i}&\boldsymbol{j}&\boldsymbol{k}\\1&-2&4\\1&1&-2\end{vmatrix}=\boldsymbol{i}(0)+\boldsymbol{j}(6)+\boldsymbol{k}3=6\boldsymbol{j}+3\boldsymbol{k}$. To find a point on the line, set $z=0$ and solve $x-2y=2,\ x+y=5$ to get $y=1,\ x=4$. Hence $(4,1,0)$ is in the line which has equations: $x=4,\ y=1+6t,\ z=3t.$

53. Let $A(1,2,3)$; $B(-1,6,2)$ and $C(2,6,3)$ are on the plane; $\boldsymbol{N}$ is parallel to $\vec{AB}\times\vec{AC}=\begin{vmatrix}\boldsymbol{i}&\boldsymbol{j}&\boldsymbol{k}\\-2&4&-1\\1&4&0\end{vmatrix}=4\boldsymbol{i}-\boldsymbol{j}-12\boldsymbol{k}$. The plane is

$4(x-1)-1\cdot(y-2)-12(x-3)=0$, or $4x-y-12z=-34.$

55. The xy-plane has equation $z=0$; $3t=0 \Rightarrow t=0 \Rightarrow x=1,\ y=-1:\ (1,-1,0)$; the yz-plane has equation $x=0$; $1+2t=0 \Rightarrow t=-\frac{1}{2} \Rightarrow y=-\frac{1}{2},\ z=-\frac{3}{2}:\ (0,-\frac{1}{2}-\frac{3}{2})$; the xz-plane has equation $y=0$; $-1-t=0 \Rightarrow t=-1 \Rightarrow x=-1,\ z=-3:\ (-1,0-3).$

57. $\boldsymbol{v}=-2\boldsymbol{i}+5\boldsymbol{j}-3\boldsymbol{k}$ is parallel to the line; $\boldsymbol{N}=2\boldsymbol{i}+\boldsymbol{j}-\boldsymbol{k}$ is perpendicular to the plane. $\boldsymbol{N}\cdot\boldsymbol{v}\neq 0 \Rightarrow$ the line and plane are not parallel. No.

11.6 SURFACES IN SPACE

1.

$x^2 + y^2 = 4$

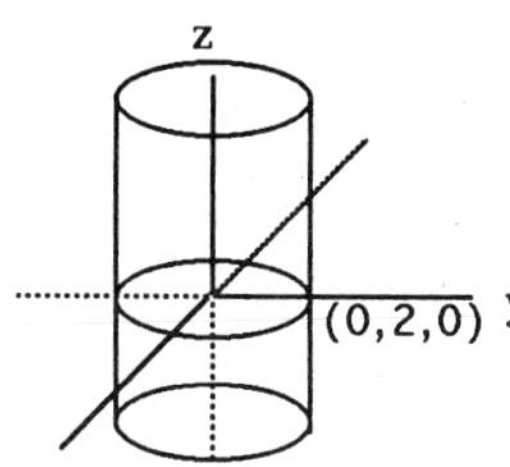

3.

$y^2 + z^2 = 1$

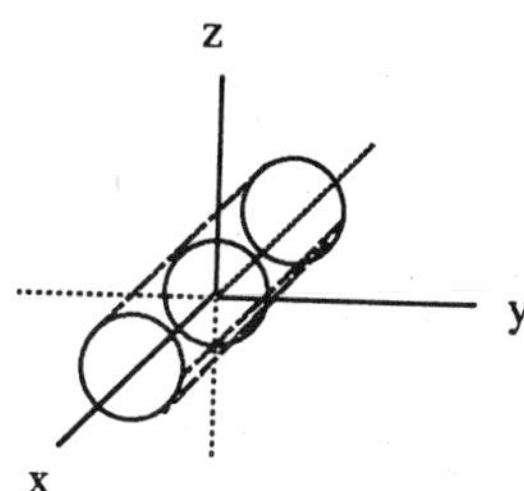

5.

$z = y^2 - 1$

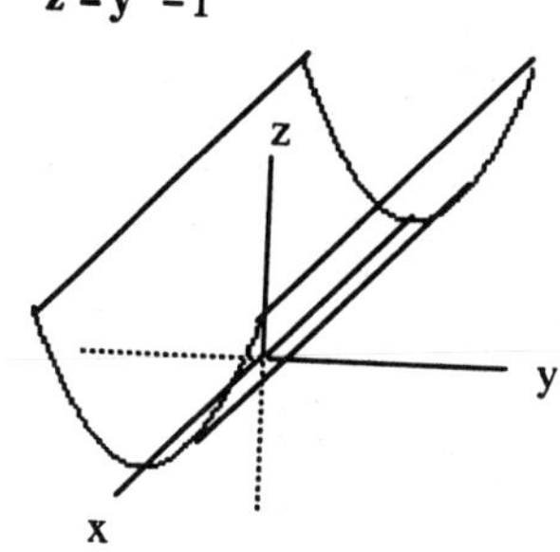

7.

$z = 4 - x^2$

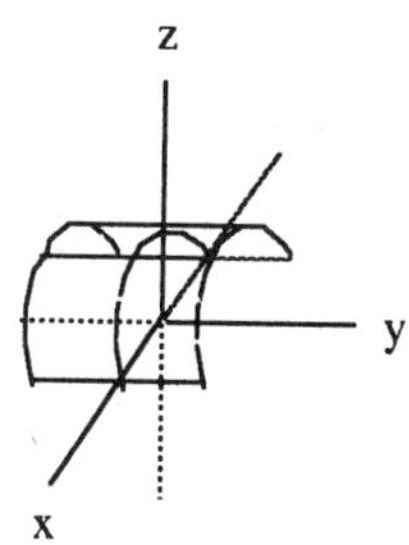

9.

$y = x^2$

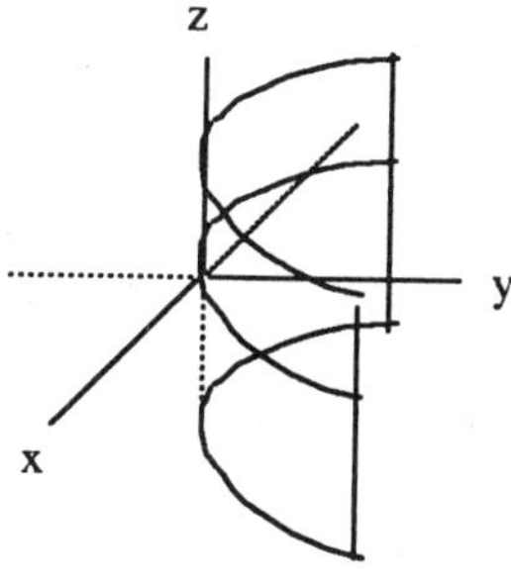

11.

$y^2 + 4z^2 = 16$

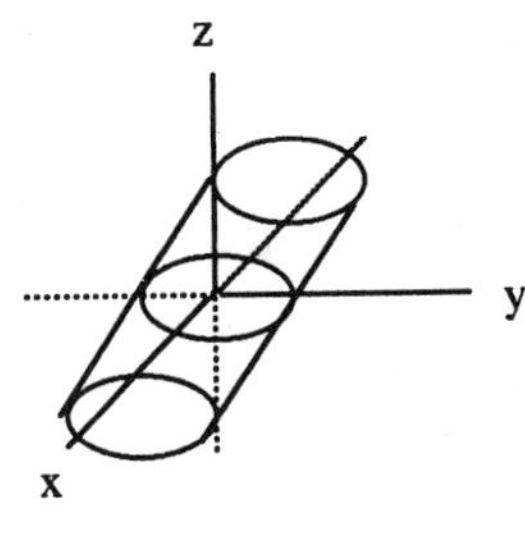

13.

$z^2 + 4y^2 = 9$

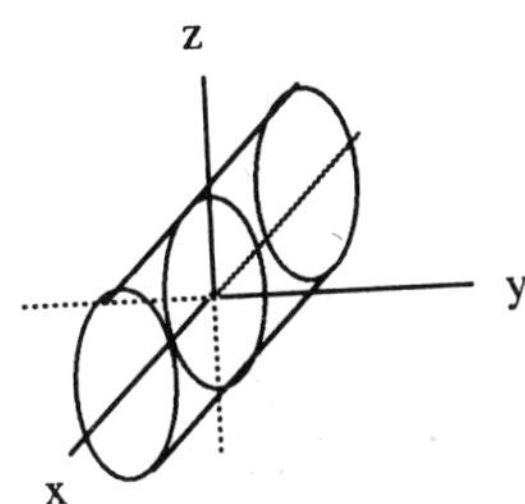

15.

$z^2 - y^2 = 1$

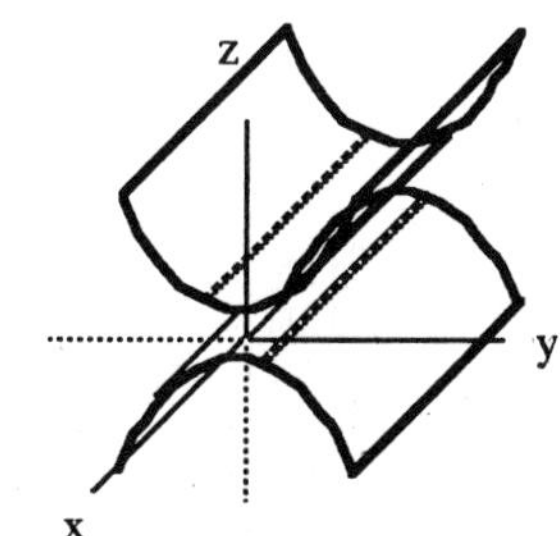

17.

$9x^2 + y^2 + z^2 = 9$

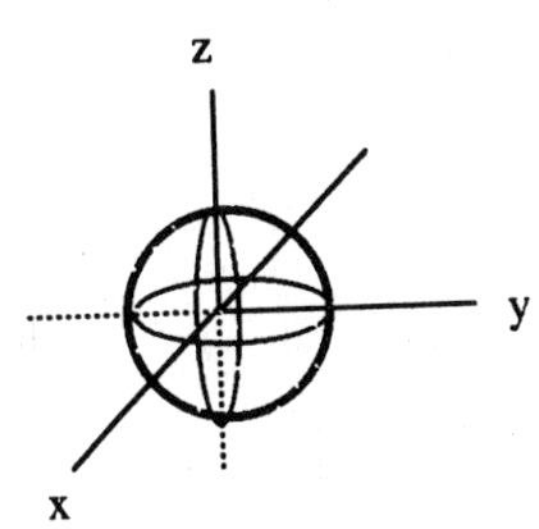

19.

$x^2 + y^2 + z^2 = 4$

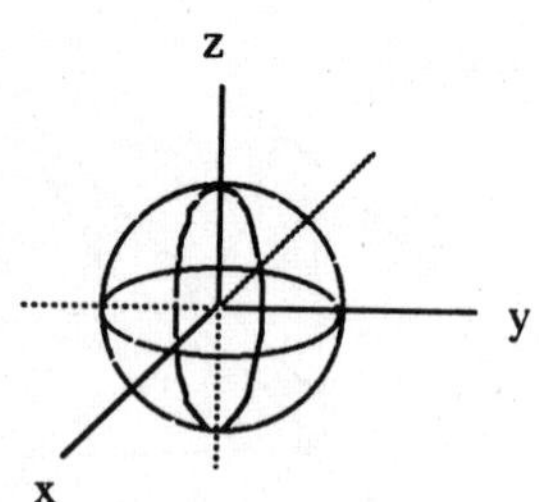

21.

$4x^2 + 9y^2 + 4z^2 = 36$

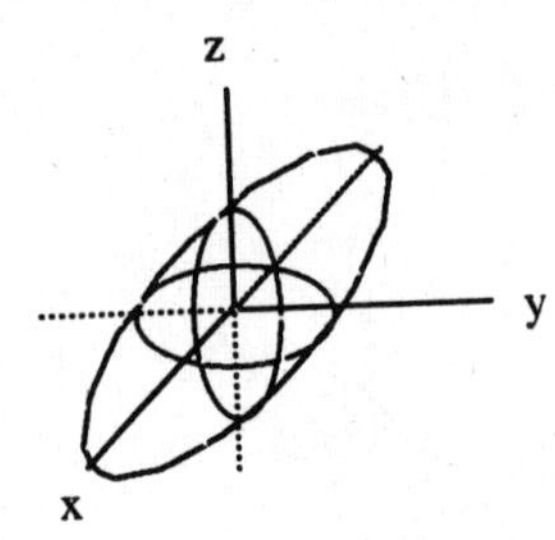

23.

$x^2 + y^2 = z$

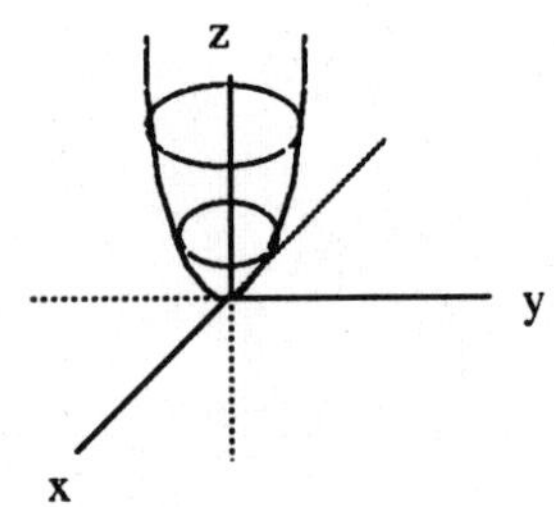

25.

$x^2 + 4y^2 = z$

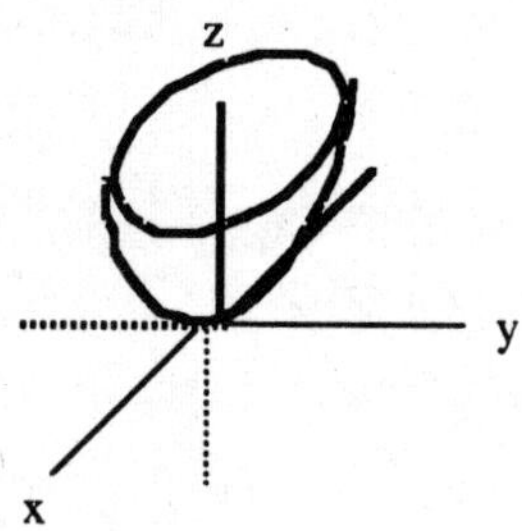

27.

$z = 8 - x^2 - y^2$

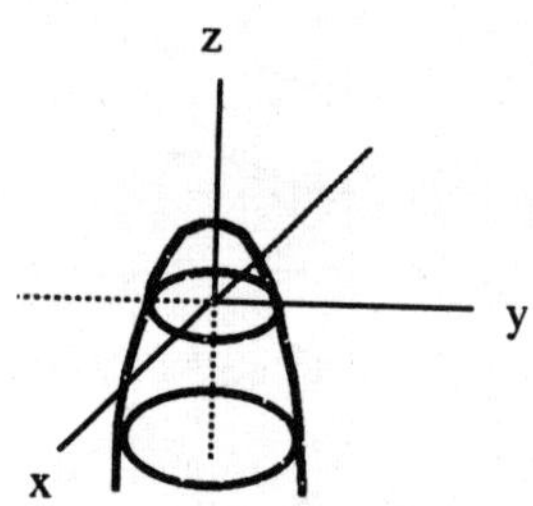

29.

$x = 4 - 4y^2 - z^2$

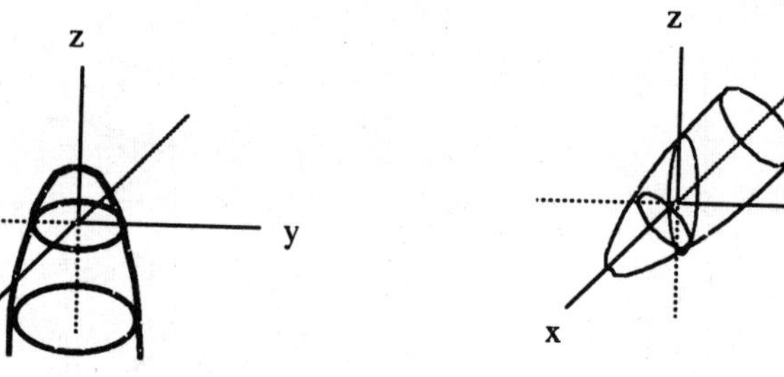

31.

$z = x^2 + y^2 + 1$

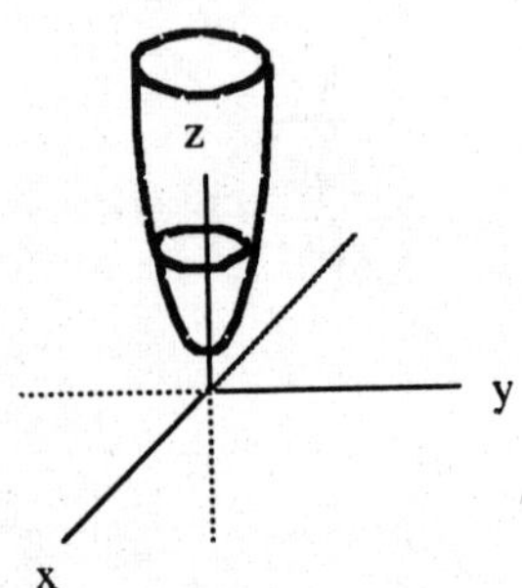

33.

$x^2 + y^2 = z^2$

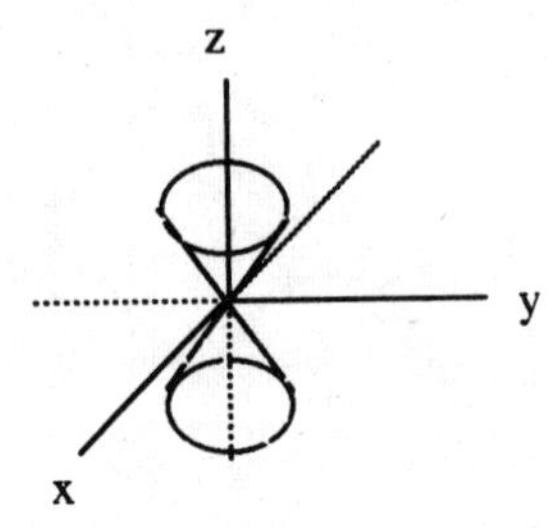

35.

$x^2 + z^2 = y^2$

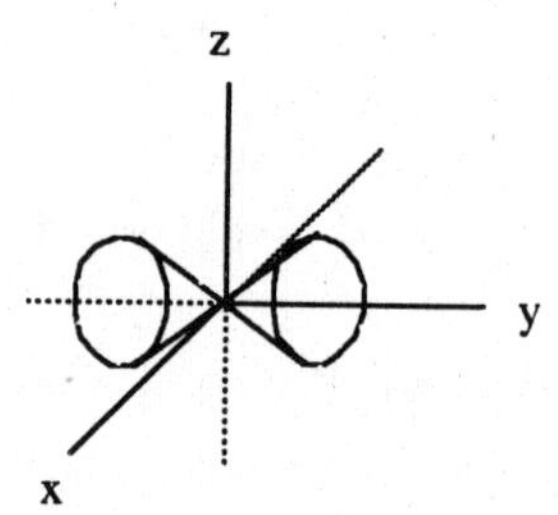

37.

$9x^2 + 4y^2 = 36z^2$

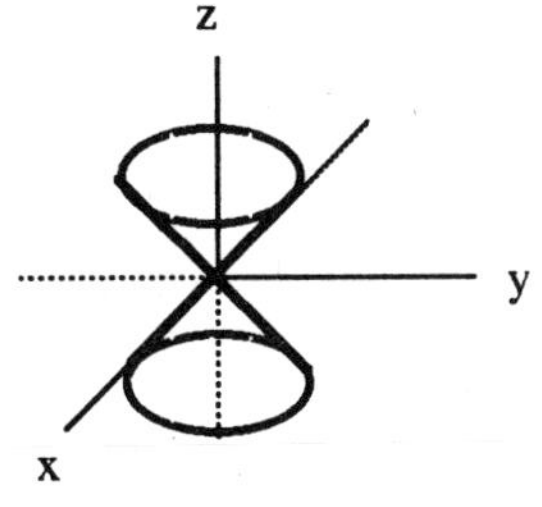

39.

$x^2 + y^2 - z^2 = 1$

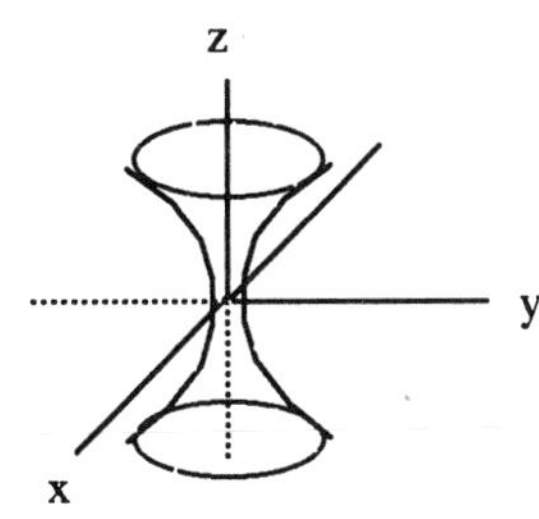

41.

$(y^2 / 4) + (z^2 / 9) - (x^2 / 4) = 1$

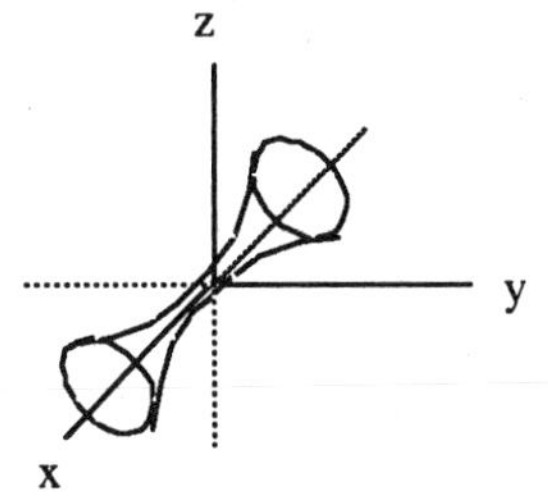

43.

$(x^2 / 4) + y^2 - z^2 = 1$

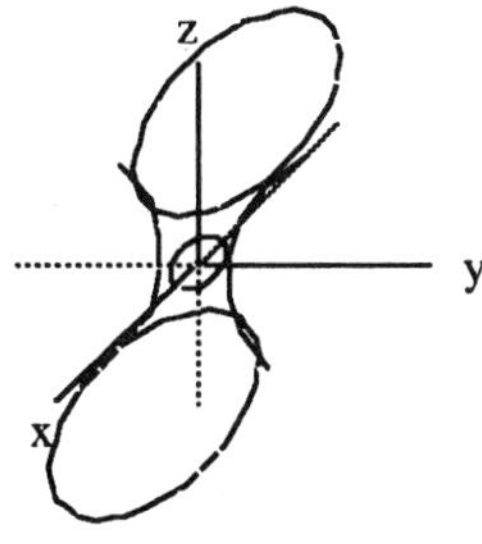

45.

$z^2 - (x^2 / 4) - y^2 = 1$

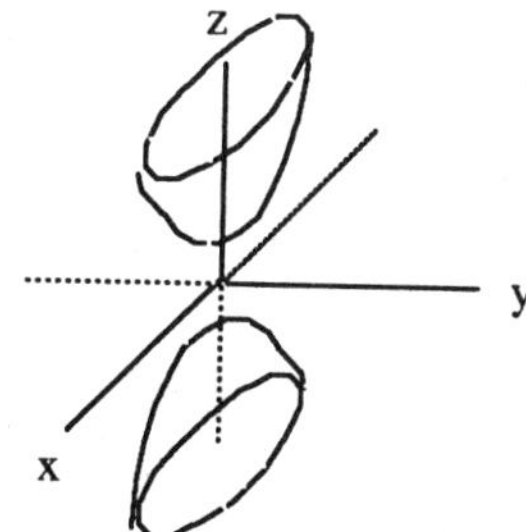

47.

$x^2 - y^2 - (z^2 / 4) = 1$

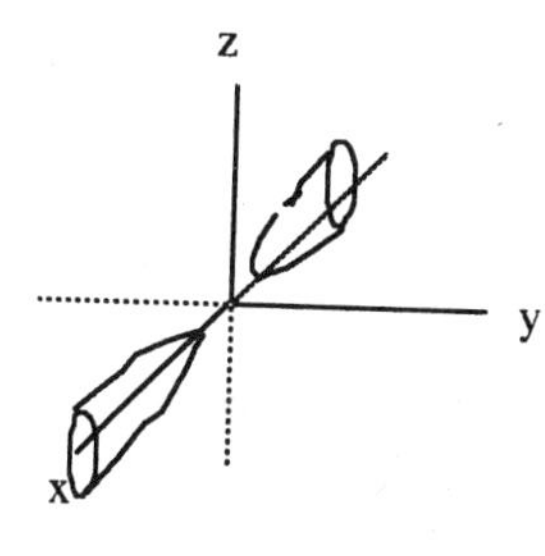

49.

$y^2 - x^2 = z$

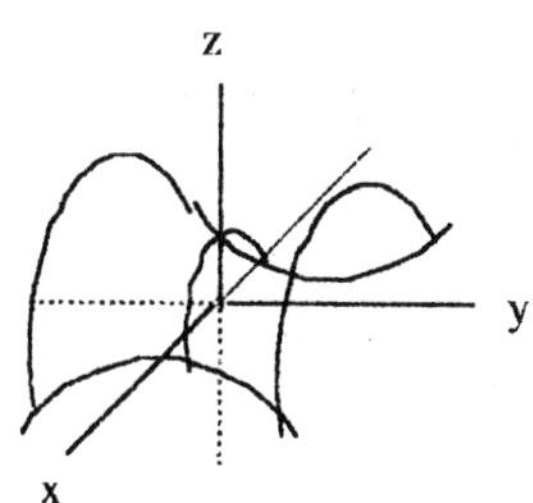

51.

Ellipsoid $9x^2 + 36y^2 + 4z^2 = 36$

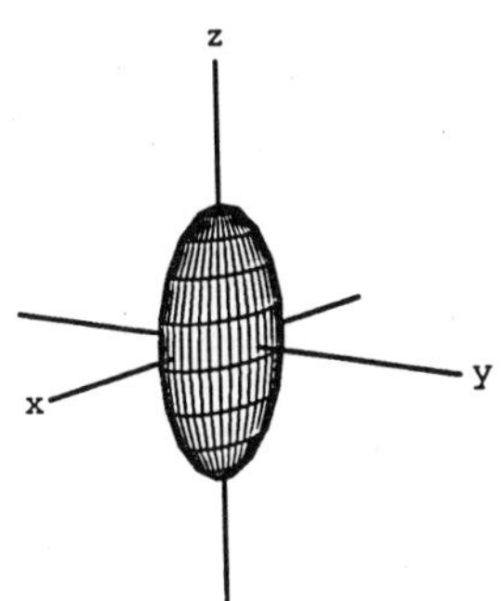

53. Cone $9x^2 + 36z^2 = 4y^2$

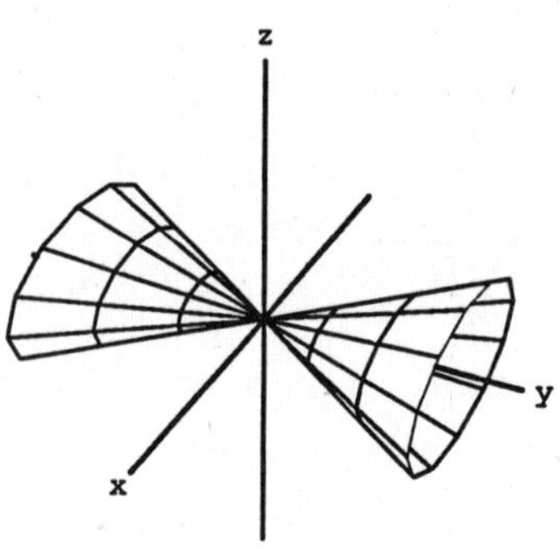

55. Paraboloid $9x^2 + 16z^2 = 72y$

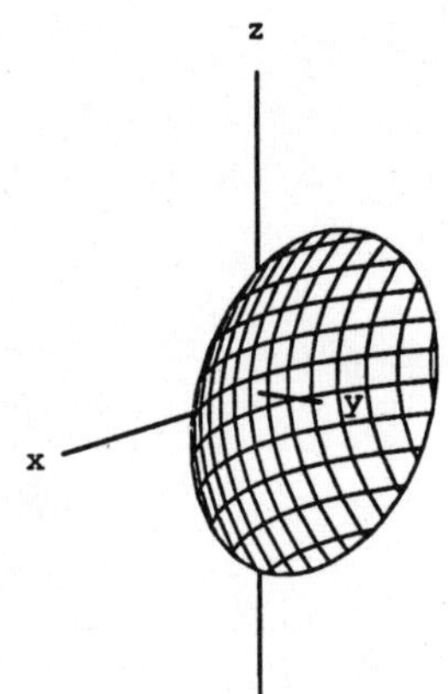

57.

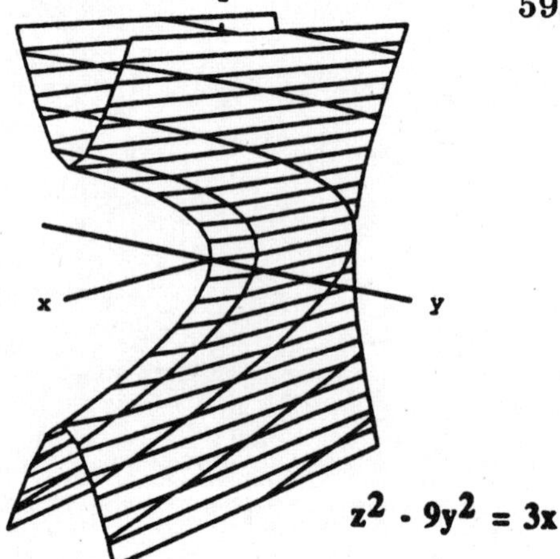

59. Paraboloid $2x^2 + 2z^2 = 3y$

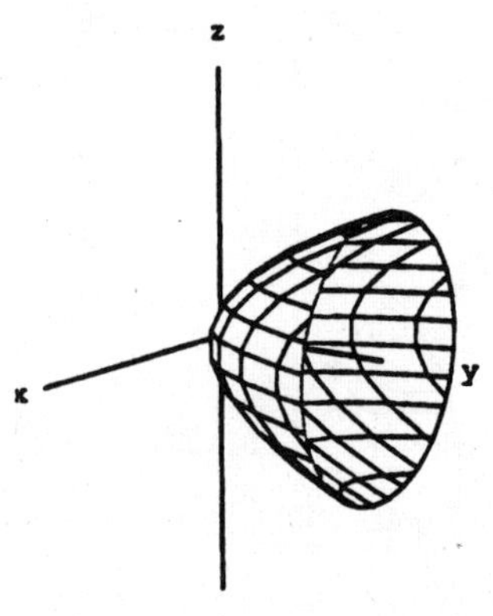

61. **Hyperboloid of two sheets $9x^2 - 36y^2 - 4z^2 = 36$**

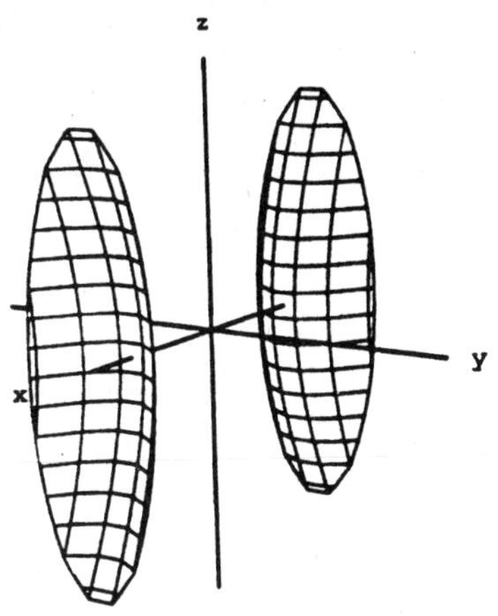

63. **Hyperboloid of one sheet $25x^2 - 100y^2 + 4z^2 = 100$**

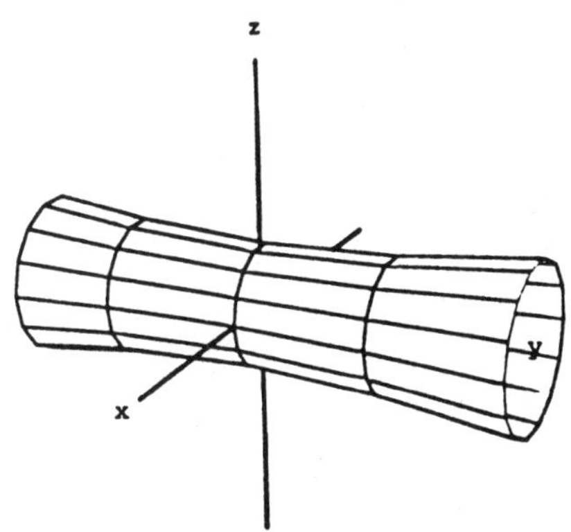

65.

$z = y^2$

67.

$z = x^2 + y^2$

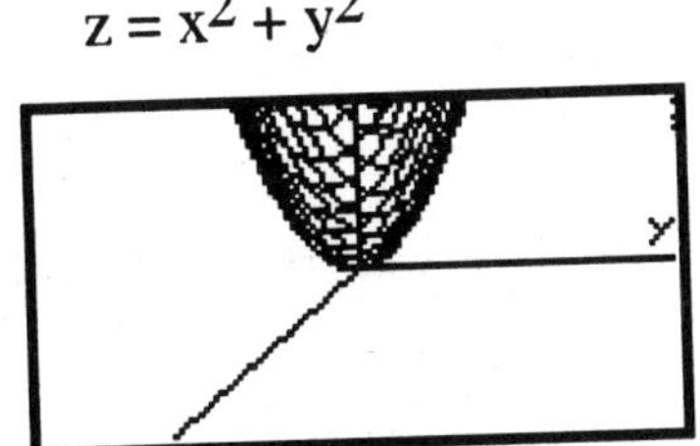

69.

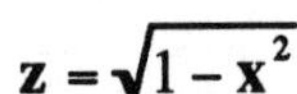

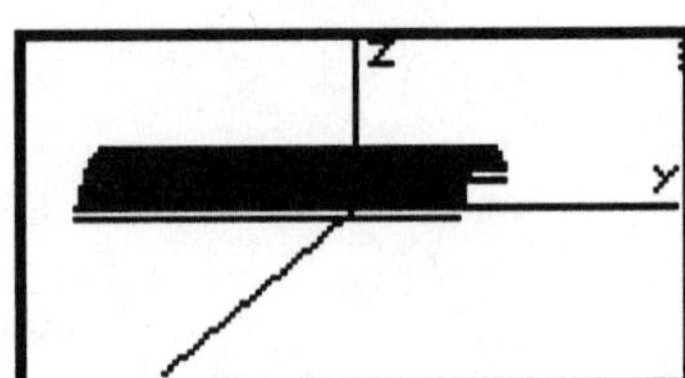

71.

$z = \sqrt{x^2 + 2y^2 + 4}$

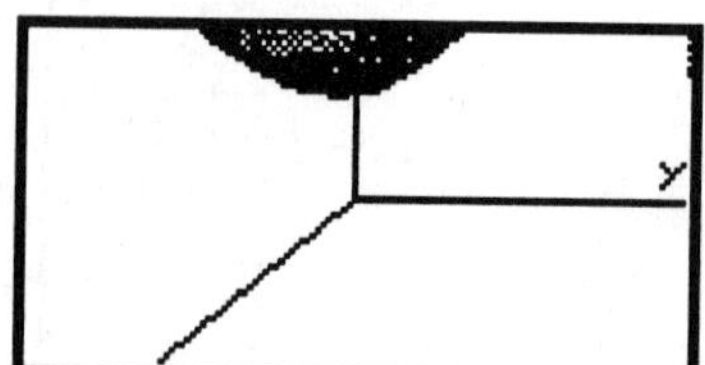

11.7 CYLINDRICAL AND SPHERICAL COORDINATES

1. – 7. Use Equations (1) and (2) to find the coordinates:

	Rectangular (x, y, z)	Cylindrical (r, θ, z)	Spherical (ρ, ϕ, θ)
1.	$(0,0,0)$	$(0,0^*,0)$	$(0,0^*,0^*)$
3.	$(0,1,0)$	$(1,\pi/2,0)$	$(1,\pi/2,\pi/2)$
5.	$(1,0,0)$	$(1,0,0)$	$(1,\pi/2,0)$
7.	$(0,1,1)$	$(1,\pi/2,1)$	$(\sqrt{2}, \cos^{-1}(1/\sqrt{2}), \pi/2)$

0^* can be any angle

9. $\phi = \pi/2 \Rightarrow$ point is in xy-plane ($z = 0$); $\theta = 3\pi/2 \Rightarrow$ point is on negative y-axis ($x = 0$); $\rho = 2\sqrt{2} \Rightarrow r^2 = 8$. Hence $(x, y, z) = (0, -\sqrt{8}, 0) = (0, -2\sqrt{2}, 0)$; $(r, \theta, z) = (2\sqrt{2}, 3\pi/2, 0)$.

11. $r = 0 \Rightarrow$ rectangular: $x = 0,\ y = 0,\ z = 0$; spherical: $\phi = 0$ or $\phi = \pi, \rho = \rho,\ \theta = \theta$. The z-axis.

13. $z = 0 \Rightarrow$ rectangular: $z = 0$; cylindrical: $z = 0$; spherical: $\phi = \pi/2$; the xy-plane.

15. $\rho\cos\varphi = 3 \Rightarrow$ rectangular, cylindrical: $z = 3$; the plane $z = 3$.

17. $\rho\sin\varphi\cos\theta = 0 \Rightarrow x = 0$; rectangular: $x = 0$; cylindrical: $\theta = \pi/2$ or $3\pi/2$; the yz-plane.

19. Spherical: $\rho^2 = 4$ or $\rho = 2$; cylindrical: $r^2 + z^2 = 4$; sphere of radius 2, center at origin.

21. $\rho = 2\sin\theta \Rightarrow \rho^2 = 2\rho\sin\theta \Rightarrow x^2+y^2+z^2 = 2y$; cylindrical: $r^2+z^2 = 2r\sin\theta$; a sphere.

23. $r = \csc\theta \Rightarrow r\sin\theta = 1 \Rightarrow$ rectangular: $y = 1$; spherical: $\rho\sin\varphi\sin\theta = 1$; the plane $y = 1$.

25. The lower half of the sphere with center at $(0,0,1)$, radius 1; spherical: $\rho^2 - 2\rho\cos\varphi = 0$, or $\rho = 2\cos\varphi$, since $\rho^2 = 2z$, $z \le 1 \Rightarrow \rho < \sqrt{2} \Rightarrow \frac{\pi}{4} < \varphi$; cylindrical: $r^2 = 2z - z^2$, $0 \le z \le 1$.

27. This is the top third of a sphere. Rectangular: $x^2 + y^2 + z^2 = 4$, $2(0.5) \le z \le 2$; cylindrical: $r^2 + z^2 = 4$, $1 \le z \le 2$.

29. $\phi = \frac{\pi}{3}$, $0 \le \rho \le 2$ is the truncated top half of a cone; rectangular: $z = \frac{1}{2}\rho \Rightarrow \rho = 2z$, $x = \frac{\sqrt{3}}{2}\rho\cos\theta$, $y = \frac{\sqrt{3}}{2}\rho\sin\theta \Rightarrow x^2 + y^2 = \frac{3}{4}\rho^2 = 3z^2$, $0 \le z \le 1$; cylindrical: $r^2 = 3z^2$, $0 \le z \le 1$.

31. A right circular cylinder whose cross-sections perpendicular to the xy-plane are circles of radius 4; The axis of the cylinder is the z-axis.

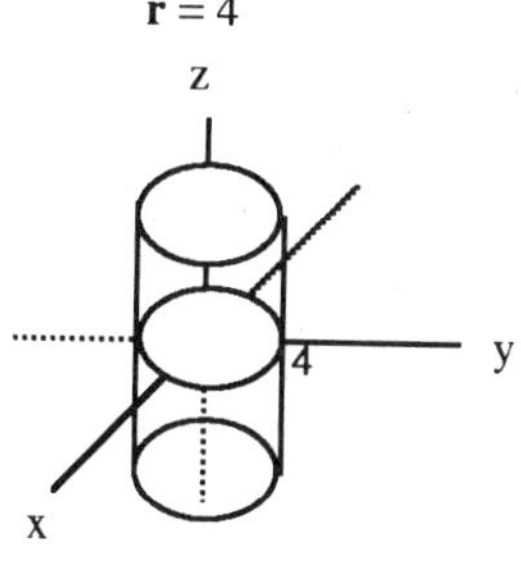

33. Right circular cylinder generated by the cardioid $r = 1 - \cos\theta$

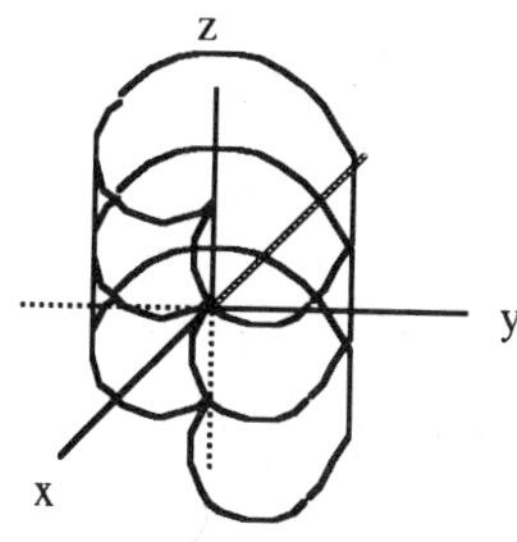

35. Circle of radius 2, center at $(0,0,3)$, parallel to the xy-plane

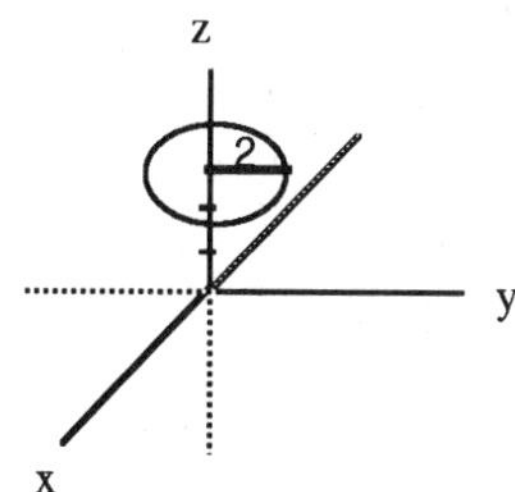

37. A spiral up the side of the cylinder $r = 3$

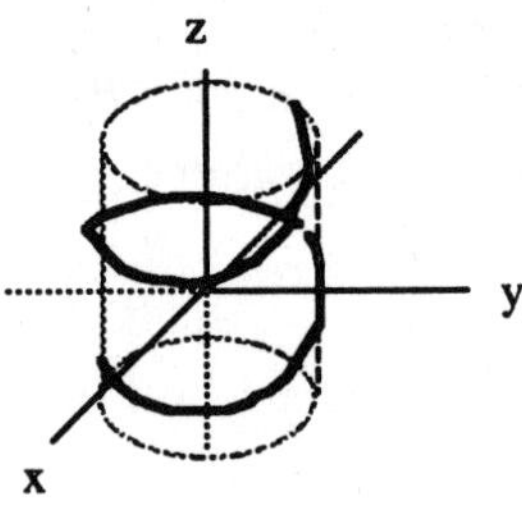

39. A circle in the xy-plane

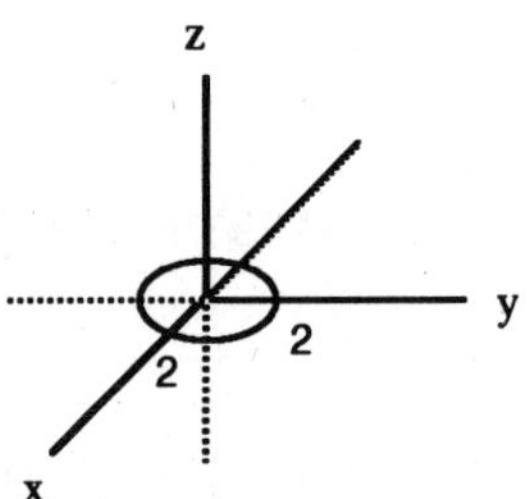

41. The intersection of the cone $\varphi = \frac{\pi}{4}$ and the plane $\theta = \frac{\pi}{4}$; intersecting lines

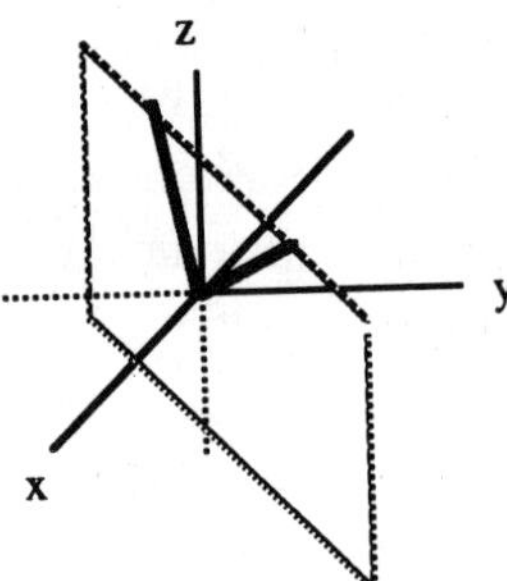

43. The curve lies in the positive half of the yz-plane $(\theta = \frac{\pi}{2})$, where $y = \rho \sin \phi$, $\rho = 4 \sin \phi \Rightarrow \rho^2 = 4\rho \sin \phi = 4y$, $y^2 + z^2 = 4y$; $(y-2)^2 + z^2 = 4$. The curve is a circle of radius 2.

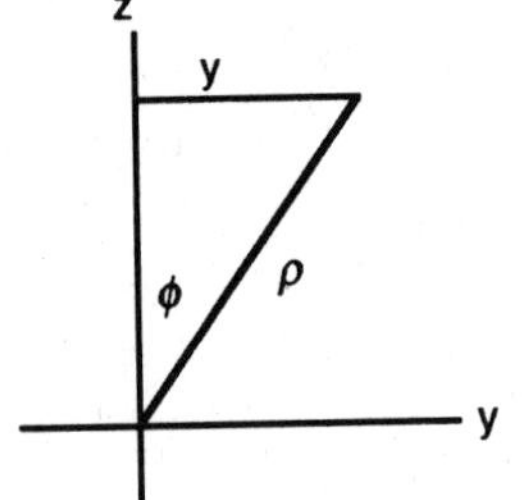

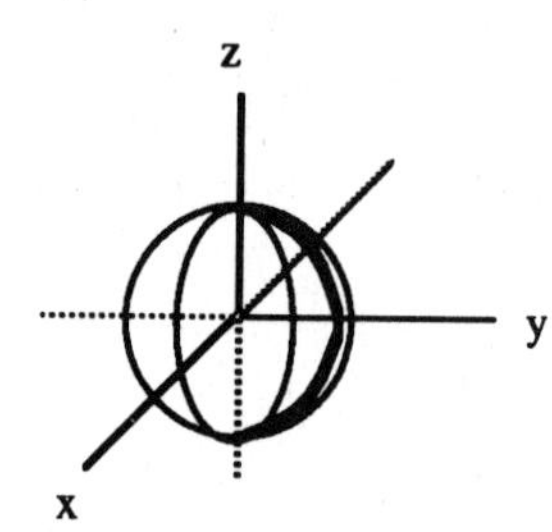

45. There will by symmetry about the z-axis. In the yz-plane, $\rho = \cos\varphi \Rightarrow \rho^2 = \rho\cos\varphi \Rightarrow y^2 + z^2 = z$, or $y^2 + (z - \frac{1}{2})^2 = \frac{1}{4}$. When revolved about the z-axis, this becomes a sphere.

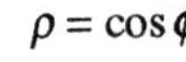

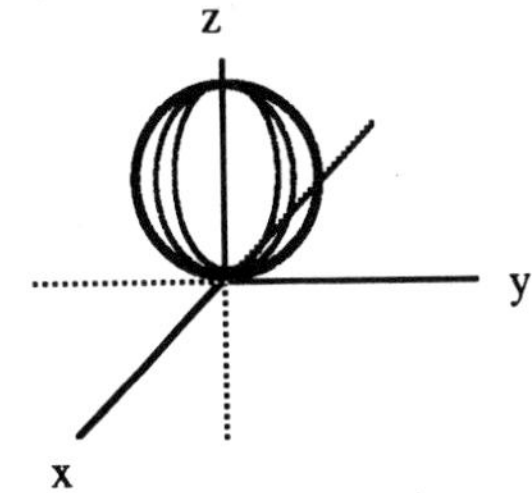

PRACTICE EXERCISES, CHAPTER 11

1.

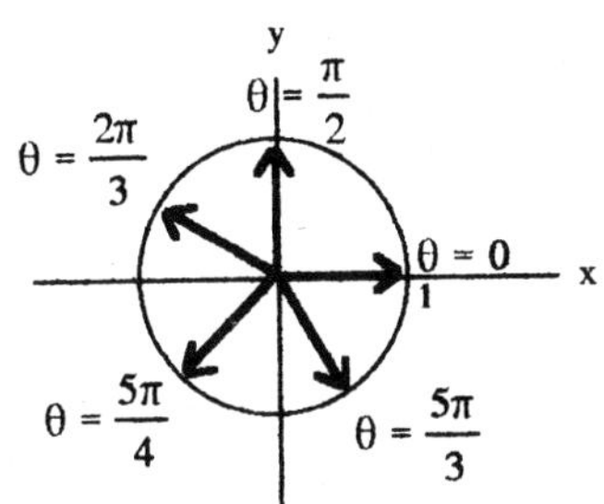

3. $y = \tan x \Rightarrow y' = \sec^2 x$; $y'(P) = (\sqrt{2}/1)^2 = 2 \Rightarrow \boldsymbol{T} = \boldsymbol{i} + 2\boldsymbol{j}$; $|\boldsymbol{T}|^2 = 5 \Rightarrow$ unit tangents are $\pm\left(\frac{1}{\sqrt{5}}\boldsymbol{i} + \frac{2}{\sqrt{5}}\boldsymbol{j}\right)$; unit normals are $\pm\left(\frac{2}{\sqrt{5}}\boldsymbol{i} - \frac{1}{\sqrt{5}}\boldsymbol{j}\right)$.

5. $\boldsymbol{v} = \sqrt{2}(\boldsymbol{i} + \boldsymbol{j}) \Rightarrow |\boldsymbol{v}| = \sqrt{2+2} = 2$; direction is $\frac{1}{|\boldsymbol{v}|}\boldsymbol{v} = \frac{1}{\sqrt{2}}\boldsymbol{i} + \frac{1}{\sqrt{2}}\boldsymbol{j}$.

7. $\boldsymbol{v} = 2\boldsymbol{i} - 3\boldsymbol{j} + 6\boldsymbol{k} \Rightarrow |\boldsymbol{v}| = \sqrt{4 + 9 + 36} = 7$; direction $= \frac{1}{|\boldsymbol{v}|}\boldsymbol{v} = \frac{2}{7}\boldsymbol{i} - \frac{3}{7}\boldsymbol{j} + \frac{6}{7}\boldsymbol{k}$.

9. $\boldsymbol{v} = c(4\boldsymbol{i} - \boldsymbol{j} + 4\boldsymbol{k})$: $|\boldsymbol{v}| = c\sqrt{16 + 1 + 16} = c\sqrt{33} = 2 \Rightarrow c = \frac{2}{\sqrt{33}}$; $\boldsymbol{v} = (\frac{2}{\sqrt{33}})(4\boldsymbol{i} - \boldsymbol{j} + 4\boldsymbol{k})$.

11. $|\boldsymbol{A}| = \sqrt{2}$; $|\boldsymbol{B}| = \sqrt{4+1+4} = 3$; $\boldsymbol{A}\cdot\boldsymbol{B} = 1\cdot 2 + 1\cdot 1 + 0\cdot(-2) = 3 = \boldsymbol{B}\cdot\boldsymbol{A}$; $\boldsymbol{A}\times\boldsymbol{B} = \begin{vmatrix} \boldsymbol{i} & \boldsymbol{j} & \boldsymbol{k} \\ 1 & 1 & 0 \\ 2 & 1 & -2 \end{vmatrix} = -2\boldsymbol{i} + 2\boldsymbol{j} - \boldsymbol{k}$; $\boldsymbol{B}\times\boldsymbol{A} = 2\boldsymbol{i} - 2\boldsymbol{j} + \boldsymbol{k}$, $|\boldsymbol{A}\times\boldsymbol{B}| = \sqrt{9} = 3$; $\cos\theta \frac{\boldsymbol{A\cdot B}}{|\boldsymbol{A}||\boldsymbol{B}|} = \frac{3}{3\sqrt{2}} = \frac{1}{\sqrt{2}} \Rightarrow \theta = \frac{\pi}{4}$; $\text{comp}_{\boldsymbol{A}}\boldsymbol{B} = \frac{\boldsymbol{A\cdot B}}{|\boldsymbol{A}|} = \frac{3}{\sqrt{2}}$; $\text{proj}_{\boldsymbol{A}}\boldsymbol{B} = \frac{\boldsymbol{A\cdot B}}{\boldsymbol{A\cdot A}}\boldsymbol{A} = \frac{3}{\sqrt{2}}(\sqrt{2}\boldsymbol{i} + \sqrt{2}\boldsymbol{j}) = 3(\boldsymbol{i} + \boldsymbol{j})$.

13. $B = \left(\frac{A\cdot B}{A\cdot A}\right)A + \left(B - \frac{A\cdot B}{A\cdot A}A\right) = \frac{8}{6}A + \left(B - \frac{8}{6}A\right) = \frac{4}{3}(2\boldsymbol{i}+\boldsymbol{j}-\boldsymbol{k})+(\boldsymbol{i}+\boldsymbol{j}-5\boldsymbol{k}-\frac{4}{3}(2\boldsymbol{i}+\boldsymbol{j}-\boldsymbol{k})) = \frac{4}{3}(2\boldsymbol{i}+\boldsymbol{j}-\boldsymbol{k})+\frac{1}{3}(-5\boldsymbol{i}-\boldsymbol{j}-11\boldsymbol{k})$.

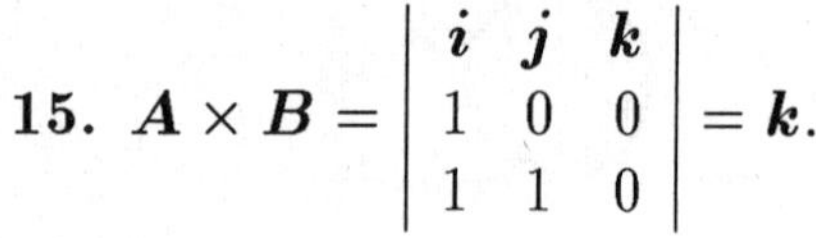

15. $\boldsymbol{A}\times\boldsymbol{B} = \begin{vmatrix} \boldsymbol{i} & \boldsymbol{j} & \boldsymbol{k} \\ 1 & 0 & 0 \\ 1 & 1 & 0 \end{vmatrix} = \boldsymbol{k}$.

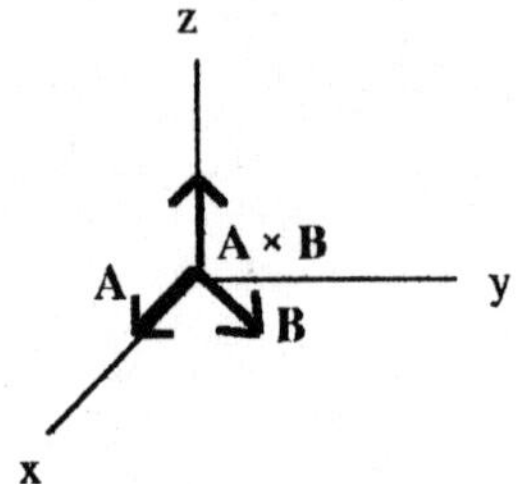

17. $P(3,2)$; $S(0,\frac{1}{2})$ is on the line, $\boldsymbol{N} = 3\boldsymbol{i}+4\boldsymbol{j}$ is normal to the line; $\vec{SP} = 3\boldsymbol{i}+\frac{3}{2}\boldsymbol{j}$;

distance P to line $= \left|\text{proj}_{\boldsymbol{N}}\ \vec{SP}\right| = \left|\frac{\boldsymbol{N}\cdot\vec{SP}}{|\boldsymbol{N}|}\right| = \frac{9+6}{\sqrt{25}} = \frac{15}{5} = 3$.

19. $P(6,0,-6)$; $S(4,0,-6)$ is on the plane; $\vec{SP} = 2\boldsymbol{i}$, $\boldsymbol{N} = \boldsymbol{i}-\boldsymbol{j}$; distance $= \left|\frac{\boldsymbol{N}\cdot\vec{SP}}{|\boldsymbol{N}|}\right| = \frac{2}{\sqrt{2}} = \sqrt{2}$.

21. $2x+y-z =$ constant; $(3,-2,1)$ on plane $\Rightarrow 6-2-1 = 3 =$ constant; $2x+y-z = 3$.

23. $\vec{PQ} = \boldsymbol{i}+2\boldsymbol{j}+\boldsymbol{k}$; $\vec{PR} = -2\boldsymbol{i}+3\boldsymbol{j}-3\boldsymbol{k}$; $\boldsymbol{N} = \vec{PQ}\times\vec{PR} = \begin{vmatrix} \boldsymbol{i} & \boldsymbol{j} & \boldsymbol{k} \\ 1 & 2 & 1 \\ -2 & 3 & -3 \end{vmatrix} = -9\boldsymbol{i}+\boldsymbol{j}+7\boldsymbol{k}$. The plane has equation $-9(x-1)+1(y+1)+7(z-2) = 0$, or $-9x+y+7z = 4$.

25. $\boldsymbol{v} = -3\boldsymbol{i}+0\boldsymbol{j}+7\boldsymbol{k} \Rightarrow x = 1-3t,\ y = 2,\ z = 3+7t$.

27. $x = 2t,\ y = -t,\ z = -t$, is the line; it intersects the plane when $3(2t)-5(-t)+2(-t) = 6$, or $t = \frac{2}{3}$. The points is $(\frac{4}{3},-\frac{2}{3},-\frac{2}{3})$.

29. Set $y-0$, then $z = 0$ to obtain points $P(10,0,-9)$ and $Q(-5,3,0)$ on the line; $\vec{PQ} = -15\boldsymbol{i}+3\boldsymbol{j}+9\boldsymbol{k}$ is the direction of the line. Using P, the line has equations $x = 10-15t,\ y = 3t,\ z = -9+9t$.

31. Minimizing $d^2 = (4-2)^2 + (4+2t+1)^2 + (4t+10)^2$, we have $0 = 2(5+2t)2 + 2(4t+10)4$ or $t = -\frac{5}{2} \Rightarrow d^2 = 4 + (0)^2 + (0)^2 \Rightarrow d = 2$.

33. Work $= \boldsymbol{F} \cdot \overrightarrow{PQ} = |\boldsymbol{F}||PQ| \cos\theta = (40 \text{ lb})(800 \text{ ft}) \cos 28° = 28254.323 \text{ ft} \cdot \text{lb}$.

35. Area $= |\boldsymbol{A} \times \boldsymbol{B}| = \begin{vmatrix} \boldsymbol{i} & \boldsymbol{j} & \boldsymbol{k} \\ 1 & 1 & -1 \\ 2 & 1 & 1 \end{vmatrix} = |2\boldsymbol{i} - 3\boldsymbol{j} - \boldsymbol{k}| = \sqrt{14}$;
Volume $\boldsymbol{C} \cdot (\boldsymbol{A} \times \boldsymbol{B}) = -2 + 6 - 3 = 1$.

37. Not always true: (b) since if: $\boldsymbol{A} = 2\boldsymbol{i}$, $\boldsymbol{A} \cdot \boldsymbol{A} = 4$, $|\boldsymbol{A}| = 2$; and (e) since: $\boldsymbol{A} \times \boldsymbol{B} = -\boldsymbol{B} \times \boldsymbol{A}$. (a), (c), (d), (f), (g), (h) are always true.

39. In plane: y-axis; in 3-space: yz-plane.

41. In plane: circle with center at origin, radius 2; in 3-space, a right circular cylinder with axis parallel to z-axis generated by the circle.

43. In the plane: parabola opening to the right with vertex at $(0, 0)$; in 3-space: the cylinder generated by the parabola with axis parallel to the z-axis.

45. In the plane: cardioid with dimple at right; in 3-space a cylinder generated by the cardioid with axis parallel to the z-axis.

47. In the plane a horizontal lemniscate (∞ sign); in 3-space a cylinder generated by the lemniscate.

49. Surface of sphere of radius 2, centered at the origin.

51. The upper nappe of a cone whose surface makes an angle of $\frac{\pi}{6}$ with the z-axis.

53. The upper hemisphere of the unit sphere.

	Rectangular	**Cylindrical**	**Spherical**
55.	$(1,0,0)$	$(1,0,0)$	$(1,\frac{\pi}{2},0)$
57.	$(0,1,1)$	$(1,\frac{\pi}{2},1)$	$(\sqrt{2},\frac{\pi}{4},\frac{\pi}{2})$
59.	$(1,1,1)$	$(\sqrt{2},\frac{\pi}{4},1)$	$(\sqrt{3},\cos^{-1}(\frac{1}{\sqrt{3}}),\frac{\pi}{4})$

61. Rectangular $z = 2 \Rightarrow$ cylindrical $z = 2 \Rightarrow$ spherical $\rho\cos\phi = 2$; this is a plane parallel to xy-plane.

63. Cylindrical: $z = r^2 \Rightarrow$ rectangular: $z = x^2 + y^2 \Rightarrow$ spherical: $\rho\cos\phi = \rho^2\sin^2\varphi\cos^2\theta + \rho^2\sin^2\varphi\sin^2\theta = \rho^2\sin^2\varphi$ or $\rho = \frac{\cos\varphi}{\sin^2\varphi}$, a paraboloid symmetric about the z-axis, opening up.

65. Spherical $\rho = 4 \Rightarrow \sqrt{r^2+z^2} = 4$ or $r^2+z^2 = 16$, cylindrical; $x^2+y^2+z^2 = 16$, rectangular; sphere of radius 4 centered at the origin.

67. Sphere

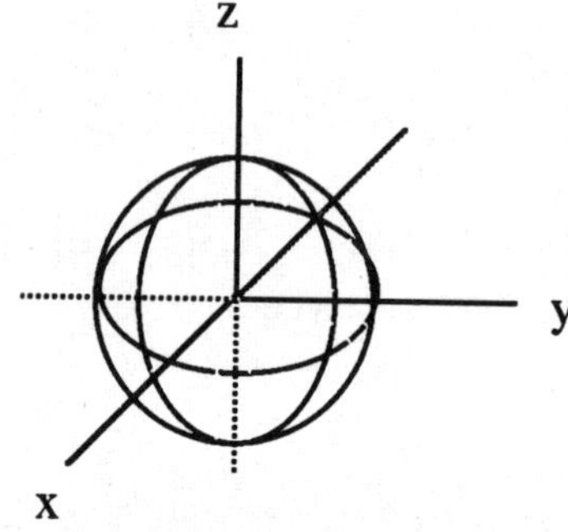

69. Ellipsoid

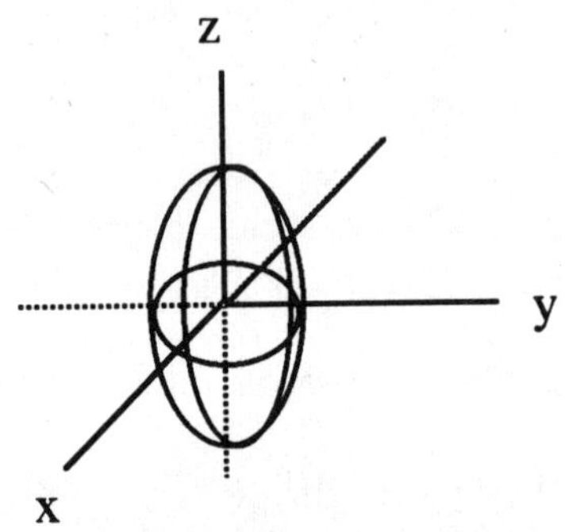

71. Circular Paraboloid

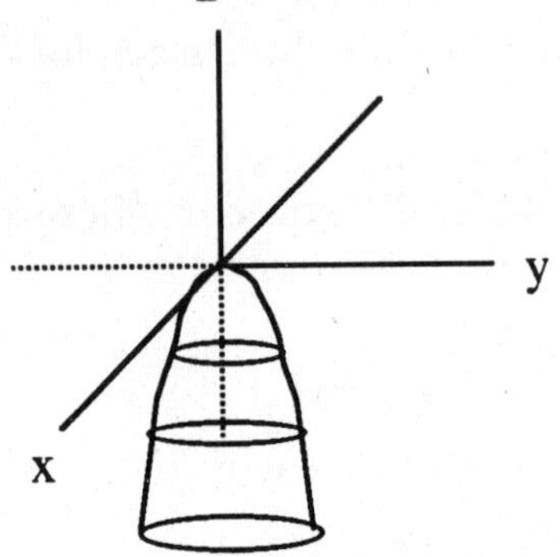

73. Cone about z-axis

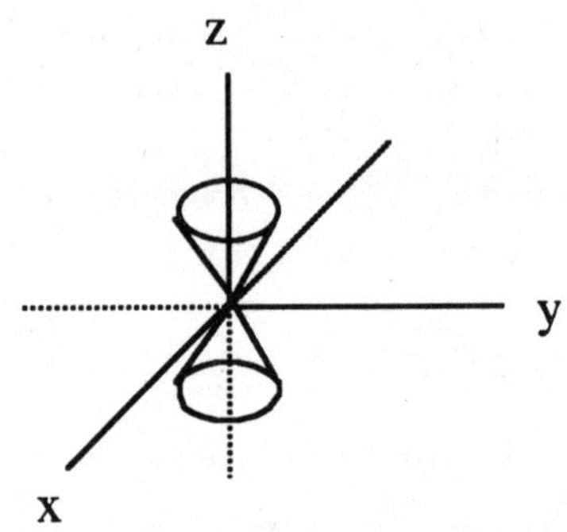

75. Hyperboloid of one sheet

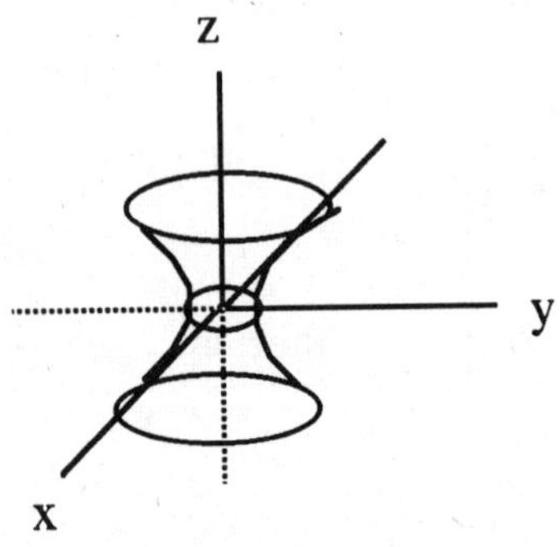

77. Hyperboloid of two sheets

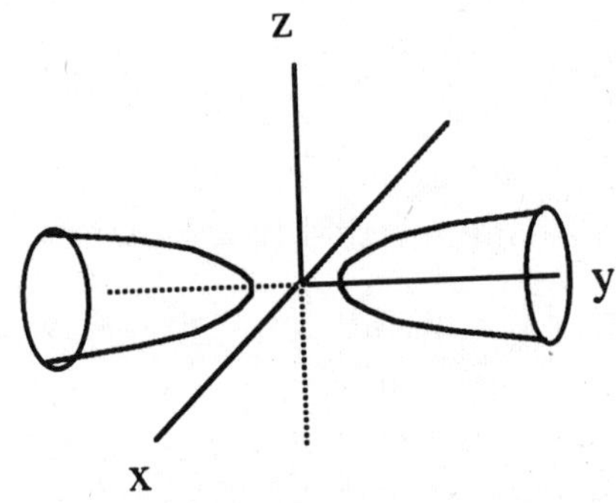

CHAPTER 12

VECTOR-VALUED FUNCTIONS, PARAMETRIZATIONS, AND MOTION IN SPACE

12.1 Vector-Valued Functions and Curves in Space; Derivatives and Integrals

1. $\boldsymbol{r} = (2\cos t)\boldsymbol{i} + (3\sin t)\boldsymbol{j} + 4t\boldsymbol{k}$, $t = \frac{\pi}{2}$. a) $\boldsymbol{v} = \frac{d\boldsymbol{r}}{dt} = (-2\sin t)\boldsymbol{i} + (3\cos t)\boldsymbol{j} + 4\boldsymbol{k}$ b) $\boldsymbol{a} = \frac{d\boldsymbol{v}}{dt} = (-2\cos t)\boldsymbol{i} + (-3\sin t)\boldsymbol{j}$ c) Speed at $t = \pi/2 = |\boldsymbol{v}(\pi/2)| = \sqrt{4\sin^2(\pi/2) + 9\cos^2(\pi/2) + 4^2} = \sqrt{20} = 2\sqrt{5}$
d) (**direction** at $t = \pi/2) = \boldsymbol{v}(\pi/2)/|\boldsymbol{v}(\pi/2)| = (-2\boldsymbol{i} + 4\boldsymbol{k})/2\sqrt{5}$
e) $\boldsymbol{v}(\pi/2) = 2\sqrt{5}\left[(-2\boldsymbol{i} + 4\boldsymbol{k})/2\sqrt{5}\right] = 2\sqrt{5}\left[-(1/\sqrt{5})\boldsymbol{i} + (2/\sqrt{5})\boldsymbol{k}\right]$.

3. a) $\boldsymbol{v} = \frac{d\boldsymbol{r}}{dt} = (-2\sin 2t)\boldsymbol{j} + (2\cos t)\boldsymbol{k}$ b) $\boldsymbol{a} = \frac{d\boldsymbol{v}}{dt} = (-4\cos 2t)\boldsymbol{j} + (-2\sin t)\boldsymbol{k}$
c) At $t = 0$: speed $= \sqrt{0^2 + 2^2} = 2$ d) **direction** $= 2\boldsymbol{k}/2 = \boldsymbol{k}$ e) $\boldsymbol{v} = 2\boldsymbol{k}$.

5. a) $\boldsymbol{v} = (\sec t\tan t)\boldsymbol{i} + (\sec^2 t)\boldsymbol{j} + (4/3)\boldsymbol{k}$ b) $\boldsymbol{a} = (\sec t\tan^2 t + \sec^3 t)\boldsymbol{i} + (2\sec^2 t\tan t)\boldsymbol{j}$ c) At $t = \pi/6$, speed $= \sqrt{(\frac{2}{\sqrt{3}}\frac{1}{\sqrt{3}})^2 + (\frac{4}{3})^2 + (\frac{4}{3})^2} = 2$
d) **direction** $= [(2/3)\boldsymbol{i} + (4/3)\boldsymbol{j} + (4/3)\boldsymbol{k}]/2 = (1/3)\boldsymbol{i} + (2/3)\boldsymbol{j} + (2/3)\boldsymbol{k}$
e) $\boldsymbol{v} = 2[(1/3)\boldsymbol{i} + (2/3)\boldsymbol{j} + (2/3)\boldsymbol{k}]$.

7. a) $\boldsymbol{v} = (-e^{-t})\boldsymbol{i} + (-6\sin 3t)\boldsymbol{j} + (6\cos 3t)\boldsymbol{k}$ b) $\boldsymbol{a} = (e^{-t})\boldsymbol{i} + (-18\cos 3t)\boldsymbol{j} + (-18\sin 3t)\boldsymbol{k}$ c) At $t = 0$: speed $= \sqrt{(-1)^2 + 0^2 + 6^2} = \sqrt{37}$
d) **direction** $= (-\boldsymbol{i} + 6\boldsymbol{k})/\sqrt{37}$ e) $\boldsymbol{v} = \sqrt{37}\left[(-\boldsymbol{i} + 6\boldsymbol{k})/\sqrt{37}\right]$.

9. $\boldsymbol{r} = (3t + 1)\boldsymbol{i} + \sqrt{3}t\boldsymbol{j} + t^2\boldsymbol{k}$, $\boldsymbol{v} = 3\boldsymbol{i} + \sqrt{3}\boldsymbol{j} + 2t\boldsymbol{k}$, $\boldsymbol{a} = 2\boldsymbol{k}$, $\boldsymbol{v}(0) = 3\boldsymbol{i} + \sqrt{3}\boldsymbol{j}$, $\boldsymbol{a}(0) = 2\boldsymbol{k}$. $\theta = \cos^{-1}[\boldsymbol{v}(0)\cdot\boldsymbol{a}(0)/|\boldsymbol{v}(0)||\boldsymbol{a}(0)|] = \cos^{-1}0 = \pi/2$.

11. $\boldsymbol{v} = \left(\frac{2t}{t^2+1}\right)\boldsymbol{i} + \left(\frac{1}{t^2+1}\right)\boldsymbol{j} + \left(\frac{t}{\sqrt{t^2+1}}\right)\boldsymbol{k}$, $\boldsymbol{a} = \frac{2(1-t^2)}{(t^2+1)^2}\boldsymbol{i} + \frac{-2t}{(t^2+1)^2}\boldsymbol{j} + \left(\frac{1}{(t^2+1)^{3/2}}\right)\boldsymbol{k}$, $\boldsymbol{v}(0) = \boldsymbol{j}$, $\boldsymbol{a}(0) = 2\boldsymbol{i} + \boldsymbol{k}$. $\theta = \cos^{-1}[\boldsymbol{v}(0)\cdot\boldsymbol{a}(0)/(|\boldsymbol{v}(0)||\boldsymbol{a}(0)|)] = \cos^{-1}0 = \pi/2$.

13. $\boldsymbol{v} = (1 - \cos t)\boldsymbol{i} + (\sin t)\boldsymbol{j}$, $\boldsymbol{a} = (\sin t)\boldsymbol{i} + (\cos t)\boldsymbol{j}$, $\boldsymbol{v}\cdot\boldsymbol{a} = 0$ yields $(1 - \cos t)\sin t + \sin t\cos t = \sin t = 0$. Answer: $t = 0, \pi, 2\pi$.

15. $\int_0^1 [t^3\boldsymbol{i} + 7\boldsymbol{j} + (t + 1)\boldsymbol{k}]\,dt = \frac{t^4}{4}\Big]_0^1\boldsymbol{i} + 7t]_0^1\boldsymbol{j} + \left[\frac{t^2}{2} + t\right]_0^1\boldsymbol{k} = \frac{1}{4}\boldsymbol{i} + 7\boldsymbol{j} + \frac{3}{2}\boldsymbol{k}$.

17. $\int_{-\pi/4}^{\pi/4} [(\sin t)\boldsymbol{i} + (1 + \cos t)\boldsymbol{j} + (\sec^2 t)\boldsymbol{k}]\,dt = [-\cos t]_{-\pi/4}^{\pi/4}\boldsymbol{i} + [t + \sin t]_{-\pi/4}^{\pi/4}\boldsymbol{j} + \tan t]_{-\pi/4}^{\pi/4}\boldsymbol{k} = (\pi/2 + \sqrt{2})\boldsymbol{j} + 2\boldsymbol{k}$.

19. $\int_1^4 \left[\frac{1}{t}\boldsymbol{i} + \frac{1}{5-t}\boldsymbol{j} + \frac{1}{2t}\boldsymbol{k}\right] dt = \ln|t|]_1^4\,\boldsymbol{i} - \ln|5-t|]_1^4\,\boldsymbol{j} + \frac{1}{2}\ln|t|\Big]_1^4\,\boldsymbol{k} = (\ln 4)\boldsymbol{i} + (\ln 4)\boldsymbol{j} + (\ln 2)\boldsymbol{k}$.

21. $\boldsymbol{r} = (\sin t)\boldsymbol{i} + (\cos t)\boldsymbol{j}$. $\boldsymbol{v} = (\cos t)\boldsymbol{i} - (\sin t)\boldsymbol{j}$, $\boldsymbol{a} = -(\sin t)\boldsymbol{i} - (\cos t)\boldsymbol{j} = -\boldsymbol{r}$. $\boldsymbol{v}(\pi/4) = (\sqrt{2}/2)(\boldsymbol{i} - \boldsymbol{j})$, $\boldsymbol{a}(\pi/4) = -(\sqrt{2}/2)(\boldsymbol{i} + \boldsymbol{j})$, $\boldsymbol{v}(\pi/2) = -\boldsymbol{j}$, $\boldsymbol{a}(\pi/2) = -\boldsymbol{i}$.

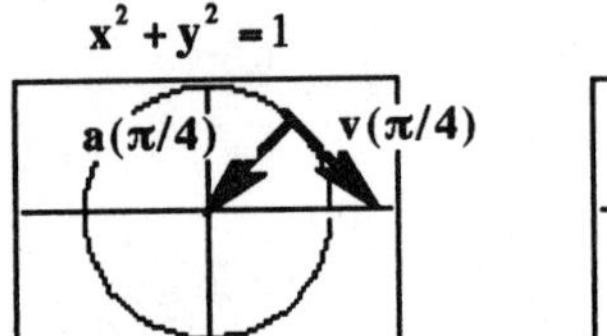

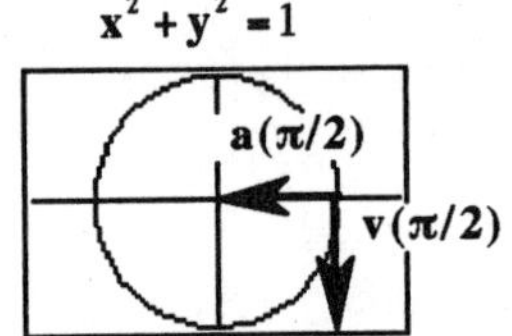

23. $\boldsymbol{v} = (1 - \cos t)\boldsymbol{i} + (\sin t)\boldsymbol{j}$, $\boldsymbol{a} = (\sin t)\boldsymbol{i} + (\cos t)\boldsymbol{j}$. $\boldsymbol{v}(\pi) = 2\boldsymbol{i}$, $\boldsymbol{a}(\pi) = -\boldsymbol{j}$, $\boldsymbol{v}(3\pi/2) = \boldsymbol{i} - \boldsymbol{j}$, $\boldsymbol{a}(3\pi/2) = -\boldsymbol{i}$.

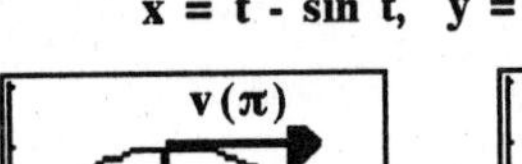

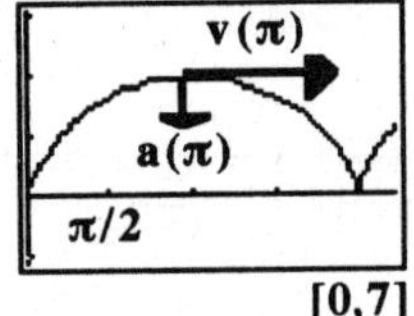

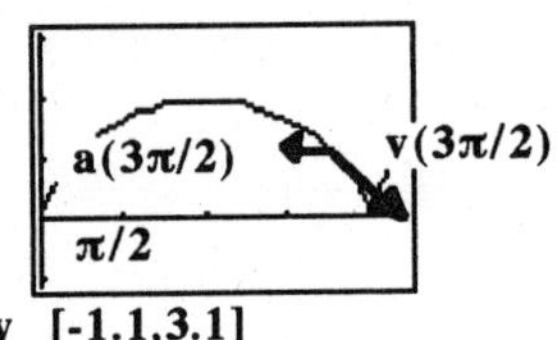

[0,7] by [-1.1,3.1]

25. Integrating, we obtain $\boldsymbol{r} = \left(-\frac{t^2}{2} + C_1\right)\boldsymbol{i} + \left(-\frac{t^2}{2} + C_2\right)\boldsymbol{j} + \left(-\frac{t^2}{2} + C_3\right)\boldsymbol{k} = -\frac{t^2}{2}(\boldsymbol{i} + \boldsymbol{j} + \boldsymbol{k}) + C_1\boldsymbol{i} + C_2\boldsymbol{j} + C_3\boldsymbol{k}$. Setting $t = 0$, we obtain $\boldsymbol{C} = C_1\boldsymbol{i} + C_2\boldsymbol{j} + C_3\boldsymbol{k} = \boldsymbol{i} + 2\boldsymbol{j} + 3\boldsymbol{k}$. Hence $\boldsymbol{r} = -\frac{t^2}{2}(\boldsymbol{i} + \boldsymbol{j} + \boldsymbol{k}) + \boldsymbol{i} + 2\boldsymbol{j} + 3\boldsymbol{k} = \left(1 - \frac{t^2}{2}\right)\boldsymbol{i} + \left(2 - \frac{t^2}{2}\right)\boldsymbol{j} + \left(3 - \frac{t^2}{2}\right)\boldsymbol{k}$.

27. $\boldsymbol{r} = \left[(t+1)^{3/2} + C_1\right]\boldsymbol{k} + (-e^{-t} + C_2)\boldsymbol{j} + [\ln(t+1) + C_3]\,\boldsymbol{k}$. When $t = 0$, $\boldsymbol{r} = (1 + C_1)\boldsymbol{i} + (-1 + C_2)\boldsymbol{j} + C_3\boldsymbol{k} = \boldsymbol{k}$ so $C_1 = -1$, $C_2 = 1$, $C_3 = 1$. $\boldsymbol{r} = \left[(t+1)^{3/2} - 1\right]\boldsymbol{i} + (1 - e^{-t})\boldsymbol{j} + [\ln(t+1) + 1]\,\boldsymbol{k}$

29. $\dfrac{d\boldsymbol{r}}{dt} = C_1\boldsymbol{i} + C_2\boldsymbol{j} + (-32t + C_3)\boldsymbol{k}$. When $t = 0$, this must equal $8\boldsymbol{i} + 8\boldsymbol{j}$ so $\dfrac{d\boldsymbol{r}}{dt} = 8\boldsymbol{i} + 8\boldsymbol{j} - 32t\boldsymbol{k}$. $\boldsymbol{r} = (8t + A_1)\boldsymbol{i} + (8t + A_2)\boldsymbol{j} + (-16t^2 + A_3)\boldsymbol{k}$. When $t = 0$: $\boldsymbol{r} = A_1\boldsymbol{i} + A_2\boldsymbol{j} + A_3\boldsymbol{k} = 100\boldsymbol{k}$ so our final answer is $\boldsymbol{r} = 8t\boldsymbol{i} + 8t\boldsymbol{j} + (100 - 16t^2)\boldsymbol{k}$.

31. $\boldsymbol{v} = (1-\cos t)\boldsymbol{i} + \sin t\boldsymbol{j}$, $\boldsymbol{a} = \sin t\boldsymbol{i} + \cos t\boldsymbol{j}$. $|\boldsymbol{v}|^2 = 1 - 2\cos t + \cos^2 t + \sin^2 t = 2(1-\cos t)$. This has maximum value when $\cos t = -1$ and minimum value when $\cos t = 1$. We conclude that the maximum and minimum values of $|\boldsymbol{v}|$ are, respectively, 2 and 0. Since $|\boldsymbol{a}| = 1$ is constant, 1 is both the maximum and minimum value of $|\boldsymbol{a}|$.

33. $\boldsymbol{f} = C_1\boldsymbol{i} + C_2\boldsymbol{j} + C_3\boldsymbol{k}$ where C_1, C_2, C_3 are constants. Hence $\dfrac{d\boldsymbol{f}}{dt} = 0i + 0j + 0k = \mathbf{0}$.

35. $\boldsymbol{u} = u_1\boldsymbol{i} + u_2\boldsymbol{j} + u_3\boldsymbol{k}$ and $\boldsymbol{v} = v_1\boldsymbol{i} + v_2\boldsymbol{j} + v_3\boldsymbol{k}$ where u_i and v_i are certain differentiable functions of t for $i = 1, 2, 3$. $\frac{d}{dt}(\boldsymbol{u}+\boldsymbol{v}) = \frac{d}{dt}\left[(u_1 + v_1)\boldsymbol{i} + (u_2 + v_2)\boldsymbol{j} + (u_3 + v_3)\boldsymbol{k}\right] = \left(\frac{du_1}{dt} + \frac{dv_1}{dt}\right)\boldsymbol{i} + \left(\frac{du_2}{dt} + \frac{dv_2}{dt}\right)\boldsymbol{j} + \left(\frac{du_3}{dt} + \frac{dv_3}{dt}\right)\boldsymbol{k} = \left(\frac{du_1}{dt}\boldsymbol{i} + \frac{du_2}{dt}\boldsymbol{j} + \frac{du_3}{dt}\boldsymbol{k}\right) + \left(\frac{dv_1}{dt}\boldsymbol{i} + \frac{dv_2}{dt}\boldsymbol{j} + \frac{dv_3}{dt}\boldsymbol{k}\right) = \dfrac{d\boldsymbol{u}}{dt} + \dfrac{d\boldsymbol{v}}{dt}$. We omit the proof of the second formula which is quite similar: it may be obtained by replacing certain +'s by −'s in the above proof.

12.2 MODELING PROJECTILE MOTION

1. $x = (v_0\cos\alpha)t$, $21 = (0.840\,\text{km/sec})(\cos 60°)t$, $t = 50\,\text{sec}$. Graph $x = 840(\cos(\pi/3))t$, $y = 840(\sin(\pi/3))t - \frac{1}{2}(9.8)t^2$, $0 \leqq t \leqq 150$, tStep $= 0.5$ in $[0, 70000]$ by $[0, 30000]$. Use TRACE to see that when $t = 50\,\text{sec}$, $x = 21000$ m. Caution: on our calculator, $\sin\pi/3$ gave 0; one can use $\sin(\pi/3)$ for the correct value.

3. a) $t = \frac{2v_0\sin\alpha}{g} = \frac{2(500\,\text{m/sec})\sin 45°}{9.8\,\text{m/sec}^2} = 72.15\,\text{sec}$. $R = \frac{v_0^2}{g}\sin 2\alpha = \frac{500^2\sin 90°}{9.8} = 25510\,\text{m} = 25.51$ km b) $x = (v_0\cos\alpha)t$, $t = (0.5\,\text{km/sec})(\cos 45°)t$, $t = 10\sqrt{2}\,\text{sec}$. $y = (v_0\sin\alpha)t - \frac{1}{2}gt^2 = 500\frac{\sqrt{2}}{2}10\sqrt{2} - \frac{1}{2}9.8(10\sqrt{2})^2 = 4020$ m $= 4.02$ km c) $y_{\text{max}} = \frac{(v_0\sin\alpha)^2}{2g} = \frac{(500\sin 45°)^2}{2(9.8)} = 6377.55$ m.

5. $R = \frac{v_0^2}{g}\sin 2\alpha$. If we replace α by $90° - \alpha$, the new range is $\frac{v_0^2}{g}\sin[2(90° - \alpha)] = \frac{v_0^2}{g}\sin(180° - 2\alpha) = \frac{v_0^2}{g}\sin 2\alpha = R$.

7. $R = \frac{v_0^2}{g}\sin 2\alpha$, $10 = \frac{v_0^2}{9.8}\sin 90°$, $v_0 = \sqrt{98} = 9.9$ m/sec. $6 = \frac{98}{9.8}\sin 2\alpha$, $\sin 2\alpha = 0.6$, $2\alpha = \sin^{-1} 0.6$ and $\pi - \sin^{-1} 0.6$. Solving for α and using degree measure, we obtain $\alpha = 18.43°$ and $71.57°$.

9. $R = \frac{v_0^2}{g}\sin 2\alpha$, $(248.8)3$ ft $= \frac{v_0^2}{32}\sin 18°$, $v_0 = \sqrt{\frac{32(248.8)3}{\sin 18°}} = 278.02$ ft/sec $= 278.02\frac{\text{ft}}{\text{sec}}\frac{1\,\text{mile}}{5280\,\text{ft}}\frac{3600\,\text{sec}}{1\,\text{hr}} = 189.56$ mph.

11. We solve $y = \frac{3}{4}y_{\max}$ for t: $(v_0 \sin\alpha)t - \frac{1}{2}gt^2 = \frac{3(v_0\sin\alpha)^2}{8g}$, (multiplying by $-8g$) $4g^2t^2 - 8gv_0\sin\alpha + 3(v_0\sin\alpha)^2 = 0$. Solving for the shorter (first) time, we get $t = \frac{8gv_0\sin\alpha - \sqrt{64g^2(v_0\sin\alpha)^2 - 48g^2(v_0\sin\alpha)^2}}{8g^2} = \frac{4gv_0\sin\alpha}{8g^2} = \frac{1}{2}\frac{v_0\sin\alpha}{g} = \frac{1}{2}$ (time for $y_{\max}$).

13. Let us find the height of the ball when $x = 369$ ft. We first have to find t such that $x(t) = 369$. $x = t(v_0\cos\alpha) = t(116\cos 45°) = 58\sqrt{2}t = 369$, $t = \frac{369}{58\sqrt{2}}$ sec. $y = (v_0\sin\alpha)t - \frac{1}{2}gt^2 = 58\sqrt{2}t - 16t^2$. $y(\frac{369}{58\sqrt{2}}) = 45.193$ ft. So in flight, the ball passes just above the pin.

15. We use the equations for launching from (x_0, y_0). $x = (v_0\cos 20°)t = 315$, $y = 3 + v_0(\sin 20°)t - 16t^2 = 37$. We solve the first equation for v_0 and substitute into the second equation. $v_0 = 315/(t\cos 20°)$, $315\tan 20° - 16t^2 = 34$. The positive solution is $t_1 = \sqrt{315\tan 20° - 34}/4 \approx 2.245$ sec. It takes t_1 seconds to get to the wall. The initial speed is $v_0 = 315/(t_1\cos 20°) \approx 149.31$ ft/sec. But $\frac{dx}{dt} = v_0\cos 20°$, $\frac{dy}{dt} = v_0\sin 20° - 32t$. The speed at t_1 is $|v(t_1)| = \sqrt{(v_0\cos 20°)^2 + (v_0\sin 20° - 32t_1)^2} \approx 141.83$ ft/sec neglecting forces other than gravity.

17. a) $x = x_0 + (v_0\cos\alpha)t = 13 + (35\cos 27°)t$ (The net will stand at $x = 25$ ft.) $y = y_0 + (v_0\sin\alpha)t - \frac{1}{2}gt^2 = 4 + (35\sin 27°)t - 16t^2$. $\boldsymbol{r} = [13 + (35\cos 27°)t]\boldsymbol{i} + [4 + (35\sin 27°)t - 16t^2]\boldsymbol{j}$. b) Graph x, y above, $0 \leqq t \leqq 1.2$, Tstep $= 0.1$ or 0.05 in $[0, 51]$ by $[-9.7, 19.7]$. Also draw in the net: Line(25, 0, 25, 6). c) i) $y_{\max} \approx 7.9$ ft occurs at $t \approx 0.5$ sec. (We are using TRACE). ii) It travels $50.4 - 13 = 37.4$ ft and hits the ground at about 1.2 sec. iii) $y = 7$ ft at $t \approx 0.25$ sec and is $x - 13 = 20.8 - 13 = 7.8$ ft from the point of impact. Also at $t \approx 0.75$ sec and $x \approx 36.4 - 13 = 23.4$ ft from the point of impact. iv) The ball hits the net by i).

19. We graph $x = \frac{152(\cos 20°)}{0.12}(1 - e^{-0.12t})$, $y = 3 + \frac{152(\sin 20°)}{0.12}(1 - e^{-0.12t}) + (\frac{32}{0.12^2})(1 - 0.12t - e^{-0.12t})$, $0 \leqq t \leqq 3.2$, tStep $= 0.01$ in $[0, 500]$ by $[0, 100]$ and we use TRACE for the approximations. a) $y_{\max} = 40.435$ ft at $t = 1.48$ sec b) It travels about 373 ft hitting the ground at $t = 3.13$ sec c) $y = 30$ ft at about $t = 0.69$ sec and $t = 2.3$ sec, 94.59 ft and 287.08 ft from home plate, respectively. d) When x is 340 ft, y is between 13.7 ft and 14.2 ft so the ball is a home run.

21. See the solution to Exercise 19 for the set up with $k = 0.12$. Here we use the TRACE function with tStep = 0.1, $0 \leqq t \leqq 3.5$.

k	$y_{\max}$	Time for y_{max}	Flight Distance	Flight Time
0.01	44.77 ft	1.6 sec	463.66 ft	3.3 sec
0.02	44.34	1.6	456.13	3.3
0.05	43.05	1.6	422.38	3.2
0.1	41.15	1.5	391.15	3.2
0.15	39.39	1.5	354.10	3.1
0.20	37.85	1.4	322.22	3.0
0.25	36.41	1.4	294.62	2.9

c) Using l'Hospital's rule, we can show $\frac{1-e^{-kt}}{k} \to t$ and $\frac{1-kt-e^{-kt}}{k^2} \to -\frac{t^2}{2}$ as $k \to 0$. Thus the limit of Eq. (13) as $k \to 0$ is Eq. (6). This makes sense: as the air density diminishes to 0, the air resistance to the motion of the projectile diminishes to 0, as was assumed in Eq. (6).

23. Marble A will be R units downrange at time t_1 where $x = (v_0 \cos\alpha)t_1 = R$, $t_1 = R/(v_0 \cos\alpha)$. At that time $y(= (v_0 \sin\alpha)t - \frac{1}{2}gt^2$ for marble A) will be $(v_0 \sin\alpha)R/(v_0\cos\alpha) - \frac{1}{2}g(R/(v_0\cos\alpha))^2$. But the height of marble B is given by $R\tan\alpha - \frac{1}{2}gt^2$ so both marbles will be at the same height at $t = t_1$ and will collide. The result is independent of the initial velocity v_0.

25. $\frac{d^2\boldsymbol{r}}{dt^2} = -g\boldsymbol{j}$ yields $\frac{d\boldsymbol{r}}{dt} = -gt\boldsymbol{j} + \boldsymbol{C}_1$. When $t = 0$, $\frac{d\boldsymbol{r}}{dt} = \boldsymbol{C}_1 = (v_0\cos\alpha)\boldsymbol{i} + (v_0\sin\alpha)\boldsymbol{j}$. So for arbitrary t, $\frac{d\boldsymbol{r}}{dt} = (v_0\cos\alpha)\boldsymbol{i} + (v_0\sin\alpha - gt)\boldsymbol{j}$. In turn this yields $\boldsymbol{r} = (v_0\cos\alpha)t\boldsymbol{i} + ((v_0\sin\alpha)t - \frac{1}{2}gt^2)\boldsymbol{j} + \boldsymbol{C}_2$. When $t = 0$, $\boldsymbol{r} = \boldsymbol{C}_2 = x_0\boldsymbol{i} + y_0\boldsymbol{j}$. So for arbitrary t, $\boldsymbol{r} = (x_0 + (v_0\cos\alpha)t)\boldsymbol{i} + (y_0 + (v_0\sin\alpha)t - \frac{1}{2}gt^2)\boldsymbol{j}$. This is equivalent to the desired parametric equations.

12.3 DIRECTED DISTANCE AND THE UNIT TANGENT VECTOR $\boldsymbol{T}$

1. $\boldsymbol{v} = \frac{d\boldsymbol{r}}{dt} = (-2\sin t)\boldsymbol{i} + (2\cos t)\boldsymbol{j} + \sqrt{5}\boldsymbol{k}$. $|\boldsymbol{v}| = \sqrt{4\sin^2 t + 4\cos^2 t + 5} = \sqrt{4+5} = 3$. $\boldsymbol{T} = \frac{\boldsymbol{v}}{|\boldsymbol{v}|} = -\frac{2}{3}\sin t\boldsymbol{i} + \frac{2}{3}\cos t\boldsymbol{j} + \frac{\sqrt{5}}{3}\boldsymbol{k}$. $L = \int_0^\pi |\boldsymbol{v}|dt = 3\pi$.

3. $\boldsymbol{v} = \frac{d\boldsymbol{r}}{dt} = \boldsymbol{i} + t^{1/2}\boldsymbol{k}$. $|\boldsymbol{v}| = \sqrt{1+t}$. $\boldsymbol{T} = \frac{1}{\sqrt{1+t}}\boldsymbol{i} + \sqrt{\frac{t}{1+t}}\boldsymbol{k}$. $L = \int_0^8 \sqrt{1+t}\,dt = \frac{2}{3}(1+t)^{3/2}]_0^8 = \frac{2}{3}[9^{3/2} - 1] = \frac{52}{3}$.

5. $\boldsymbol{v} = \boldsymbol{i} - \boldsymbol{j} + \boldsymbol{k}$, $|\boldsymbol{v}| = \sqrt{3}$. $\boldsymbol{T} = \frac{1}{\sqrt{3}}(\boldsymbol{i} - \boldsymbol{j} + \boldsymbol{k})$. $L = \int_0^3 \sqrt{3}dt = 3\sqrt{3}$.

7. $\boldsymbol{v} = (\cos t - t\sin t)\boldsymbol{i} + (\sin t + t\cos t)\boldsymbol{j} + \sqrt{2}t^{1/2}\boldsymbol{k}$. $|\boldsymbol{v}| = \sqrt{\cos^2 t - 2t\cos t\sin t + t^2\sin^2 t + \sin^2 t + 2t\sin t\cos t + t^2\cos^2 t + 2t} = \sqrt{t^2 + 2t + 1} = t + 1$. $\boldsymbol{T} = (\frac{\cos t - t\sin t}{t+1})\boldsymbol{i} + (\frac{\sin t + t\cos t}{t+1})\boldsymbol{j} + (\frac{\sqrt{2t}}{t+1})\boldsymbol{k}$. $L = \int_0^\pi (t+1)dt = \frac{t^2}{2} + t]_0^\pi = \frac{\pi^2}{2} + \pi$.

9. $\boldsymbol{v} = -4\sin t\boldsymbol{i} + 4\cos t\boldsymbol{j} + 3\boldsymbol{k}$, $|\boldsymbol{v}| = \sqrt{16\sin^2 t + 16\cos^2 t + 9} = 5$. $s(t) = \int_{t_0}^t 5d\tau = 5(t - t_0)$. $L = \frac{5\pi}{2}$.

11. $\boldsymbol{v} = (e^t\cos t - e^t\sin t)\boldsymbol{i} + (e^t\sin t + e^t\cos t)\boldsymbol{j} + e^t\boldsymbol{k}$. $|\boldsymbol{v}| = \sqrt{e^{2t}(\cos^2 t + \sin^2 t) + e^{2t}(\sin^2 t + \cos^2 t) + e^{2t}} = \sqrt{3}e^t$. $s(t) = \int_{t_0}^t \sqrt{3}e^\tau d\tau = \sqrt{3}(e^t - e^{t_0})$. $L = \sqrt{3}(e^{\ln 4} - e^0) = \sqrt{3}(4-1) = 3\sqrt{3}$.

13. $\boldsymbol{v} = \sqrt{2}\boldsymbol{i} + \sqrt{2}\boldsymbol{j} = 2t\boldsymbol{k}$. $|\boldsymbol{v}| = \sqrt{2+2+4t^2} = 2\sqrt{1+t^2}$. The points correspond to $t = 0$ to $t = 1$. From a table of integrals we use $\int \sqrt{x^2+a^2}dx = \frac{x}{2}\sqrt{x^2+a^2} + \frac{a^2}{2}\ln|x + \sqrt{x^2+a^2}| + C$. $L = 2\int_0^1 \sqrt{t^2+1}dt = 2[\frac{t}{2}\sqrt{t^2+1} + \frac{1}{2}\ln|t + \sqrt{t^2+1}|]_0^1 = \sqrt{2} + \ln(1+\sqrt{2})$.

15. a) $\boldsymbol{v} = -4\sin 4t\boldsymbol{i} + 4\cos 4t\boldsymbol{j} + 4\boldsymbol{k}$. $|\boldsymbol{v}| = \sqrt{16+16} = 4\sqrt{2}$. $L = \int_0^{\pi/2} 4\sqrt{2}dt = 2\sqrt{2}\pi$. b) $\boldsymbol{v} = -\frac{1}{2}\sin(\frac{t}{2})\boldsymbol{i} + \frac{1}{2}\cos(\frac{t}{2})\boldsymbol{j} + \frac{1}{2}\boldsymbol{k}$. $|\boldsymbol{v}| = \sqrt{\frac{1}{4} + \frac{1}{4}} = \frac{\sqrt{2}}{2}$. $L = \int_0^{4\pi} \frac{\sqrt{2}}{2}dt = 2\sqrt{2}\pi$. c) $\boldsymbol{v} = -\sin t\boldsymbol{i} - \cos t\boldsymbol{j} - \boldsymbol{k}$. $|\boldsymbol{v}| = \sqrt{2}$. $L = \int_{-2\pi}^0 \sqrt{2}dt = 2\sqrt{2}\pi$.

12.4 CURVATURE, TORSION, AND THE *TNB* FRAME

1. $\boldsymbol{v} = \boldsymbol{i} - \tan t\boldsymbol{j}$. $|\boldsymbol{v}| = \sqrt{1+\tan^2 t} = \sec t$ (because $\sec t > 0$ for $-\pi/2 < t < \pi/2$). $\boldsymbol{T} = \frac{\boldsymbol{v}}{|\boldsymbol{v}|} = \frac{1}{\sec t}\boldsymbol{i} - \frac{\tan t}{\sec t}\boldsymbol{j} = \cos t\boldsymbol{i} - \sin t\boldsymbol{j}$. $\dfrac{d\boldsymbol{T}}{dt} = -\sin t\boldsymbol{i} - \cos t\boldsymbol{j}$. $\boldsymbol{N} = \dfrac{(d\boldsymbol{T}/dt)}{|d\boldsymbol{T}/dt|} = -\sin t\boldsymbol{i} - \cos t\boldsymbol{j}$. $\kappa = \left|\dfrac{d\boldsymbol{T}}{ds}\right| = \left|\dfrac{d\boldsymbol{T}}{dt}\dfrac{dt}{ds}\right| = \left|\dfrac{d\boldsymbol{T}}{dt}\right|\dfrac{1}{|\boldsymbol{v}|} = \dfrac{1}{\sec t} = \cos t$.

3. $\boldsymbol{v} = 2\boldsymbol{i} - 2t\boldsymbol{j}$, $|\boldsymbol{v}| = \sqrt{4+4t^2} = 2\sqrt{1+t^2}$. $\boldsymbol{T} = \dfrac{2\boldsymbol{i} - 2t\boldsymbol{j}}{2\sqrt{1+t^2}} = \dfrac{\boldsymbol{i} - t\boldsymbol{j}}{\sqrt{1+t^2}}$. $\dfrac{d\boldsymbol{T}}{dt} = -(1+t^2)^{-3/2}t\boldsymbol{i} - \frac{\sqrt{1+t^2} - t^2/\sqrt{1+t^2}}{1+t^2}\boldsymbol{j} = \frac{-t}{(1+t^2)^{3/2}}\boldsymbol{i} - \frac{1}{(1+t^2)^{3/2}}\boldsymbol{j}$. $\left|\dfrac{d\boldsymbol{T}}{dt}\right| =$

$\sqrt{\frac{t^2}{(1+t^2)^3}+\frac{1}{(1+t^2)^3}} = \frac{1}{1+t^2}$. $\boldsymbol{N} = \dfrac{d\boldsymbol{T}/dt}{|d\boldsymbol{T}/dt|} = -\dfrac{t}{\sqrt{1+t^2}}\boldsymbol{i} - \dfrac{1}{\sqrt{1+t^2}}\boldsymbol{j}$. $\kappa = \left|\dfrac{d\boldsymbol{T}}{ds}\right| = \left|\dfrac{d\boldsymbol{T}}{dt}\dfrac{dt}{ds}\right| = \dfrac{1}{1+t^2}\dfrac{1}{|\boldsymbol{v}|} = \dfrac{1}{2(1+t^2)^{3/2}}$.

5. $\boldsymbol{v} = (3\cos t)\boldsymbol{i} - 3\sin t\boldsymbol{j} + 4\boldsymbol{k}$, $\boldsymbol{a} = -3\sin t\boldsymbol{i} - 3\cos t\boldsymbol{j}$, $|\boldsymbol{v}| = \sqrt{9\cos^2 t + 9\sin^2 t + 16} = 5$. $\boldsymbol{T} = \dfrac{\boldsymbol{v}}{|\boldsymbol{v}|} = \dfrac{3}{5}\cos t\boldsymbol{i} - \dfrac{3}{5}\sin t\boldsymbol{j} + \dfrac{4}{5}\boldsymbol{k}$. $\dfrac{d\boldsymbol{T}}{dt} = -\dfrac{3}{5}\sin t\boldsymbol{i} - \dfrac{3}{5}\cos t\boldsymbol{j}$, $\left|\dfrac{d\boldsymbol{T}}{dt}\right| = \dfrac{3}{5}$, $\boldsymbol{N} = -\sin t\boldsymbol{i} - \cos t\boldsymbol{j}$. $\boldsymbol{B} = \boldsymbol{T}\times\boldsymbol{N} = \begin{vmatrix} \boldsymbol{i} & \boldsymbol{j} & \boldsymbol{k} \\ \frac{3}{5}\cos t & -\frac{3}{5}\sin t & \frac{4}{5} \\ -\sin t & -\cos t & 0 \end{vmatrix} = \dfrac{4}{5}\cos t\boldsymbol{i} - \dfrac{4}{5}\sin t\boldsymbol{j} - \dfrac{3}{5}\boldsymbol{k}$. $\kappa = \left|\dfrac{d\boldsymbol{T}}{ds}\right| = \left|\dfrac{d\boldsymbol{T}}{dt}\dfrac{dt}{ds}\right| = \dfrac{3}{5}\dfrac{1}{|\boldsymbol{v}|} = \dfrac{3}{25}$. $\tau = \left\|\begin{matrix} 3\cos t & -3\sin t & 4 \\ -3\sin t & -3\cos t & 0 \\ -3\cos t & 3\sin t & 0 \end{matrix}\right\| \div |\boldsymbol{v}\times\boldsymbol{a}|^2 = 36/|\boldsymbol{v}\times\boldsymbol{a}|^2$. $\boldsymbol{v}\times\boldsymbol{a} = \begin{vmatrix} \boldsymbol{i} & \boldsymbol{j} & \boldsymbol{k} \\ 3\cos t & -3\sin t & 4 \\ -3\sin t & -3\cos t & 0 \end{vmatrix} = 12\cos t\boldsymbol{i} - 12\sin\boldsymbol{j} - 9\boldsymbol{k}$. $|\boldsymbol{v}\times\boldsymbol{a}|^2 = 12^2 + 9^2 = 225$. $\tau = 36/225 = 4/25$.

7. $\boldsymbol{v} = (e^t\cos t - e^t\sin t)\boldsymbol{i} + (e^t\cos t + e^t\sin t)\boldsymbol{j}$, $\boldsymbol{a} = (-e^t\sin t + e^t\cos t - e^t\cos t - e^t\sin t)\boldsymbol{i} + (e^t\cos t - e^t\sin t + e^t\sin t + e^t\cos t)\boldsymbol{j} = -2e^t\sin t\boldsymbol{i} + 2e^t\cos t\boldsymbol{j}$. $|\boldsymbol{v}| = \sqrt{e^{2t}(\cos t - \sin t)^2 + e^{2t}(\cos t + \sin t)^2} = \sqrt{2e^{2t}} = \sqrt{2}e^t$. $\boldsymbol{T} = \frac{\boldsymbol{v}}{|\boldsymbol{v}|} = \left(\frac{\cos t - \sin t}{\sqrt{2}}\right)\boldsymbol{i} + \left(\frac{\cos t + \sin t}{\sqrt{2}}\right)\boldsymbol{j}$. $\dfrac{d\boldsymbol{T}}{dt} = \dfrac{-\sin t - \cos t}{\sqrt{2}}\boldsymbol{i} + \dfrac{-\sin t + \cos t}{\sqrt{2}}\boldsymbol{j}$. $\left|\dfrac{d\boldsymbol{T}}{dt}\right| = \dfrac{1}{\sqrt{2}}\sqrt{(\sin t + \cos t)^2 + (\cos t - \sin t)^2} = 1$. $\boldsymbol{N} = -\dfrac{(\sin t + \cos t)}{\sqrt{2}}\boldsymbol{i} + \dfrac{(\cos t - \sin t)}{\sqrt{2}}\boldsymbol{j}$. $\boldsymbol{B} = \boldsymbol{T}\times\boldsymbol{N} = \begin{vmatrix} \boldsymbol{i} & \boldsymbol{j} & \boldsymbol{k} \\ \frac{\cos t - \sin t}{\sqrt{2}} & \frac{\cos t + \sin t}{\sqrt{2}} & 0 \\ -\frac{(\cos t + \sin t)}{\sqrt{2}} & \frac{\cos t - \sin t}{\sqrt{2}} & 0 \end{vmatrix} = \frac{1}{2}[(\cos t - \sin t)^2 + (\cos t + \sin t)^2]\boldsymbol{k} = \boldsymbol{k}$. $\kappa = \left|\dfrac{d\boldsymbol{T}}{ds}\right| = \left|\dfrac{d\boldsymbol{T}}{dt}\dfrac{dt}{ds}\right| = 1\left(\dfrac{1}{|\boldsymbol{v}|}\right) = \dfrac{1}{\sqrt{2}e^t}$. $\tau = 0$ because the last column of the determinant is all 0's.

9. $\boldsymbol{v} = 2\boldsymbol{i} + 2t\boldsymbol{j}$, $|\boldsymbol{v}| = \sqrt{4+4t^2} = 2\sqrt{1+t^2}$. $a_T = \frac{d}{dt}|\boldsymbol{v}| = \frac{2t}{\sqrt{1+t^2}}$. $\boldsymbol{a} = 2\boldsymbol{j}$, $|\boldsymbol{a}| = 2$. $a_N = \sqrt{|\boldsymbol{a}|^2 - a_T^2} = \sqrt{4 - \frac{4t^2}{1+t^2}} = \frac{2}{\sqrt{1+t^2}}$. $\boldsymbol{a} = a_T\boldsymbol{T} + a_N\boldsymbol{N} = \frac{2t}{\sqrt{1+t^2}}\boldsymbol{T} + \frac{2}{\sqrt{1+t^2}}\boldsymbol{N}$.

11. $\boldsymbol{v} = -a\sin t\boldsymbol{i} + a\cos t\boldsymbol{j} + b\boldsymbol{k}$, $|\boldsymbol{v}| = \sqrt{a^2+b^2}$. $a_T = \frac{d}{dt}|\boldsymbol{v}| = 0$. $\boldsymbol{a} = -a\cos t\boldsymbol{i} - a\sin t\boldsymbol{j}$, $|\boldsymbol{a}| = a$. $a_N = \sqrt{|\boldsymbol{a}|^2 - a_T^2} = \sqrt{a^2} = a$, assuming $a > 0$. $\boldsymbol{a} = a_T\boldsymbol{T} + a_N\boldsymbol{N} = a\boldsymbol{N}$.

13. $\boldsymbol{v} = \boldsymbol{i} + 2\boldsymbol{j} + 2t\boldsymbol{k}$, $\boldsymbol{a} = 2\boldsymbol{k}$, $|\boldsymbol{v}| = \sqrt{1+4+4t^2} = \sqrt{5+4t^2}$, $\frac{d}{dt}|\boldsymbol{v}| = \frac{4t}{\sqrt{5+4t^2}}$. When $t = 1$, $a_T = \frac{4}{\sqrt{9}} = \frac{4}{3}$. $a_N = \sqrt{|\boldsymbol{a}|^2 - a_T^2} = \sqrt{4 - \frac{16}{9}} = \frac{\sqrt{20}}{3} = \frac{2\sqrt{5}}{3}$. $\boldsymbol{a} = \frac{4}{3}\boldsymbol{T} + \frac{2\sqrt{5}}{3}\boldsymbol{N}$.

15. $\boldsymbol{v} = 2t\boldsymbol{i} + (1+t^2)\boldsymbol{j} + (1-t^2)\boldsymbol{k}$, $\boldsymbol{a} = 2\boldsymbol{i} + 2t\boldsymbol{j} - 2t\boldsymbol{k}$. $|\boldsymbol{v}| = \sqrt{4t^2+1+2t^2+t^4+1-2t^2+t^4} = \sqrt{2t^4+4t^2+2} = \sqrt{2}(t^2+1)$. $\frac{d|\boldsymbol{v}|}{dt} = 2\sqrt{2}t$. At $t = 0$, $a_T = 0$, $\boldsymbol{a} = 2\boldsymbol{i}$, $a_N = \sqrt{|a|^2 - a_T^2} = \sqrt{4-0} = 2$, $\boldsymbol{a} = 0\boldsymbol{T} + 2\boldsymbol{N} = 2\boldsymbol{N}$.

17. $\boldsymbol{r} = (\cos t)\boldsymbol{i} + (\sin t)\boldsymbol{j} - \boldsymbol{k}$. $\boldsymbol{r}(\frac{\pi}{4}) = \frac{\sqrt{2}}{2}\boldsymbol{i} + \frac{\sqrt{2}}{2}\boldsymbol{j} - \boldsymbol{k}$. $\boldsymbol{v} = -\sin t\boldsymbol{i} + \cos t\boldsymbol{j}$, $|\boldsymbol{v}| = 1$, so $\boldsymbol{T} = \boldsymbol{v}$. $\boldsymbol{T}(\frac{\pi}{4}) = -\frac{\sqrt{2}}{2}\boldsymbol{i} + \frac{\sqrt{2}}{2}\boldsymbol{j}$. $\frac{d\boldsymbol{T}}{dt} = \frac{d\boldsymbol{v}}{dt} = -\cos t\boldsymbol{i} - \sin t\boldsymbol{j}$, a unit vector, so $\boldsymbol{N} = -\cos t\boldsymbol{i} - \sin t\boldsymbol{j}$, $\boldsymbol{B}(\frac{\pi}{4}) = \boldsymbol{T}(\frac{\pi}{4}) \times \boldsymbol{N}(\frac{\pi}{4}) = \begin{vmatrix} \boldsymbol{i} & \boldsymbol{j} & \boldsymbol{k} \\ -\frac{\sqrt{2}}{2} & \frac{\sqrt{2}}{2} & 0 \\ -\frac{\sqrt{2}}{2} & -\frac{\sqrt{2}}{2} & 0 \end{vmatrix} = \boldsymbol{k}$. The position vector, $\boldsymbol{r}(\frac{\pi}{4})$, has terminal point $(\frac{\sqrt{2}}{2}, \frac{\sqrt{2}}{2}, -1)$. For the equation of a plane we use $A(x-x_0) + B(y-y_0) + C(z-z_0) = 0$. Osculating plane: normal $\boldsymbol{B}$, $1 \cdot (z+1) = 0$ or $z = -1$. Normal plane: normal vector $\boldsymbol{T}$, $-\frac{\sqrt{2}}{2}(x - \frac{\sqrt{2}}{2}) + \frac{\sqrt{2}}{2}(y - \frac{\sqrt{2}}{2}) = 0$ or $y = x$. Rectifying plane: normal vector $\boldsymbol{N}$, $-\frac{\sqrt{2}}{2}(x - \frac{\sqrt{2}}{2}) - \frac{\sqrt{2}}{2}(y - \frac{\sqrt{2}}{2}) = 0$ which reduces to $y = -x$.

19. If $|\boldsymbol{v}|$ is constant, so is $|\boldsymbol{v}|^2 = \boldsymbol{v} \cdot \boldsymbol{v}$. Hence $\frac{d}{dt}(\boldsymbol{v} \cdot \boldsymbol{v}) = \frac{d\boldsymbol{v}}{dt} \cdot \boldsymbol{v} + \boldsymbol{v} \cdot \frac{d\boldsymbol{v}}{dt} = 2(\boldsymbol{a} \cdot \boldsymbol{v}) = 0$. Therefore $\boldsymbol{a}$ is normal to $\boldsymbol{v}$ whose direction is the direction of the path. Alternatively, $a_T = \frac{d}{dt}|\boldsymbol{v}| = 0$ so $\boldsymbol{a} = a_T\boldsymbol{T} + a_N\boldsymbol{N} = a_N\boldsymbol{N}$ which is normal to $\boldsymbol{T}$.

21. $\boldsymbol{r} = t\boldsymbol{i} + \sin t\boldsymbol{j}$, $\boldsymbol{v} = \boldsymbol{i} + \cos t\boldsymbol{j}$, $\boldsymbol{a} = -\sin t\boldsymbol{j}$. At $t = \frac{\pi}{2}$, $\boldsymbol{v} = \boldsymbol{i}$, $|\boldsymbol{v}| = 1$, $\boldsymbol{a} = -\boldsymbol{j}$. $\kappa = \frac{|\boldsymbol{v} \times \boldsymbol{a}|}{|\boldsymbol{v}|^3} = \left\| \begin{matrix} \boldsymbol{i} & \boldsymbol{j} & \boldsymbol{k} \\ 1 & 0 & 0 \\ 0 & -1 & 0 \end{matrix} \right\| = |-\boldsymbol{k}| = 1$. Radius $= 1/\kappa = 1$. Thus the center of the circle of curvature is one unit below $(\pi/2, 1)$ so is $(\pi/2, 0)$. The circle has equation $(x - \pi/2)^2 + y^2 = 1$. Graph $x_1(t) = t$, $y_1(t) = \sin t$ and $x_2(t) = \frac{\pi}{2} + \cos t$, $y_2(t) = \sin t$, $-2\pi \leqq t \leqq 2\pi$, tstep $= 0.1$ in $[-2\pi, 2\pi]$ by $[-4.2, 3.2]$.

23. $\boldsymbol{v}\times\boldsymbol{a} = (ds/dt)\boldsymbol{T}\times\left[\frac{d^2s}{dt^2}\boldsymbol{T} + \kappa(ds/dt)^2\boldsymbol{N}\right] = (\frac{ds}{dt})(\frac{d^2s}{dt^2})\boldsymbol{T}\times\boldsymbol{T} + \kappa(\frac{ds}{dt})^3\boldsymbol{T}\times\boldsymbol{N} = \boldsymbol{0} + \kappa(\frac{ds}{dt})^3\boldsymbol{T}\times\boldsymbol{N} = \kappa(\frac{ds}{dt})^3\boldsymbol{T}\times\boldsymbol{N}$. $|\boldsymbol{v}\times\boldsymbol{a}| = \kappa(\frac{ds}{dt})^3|\boldsymbol{T}||\boldsymbol{N}| = \kappa(\frac{ds}{dt})^3 = \kappa|\boldsymbol{v}|^3$. Therefore $\kappa = \dfrac{|\boldsymbol{v}\times\boldsymbol{a}|}{|\boldsymbol{v}|^3}$.

25. $\kappa = \frac{a}{a^2+b^2}$. We wish to find the maximum value of the function $f(x) = \frac{x}{x^2+b^2}$, $x \geqq 0$. $f'(x) = \frac{(x^2+b^2)-x(2x)}{(x^2+b^2)^2} = \frac{b^2-x^2}{x^2+b^2}$. $f'(x) = 0$, $x \geqq 0$ yields $x = b$. Since $f'(x) > 0$ for $0 \leqq x < b$ and $f'(x) < 0$ for $x > b$, $f(b) = \frac{b}{2b^2} = \frac{1}{2b}$ is the absolute maximum of f and of κ. $b > 0$ or else the helix degenerates into a circle.

12.5 PLANETARY MOTION AND SATELLITES

1. $T^2 = \frac{4\pi a^3}{GM}$. $T = \sqrt{\frac{4(6.808\times 10^6)^3\pi^2}{(6.6720\times 10^{-11})(5.975\times 10^{24})}} = 5590.00524462\,\text{sec} = 93.17\,\text{min}$ compared to 93.11 min. in the Table.

3. $\frac{T^2}{a^3} = \frac{4\pi^2}{GM}$, $\frac{a^3}{T^2} = \frac{GM}{4\pi^2}$, $a = (\frac{GMT^2}{4\pi^2})^{1/3} = ((6.6720\times 10^{-11})(5.975\times 10^{24})((92.25)60)^2/(4\pi^2))^{1/3} = 6.763\times 10^6$ m $=$ 6763 km. $a =$ (diameter of Earth + perigee height + apogee height)/2 $=$ $(2(6378.533) + 183 + 589)/2 = 6765$ km.

5. $T = (4\pi^2a^3/(GM))^{1/2} = (4\pi^2(2.2030\times 10^7)^3/((6.6720\times 10^{-11})(6.418\times 10^{23}))^{1/2} = 9.9283\times 10^4\,\text{sec} = 1655\,\text{min}$.

7. $\frac{T^2}{a^3} = \frac{4\pi^2}{GM}$, $\frac{a^3}{T^2} = \frac{GM}{4\pi^2}$, $a = (GMT^2/(4\pi^2))^{1/3} = ((6.6720\times 10^{-11})(6.418\times 10^{23})(1477.4(60))^2/(4\pi^2))^{1/3} = 2.042\times 10^7$ m$=$20420 km.

9. From Eq. (24), $|\boldsymbol{v}| = \sqrt{GM/r} = ((6.6720\times 10^{-11}(5.975\times 10^{24})/r)^{1/2} = (1.9967\times 10^7)/\sqrt{r}$ m/sec.

11. $e = \frac{r_0v_0^2}{GM} - 1$. Circle: $e = 0$, $\frac{r_0v_0^2}{GM} = 1$, $v_0 = \sqrt{\frac{GM}{r_0}}$. Ellipse: $e < 1$, $\frac{r_0v_0^2}{GM} - 1 < 1$, $\frac{r_0v_0^2}{GM} < 2$, $v_0 < \sqrt{\frac{2GM}{r_0}}$. Hyperbola: $e > 1$, $v_0 > \sqrt{\frac{2GM}{r_0}}$. Since $e \geq 0$, $v_0 \geqq \sqrt{\frac{GM}{r_0}}$ in all cases.

13. $\boldsymbol{u_r} = \cos\theta\boldsymbol{i} + \sin\theta\boldsymbol{j}$. The $\boldsymbol{u_r}$-component of $\boldsymbol{a} = \frac{d^2r}{dt^2} - r(\frac{d\theta}{dt})^2$. $\frac{dr}{dt} = (\sinh\theta)\frac{d\theta}{dt} = 2\sinh\theta$, $\frac{d^2r}{dt^2} = 2(\cosh\theta)\frac{d\theta}{dt} = 4\cosh\theta$. Hence $\frac{d^2r}{dt^2} - r(\frac{d\theta}{dt})^2 = 4\cosh\theta - (\cosh\theta)(2)^2 = 0$.

PRACTICE EXERCISES, CHAPTER 12

1. $\boldsymbol{r} = (4\cos t)\boldsymbol{i} + (\sqrt{2}\sin t)\boldsymbol{j}$, $\boldsymbol{v} = (-4\sin t)\boldsymbol{i} + (\sqrt{2}\cos t)\boldsymbol{j}$, $\boldsymbol{a} = (-4\cos t)\boldsymbol{i} + (-\sqrt{2}\sin t)\boldsymbol{j}$. $\boldsymbol{r}(0) = 4\boldsymbol{i}$, $\boldsymbol{v}(0) = \sqrt{2}\boldsymbol{j}$, $\boldsymbol{a}(0) = -4\boldsymbol{i}$. $\boldsymbol{r}(\pi/4) = 2\sqrt{2}\boldsymbol{i} + \boldsymbol{j}$, $\boldsymbol{v}(\pi/4) = -2\sqrt{2}\boldsymbol{i} + \boldsymbol{j}$, $\boldsymbol{a}(\pi/4) = -2\sqrt{2}\boldsymbol{i} - \boldsymbol{j}$.

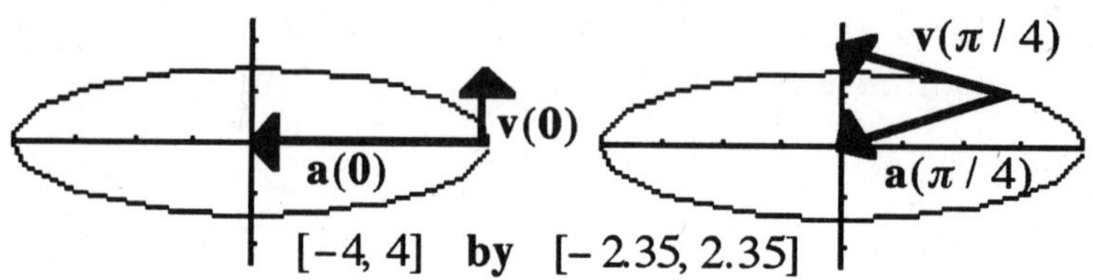

3. $\int_0^1 [(3+6t)\boldsymbol{i} + (4+8t)\boldsymbol{j} + (6\pi\cos\pi t)\boldsymbol{k}]dt = [3t+3t^2]_0^1\boldsymbol{i} + [4t+4t^2]_0^1\boldsymbol{j} + [6\sin\pi t]_0^1\boldsymbol{k} = 6\boldsymbol{i} + 8\boldsymbol{j}$.

5. $\frac{d\boldsymbol{r}}{dt} = -(\sin t)\boldsymbol{i} + (\cos t)\boldsymbol{j} + \boldsymbol{k}$, $\boldsymbol{r} = \boldsymbol{j}$ when $t = 0$. $\boldsymbol{r} = \cos t\boldsymbol{i} + \sin t\boldsymbol{j} + t\boldsymbol{k} + \boldsymbol{C}$. $\boldsymbol{r}(0) = \boldsymbol{i} + \boldsymbol{C} = \boldsymbol{j}$, $\boldsymbol{C} = -\boldsymbol{i} + \boldsymbol{j}$, $\boldsymbol{r} = (\cos t - 1)\boldsymbol{i} + (\sin t + 1)\boldsymbol{j} + t\boldsymbol{k}$.

7. $\frac{d^2\boldsymbol{r}}{dt^2} = 2\boldsymbol{j}$, $\frac{d\boldsymbol{r}}{dt} = \boldsymbol{k}$ and $\boldsymbol{r} = \boldsymbol{i}$ when $t = 0$. $\frac{d\boldsymbol{r}}{dt} = 2t\boldsymbol{j} + \boldsymbol{C}_1$, $\frac{d\boldsymbol{r}}{dt}(0) = \boldsymbol{C}_1 = \boldsymbol{k}$, $\frac{d\boldsymbol{r}}{dt} = 2t\boldsymbol{j} + \boldsymbol{k}$. $\boldsymbol{r} = t^2\boldsymbol{j} + t\boldsymbol{k} + \boldsymbol{C}_2$, $\boldsymbol{r}(0) = \boldsymbol{C}_2 = \boldsymbol{i}$, $\boldsymbol{r} = \boldsymbol{i} + t^2\boldsymbol{j} + t\boldsymbol{k}$.

9. $L = \int_0^{\pi/4} \sqrt{(\frac{dx}{dt})^2 + (\frac{dy}{dt})^2 + (\frac{dz}{dt})^2} = \int_0^{\pi/4} \sqrt{(-2\sin t)^2 + (2\cos t)^2 + (2t)^2}dt = \int_0^{\pi/4} 2\sqrt{1+t^2} = t\sqrt{1+t^2} + \ln|t + \sqrt{t^2+1}|]_0^{\pi/4} = \frac{\pi}{4}\frac{\sqrt{16+\pi^2}}{4} + \ln(\frac{\pi}{4} + \frac{\sqrt{16+\pi^2}}{4}) = \frac{\pi\sqrt{16+\pi^2}}{16} + \ln(\frac{\pi+\sqrt{16+\pi^2}}{4}) = 1.7199\ldots$.

11. $\boldsymbol{r} = \frac{4}{9}(1+t)^{3/2}\boldsymbol{i} + \frac{4}{9}(1-t)^{3/2}\boldsymbol{j} + \frac{1}{3}t\boldsymbol{k}$, $\boldsymbol{v} = \frac{2}{3}(1+t)^{1/2}\boldsymbol{i} - \frac{2}{3}(1-t)^{1/2}\boldsymbol{j} + \frac{1}{3}\boldsymbol{k}$, $\boldsymbol{a} = \frac{1}{3}(1+t)^{-1/2}\boldsymbol{i} + \frac{1}{3}(1-t)^{-1/2}\boldsymbol{j}$. $|\boldsymbol{v}| = \sqrt{\frac{4}{9}(1+t) + \frac{4}{9}(1-t) + \frac{1}{9}} = 1$ so $\boldsymbol{T} = \boldsymbol{v}/|\boldsymbol{v}| = \boldsymbol{v}$. $\boldsymbol{T}(0) = \frac{2}{3}\boldsymbol{i} - \frac{2}{3}\boldsymbol{j} + \frac{1}{3}\boldsymbol{k}$. $d\boldsymbol{T}/dt = \boldsymbol{a}$, $|\frac{d\boldsymbol{T}}{dt}| = \sqrt{\frac{1}{9}\frac{1}{1+t} + \frac{1}{9}\frac{1}{1-t}} = \frac{1}{3}\sqrt{\frac{2}{1-t^2}}$. $\boldsymbol{N} = (d\boldsymbol{T}/dt)/|d\boldsymbol{T}/dt|$, $\boldsymbol{N}(0) = (\frac{1}{3}\boldsymbol{i} + \frac{1}{3}\boldsymbol{j})/(\sqrt{2}/3) = (\boldsymbol{i} + \boldsymbol{j})/\sqrt{2}$.

$$\boldsymbol{B} = \boldsymbol{T} \times \boldsymbol{N},\ \boldsymbol{B}(0) = \begin{vmatrix} \boldsymbol{i} & \boldsymbol{j} & \boldsymbol{k} \\ \frac{2}{3} & -\frac{2}{3} & \frac{1}{3} \\ \frac{1}{\sqrt{2}} & \frac{1}{\sqrt{2}} & 0 \end{vmatrix} = -\frac{1}{3\sqrt{2}}\boldsymbol{i} + \frac{1}{3\sqrt{2}}\boldsymbol{j} + \frac{4}{3\sqrt{2}}\boldsymbol{k}.$$ At $t = 0$, $\kappa =$

$$|\boldsymbol{v}(0) \times \boldsymbol{a}(0)|/|\boldsymbol{v}(0)|^3 = |\boldsymbol{v}(0) \times \boldsymbol{a}(0)| = \left\| \begin{matrix} \boldsymbol{i} & \boldsymbol{j} & \boldsymbol{k} \\ \frac{2}{3} & -\frac{2}{3} & \frac{1}{3} \\ \frac{1}{3} & \frac{1}{3} & 0 \end{matrix} \right\| = |-\tfrac{1}{9}\boldsymbol{i} + \tfrac{1}{9}\boldsymbol{j} + \tfrac{4}{9}\boldsymbol{k}| =$$

$\frac{1}{9}|-\boldsymbol{i}+\boldsymbol{j}+4\boldsymbol{k}| = \sqrt{18}/9 = \sqrt{2}/3$. $\tau(0) = \pm\begin{vmatrix} \dot{x} & \dot{y} & \dot{z} \\ \ddot{x} & \ddot{y} & \ddot{z} \\ \dddot{x} & \dddot{y} & \dddot{z} \end{vmatrix} \Big/ |\boldsymbol{v}\times\boldsymbol{a}|^2_{t=0} =$

$\pm\begin{vmatrix} \frac{2}{3} & -\frac{2}{3} & \frac{1}{3} \\ \frac{1}{3} & \frac{1}{3} & 0 \\ -\frac{1}{6} & \frac{1}{6} & 0 \end{vmatrix} /(\sqrt{2}/3)^2 = (1/27)/(2/9) = 1/6.$

13. $\boldsymbol{v} = (3+6t)\boldsymbol{i} + (4+8t)\boldsymbol{j} + 6\sin t\boldsymbol{k}$, $|\boldsymbol{v}| = \sqrt{[3(1+2t)]^2 + [4(1+2t)]^2 + 36\sin^2 t} = \sqrt{25(1+2t)^2 + 36\sin^2 t}$. $a_T = \frac{d|\boldsymbol{v}|}{dt} = \frac{100(1+2t)+72\sin t\cos t}{2\sqrt{25(1+2t)^2+36\sin^2 t}}$. When $t=0$, $a_T = \frac{100}{10} = 10$. $\boldsymbol{a} = 6\boldsymbol{i} + 8\boldsymbol{j} + 6\cos t\boldsymbol{k}$, $\boldsymbol{a}(0) = 6\boldsymbol{i} + 8\boldsymbol{j} + 6\boldsymbol{k}$. $a_N = \sqrt{|\boldsymbol{a}|^2 - a_T^2} = \sqrt{6^2+8^2+6^2-10^2} = 6$. $\boldsymbol{a}(0) = 10\boldsymbol{T}(0) + 6\boldsymbol{N}(0)$.

15. $\boldsymbol{v} = -\frac{1}{2}(1+t^2)^{-3/2}2t\boldsymbol{i} + \frac{\sqrt{1+t^2} - t[\frac{1}{2}(1+t^2)^{-1/2}2t]}{1+t^2}\frac{\sqrt{1+t^2}}{\sqrt{1+t^2}}\boldsymbol{j} = -\frac{t}{(1+t^2)^{3/2}}\boldsymbol{i} + \frac{1+t^2-t^2}{(1+t^2)^{3/2}}\boldsymbol{j}$. $|\boldsymbol{v}|^2 = \frac{t^2+1}{(1+t^2)^3} = \frac{1}{(1+t^2)^2}$, $|\boldsymbol{v}| = \frac{1}{1+t^2}$. Maximal speed $= |v(0)| = 1$, when denominator is smallest.

17. $\boldsymbol{v} = -5\sin t\boldsymbol{j} + 3\cos t\boldsymbol{k}$, $\boldsymbol{a} = -5\cos t\boldsymbol{j} - 3\sin t\boldsymbol{k}$. $\boldsymbol{v}\cdot\boldsymbol{a} = 25\sin t\cos t - 9\sin t\cos t = 8\sin 2t = 0$ when $2t = n\pi$, $t = \frac{n\pi}{2}$. In $0 \leqq t \leqq \pi$ when $t = 0, \frac{\pi}{2}, \pi$.

19. $x = (v_0\cos\alpha)t = (44\cos 45°)3 = 66\sqrt{2}$ ft. $y = 8 + (v_0\sin\alpha)t - 16t^2 = 8 + 66\sqrt{2} - 144 = -42.66$ ft. So it must be on the level ground.

21. $x = (v_0\cos 45°)t = (4325)3$ ft. $y = (v_0\sin 45°)t - 16t^2 = 0$. From the first equation $t = 3(4325)\sqrt{2}/v_0$. From the second equation $v_0 = 16\sqrt{2}t = 16(6)(4325)/v_0$, $v_0^2 = 16(2)(4325)$, $v_0 = 4\sqrt{6(4325)} = 644.36$ ft/sec. Replacing 4325 by 4752, we get $v_0 = 675.42$ ft/sec.

23. a) $\frac{8\text{ mi}}{\text{hr}} = \frac{8\text{ mi}}{\text{hr}}\frac{1\text{ hr}}{3600\text{ sec}}\frac{5280\text{ ft}}{1\text{mi}} = \frac{176}{15}$ ft/sec (to avoid round off error). $x = (155\cos 18°)t - (176/15)t$, $y = 4 + (155\sin 18°)t - 16t^2$ and with the same $x, y, \boldsymbol{r} = x\boldsymbol{i} + y\boldsymbol{j}$. b) Graph x, y in a), $0 \leqq t \leqq 4$, tstep = 0.05 in $[0, 500]$ by $[0, 100]$. Also include LINE $(380, 0, 380, 10)$. c) With the setup in b) we use the TRACE function, (c-i) $y_{\max} = 39.85$ ft at $t = 1.5$ sec (c-ii) $y = 0$ at about $t = 3.1$ sec and $x = 420.61$ ft. (c-iii) $y = 25$ ft at about $t = 0.55$ sec and $t = 2.45$ sec, when $x = 74.62$ ft and 332.42 ft, respectively. (c-4) Zooming-in and again including LINE $(380, 0, 380, 10)$ shows it is a home run.

25. $\boldsymbol{r} = e^t\boldsymbol{i} + \sin t\boldsymbol{j} + \ln(1-t)\boldsymbol{k}$. When $t = 0$, we get the point $(1,0,0)$. $\boldsymbol{v} = e^t\boldsymbol{i} + \cos t\boldsymbol{j} - \frac{1}{1-t}\boldsymbol{k}$, $\boldsymbol{v}(0) = \boldsymbol{i}+\boldsymbol{j}-\boldsymbol{k}$. The tangent line has direction numbers $1, 1, -1$ and parametric equations $x = 1+t,\ y = t,\ z = -t$.

27. $\boldsymbol{v}\times\boldsymbol{a} = \begin{vmatrix} \boldsymbol{i} & \boldsymbol{j} & \boldsymbol{k} \\ 3 & 4 & 0 \\ 5 & 15 & 0 \end{vmatrix} = 25\boldsymbol{k}$. $|\boldsymbol{v}| = \sqrt{3^2+4^2} = 5$. $\kappa = \frac{|\boldsymbol{v}\times\boldsymbol{a}|}{|\boldsymbol{v}|^3} = \frac{25}{125} = \frac{1}{5}$.

29. $\begin{vmatrix} -a\sin t & a\cos t & b \\ -a\cos t & -a\sin t & 0 \\ a\sin t & -a\cos t & 0 \end{vmatrix} = b(a^2\cos^2 t + a^2\sin^2 t) = ba^2$. $\boldsymbol{v}\times\boldsymbol{a} =$

$\begin{vmatrix} \boldsymbol{i} & \boldsymbol{j} & \boldsymbol{k} \\ -a\sin t & a\cos t & b \\ -a\cos t & -a\sin t & 0 \end{vmatrix} = ab\sin t\boldsymbol{i} - ab\cos t\boldsymbol{j} + a^2\boldsymbol{k}$. $|\boldsymbol{v}\times\boldsymbol{a}|^2 = a^2b^2(\sin^2 t + \cos^2 t) + a^4 = a^2(a^2+b^2)$. $\tau = ba^2/[a^2(a^2+b^2)] = b/(a^2+b^2)$. For a given value of a this has maximum value $1/(2a)$. (See the solution of 12.4 #25 and interchange a and b.)

31. a) Let R be the point $(0, y_0)$ and let $\alpha = \angle ROT$. In triangle ROT, $\cos\alpha = y_0/6380$. In triangle SOT, $\cos\alpha = 6380/(6380+437)$. Hence $y_0 = \frac{6380^2}{(6380+437)} = 5971$ km.

b) $ds = \sqrt{1+(\frac{dx}{dy})^2}dy$. $x = \sqrt{(6380)^2 - y^2}$, $\frac{dx}{dy} = \frac{-y}{\sqrt{(6380)^2-y^2}}$, $1+(\frac{dx}{dy})^2 = \frac{6380^2}{6380^2-y^2}$. $VA = \int_{5971}^{6380} 2\pi x\,ds = 2\pi\int_{5971}^{6380} 6380\,dy = 2\pi 6380(6380-5971) = (5.2188\times 10^6)\pi$ km^2.

c) $(VA/4\pi r^2)100\% = (5.2188\times 10^8)/(4(6380)^2)\% = 3.21\%$.

33. a) $f(0) = -1$, $f(2) = 2-1-\frac{1}{2}\sin 2 = 1-\frac{1}{2}\sin 2 > 1-\frac{1}{2} = \frac{1}{2} > 0$. By the Intermediate Value Property $f(x) = 0$ has a solution between $x = 0$ and $x = 2$. b) $f'(x) = 1 - 0.5\cos x$. We use $x - \frac{f(x)}{f'(x)} = x - \frac{x-1-0.5\sin x}{1-0.5\cos x}$ for our recursion formula, starting with $x_0 = 0$. The sequence of numbers we obtain converges to 1.49870113352. Graph f in the window $[-2,3]$ by $[-2,2]$.